AF678027

STRUCTURE AND FUNCTIONING OF CLUSTER ROOTS AND PLANT RESPONSES TO PHOSPHATE DEFICIENCY

Structure and Functioning of Cluster Roots and Plant Responses to Phosphate Deficiency

Edited by

H. LAMBERS

and

P. POOT

Reprinted from *Plant and Soil*, Volume 248, Nos. 1–2 (2003).

KLUWER ACADEMIC PUBLISHERS
DORDRECHT / BOSTON / LONDON

Library of Congress Cataloging-in-Publication Data

A C.I.P. Catalogue record for this book is available from the library of Congress.

ISBN 1-4020-0434-6

Published by Kluwer Academic Publishers,
P.O. Box 17, 3300 AA Dordrecht, The Netherlands

Sold and distributed in North, Central and South America
by Kluwer Academic Publishers,
101 Philip Drive, Norwell, MA 02061, U.S.A.

In all other countries, sold and distributed
by Kluwer Academic Publishers,
P.O. Box 322, 3300 AH Dordrecht, The Netherlands

Printed on acid-free paper

All Rights reserved
© 2003 Kluwer Academic Publishers
No part of the material protected by this copyright notice
may be reproduced or utilized in any form or by any means,
electronic or mechanical, including photocopying,
recording or by any information storage and retrieval
system, without written permission from the copyright owner.

Printed in the Netherlands

Contents

Cover Photo: Cluster roots of *Hakea prostrata* R.Br., grown in nutrient solution with 1 μM phosphate. Cluster roots occur in most species of the Proteaceae and in a range of other species. They release inorganic anions, such as citrate and malate, into the rhizosphere and mobilise and absorb sparingly available phosphate. Cluster roots allow plants to grow in nutrient-impoverished or phosphate-fixing soils. Photograph by Michael W. Shane.

Plant and Soil **248:** ix–xix, 2003.
© 2003 *Kluwer Academic Publishers. Printed in the Netherlands.*

Introduction

Structure and functioning of cluster roots and plant responses to phosphate deficiency

Hans Lambers[1], Michael D Cramer[1,2], Michael W Shane[1], Madeleine Wouterlood[1], Pieter Poot[1] and Erik J Veneklaas[1]
[1]*School of Plant Biology, Faculty of Natural and Agricultural Sciences, The University of Western Australia, Crawley, WA 6009, Australia, and* [2]*Botany Department, University of Stellenbosch, Private Bag XI, Matieland, 7602, South Africa*

Introduction

Phosphate readily binds to cations and particles in the soil forming sparingly soluble precipitates and making P the least accessible macronutrient for plants. Its availability frequently limits plant growth in both natural and agricultural systems. Fertiliser P is acquired from non-renewable resources, and it is anticipated that readily available sources of P sediments will be depleted by the end of this century. This will cause P-fertiliser prices to rise, increasing the incentive to develop new crops that are better able to access P in soil. This is particularly relevant in large parts of the world where the major part of applied fertiliser is extensively immobilised, and, therefore, unavailable to most traditional crops, e.g., wheat.

Plant roots are now viewed as active "miners" rather than simply passive accumulators of soil minerals. This 'mining' activity is realised through symbiotic associations with mycorrhizal fungi (Fransson et al., 2003; Solaiman and Abbott, 2003), root-hair formation (Bates and Lynch, 1996) and other morphological responses (De Groot et al., 2003; He et al., 2003), and chemical and physical modification of the substrate (Jones et al., 2003). There is also evidence that crops that are very good at accessing sparingly available P can have a favourable effect on plants with which they are intercropped (Horst and Waschkies, 1987; Li et al., 2003) and on the following crop (Kamh et al., 1999). The mechanisms that are responsible for these effects (Zhang and Li, 2003) remain to be further explored.

Plant roots have a number of chemical mechanisms for liberating soil minerals, such as P (Hinsinger et al., 2003; Jones et al., 2003), and for defending themselves against toxic minerals such as Al (Ryan et al., 2003). Although all plant roots are, to some extent, capable of chemical and physical 'mining' of the soil, the highly branched and specialised roots of certain plant species, known as 'proteoid' or 'cluster' roots, are extremely effective in mobilising sparingly soluble minerals (Watt and Evans, 2003). It has been proposed that the present-day soil characteristics of some regions are, in fact, the result of millennia of cluster-root 'mining' activity (Pate et al., 2001; Verboom and Pate, 2003).

Although the morphology of cluster roots has been known since Engler described them at the end of the 19th century for Proteaceae growing in the Leipzig Botanic Gardens (cited in Skene, 1998), there has recently been renewed interest in cluster-rooted plant species. This renewed interest has, to some extent, been driven by the recognition of the role of cluster roots in P acquisition in some crop plants (*e.g.*, *Lupinus albus*) (Gardner et al., 1982). The importance of cluster roots to the survival of P-limited native flora, especially the Proteaceae growing on highly leached soils (e.g., regions of Australia and Southern Africa), has also been increasingly recognised (Lamont, 2003). Proteaceae occur at their highest frequency in the southwest of Western Australia and South Africa (Pate et al., 2001). Almost all Proteaceae have 'proteoid' or 'cluster' roots. The enormous diversity of plant species in the flora of the southwest of Western Australia and South Africa (Myers et al., 2000) occurs on the most nutrient-impoverished soils in the world (Pate and Dell, 1984; Specht and Specht, 1999), suggesting that species with 'proteoid' or 'cluster' roots are particularly well adapted to such environments.

The importance of understanding the functioning of these roots to the conservation of native flora and the increasing cultivation of cluster-rooted native plants for floral, beverage (e.g., tea, *Aspalathus linearis*, F. Dakora, pers. comm.), food (*Macadamia*), timber (*Grevillea robusta*) and medicinal products has sparked further interest in these specialised roots. Increased understanding of the form and function of cluster roots has also led to further research into mechanisms of P acquisition of non-cluster rooted plants. For example, the discovery that the carboxylates that are exuded from cluster roots enhance a plant's capacity to acquire P has led to studies of carboxylate exudation from non-cluster rooted crop species, such as chickpea (Li et al., 2003; Veneklaas et al., 2003). In addition, studies that examine the role of cluster roots for acquiring N provide strong indications that they can access complex forms of organic N, but this remains to be confirmed (Schmidt et al., 2003).

The ability to mobilise sparingly soluble nutrients such as P and Fe, and to access organic forms of N and P is a highly desirable trait. It would be naïve to think that crops that can mobilise sparingly soluble nutrients no longer require an input of (organic or inorganic) fertiliser P. Like infection with mycorrhizal fungi, cluster roots enable more efficient utilisation of soil P (mobilisation), and the optimum soil P concentration for plant productivity is lower. Use of species that have fast rates of carboxylate exudation, and, therefore, locally enhance the carboxylate concentration to a great extent, would result in less immobilisation of added P by soil particles, and hence provide greater P availability for plant growth. In some soils, more than 80% of the P applied as fertiliser may be fixed to soil particles; therefore, using species with cluster roots would allow a greater proportion of P added as fertiliser to be used by the crop. Consequently, with lower applications of P fertilisers, over time there will be less P leaching from soils providing benefits for conservation, particularly of river systems and wetlands.

A more widespread use of crops with cluster roots, or with alternative structures with similar function, such as the third-order laterals described for *L. luteus* (Brennand and Bolland, 2003), will require a better understanding of the interactions between plant functioning and soil properties. The third-order lateral roots of, *L. luteus* may function in a similar manner to cluster roots, although this remains to be established. *L. luteus* is better at accessing P than *L. angustifolius*, which makes neither cluster roots nor third-order laterals (Brennan and Bolland, 2003). Furthermore, for cluster roots to be effective, management practices should be adopted that protect the cluster roots, because these often develop in the drought-prone surface soils, and are also highly sensitive to fertiliser inputs and mechanical damage.

Cluster-root morphology

Purnell (1960) coined the term 'proteoid' roots, because of their widespread occurrence within the Proteaceae she had studied. Proteoid roots have since been described in many other families, first in Fabaceae: *Viminaria juncea* (Lamont, 1972a), *L. albus* (Trinick, 1977) and a range of Australian species (Adams et al., 2002). Subsequently, similar root structures have been found in many other species from a range of families (Table 1), and 'proteoid' roots are consequently frequently referred to as 'cluster' roots (Lamont, 2003).

Cluster roots are morphologically distinct structures that are specialised for nutrient mobilisation and consist of very large numbers of short, determinate, unbranched, lateral rootlets that develop as discrete units in regions proximal from the tip of primary or secondary lateral roots (Skene, 2003). Each of these clustered regions may be several centimetres in length (Figure 1). A lateral root axis may have several clustered regions along its length, each separated by unbranched regions (Lamont, 1972b). The density of lateral rootlets in a cluster root depends upon the species. In the genus *Lupinus*, e.g., *L. albus*, a cluster root generally has 1–4 rootlets mm^{-1} root axis (Dinkelaker et al., 1989; Johnson et al., 1994). The Proteaceae, however, provide "probably the most spectacular example of clumping of lateral roots" (McCully, 1975) which may reach 100 rootlets mm^{-1} root axis in *Hakea prostrata* (Lamont, 1972b). The rootlets have their origin in pericycle cells located over protoxylem poles, and, therefore, a greater number of protoxylem poles leads to greater rootlet density.

The dense packing of rootlets in a longitudinal file that is typical for cluster roots (Lamont, 2003; Peek et al., 2003) has a remarkable similarity to that of the adventitious roots in the adhesive root pads used for climbing, which develop on the stems of the climbing fig, *Ficus pumila* (Groot et al., 2003).

The pattern of cluster-root development is either 'simple' and bottlebrush-like, as in *Hakea* (Lamont, 1972b; Purnell, 1960; Figure 1B), *Lupinus* (Peek et al., 2003; Figure 1C) and *Protea* (Figure 1D), or 'compound' and mat-forming, as in *Banksia* species (Grierson, 1992; Figure 1A, F). In the South African Proteaceae both 'simple' and 'compound' cluster-roots are typically found on individual plants (Lamont,

Table 1. Families with specialised root structures. The species belonging to the Cyperaceae have 'dauciform' (carrot-shaped) roots, as shown in Figure 1E. There is no information on their capacity to release exudates. Species belonging to the other families have 'proteoid' or 'cluster' roots. *Ficus pumila* has adhesive pads, which have a similar anatomy and morphology as cluster roots, and may even be similar in terms of exudation of organic substances (Groot et al., 2003).

Family	Representative species	Reference(s)
Betulaceae	*Alnus incana*	Hurd and Schwintzer, 1996
Casuarinaceae	*Casuarina cunninghamiana*	Diem et al., 1981; Reddell et al., 1997; Zaïd et al., 2003
	C. glauca	Arahou and Diem, 1997; Zaïd et al., 2003
	Gymnostoma papuanum	Racette et al., 1990
Cucurbitaceae	*Cucurbita pepo*	Waters and Blevins, 2000
Cyperaceae	*Carex flacca*	Davies et al., 1973
	Cyathochaete avenacea	Lamont and McComb, 1974
Eleagnaceae	*Hippophae rhamnoides*	Skene, 1998
Fabaceae	*Lupinus albus*	Kania et al., 2003; Shen et al., 2003; Trinick, 1977; Veneklaas et al., 2003; Watt and Evans, 2003
	L. cosentinii	Clements et al., 1993
	L. luteus	Brennan and Bolland, 2003
	Viminaria juncea	Lamont, 1972a
Moraceae	*Ficus benjamina*	Rosenfield et al., 1991
Myricaceae	*Comptonia peregrina*	Hurd and Schwintzer, 1997
	Myrica cerifera	Louis et al., 1990
	M. gale	Crocker and Schwintzer, 1993
Proteaceae	*Banksia grandis**	Lambers et al., 2002
	*Banksia integrifolia**	Grierson, 1992
	Dryandra sessilis	Roelofs et al., 2001
	Hakea obliqua	Lamont, 1972b
	Hakea prostrata	Shane et al., 2003
	Leucadendron laureolum	Lamont et al., 1984
	Telopea speciosissima	Grose, 1989

*Compound cluster roots in *Banksia* and *Dryandra* species. Simple cluster roots in *Hakea* species

1983). Table 1 lists some representative species from several families that develop simple or compound cluster roots.

Cluster-root strategies of nutrient acquisition

The functions of the vast array of components that have been reported in root exudates and their significance in nutrient (e.g., P) mobilisation in the rhizosphere are of great importance to understanding plant mineral nutrition, if we are to find ways of improving it. In general, root exudates are a blend of many compounds including enzymes, water, carboxylates, sugars, phenolic compounds and mucilage (Kania et al., 2003; Marschner, 1995; Wasaki et al., 2003). Of particular importance to inorganic P nutrition is any component that increases the plant availability of sparingly soluble P. These components include carboxylic acid anions (carboxylates), such as citrate and malate (Gardner et al., 1982), that are capable of solubilising complexed metals by chelation and by displacing P bound to positively charged soil particles, thus increasing the availability of P (Gerke et al., 2000a). For example, the influence of citrate exudation from proteoid roots of *L. albus* on P mobilisation was elegantly demonstrated by precipitation of calcium citrate in the cluster-root rhizosphere (Dinkelaker et al., 1989). The results illustrate the chemical nature of enhanced P mobilisation through citrate exudation by cluster roots, and that uptake of P was from a previously insoluble P source that would be largely inaccessible to most plant roots.

Morphological components

In P-limited *L. albus* the exudation rate of citrate from the cluster roots is considerably faster than that from unbranched roots of similar age (Keerthisinghe et al., 1998). Citrate-exudation rates are even faster from roots of P-limited *L. angustifolius*, which does not make cluster roots (Römer et al., 2000). However, *L. angustifolius* is far less efficient at acquiring P from low-P soils than *L. albus* (Bolland, 1997). This indicates that citrate exudation from non-cluster roots is not as efficient in mobilising P as that from dense clusters of rootlets. In light of these findings it seems that the morphological component of the cluster-root strategy for P-acquisition is an important adjunct to the chemical component. It is hypothesised that the tightly packed rootlets covered with an enormous number of root hairs greatly enhance the carboxylate concentration in the rhizosphere, and also ensure an efficient recovery of solubilised P (Gerke et al., 2000b). For plants without cluster roots, carboxylate exudation rates have to be much faster in order to reach local concentrations that are effective in mobilising P, and even then the recovery of solubilised P will not be as efficient because of diffusion away from the root surface.

Physiological components

Once P has been mobilised in the soil, it needs to be taken up, using P_i transporters. The system responsible for P_i uptake by cluster roots appears to be similar to that used in non-cluster roots (Smith et al., 2003). The systems responsible for P_i uptake from the root environment are $H_2PO_4^-/H^+$ symporters, belonging to the Pht1 family. In *Arabidopsis thaliana* and many other species, most of the genes of the Pht1 family that are expressed in roots are up-regulated in P-deficient plants (Smith et al., 2003). The efficient acquisition of P_i by cluster roots arises primarily from their capacity to increase the availability of soil P_i immediately adjacent to the rootlets by excretion of carboxylates, protons and phosphatases (Smith et al., 2003). Shane et al. (2003) provided evidence that *H. prostrata*, a P-sensitive species belonging to the Proteaceae, has a very low capacity to down-regulate its P_i transporters. This would account, in part, for the P-toxicity symptoms that are found in many species of this family (Grose, 1989; Lambers et al., 2002; Parks et al., 2000). The sensitivity to P fertilization is also partly due to a low tolerance of increased P concentrations in leaf tissues (Lambers et al., 2002; Parks et al., 2000).

Factors influencing cluster-root development

Phenotypic effects

The contribution of cluster-root biomass to the total root system can be increased or decreased by variation in root aeration (Crocker and Schwintzer, 1993; Lamont, 1976; Racette et al., 1990) and pH, and may also be affected to some extent by the presence of specific micro-organisms (Dinkelaker et al., 1995; Lamont, 2003). This variation in the proportion of cluster-root biomass suggests that factors other than plant nutritional status alter their growth. However, aeration and pH both influence nutritional status, as might micro-organisms, which in turn influences growth rates, but the relationship between factors which influence plant growth rate and the development of cluster roots has yet to be thoroughly investigated. For the Proteaceae native to the Mediterranean region of the southwest of Western Australia, the presence of moisture in the upper soil/litter layers is a key requirement for the seasonal surge of cluster-root growth (Lamont, 2003). The length of time during which cluster roots are produced in the upper soil layers depends on the length of the wet season, which may be only 2–4 months in Mediterranean regions, compared with wetter regions where functional dense root mats may persist for longer periods. Cluster roots occur in surface layers (Dinkelaker et al., 1995; Purnell, 1960), and have recently been found also in deeper soil layers that are rich in $FePO_4$ (Pate et al., 2001). This indicates, firstly, that even after the surface layers dry out, growth of cluster roots may continue deeper in the soil profile. Secondly, this supports the contention that cluster roots are more likely to form where nutrients can be made available (Dinkelaker et al., 1995; Pate et al., 2001; Purnell, 1960).

Once the environmental and local conditions for cluster-root development are met, the plant's nutritional status has the next greatest influence on cluster-root initiation, growth and carboxylate exudation. P has been repeatedly demonstrated to be the key element having the greatest influence on root development that leads to cluster-root initiation and growth (Marschner, 1995). In *L. albus*, P deficiency (Keerthisinghe et al., 1998; Watt and Evans, 1999) and to a lesser extent Fe deficiency (Hagström et al., 2001; Liang and Li, 2003; White and Robson, 1989) enhance cluster-root development. For plants equally supplied with P, a low supply of N (*H. prostrata*; Lamont,

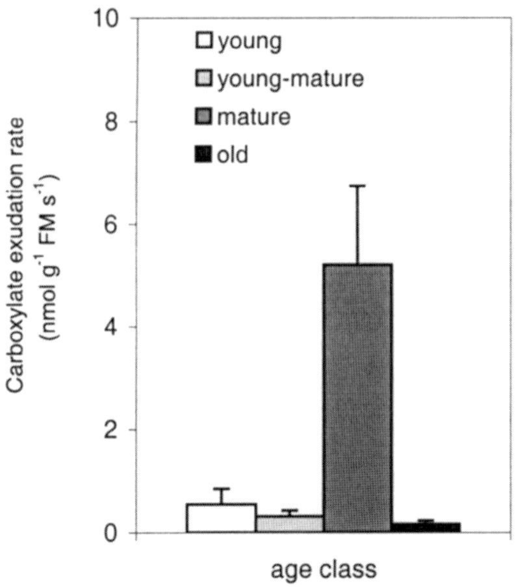

Figure 2. Carboxylate exudation by cluster roots of *Hakea prostrata* at four developmental stages. The 'exudative burst' lasts for a few days. Plants were grown in nutrient solution, with a low P concentration (4 μm P was added every second week). Data are mean values based on 8, 4, 9 and independent experiments, for young, young-mature, mature and old cluster roots, respectively; error bars give SE.

1973) can also account for some enhanced cluster-root growth.

Foliar feeding of P to the shoots of *L. albus* suppresses cluster-root development to a similar degree as when P is supplied in the soil (Gilbert et al., 1997; Marschner et al., 1987). This indicates that the internal P concentration regulates cluster-root development (Gilbert et al., 1997). Evidence from split-root studies with *H. prostrata* and *L. albus* suggests that shoot P concentration has a greater influence on cluster-root growth and carboxylate exudation than root P concentration (Shane et al., 2002; 2003). This requires the 'sensing' of the plant P status, in a manner that still needs to be further explored (Abel et al., 2002). The exogenous application of synthetic auxins to the shoots of P-sufficient *L. albus* plants overrides the suppression of cluster root growth caused by high P supply to roots (Gilbert et al., 2000). Similarly, auxin-transport inhibitors suppress cluster-root growth in P-stressed *L. albus*. This suggests that auxins, which are known to influence lateral root development (Salisbury and Ross, 1992) are involved in cluster-root initiation, but they probably play a role in the final stage(s) in the signal-transduction pathway (Gilbert et al., 2000).

When grown in solution culture low in P, cluster roots of *L. albus* stop elongating when the lateral rootlets are approximately three to four days old, coinciding with a burst of citrate exudation (Hagström et al., 2001; Watt and Evans, 1999). In *L. albus*, a low Fe supply also results in enhanced exudation of citrate (Hagström et al., 2001). In the Proteaceae, the exudation of phenolic compounds (soil-grown *H. undulata*; Dinkelaker et al., 1995) and carboxylates (nutrient solution-grown *H. prostrata*; Figure 2) also follow a developmental time course with a brief exudative burst clearly demarcated during rootlet development.

Genotypic effects

Since their initial discovery in the Proteaceae, cluster roots have been found in many, taxonomically distant families (Table 1). Moreover, in the genus *Lupinus*, cluster roots are only found in two out of five clades (Skene, 1998). It would seem that none of the *Lupinus* species that belong to the two clades from the New World (Ainouche and Bayer, 1999) are capable of making cluster roots. This raises a fascinating question as to the genetic control of cluster-root formation in closely related species, and presents challenging opportunities to possibly genetically modify non-cluster forming *Lupinus* species, to provide them with the capacity to make cluster roots.

Dinkelaker et al. (1995) performed reciprocal grafting of rootstocks of *L. albus*, which develops cluster roots, with *L. angustifolius*, which does not. Their results demonstrated that cluster-root development depends on the rootstock, and, therefore, the roots must belong to a cluster-root forming species (Figure 3A). These reciprocal grafting studies can be interpreted as follows: 1) that *L. angustifolius* roots were not capable of responding to some putative 'signal' from the shoots of *L. albus*; or, 2) that this 'signal' is not perceived by the roots of *L. angustifolius*; or, 3) at least part of the signal-transduction pathway for cluster-root development originates from the cluster root-forming rootstock. Grafting experiments by us (M.W. Shane, M.D. Cramer and C. Szota), where an individual shoot has two rootstocks, suggest that a 'signal' originating from the cluster-root forming rootstock need not be invoked as an explanation. We followed an approach (C. Beveridge, pers. comm.; Beveridge et al., 1994) in which two shoots, e.g., one from *L. albus* and another one from *L. angustifolius* (and controls), were grafted together at the epicotyl (Figure 3B–E). After approximately two weeks, one

Figure 1.

Figure 3.

←

Figure 1. Specialised root structures: Compound cluster roots of (A) *Banksia grandis* and (F) *B. oreophila* (Proteaceae), grown in nutrient solution with less than 1 μm P. Simple cluster roots of (B) *Hakea prostrata* (Proteaceae), (C) *Lupinus albus* (Fabaceae) grown in nutrient solution with 1 μm P, and (D) *Protea repens* (Proteaceae) excavated in its natural habitat. Very dense and long root hairs give the appearance of a cotton ball to (E) dauciform roots of *Cyathochaete avenacea* (Cyperaceae). Scale bars: 30 mm (A), 15 mm (B), 12.5 mm (C), 7.5 mm (E) and 20 mm (F).

Figure 3. Grafting of lupins. (A) Graft of a shoot of *Lupinus angustifolius* onto the rootstock of *L. albus*. Note the cluster roots (arrow). (B) Two plants of *L. albus*, grafted together at the epicotyl (arrow) when they were two weeks old. (C) One plant of *L. albus* and one of *L. angustifolius*, grafted together (arrow pointing to the right) at two weeks. Once the graft was established, the shoot of *L. albus* was severed, and the grafted plant was allowed to continue to grow for several weeks, until flowering. Note the presence of cluster roots on the roots of *L. albus* (left arrow), and the absence of cluster roots on those of *L. angustifolius* (right arrow). (D) A graft as in (C), but now with the shoot of *L. albus* left intact, and that of *L. angustifolius* severed. The arrow points towards the union. (E) The root systems of the plant shown in (D), showing several cluster roots (right arrow) on the roots of *L. albus*, but none on those of *L. angustifolius* (left arrow).

of the shoots, either *L. albus* or *L. angustifolius* was removed and the growth of the two root systems monitored over the course of the following weeks. The results confirmed those of Dinkelaker et al. (1995), showing that cluster-root development is restricted to the rootstocks of *L. albus*, regardless of the scion or adjacent rootstock. Therefore, the absence of cluster roots on *L. angustifolius* roots is not due to the absence of a putative root signal that would trigger a shoot signal. Applying Occam's razor, it may be concluded that the *L. angustifolius* root is genetically not capable of forming cluster roots or does not perceive the shoot 'signal (s)' as does the *L. albus* root.

The identity of the 'signal (s)' remains a mystery. Auxins and cytokinins are potent regulators of lateral root formation (Salisbury and Ross, 1992), and it is, therefore, likely that they have a role in the transduction pathway to initiate development of the dense clusters of rootlets. The lack of auxin-induced development of cluster roots in root systems of plants that cannot develop cluster roots suggests that the mere presence of auxins is not responsible for the localised proliferation of dense clusters of lateral rootlets (Skene and James, 2000). The sensing of low internal P status and the subsequent initiation of cluster roots may also involve cytokinins (Abel et al., 2002; De Groot et al., 2003), but this has not been determined for cluster-root forming species.

Carboxylate synthesis and exudation

The physiology, biochemistry and molecular biology of citrate exudation from the cluster roots of the agriculturally important *L. albus* have been studied in greater detail than those of any other cluster-root forming species. We are rapidly accumulating knowledge of the genes involved in carbon metabolism that are up- or down-regulated at the time before citrate exudation peaks during cluster-root development (Kania et al., 2003; Uhde-Stone et al., 2003). However, it is very likely that transport, rather than synthesis, is the rate-determining step for citrate exudation (Watt and Evans, 1999). Citrate exudation probably involves at least an anion channel (Neumann et al., 2000). This conclusion, which was based on evidence obtained using anion-channel blockers, needs independent confirmation, perhaps using the patch-clamp technique (Ryan et al., 1997; 2003). We also need to assess which rootlet cells are involved in carboxylate synthesis and exudation. Are the root hairs important for P uptake, or are these also responsible for carboxylate exudation?

Many species, including *L. albus* and *Cicer arietinum*, show enhanced carboxylate exudation when they become P deficient (Neumann and Römheld, 1999). For *H. prostrata*, it was found that this is, at least partly, a response to the internal P status (Shane et al., 2003). However, as with the response to Al (Ryan et al., 1997), there may well be additional responses to external P, which remain to be further explored.

Carboxylates are not only important in the mobilisation of P, but also for the detoxification of Al in the rhizosphere, stimulating growing interest in genotypes with altered carboxylate metabolism. Mutants with a reduced activity of NADP-isocitrate dehydrogenase, which would reduce citrate catabolism and potentially enhance citrate exudation, offer tools to explore to what extent the metabolism of citrate affects its exudation (Kihara et al., 2002). In *Nicotiana tabacum*, transformants have been produced that have a modified citrate metabolism. Such transformants allow testing the hypothesis that increased or decreased citrate synthesis or breakdown affect a plant's Al resistance. In tobacco, internal citrate concentrations are largely insensitive to large changes in mitochondrial citrate synthase and cytosolic isocitrate dehydrogenase activities. Other factors, such as transport out of the roots, are more likely to the rate control citrate exudation (Delhaize et al., 2003).

Amongst the many exciting recent developments is the observation that roots of *C. arietinum* and *L. albus* release malate as well as citrate, and that this ratio depends on soil conditions including soil pH (Veneklaas et al., 2002). This observation offers a tool to assess which genes are involved in the switch from citrate exudation to malate exudation. If *Arabidopsis thaliana* shows a similar response to soil conditions and pH, this would offer a tool to explore genes that are up- or down-regulated depending on soil conditions. Some genotypes that are very efficient at accessing sparingly soluble P do, indeed, release carboxylates (Narang et al, 2000), but it is not known if the composition of the released carboxylates depends on soil conditions. Studies with Proteaceae are lacking, but it appears that they release a wider spectrum of carboxylates (Roelofs et al., 2001), and that the composition also responds to the chemical nature of P-containing compounds in the soil (Lambers et al., 2002).

Concluding Remarks

Bringing together a group of scientists at the University of Western Australia, Perth, in November 2001, to discuss the structure and functioning of cluster roots and plant responses to P deficiency, has led to a review of the state of the art and identified major challenges for future research. Such challenges include the identification of the genes that are involved in the production of cluster roots in *Lupinus* species. That knowledge may allow us to, firstly, breed or engineer other *Lupinus* species with cluster roots, and, secondly, produce species from other genera with cluster roots. Such new crops could be more efficient at acquiring P from sparingly available sources, and reduce P losses due to run-off and leaching.

Another challenge is to identify the mechanisms that are responsible for the switch from predominantly citrate exudation to mainly malate exudation, as dependent on soil conditions. That switch must involve 'sensing' of the initial soil conditions, which subsequently allows the 'design' of a rhizosphere most amenable to acquiring P at the lowest metabolic cost. That is, small carboxylates should be preferred when these are as effective as larger ones, whereas larges ones should be preferred when these are much more effective. We also need further research on the sensing of the internal P status, and on the signal-transduction pathway that leads to numerous adaptive responses, including the development of cluster roots, synthesis and release of exudates, and the expression of P transporters. *L. albus* is a great model plant when it comes to a crop species, but we have to be aware that many species belonging to the Proteaceae are much better at accessing P from a wider range of P-impoverished soil types. These species may, therefore, offer important information that we cannot obtain from *L. albus*. Also, the intricate relationship between species of the Proteaceae and the soil in their natural habitat offers further opportunities to study how soil conditions are sensed and account for plant distribution. Moreover, an analysis of the extreme sensitivity to P supply in many species of the Proteaceae may lead to a better understanding of P metabolism in general.

Proteoid or cluster roots may once have been thought of as a curiosity in slow-growing woody species. However, their discovery in species outside the Proteaceae and the growing knowledge on their functioning in P acquisition from sparingly available sources has, quite rightly, triggered renewed interest. Clearly, with non-renewable P reserves being rapidly depleted, efficient P-acquisition mechanisms must become a major research focus.

Acknowledgements

We thank Chris Szota for his help in the grafting experiments, and for his assistance in making the plate for Figures 1 and 3. We gratefully acknowledge the financial support provided by the University of Western Australia, the Grains Research and Development Corporation, Kluwer Academic Publishers, the Australian Society for Plant Physiologists, Kings Park & Botanic Gardens, and CSBP Futurefarm.

References

Abel S A, Ticconi C A and Delatorre C A 2002 Phosphate sensing in plants. Physiol. Plant. 115, 1–8.

Adams m A, Bell T L and Pate J S 2002 Phosphorus sources and availability modify growth and distribution of root clusters and nodules of native Australian legumes. Plant, Cell Environ. 25, 837–850.

Ainouche A-K and Bayer R J 1999 Phylogenetic relationships in *Lupinus* (Fabaceae: Papilionoideae) based on internal transcribed spacer sequences (its) nuclear ribosomal DNA. Am. J. Bot. 86: 590–607.

Arahou M and Diem H G 1997 Iron deficiency induces cluster (proteoid) root formation in *Casuarina glauca*. Plant Soil 196, 71–79.

Bates T R and Lynch J P 1996 Stimulation of root hair elongation in *Arabidopsis thaliana* by low phosphorus availability. Plant Cell Environ. 19, 529–538.

Bolland m D A 1997 Comparative phosphorus requirement of four lupin species. J. Plant Nutr. 20, 1239–1253.

Beveridge C A, Ross J J and Murfet I C 1994 Branching mutant rms-2 in *Pisum sativum*. Grafting studies and endogenous indole-3-acetic acid levels. Plant Physiol. 104, 953–959.

Brennan R F and Bolland M D A 2003 *Lupinus luteus* cv. Wodjil takes up more phosphorus and cadmium than *Lupinus angustifolius* cv. Kalya. Plant Soil 248, 167–185.

Clements J C, White P F and Buirchell B J 1993 The root morphology of *Lupinus angustifolius* in relation to other *Lupinus* species. Aust. J. Agric. Res. 44, 1367–1375.

Crocker L J and Schwintzer C R 1993 Factors affecting formation of cluster roots in *Myrica gale* seedlings in water culture. Plant Soil 152, 287–298.

Davies J, Briarty L g and Rieley J O 1973 Observations on the swollen lateral roots of the Cyperaceae. New Phytol. 72, 167–174.

De Groot C, Marcelis L F M, Van den Boogaard R and Lambers H 2003 Interaction of nitrogen and phosphorus nutrition in determining growth. Plant Soil 248, 257–268.

Delhaize E, Ryan P R, Hocking P J, and Richardson A E 2003 Effects of altered citrate synthase and isocitrate dehydrogenase expression on internal citrate concentrations and citrate efflux of tobacco (*Nicotiana tabacum* L.) roots. Plant Soil 248, 137–144.

Diem H G, Gueye I, Gianinazzi-Pearson V, Fortin J A and Dommergues Y R 1981 Ecology of VA mycorrhizae in the tropics: the semi-arid zone of Senegal. Oecol. Plant. 2, 53–62.

Dinkelaker B, Römheld V and Marschner H 1989 Citric acid excretion and precipitation of calcium citrate in the rhizosphere of white lupin (*Lupinus albus* L.). Plant, Cell Environ. 12, 285–292.

Dinkelaker B, Hengeler C and Marschner H 1995 Distribution and function of proteoid root clusters and other root clusters. Bot. Acta 108, 183–200.

Fransson A-M, Van Aarle I, Olsson P A and Tyler G 2003 *Plantago lanceolata* L. and *Rumex acetosella* L. differ in their utilisation of soil phosphorus fractions. Plant Soil 248, 285–295.

Gardner W K, Parbery D G and Barber D A 1982 The acquisition of phosphorus by *Lupinus albus* L. I. Some characteristics of the soil/root interface. Plant Soil 67, 19–32.

Gerke J, Beißner L and Römer W 2000a The quantitative effect of chemical phosphate mobilization by carboxylate anions on P uptake by a single root. I. The basic concept and determination of soil parameters. J. Plant Nutr. Soil Sci. 163, 207–212.

Gerke J, Römer W and Beißner L 2000b The quantitative effect of chemical phosphate mobilization by carboxylate anions on P uptake by a single root. II. The importance of soil and plant parameters for uptake of mobilized P. J. Plant Nutr. Soil Sci. 163, 213–219.

Gilbert G A, Allan D A and Vance C P 1997 Phosphorus deficiency in white lupin alters root development and metabolism. *In* Radical Biology: Advances and perspectives on the function of plant roots. Eds. Flores H E, Lynch D and Eissenstat D pp. 92–103, American Society of Plant Physiologists, Rockville.

Gilbert G A, Knight J D, Vance C P and Allan DL 2000 Proteoid root development of phosphorus deficient white lupin is mimicked by auxin and phosphonate. Ann. Bot. 85, 921–928.

Grierson P F 1992 Organic acids in the rhizosphere of *Banksia integrifolia* L.f.. Plant Soil 144, 259–265.

Groot E P, Sweeney E J and Rost T L 2003 Development of the adhesive pad on climbing fig (*Ficus pumila*) stems from clusters of adventitious roots. Plant Soil 248, 85–96.

Grose M J 1989 Phosphorus nutrition of seedlings of the Waratah *Telopea speciosissima* (Sm.) R. Br. (Proteaceae). Aust. J. Bot. 37, 313–320.

Hagström J, James W M and Skene K R 2001 A comparison of structure, development and function in cluster roots of *Lupinus albus* L. under phosphate and iron stress. Plant Soil 232, 81–90.

He Y, Liao H and Yan X 2003 Localized supply of phosphorus induces root morphological and architectural changes of rice in split and stratified soil cultures. Plant Soil 248, 247–256.

Hinsinger P, Plassard C, Tang C and Jaillard B 2003 Origins of root-mediated pH changes in the rhizosphere and their responses to environmental constraints – a review. Plant Soil 248, 43–59.

Horst W J and Waschkies C 1987 Phosphatversorgerung von Sommerweizen (*Triticum aestivum* L.) in Mischkultur mit Weiszer Lupine (*Lupinus albus* L.). Z. Pflanzenernähr. Bodenk. 150, 1-8.

Hurd T M and Schwintzer C R 1996 Cluster roots of *Alnus incana* ssp. *Rugosa* in the field and four *Alnus* species in water culture with phosphorus and iron deficiency. Can. J. Bot. 74, 1684–1686.

Hurd T M and Schwintzer C R 1997 Formation of cluster roots and mycorrhizal status of *Comptonia peregrina* and *Myrica pensylvanica* in Maine, USA. Physiol. Plant. 99, 680–689.

Jones D L, Dennis P G, Owen A G and van Hees, P A W 2003 Organic acid behavior in soils – misconceptions and knowledge gaps. Plant Soil 248, 31–41.

Johnson J F, Allan D L and Vance C P 1994. Phosphorus stress-induced proteoid roots show altered metabolism in *Lupinus albus*. Plant Physiol. 104, 757–665.

Kamh M, Horst W J, Amer F, Mostafa H and Maier P 1999 Mobilization of soil and fertilizer phosphate by cover crops. Plant Soil 211, 19–27.

Kania, A, Langlade N, Martinoia E and Neumann G 2003 Phosphorus deficiency-induced modifications in citrate catabolism and in cytosolic pH as related to citrate exudation in cluster roots of white lupin. Plant Soil 248, 117–127.

Keerthisinghe G, Hocking P, Ryan P R and Delhaize E 1998 Proteoid roots of lupin (*Lupinus albus* L.): Effect of phosphorus supply on formation and spatial variation in citrate efflux and enzyme activity. Plant Cell Environ. 21, 467–478.

Kihara T, Ohono T, Koyama H, Sawafuji T and Hara T 2003 Characterization of NADP-isocitrate dehydrogenase expression in a carrot mutant cell line with enhanced citrate excretion. Plant Soil 248, 145–153.

Lambers H, Juniper D, Cawthray G R, Veneklaas E J and Martinez E 2002. The pattern of carboxylate exudation in *Banksia grandis* (Proteaceae) is affected by the form of phosphate added to the soil. Plant Soil 238, 111–122.

Lamont B 1972a 'Proteoid' roots in the legume *Viminaria juncea*. Search 3, 90–91.

Lamont B 1972b The morphology and anatomy of proteoid roots in the genus *Hakea*. Aust. J. Bot. 20, 155–174.

Lamont B 1973 Factors affecting the distribution of proteoid roots within the root systems of two *Hakea* species. Aust. J. Bot. 21, 165–187.

Lamont B B and McComb A J 1974 Soil microorganisms and the formation of proteoid roots. Aust. J. Bot. 22, 681–688.

Lamont B 1976 The effects of seasonality and waterlogging on the root systems of a number of *Hakea* species. Aust. J. Bot. 24, 691–702.

Lamont B 1983 Proteoid roots in the South African Proteaceae. S. Afr. J. Bot. 49, 103–123.

Lamont B 2003 Structure, ecology and physiology of root clusters – a review. Plant Soil 248, 1–19.

Lamont B, Brown G. and Mitchell D T 1984 Structure, environmental effects on their formation and function of proteoid roots in *Leucadendron laureolum* (Proteaceae). New Phytol. 97, 381–390.

Li L, Tang C, Rengel Z and Zhang F S 2003 Chickpea facilitates phosphorus uptake by intercropped wheat from an organic phosphorus source. Plant Soil 248, 297–303.

Liang R and Li C 2003 Differences in cluster-root formation and carboxylate exudation in *Lupinus albus* L. under different nutrient deficiencies. Plant Soil 248, 221–227.

Louis I, Racette S and Torrey J g 1990 Occurrence of cluster roots in *Myrica cerifera* L. (Myricaceae) in water culture in relation to phosphorus nutrition. New Phytol. 115, 311–318.

Marschner H 1995 Mineral Nutrition of Higher Plants. 2nd edn. Academic Press, London.

Marschner H, Römheld V and Cakmak I 1987 Root-induced changes of nutrient availability in the rhizosphere. J. Plant Nutr. 10, 1175–1184.

McCully M E 1975 The development of lateral roots. *In* The development and function of roots Third Cabot Symposium. Eds. Torrey J G and Clarkson D T pp. 105–124. Academic Press, New York.

Myers N, Mittermeier R A., Mittermeier C G, da Fonseca G A B and Kent J 2000 Biodiversity hotspots for conservation priorities. Nature 403, 853–858.

Narang R, Bruene A, and Altmann T 2000 Analysis of phosphate acquisition efficiency in different *Arabidopsis* accessions. Plant Physiol. 124, 1786–1799.

Neumann G and Römheld V 1999 Root excretion of carboxylic acids and protons in phosphorus-deficient plants. Plant Soil 211, 121–130.

Neumann G, Massonneau A, Langlade N, Dinkelaker B, Hengeler C, Römheld V and Martinoia E 2000 Physiological aspects of cluster root function and development in phosphorus-deficient white lupin (*Lupinus albus* L.). Ann. Bot. 85, 909–919.

Parks S E, Haigh A M and Creswell G C 2000 Stem tissue phosphorus as an index of the phosphorus status of *Banksia ericifolia* L. f. Plant Soil 227, 59–65.

Pate J S and Dell B 1984 Economy of nutrients in sandplain species. *In* Kwongan, Plant Life of the Sandplain. Eds. J S Pate and J S Beard. pp. 227–252. The University of Western Australia Press, Nedlands.

Pate J S, Verboom W H and Galloway P D 2001 Co-occurrence of Proteaceae, laterite and related oligotrophic soils: coincidental associations or causative inter-relationships? Aust. J. Bot. 49, 529–560.

Peek C S, Robson A D and Kuo J 2003 The formation, morphology and anatomy of cluster root of *Lupinus albus* L. as dependent on soil type and phosphorus supply. Plant Soil 248, 237–246.

Purnell H M 1960 Studies of the family Proteaceae. I. Anatomy and morphology of the roots of some Victorian species. Aust. J. Bot. 8, 38–50.

Reddell P, Yun Y & Shipton WA 1997 Cluster roots and mycorrhizae in *Casuarina cunninghamiana*: their occurrence and formation in relation to phosphorus supply. Aust. J. Bot. 45: 41–51.

Römer W, Kang D K, Egle K Gerke J and Keller H 2000 The acquisition of cadmium by *Lupinus albus* L., *Lupinus angustifolius* L., and *Lolium multiflorum* Lam.. J. Plant Nutr. Soil Sci. 163, 623–628.

Racette S, Louis I, and Torrey J g 1990 Cluster roots formation by *Gymnostoma papuanum* (Casuarinaceae) in relation to aeration and mineral nutrient availability in water culture. Can. J. Bot. 68, 2564–2570.

Roelofs R F R, Rengel Z., Cawthray G R, Dixon K W and Lambers H 2001. Exudation of carboxylates in Australian Proteaceae: chemical composition. Plant, Cell Environ. 24, 891–904.

Rosenfield C-L, Reed D W and Kent M W 1991. Dependency of iron reduction on development of a unique root morphology in *Ficus benjamina* L. Plant Physiol. 95: 1120–1124.

Ryan P R, Skerrett M, Findlay G P, Delhaize E and Tyerman S D 1997 Aluminium activates an anion channel in the apical cells of wheat roots. Proc. Natl. Acad. Sci. USA 94, 6547–6552.

Ryan P R, Dong B, Watt M, Kataoka T and Delhaize E 2003 Strategies to isolate transporters that facilitate organic anion efflux from plant roots. Plant Soil 248, 61–69.

Salisbury F B and Ross C W 1992 Plant Physiology. Fourth edition. Belmont, CA: Wadsworth, Inc.

Schmidt S, Mason M, Sangtiean T and Stewart G R 2002 Do cluster roots of *Hakea actities* (Proteaceae) acquire complex organic nitrogen? Plant Soil 248, 157–165.

Shane M W, De Vos M, De Roock S and Lambers H. 2003 Shoot phosphorus status regulates citrate exudation and cluster root growth in divided root systems of *Lupinus albus* Plant, Cell Environ., in press.

Shane M W, De Vos M, De Roock S, Cawthray G R and Lambers H 2003 Effects of external phosphorus supply on internal phosphorus concentrations and the initiation, growth and exudation of cluster roots in *Hakea prostrata* R. Br. Plant Soil 248, 209–219.

Shen J, Rengel Z, Tang C and Zhang F 2003 Role of phosphorus nutrition in development of cluster roots and release of carboxylates in soil-grown *Lupinus albus.* Plant Soil 248, 199–206.

Skene K R 1998 Cluster roots: some ecological considerations. J. Ecol. 86, 1060–1064.

Skene K R 2003 The evolution of physiology and development in the cluster root: teaching an old dog new tricks? Plant Soil 248, 21–30.

Skene K R and James W M 2000 A comparison of the effects of auxin on cluster root initiation and development in *Grevillea robusta* Cunn. ex R. Br. (Proteaceae) and in the genus *Lupinus* (Leguminosae). Plant Soil 219, 221–229.

Smith F W, Mudge S R, Rae A L and Glassop D 2003 Phosphate transport in plants. Plant Soil 248, 71–83.

Solaiman M Z and Abbott L K 2003 Phosphorus uptake by a community of arbuscular mycorrhizal fungi in jarrah forest. Plant Soil 248, 313–320.

Specht R L and Specht A 1999 Australian plant communities. Dynamics of structure, growth and biodiversity Oxford University Press, Oxford.

Trinick M J 1977 Vesicular-arbuscular infection and soil phosphorus utilization in *Lupinus* spp. New Phytol. 78, 297–304.

Uhde-Stone C, Gilbert G, Johnson J M-F, Litjens R, Zinn K E, Temple S J, Vance C P and Allan D L 2003 Acclimation of white lupin to phosphorus deficiency involves enhanced expression of genes related to organic acid metabolism. Plant Soil 248, 99–116.

Veneklaas E J, Stevens J, Cawthray G R, Turner S, Grigg A M and Lambers H 2003 Chickpea and white lupin rhizosphere carboxylates vary with soil properties and enhance phosphorus uptake. Plant Soil 248, 187–197.

Verboom W H and Pate J S 2003 Relationships between cluster root-bearing taxa and laterite across landscapes in south west Western Australia: an approach using airborne radiometric and digital elevation models. Plant Soil 248, 321–333.

Wasaki J, Yamamura T, Shinano T and Osaki M 2003 Secreted acid phosphatase is expressed in cluster roots of lupin in response to phosphorus deficiency. Plant Soil 248, 129–136.

Waters B m and Blevins D g 2000 Ethylene production, cluster root formation, and localization of iron (III) reducing capacity in Fe deficient squash roots. Plant Soil 225, 21–31.

Watt M, Evans J R 1999 Linking development and determinacy with organic acid efflux from proteoid roots of *Lupinus albus* L. grown with low phosphorus and ambient or elevated atmospheric CO_2 concentration. Plant Physiol 120, 705–716.

Watt M and Evans J R 2003 Phosphorus acquisition from soil by white lupin (*Lupinus albus* L.) and soybean (*Glycine max* L.)-species with contrasting root development. Plant Soil 248, 271–283.

White D and Robson A 1989 Rhizosphere acidification and Fe^{3+} reduction in lupinus and pea: Iron deficiency is not due to a poor ability to reduce Fe^{3+}. Plant Soil 119, 163–175.

Zaïd E H, Arahou M, Diem H G and El Morabet R 2003 Is Fe deficiency rather than P deficiency the cause of cluster root formation in *Casuarina* species? Plant Soil 248, 229–235.

Zhang F and Li L 2002 Using competitive and facilitative interactions in intercropping systems enhances crop productivity and nutrient-use efficiency. Plant Soil 248, 305–312.

The biology of cluster roots and the acquisition of P from the rhizosphere

Plant and Soil **248:** 1–19, 2003.
© 2003 *Kluwer Academic Publishers. Printed in the Netherlands.*

Structure, ecology and physiology of root clusters – a review

Byron B. Lamont*
Department of Environmental Biology, Curtin University of Technology, GPO Box U1987, Perth, WA 6845, Australia

Received 5 February 2002. Accepted in revised form 12 July 2002

Key words: auxin, cluster roots, proteoid roots, root function, root structure, soil bacteria, soil nutrients, soil water

Abstract

Hairy rootlets, aggregated in longitudinal rows to form distinct clusters, are a major part of the root system in some species. These root clusters are almost universal (1600 species) in the family Proteaceae (proteoid roots), with fewer species in another seven families. There may be 10–1000 rootlets per cm length of parent root in 2–7 rows. Proteoid roots may increase the surface area by over 140× and soil volume explored by 300× that per length of an equivalent non-proteoid root. This greatly enhances exudation of carboxylates, phenolics and water, solubilisation of mineral and organic nutrients and uptake of inorganic nutrients, amino acids and water per unit root mass. Root cluster production peaks at soil nutrient levels (P, N, Fe) suboptimal for growth of the rest of the root system, and may cease when shoot mass peaks. As with other root types, root cluster production is controlled by the interplay between external and internal nutrient levels, and mediated by auxin and other hormones to which the process is particularly sensitive. Proteoid roots are concentrated in the humus-rich surface soil horizons, by 800× in *Banksia* scrub-heath. Compared with an equal mass of the B horizon, the A_1 horizon has much higher levels of N, P, K and Ca in soils where species with proteoid root clusters are prominent, and the concentration of root clusters in that region ensures that uptake is optimal where supply is maximal. Both proteoid and non-proteoid root growth are promoted wherever the humus-rich layer is located in the soil profile, with 4× more proteoid roots per root length in *Hakea laurina*. Proteoid root production near the soil surface is favoured among hakeas, even in uniform soil, but to a lesser extent, while addition of dilute N or P solutions in split-root system studies promotes non-proteoid, but inhibits proteoid, root production. Local or seasonal applications of water to hakeas initiate non-proteoid, then proteoid, root production, while waterlogging inhibits non-proteoid, but promotes proteoid, root production near the soil surface. A chemical stimulus, probably of bacterial origin, may be associated with root cluster initiation, but most experiments have alternative interpretations. It is possible that the bacterial component of soil pockets rich in organic matter, rather than their nutrient component, could be responsible for the proliferation of proteoid roots there, but much more research on root cluster microbiology is needed.

Abbreviations: FeEDDHA – Fe ethylenediamine dihydroxyacetic acid; 1AA – indole-3-acetic acid; NAA – α naphthaleneacetic acid; OM – organic matter; P_i – inorganic P; 2,4,5–T – 2,4,5 trichlorophenoxyacetic acid

Introduction

Clusters of hairy rootlets in longitudinal rows were called proteoid roots by Purnell (1960) in reference to their widespread occurrence in the Proteaceae (Figure 1). All species examined in this essentially Southern Hemisphere family produce these root clusters, probably at least 1600 species in 73 genera, except for the primitive Persoonioideae (Lamont, 1982, 1983a). This includes 23 genera of rainforest Proteaceae (P. Reddell, pers. comm.). Almost identical structures have been identified in the pea legume, *Viminaria juncea* (Lamont, 1972a): formed in waterlogged soils, they lack root hairs. Related structures, with a much lower density and number of rootlets are now recognised

* FAX No: +61-9266-2495.
E-mail: b.lamont@curtin.edu.au

Figure 1. Morphology of proteoid root clusters. (a) New-season lateral of *Hakea lissocarpha* (Proteaceae) with three proteoid roots of decreasing age from left (dried out) to right (yet to mature). (b) Rootlets just emerging in double rows from parent root of *H. prostrata*. (c) Young proteoid root of *Leucadendron laureolum* (Proteaceae) before root hair development on rootlets. (d) Lateral branch of *Viminaria juncea* (Fabaceae) showing two mature proteoid roots and nodules (*n*). Scale: a, d = 4 cm; b, c = 1 mm. (a, b, d) floated in water; (a, d) conventional photography; (b) dissecting microscope view; (c) freeze-dried and scanning electron microscopy.

in several other genera of peas – notably *Lupinus* (Adams et al., 2002; Fabaceae, Trinick, 1977), *Acacia* (Mimosaceae, Sward, 1978), Casuarinaceae (Reddell et al., 1986), Myricaceae (Louis et al., 1990), Elaeagnaceae (Skene, 1998), Betulaceae (Hurd and Schwintzer, 1996), *Ficus* (Moraceae, Rosenfield et al., 1990) and Curcurbitaceae (Waters and Blevins, 2000). No doubt even more species, genera and families will be shown to produce these fascinating root structures, now considered one of the three major root modifications, along with mycorrhizas and nodules, among flowering plants (Lamont, 1982; Skene, 1998).

Lamont (1982) introduced the term 'root clusters' to cover all the above, calling them proteoid when the rootlets were in longitudinal rows. The subsequent trend has been to describe the clusters of lupins and Proteaceae as proteoid and those of other taxa as cluster roots. Dinkelaker et al. (1995) separated 'proteoid-like root clusters' into proteoid root clusters, produced by the Proteaceae, and 'other root clusters' for all other families. I will follow this terminology, using root clusters in the original sense to include proteoid roots. In all cases, the clumping of hairy rootlets is the key feature, with Proteaceae showing the most elaborate structures.

In this review, I first describe the structure and development of root clusters. I then briefly review their function (exudation, solubilisation and uptake). This is followed by an account of their contribution to the total root system, including the importance of plant age, mineral nutrients and water, and role of plant hormones, in their formation. I then consider distribution of root clusters within the root system, especially the possible role of organic matter in accounting for their proliferation at or near the soil surface. The review finishes with an account of the possible role of microorganisms in root cluster initiation. Nomenclature for all species follows that used in the original papers.

Structure and development

In a study of 54 South African Proteaceae in 13 genera, Lamont (1983a) showed that their individual clusters ranged 2–75 mm in length and 1–34 mm in width. Almost all other species have clusters in this range (Dinkelaker et al., 1995), except that those in mature plants of *Hakea prostrata* may reach 200 mm long and 70 mm wide (Lamont, 1972b). Purnell (1960) recognised 'compound' proteoid roots as clusters of 'simple' proteoid roots (Figure 2). They are common in many Proteaceae (Lamont, 1983a) reaching their peak complexity in the major Australian genera, *Banksia* and *Dryandra*, where the ultimate parent roots are determinate as well as their rootlets. In 'open' compound roots, adjacent clusters arise from parent roots attached to a lower-order parent root lacking rootlets (Figure 2). In 'closed' compound roots, numerous small clusters extend from the confines of a larger lower-order proteoid root active at the same time. Thus, for compound root clusters, any bare parent root segments within the confines of the ultimate ends of the component clusters are included. This overestimates the contribution of root clusters to the root system, but, for practical purposes as well as conceptual, is unavoidable. A recent attempt to include all parts of parent roots outside the area with clusters in the definition of proteoid roots (Watt and Evans, 1999), however, does not do justice to the structural or historical basis of the term, and cannot be adopted.

The rootlets arise simultaneously opposite every protoxylem pole so that for all species there are 2–7 longitudinal rows per cluster (Figure 3). Watt and Evans (1999) provide a detailed account of the structure and development of the rootlets of *Lupinus albus*. Some rootlets are so tightly packed that they emerge in double rows (Figure 1b) to give up to 1000 rootlets per cm length of parent root. Those with the more typical single rows (Figure 1d) and 4–5 poles have 250–500 rootlets per cm. Non-proteoid root clusters often have gaps between the rootlets in a row (non-contiguous) and 2–3 poles with < 50 rootlets per cm (Dinkelaker et al., 1989). The limit is 10 rootlets per cm recognised in *Lupinus albus* (Gilbert et al., 1999; Watt and Evans, 1999) – clearly, with time, the concept has become more generous! The cortex is split by the emerging rootlets where they are contiguous (Figures 1d, 3), but whether this leads to increased leakage or uptake of substances is unknown.

The rootlets measured so far are 0.6–35 mm long and 8–250 μm wide (Lamont, 1983a). Their root hairs are 100–2000 μm long and 7–11 μm wide (Lamont, 1972b, 1983a; Skene et al., 1998). These are often longer and denser but similar in diameter to those of the parent root. For the clusters of *Leucadendron laureolum*, root hairs are 3× longer and 1.5× denser (Table 1). The presence of these relatively short (7 mm long) rootlets is sufficient to increase the surface area per unit length of parent root by over 140×, mainly due to the contribution made by the root hairs. The volume of soil explored within the confines of these rootlets is increased by 288× that of an equivalent

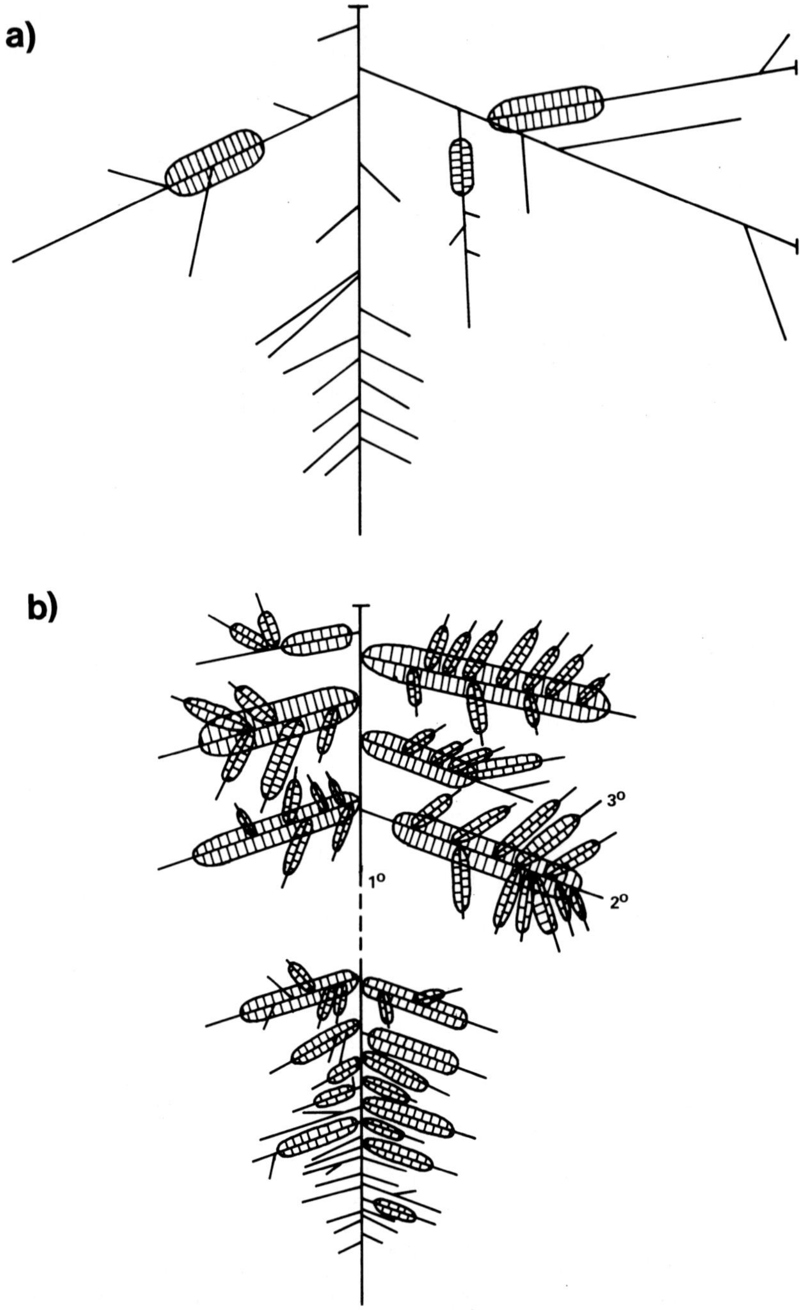

Figure 2. Arrangement of proteoid roots. (a) Three 'simple' proteoid roots of *Serruria floribunda* (Proteaceae) on a new season's lateral branch. (b) 'Incompletely open compound' proteoid root of *Leucospermum formosum* (Proteaceae) comprising 111 simple proteoid roots (midportion omitted). Note that most of the third order (3°) parent roots are 'closed', i.e. they arise from within an older proteoid root. From Lamont (1983).

Figure 3. Anatomy of proteoid roots. (a) Transection through parent root of *Hakea prostrata* showing rootlets arising from pericycle and splitting of the cortex. (b) Transection of *Viminaria juncea* as in (a). Note aerenchyma nature of cortex. (c) Transection through mature proteoid root of *H. prostrata*. Secondary growth has begun in the vascular core. Scale = 1 mm. (a, b, c) wax-embedded and transmission light microscopy; (a, b) stained with safranin and fast green; (c) stained with $FeCl_3$–$K_3Fe(CN)_6$ to highlight the presence of phenolics.

Table 1. Dimensions of proteoid and non-proteoid roots of *Leucadendron laureolum* and associated soil volume explored. Adapted from Lamont (1983b)

	Proteoid rootlet	Parent root	Ratio
Width (mm)	0.16	0.66	
Length (mm)	7.00	Indefinite	
No. per cm parent root	267	–	
Root hair width (μm)	10.4	11.3	
Root hair length (mm)	0.30	0.10	3.0×
No. root hairs mm^{-2} root surface	806	528	1.5×
	Proteoid root	**Parent root**	**Ratio**
Surface area/length (mm^2 mm^{-1})	842.1	6.0	141.3×
Proportion due to root hairs	89%	65%	
Surface/volume (mm^2 mm^{-3})	183.6	17.3	10.6×
Surface/mass (mm^2 mg^{-1})	4658	295	15.8×
Soil volume/root length (mm^3 mm^{-1})	168.9	0.6	288.2×
Soil volume/root mass (mm^3 mg^{-1})	934	29	32.5×

length of non-proteoid root. It is unknown what is the relative nutrient cost of producing the two types of roots, although the total respiratory cost appears to be similar in *Lupinus albus* (Keerthisinghe et al., 1998). On an equal mass basis, these proteoid roots have 16× the surface area and explore 33× as much soil volume (Table 1).

Function

What are the functional implications of this vast increase in effective root surface and soil volume explored? The first effect is on enhancing rates of exudation. Assuming that rates of exudation are limited by the availability of release sites at the root surface (i.e., they are area-limited) rather than supply, then excretion of carboxylates (Gardner et al., 1983; Johnson et al., 1996; Roelofs et al., 2001), protons (Grierson and Attiwill, 1989), phenolics (Lamont, 1972b; Neumann et al., 2000) and water (Burgess et al., 2000; Lamont, 1983a) will be enhanced. Total citrate exuded by lupin root clusters may reach up to 23% of total plant mass (Dinkelaker et al., 1989), which implies that they may be metabolically more active than other roots as well. There is conflicting evidence about that (Johnson et al., 1994; Malajczuk and Bowen, 1974; Vorster and Jooste, 1986a) and some biochemical processes appear to be enhanced while others are reduced (Neumann et al., 2000). Abundant starch grains in the rootlets (Lamont, 1972b) and mitochondria in the root hairs (Skene et al., 1996) support the potential for substantial metabolic activity. The 'pulse' of carboxylate exudation lasts 2–3 days in mature clusters of *Lupinus albus*, and the rate shows a strong response to light but not atmospheric CO_2 supply (Watt and Evans, 1999).

Root clusters also secrete certain enzymes into the rhizosphere. Enhanced exudation of acid phosphatases is notable, especially at low available P (Dinkelaker et al., 1997; Gilbert et al., 1999). Acid phosphatases hydrolyse poorly soluble organic (humic) Ca/Fe/Al phosphates (Gilbert et al., 1999). Since these studies have not been conducted under sterile conditions it is possible that microbes in the rhizosphere (Figure 4) contribute enzymes and protons to the solubilisation process. However, Grierson and Adams (2000) were unable to attribute enhanced phosphatase activity of surface soil over the wet season to greater microbial activity in the presence of active proteoid roots. Clusters in *Cucurbita pepo* have strong Fe^{3+} reductase activity at low environmental Fe (Waters and Blevins, 2000). For inorganic sources of P, Wenzel et al. (1994) showed that bacteria isolated from the rhizosphere of *Telopea speciosissima* (Proteaceae) reduced pH of the culture medium and solubilised $Ca_3(PO_4)_2$, especially when the N source was NH_4 rather than NO_3.

The second effect of the large surface-volume ratio is on enhancing solubilisation processes. The concentration of these exudates in a confined space promotes release of Fe, Ca, P, Mn and Zn from insoluble or-

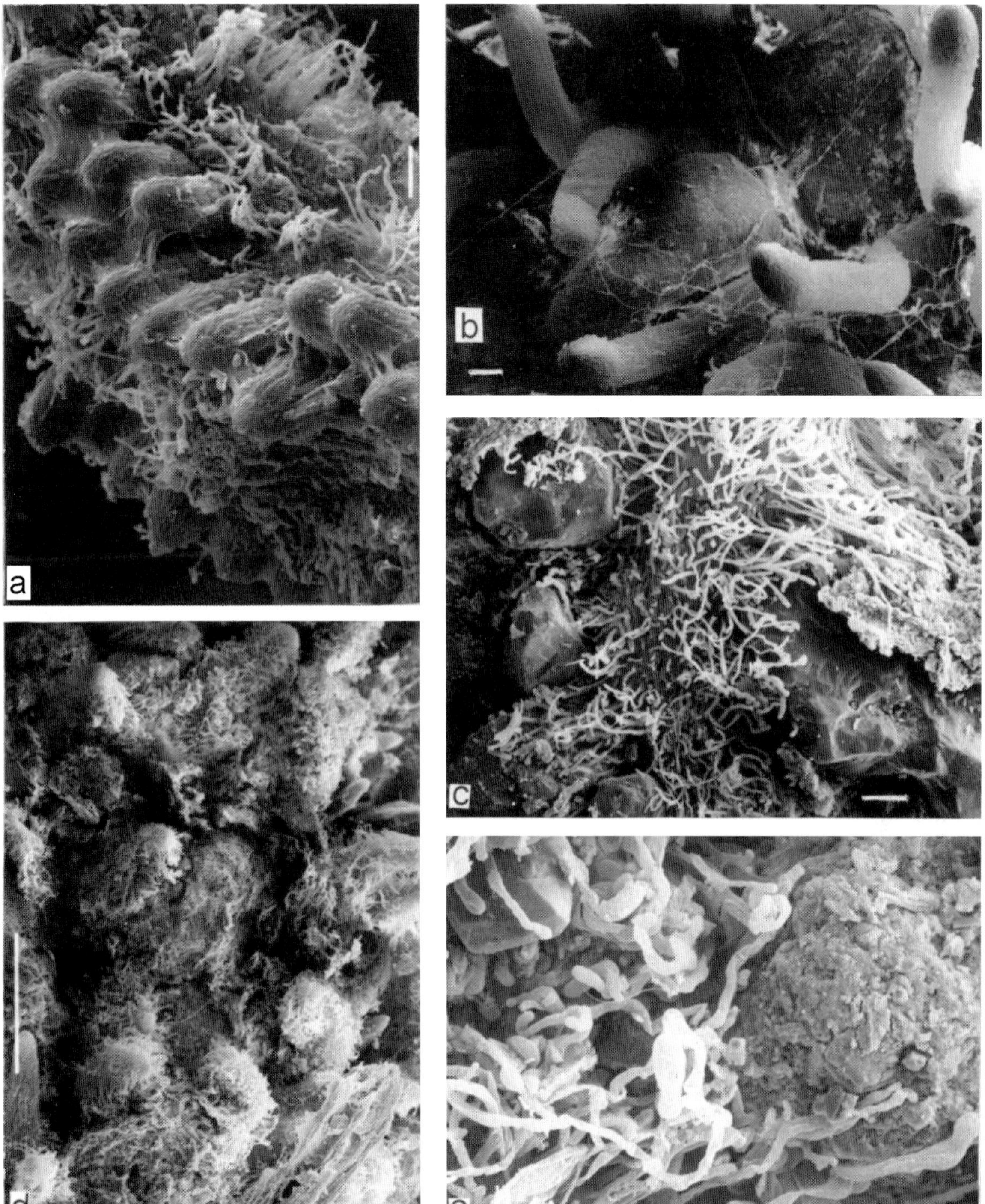

Figure 4. Relationship of *Leucadendron laureolum* proteoid roots with soil. (a) Blanket of mucilage over tips of elongating rootlets. (b) Sand grains held in place by 'pincushion' effect of rootlets even in absence of root hairs. Note abundant fungal hyphae. (c) Rootlet squeezing between soil particles with root hairs stretching over soil films. (d) Surface of proteoid root as removed from the soil. About two rootlets per mm^2 with root hairs enmeshed among soil particles. (e) Root hairs clinging to sand grain and surface colloids. Scale: 100 μm. Removed from soil without washing, freeze-dried and scanning electron microscopy.

ganic and inorganic forms (Dinkelaker et al., 1989, 1995; Gardner et al., 1982b, 1983; Gerke et al., 1994; Pate et al., 2001). Toxic Al ions are also released but tend to be immobilised by carboxylates (Dinkelaker et al., 1989) as is Ca (Gardner et al., 1983). Nevertheless, Proteaceae is a highly Al-tolerant family (Webb, 1954), though the mechanism is unknown. Solubilisation of MnO_2 and organic P may even extend up to 5 mm beyond the confines of the cluster in agar and soil (Gardner et al., 1982b; Li et al., 1997). This is no doubt promoted by root hairs enveloping tips of the rootlets at maturity in many species (Skene et al., 1996). Pate et al. (2001) have identified a major role in Proteaceae-dominated vegetation for the Fe-Al-P-solubilising function of proteoid roots in soil-profile development and lateritisation processes.

The third effect of the large surface-volume ratio is on enhancing nutrient uptake. Root clusters have strong soil 'interception' due to their constant exploration of new parts of the soil profile during the growing season, large volume explored (Table 1), and intimate association of the rootlets and root hairs with soil particles and films (Figure 4; Gould, 1998). Such 'contact exchange' is important for Ca and Mg (Barber, 1995). It will also maximise the soil-root water potential gradient pathway for nutrients, such as Ca, Mg, S and NO_3, whose uptake is largely controlled by 'mass flow'. It will also minimise the path length for nutrients, such as K, NH_4 and P, whose uptake is dependent on diffusion (Voorster and Jooste, 1986a).

These acquisition mechanisms are further enhanced by the mucopolysaccharides exuded by the root caps, epidermis and root hairs that 'glue' the root cluster to the soil particles (Figure 4a; Dell et al., 1980; Skene et al., 1996). This may be particularly important in the surface soil with low bulk density, where most proteoid root clusters are located, and in drying soil, both of which favour root hair development. Gould (1998) highlighted the role of the proteoid root mat of *Banksia serrata* in reducing soil erosion in steep terrain after fire when the loose litter layer was consumed. The mucilage components may also immobilise released Al and other cations (Horst et al., 1982). These properties of root clusters will also enhance water uptake (Burgess et al., 2000) but comparisons with non-clustered roots are needed.

Early research on proteoid root function emphasised their uptake of inorganic P (P_i). Jeffrey (1967) considered that the rapid conversion of orthoP_i as it was released from decomposing litter to inorganic polyP_i inside the root was the key, but subsequent research has yet to confirm that higher plants, as distinct from microbes, can synthesise polyP_i. Bound or free radioactive P_i are usually absorbed more readily by root clusters than non-clustered roots on a mass or soil volume, or even surface area, basis (Braum and Helmke, 1995; Keerthisinghe et al., 1998; Lamont, 1982; Malajczuk and Bowen, 1974; Neumann et al., 1999; Vorster and Jooste, 1986a). Diffusional limitations or use of post-pulse clusters might explain why some studies of root systems suspended in water culture show little difference in rates of P_i uptake between clusters and their associated laterals (Dinkelaker et al., 1995; Roelofs et al., 2001). Nevertheless, rates may be markedly different in others where enhanced active uptake has been demonstrated (Malajczuk and Bowen, 1974; Neumann et al., 2000).

Solubilised Fe, P, Ca and Mn appear to be readily absorbed by root clusters, but there is little direct evidence (Braum and Helmke, 1995; Dinkelaker et al., 1989; Gardner et al., 1982b; Lamont et al., 1984; Waters and Blevins, 2000). In xylem sap of proteoid roots of *Banksia prionotes*, N, K and Cl concentrations are higher than in non-proteoid roots, but those of Mg, Ca and S are similar (Pate and Jeschke, 1993). Vorster and Jooste (1986b) followed translocation of labelled K and P from proteoid roots to the shoots. It has now been shown that root clusters can absorb organic N (Schmidt and Stewart, 1997). ^{15}N-labelled glycine was more readily absorbed per unit weight by isolated proteoid rootlets of *Hakea* aff. *gibbosa* than by non-proteoid rootlets in this and seven other species from waterlogged soils. Studies have yet to be undertaken on whether root clusters can solubilise organic N, such as polypeptides. Research has yet to show an increase in shoot growth or nutrient concentration specifically due to the presence of root clusters. Part of the problem is the inability to produce root systems lacking root clusters other than by altering nutrient supply. Overall schemes on the biochemistry of root-cluster function are now available (Neumann et al., 1999; Pate et al., 2001).

Contribution of root clusters to the root system

Importance of mineral nutrients

Species capable of producing root clusters invariably possess them in abundance under field conditions (Crocker and Schwintzer, 1994; Keerthisinghe et al., 1998; Lamont, 1972b, 1983; Purnell, 1960; Reddell

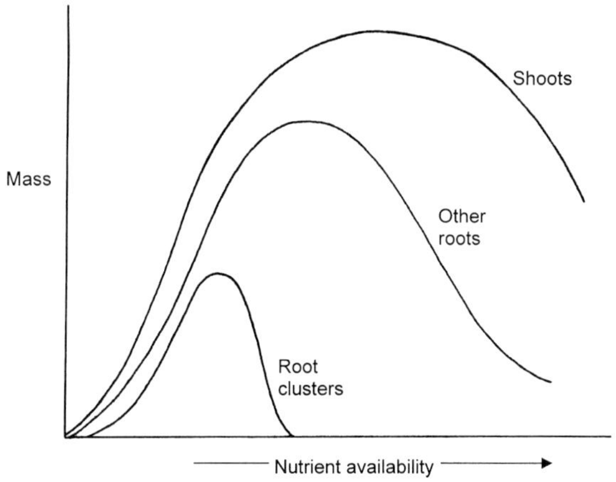

Figure 5. Idealised relationship between soil nutrient (N, P) availability and production of root clusters, other roots, and shoots. Note the absence of root clusters at extremely low and moderately high nutrient levels. Adapted from Lamont (1982).

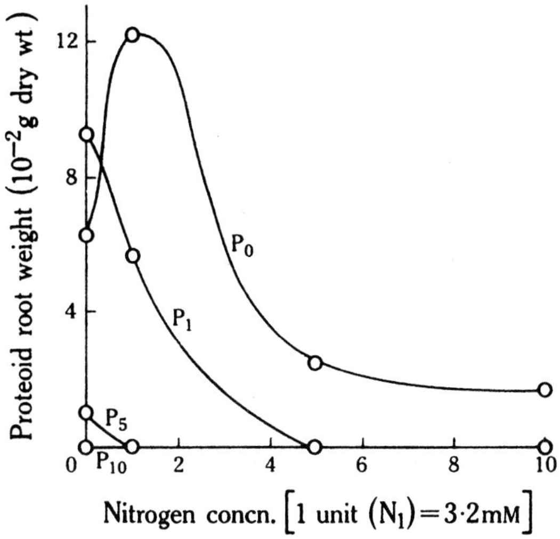

Figure 6. Proteoid root production by 6-month-old *Hakea prostrata* plants in sand culture at different levels of applied NH_4NO_3 and H_3PO_4 (1 unit (P_1) = 2 mM P nutrient solution). Five replicates per treatment, with $P < 0.05$ for analysis of variance between N levels for P_0, P_1 and P_5 treatments. Adapted from Lamont (1972b).

et al., 1997). This is despite the fact that, under experimental conditions optimal for growth of the rest of the root system, they may be rare or absent altogether (Keerthisinghe et al., 1998; Lamont, 1972b; Louis et al., 1990; Reddell et al., 1997). This is significant on two counts: plant growth is (not unexpectedly) suboptimal under 'natural' conditions, and root clusters can be viewed as facultative structures.

Root cluster production peaks at soil levels of P, N and/or Fe suboptimal for other roots (Figure 5; Crocker and Schwintzer, 1993; Gardner et al., 1982c; Keertsinghe et al., 1998; Lamont, 1972c; Louis et al., 1990; Racette et al., 1990; Reddell et al., 1997; Rosenfield et al., 1991). Proteoid root production is promoted by applications to *Banksia ericifolia* seedlings of up to 10 mg P_i L^{-1} potting mix and 430 mg FeEDDHA L^{-1} but the results are not available relative to the rest of the root system (Handreck, 1991). Seedlings of *Hakea prostrata* lacking root clusters due to artificially high soil levels of N or P have a high shoot:root ratio (Figure 5; Lamont, 1986), and this presumably increases their susceptibility to drought when well fertilised under cultivation. Failure to detect a promotory range of some nutrients (i.e., production never increases from the lowest concentration used) can be attributed to too high a base level of other nutrients, especially P (Figure 6). This is highlighted by inoculation of *Casuarina cunninghamiana* with the mycorrhizal fungus, *Glomus*, which enhances P uptake (Reddell et al., 1997). For any level of applied P, root cluster production was reduced compared with the uninoculated plants, with a peak no longer attained, only a fall.

That there is a threshold nutrient level required for root-cluster formation (actually a threshold plant size, Figure 5) is indicated by failure of *Viminaria juncea* and *C. cunninghamiana* to produce root clusters, in the absence of applied N, until they are nodulated (Pathmaranee, 1974; Reddell et al., 1997). Interestingly, peak root-nodule production for *V. juncea* occurs at 4× the P level required for peak root-cluster production. Source of N may also affect root-cluster formation, with urea more effective than NO_3 in *Myrica gale*, although cluster size is reduced (Crocker and Schwintzer, 1993). However, NO_3 is much more effective than NH_4 in promoting cluster production in *Gymnostoma papuanum* (Casuarinaceae), possibly because NH_4 is more readily absorbed, raising the plant's N status (Racette et al., 1990).

Seedlings of different species vary in their ability to initiate root clusters. Lupins may take several days, while Proteaceae species take several months, or longer, partly dependent on presence and size (nutrient status) of the cotyledons (Keerthisinghe et al., 1998; Lamont, 1973, 1983a; Lamont and Groom, 2002). As they produce more higher-order lateral roots, so their capacity to produce root clusters increases (Lamont, 1972a). At low base levels of nutrients, root clusters of seedlings in their first season may contribute 15–80% of total root mass (down to 0% at high base levels), and

Table 2. Growth of *Hakea prostrata* seedlings originally grown in sand receiving 4% Hoaglands solution with (+N) or without (−N) 640 μM NH_4NO_3 for 12 weeks. Root systems were washed free of sand, wrapped tightly in cotton gauze, repotted into sand and given tapwater for a further 10 weeks. New root growth and N concentration of the shoots were then analysed. *P* values refer to 1-way ANOVA. Adapted from Lamont (1973)

	+N	−N	*P*
Initial N (mg/g leaf)	9.4	3.2	< 0.001
New root mass (g)	0.17	0.11	< 0.01
Proteoid/non-proteoid roots	0.05	0.16	< 0.01
No. proteoid roots/g	18	79	< 0.01

up to 134 clusters per g root mass (Lamont, 1972b, 1983a; Louis et al., 1990; Reddell et al., 1997). As plant size and age can affect the ability to produce root clusters it is sometimes difficult to distinguish a maturity/size effect from a direct effect of the treatment on root cluster production. Thus, there may be advantages in using older seedlings (and applying new treatments), rooted cuttings or split-root systems in experiments (Lamont, 1973, 1983a; Lamont et al., 1984; Shane et al., 2003). Most of the new season's fine roots, especially near the soil surface, may be proteoid in mature plants (Figure 1; Lamont, 1983b; Low and Lamont, 1990; Pate et al., 2001).

It is expected that root-cluster formation is controlled by the plant rather than by a direct response of the parent root to soil nutrients. Shoot, leaf and root concentrations of P or N are closely correlated with root cluster production (Keerthisinghe et al., 1998; Lamont, 1973; Louis et al., 1990). Foliar applications of P to *Myrica cerifera* and *Lupinus albus* suppress root-cluster formation (Louis et al., 1990; Marschner et al., 1987). Results of Handreck (1991) for *Banksia ericifolia* show that proteoid root production ceases at about 5 mg g^{-1} P in the dry shoots. *Banksia grandis* also failed to produce proteoid roots at > 5 mg g^{-1} leaf P (supplied as KH_2PO_4), whereas they were abundant at < 1 mg g^{-1} (supplied as Fe or Al PO_4) (Lambers et al., 2002). Application of phosphorous acid, which reduces plant sensitivity to internal P levels, increases root-cluster production in high-P plants of *Lupinus albus* (Gilbert et al., 2000). After repotting into N-free quartz sand, seedlings of *Hakea prostrata* originally at low N had one-third the N concentration and 4.4× the frequency of proteoid roots as those originally at high N (Table 2).

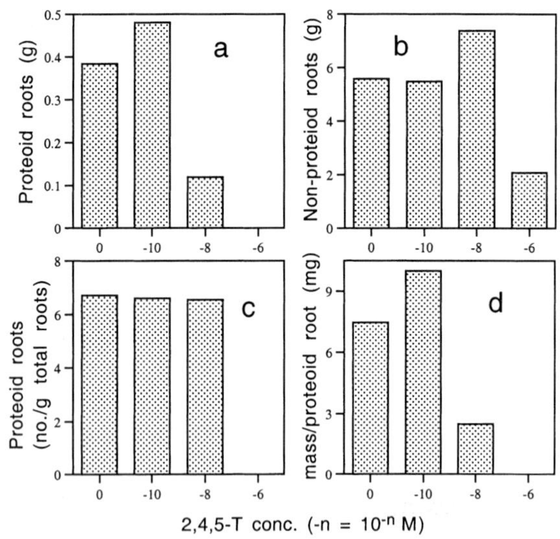

Figure 7. Production of proteoid and non-proteoid roots by 12-month-old *Hakea laurina* repotted into perlite (containing 0.2% Hoagland's solution per L) kept moist with tapwater for 8 weeks. Based on five bulked plants per treatment. Concentrations of 2,4,5–T are per L perlite, half added initially and half at 3 weeks. B.B. Lamont (unpublished).

Role of hormones

Mineral nutrient status affects the level of plant hormones (Kuiper et al., 1989). While there have been no attempts at showing that nutrient levels control hormone concentrations in species with root clusters, their exogenous application may mimic the pattern for nutrient supply (Watt and Evans, 1999). The auxin analogue, 2,4,5–T, causes a peak in proteoid root production when applied at 10^{-10} M in perlite growth medium to the roots of *Hakea laurina* watered with 0.2% Hoagland's initially and tapwater thereafter (Figure 7). These plants had 295 rootlets per cm of parent root. Cluster production is strongly suppressed at 10^{-8} M 2,4,5–T, the peak level for non-proteoid roots. The effect was on rootlet number per cluster rather than on cluster production itself. Non-proteoid root growth is severely reduced at 10^{-6} M, and all plants died soon after receiving 10^{-4} M 2,4,5–T. However, at 10^{-6} M, the number of non-proteoid rootlets had increased from 8 to 15 per cm parent root (enough to be considered proteoid among non-Proteaceae) with an increase in abundance of root hairs but the rootlets were longer and more variable in length (B.B. Lamont, unpublished). This indicates that 2,4,5–T has similar properties to the endogenous hormone, at first adding

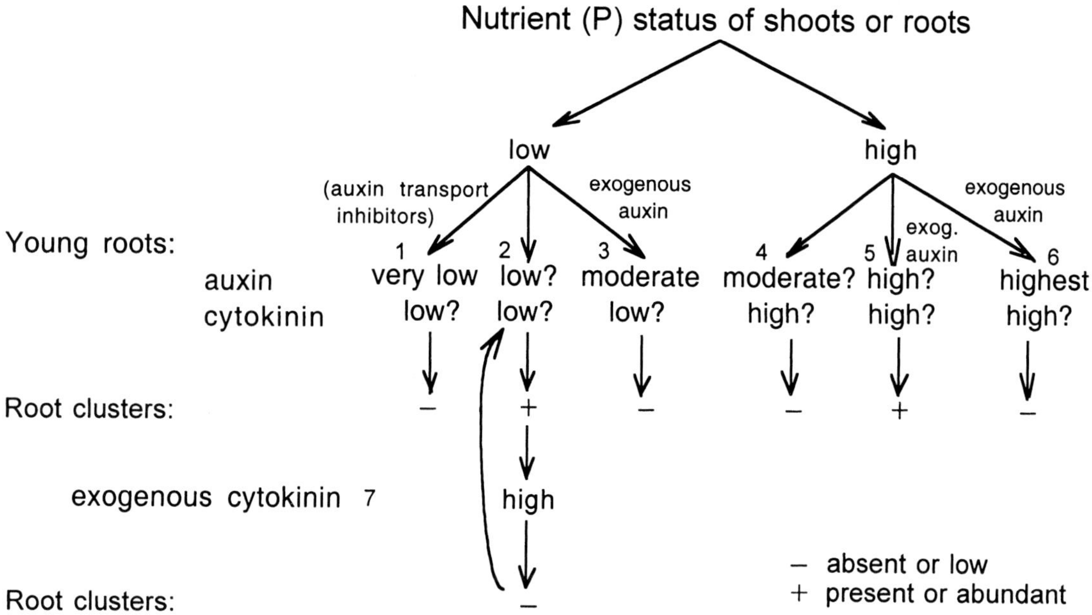

Figure 8. Model of role of plant hormones in root-cluster production as controlled by mineral nutrient status. Treatments 1, 2, 3, 5, 7: *Lupinus albus* (Gilbert et al., 2000; Neumann et al., 2000; Skene and James, 2000); 1, 2, 3: *Hakea laurina* (Figure 7); *Grevillea robusta* (Skene and James, 2000). ? refers to expectation from the literature but not yet demonstrated on species producing root clusters. See text for more details.

to its effect and then replacing it as its concentration increased.

For P levels at which root clusters are normally suppressed, Skene and James (2000) were able to induce them to form on *Grevillea robusta* (peak at 10^{-8} M solution) and *L. albus* (peak at 10^{-9} M) with the synthetic auxin, NAA, although production was well below that in the absence of P (−P) in the absence of NAA. Neumann et al. (2000) induced maximum root-cluster formation on young *Lupinus albus* plants in water culture with 250 μM P (normally suppressive to cluster formation) at $7{\cdot}10^{-7}$–$7{\cdot}10^{-6}$ M of the natural auxin, IAA. At −P, added kinetin suppressed root-cluster development until, at $5{\cdot}10^{-6}$ M, their production ceased. They suggested that cytokinin production by rootlet tips of the auxin-induced clusters could suppress their ongoing elongation and explain their determinate growth.

Figure 8 summarises present knowledge on internal control of root-cluster development within the context of general controls of root growth. Low internal levels of N and P (shoots or roots) are associated with low levels of auxin and cytokinins (Kuiper et al., 1989). This promotes root growth rather than shoot growth (van der Werf and Nagel 1996) and clustered rootlets rather than non-clustered (Figure 7). Auxin received from the shoots induces prolific rootlet initiation in the recently differentiated zone of the parent root (Figure 1a; Celenza et al., 1995). IAA at first maintains rootlet elongation. Increasing cytokinin production from their apices finally inhibits further elongation and possibly suppresses new root cluster formation for a set distance basally when root clusters can again be initiated (Figure 1a). High levels of N and P in the shoots or roots appear to promote auxin synthesis to levels inhibitory for root-cluster initiation but promotory for other roots (Figures 6, 7). This is confounded by the fact that clusters can be induced at high P with exogenous auxin. In that case, it might be the ratio of auxin to other hormones, e.g., cytokinins, that is important (Neumann et al., 2000).

Ethylene might also have a role in root-cluster formation, for, unlike IAA, it promotes root-hair elongation and possibly reduces root gravitropism, and some studies have shown ethylene production is promoted by low P or Fe (Lynch and Brown, 1997; Waters and Blevins, 2000). However, while enhanced ethylene production is associated with increased root-cluster production at low P by *Lupinus albus*, ethylene inhibitors have no effect on their production (Gilbert et al., 2000). The hypersensitivity of root-cluster formation to hormones generally is further supported by their production halving in the presence of gibberellin (GA_3) sprayed on the foliage of *Hakea laurina*

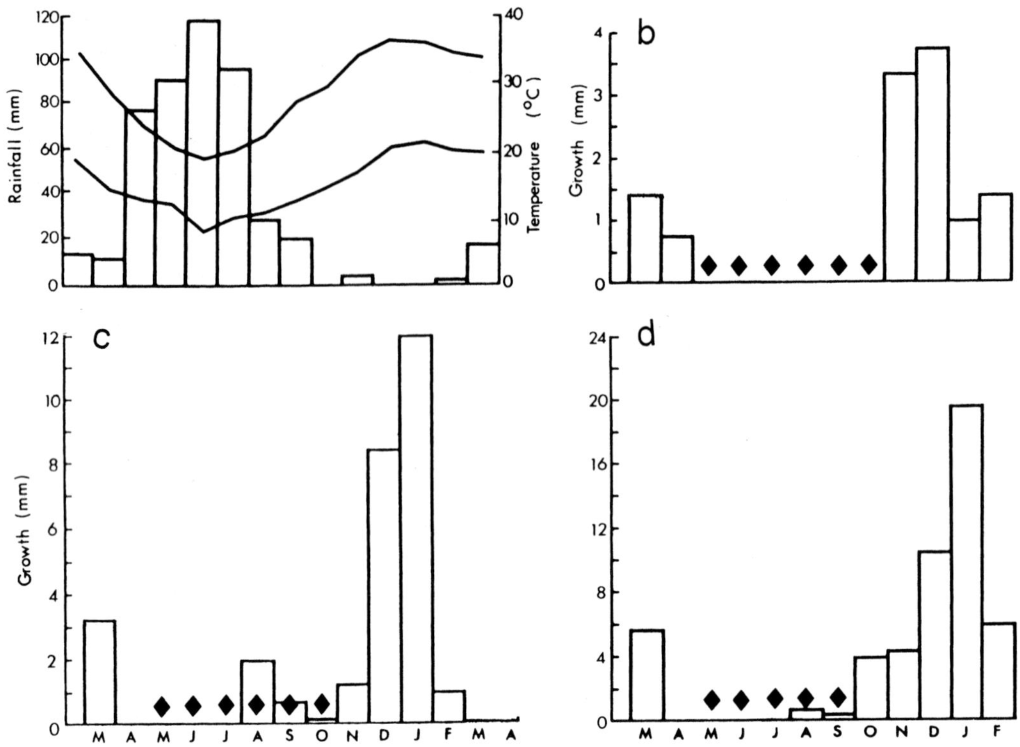

Figure 9. (a) Monthly rainfall and minimum and maximum temperatures at Eneabba, southwestern Australia, and presence of fresh proteoid roots (♦) and extension shoot growth of (b) *Banksia menziesii*, (c) *B. attenuata* and (d) *B. hookeriana* (Proteaceae). Note how root and shoot growth are out of phase. Adapted from Lamont and Bergl (1991).

without affecting total root growth (B.B. Lamont, unpublished). Roots are a major source of GA, which may inhibit root initiation when applied to cuttings (Salisbury and Ross, 1992). While the research to date highlights the high sensitivity of the root-cluster formation process to exogenous hormones, data are lacking on the actual levels at different stages and growing conditions, or on their mode of action at the biochemical level. In addition, to what extent pulsing of hormone availability from the shoots and roots explains the synchronised production of clusters in the root system (Watt and Evans 1999) has yet to be explored.

Importance of water

Water is essential for parent root growth, and thus is a prerequisite for root-cluster formation. Watering potted seedlings of three *Banksia* species at short intervals, rather than every 2–4 days, was optimal for proteoid root production (Siddiqi and Carolin, 1976). In the highly seasonal Mediterranean climates where most cluster-bearing species occur, root cluster production is confined to the wet (winter–spring) season (Figure 9; Lamont, 1976, 1983a; Lamont and Bergl, 1991). Proteoid (and non-proteoid) root growth accompanies or (more typically) precedes shoot growth, consistent with their role in maximising nutrient uptake in impoverished soils (Specht et al., 1983). Most rootlets shrivel after 6–10 weeks and are replaced further along the parent root during the wet season (Figure 1a). It is possible to induce local proteoid root formation during the summer drought if that part of the root system receives sufficient water (Lamont, 1976). There is also another mechanism: water absorbed by the sinker and tap roots is secreted by the surface laterals into soil and creates conditions suitable for proteoid root initiation (Burgess et al., 2000; Lamont 1983a).

Many species that produce root clusters grow under permanently or seasonally waterlogged conditions – *Viminaria juncea* (Figure 1b), *Brabejum stellatifolium*, some species of *Leucadendron, Hakea, Banksia*, Casuarinaceae and *Myrica* (Lamont 1976; Schmidt and Stewart, 1997; pers. observ.). Reddell et al. (1997) have suggested that, under these conditions, mycor-

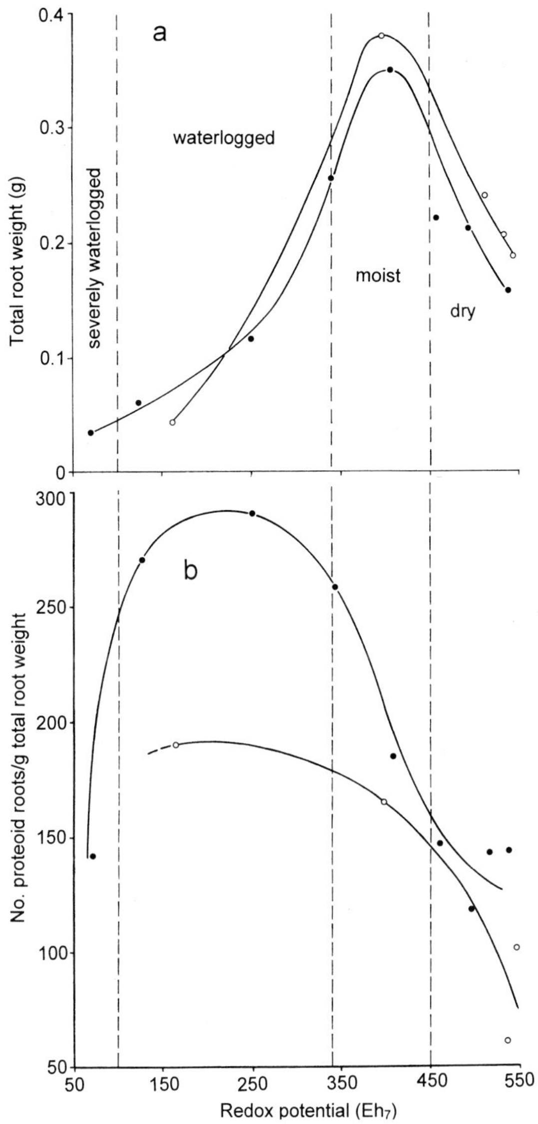

Figure 10. Total root mass (upper) and number of proteoid roots per unit root mass (lower) of *Hakea laurina* (○) and *H. multilineata* (●) grown for 22 weeks in soil varying in water availability (redox potential). Adapted from Lamont (1976).

rhizal development is poor, and that root clusters are a suitable alternative for enhancing P uptake in impoverished soils. In addition, root clusters readily form in water culture, especially if aerated (Crocker and Schwintzer, 1993; Racette et al., 1990; Roelofs et al., 2001). I showed (Lamont, 1976) that contribution on a mass basis of proteoid roots to the root systems of three *Hakea* species at field capacity (FC, redox potential of 450 mV) to 2× FC (360 mV) was much higher than near wilting point (550 mV) and at 3× FC (110 mV). While the rest of the root system was severely retarded at 2–3× FC, number of proteoid roots per unit total root mass actually peaked (Figure 10). This suggests that root clusters probably have a special contribution to make to nutrient uptake even under waterlogged conditions.

Distribution of root clusters through soil

Under field conditions, root clusters in all Proteaceae and *Viminaria* are concentrated in the humus-rich surface soil horizons. Data are not available for other taxa, although new roots anywhere in the soil appear to have the propensity to produce clusters. When a thick layer of litter is present, the concentration of proteoid roots is greatest in the actively decomposing layer (A_0) with proteoid root clusters growing up into the raw litter (Gould, 1998; Lamont, 1973). When the A_0 horizon is absent then the clusters are concentrated in the A_1 humic layer. Combining the two layers, proteoid roots usually form a mat 3–15 cm thick that is especially dense under the plant crown and for species with compound proteoid roots (Jeffrey, 1967; Lamont 1973, 1983a; Lamont et al., 1984; Low and Lamont, 1990). Nevertheless, occasional proteoid roots have been detected to a depth of 5 m associated with sesquioxide deposits (Pate et al., 2001).

Part of the explanation for this pattern could be that the surface root system has a greater propensity for root-cluster production. One-year-old plants grown in 44 L containers of quartz sand had 4.44× (*Hakea laurina*) and 3.10× (*H. prostata*) the proteoid/non-proteoid root ratio at 0–8.5 cm depth compared with 34–42.5 cm depth (Lamont, 1973). This may be associated with the greater abundance of higher-order laterals at the surface as these produce more root clusters (Lamont, 1972b). Nutrient and water levels were the same, or slightly lower, at the surface. When the A_{0-1} and B_{1-2} (depositional, non-humic) horizons are at similar depths to the above, field plants have a ratio of 6.86× (*H. laurina*) and 12.21× (*H. prostrata*). This is even more marked for *Banksia*-dominated scrub-heath in southwestern Australia where 81% of fine roots in the uppermost 15 cm are proteoid (Gould 1998; Low and Lamont, 1990). Living fine roots at a depth of 75–180 cm have 20% of the mass of those in the surface 15 cm but lack proteoid roots altogether. These higher ratios for field plants could be because the plants are much older and/or include responses to differences in soil properties of the A and B horizons.

Table 4. Growth of 18-month-old *Hakea prostrata* in cotton-wool-plugged bottles of autoclaved (A) or non-autoclaved sand. Adapted from Lamont and McComb (1974)

	n	Shoots (g)	Roots (mg)	No. proteoid roots/g root
Non-autoclaved	10	0.19	79	5.1
A	10	0.16	45	0.0
A+autoclaved soil extract	12	0.14	29	0.0
A+non-autoclaved soil extract	11	0.13	25	160.0
A+proteoid root wash	20	0.16	31	9.7
A+proteoid root and soil microbes	26	0.12	30	0.0

similar functions, especially in enhancing P uptake (Lamont, 1982).

The role of non-invasive microbes in root cluster initiation is less clear. Malajczuk and Bowen (1974) showed that 4-month-old plants of *Banksia grandis* grown in two γ-irradiated soil types produced proteoid roots after inoculation with a proteoid root extract while the controls did not. Since shoots of the inoculated plants had twice the mass of the non-inoculated it is possible that proteoid root production was only an indirect response to the presence of microbes. The results of Lamont and McComb (1974) were more definitive, as 12-month-old plants of *Hakea prostrata* in autoclaved, washed sand (inoculated with an autoclaved soil extract) grew as well as the controls (inoculated with a non-autoclaved soil extract), but in contrast, lacked proteoid roots (Table 4).

The bactericide, chloromycetin (1.5 μM) was sprayed in aqueous solution onto the foliage of *H. laurina*: number of proteoid roots per unit total root weight was reduced by 25% in the antibiotic-treated plants compared with those sprayed with water only, without any change in shoot weight (B.B. Lamont, unpublished). A more detailed study involved a split-root design with *Leucadendron laureolum* one side of which was watered with 0.3 mM chloromycetin (Lamont et al., 1984). After 9 months, the chloromycetin-treated part of the root system had a 63% reduction in number of proteoid roots per gram of total roots compared with the controls. However, this was accompanied by a 36% reduction in total root growth making interpretation difficult: either chloromycetin/bacteria in the soil also affected general root growth, or reduced root growth was a result of the reduction in proteoid roots. Unfortunately, in neither study was microbial status of the soil or rhizosphere assessed.

Problems with the bacterial stimulus hypothesis have come from two sources: attempts to induce production of proteoid roots with a selection of soil microbes, and apparently conflicting evidence for root-cluster formation in non-Proteaceae. Washings and cultured microbes from young proteoid roots and various soil microbes, including two bacteria, failed to induce proteoid roots (Table 4). Some exudates in the washings (e.g., phenolics) might have suppressed the causative organisms (Neumann et al., 2000), but there is no evidence that bacterial activity in proteoid roots is low (Lamont and McComb, 1974; Malajczuk and Bowen, 1974). A wide range of growth-inhibiting rhizobacteria exists and these may have masked a positive response (Barazini and Friedman, 2001). Thus, the identity of any initiating microorganisms remains elusive.

In a similar study on *Casuarina equisetifolia*, but grown in non-sterile soil, Sabba Rao and Rodríquez-Barrueco (1995) noted no significant increase in root-cluster formation when inoculated with four microbial types. Plants receiving aerobic microbes from soil extracts had 4× cluster weight of the controls, although total root weight was 38% higher. *Alnus glutinosa* (Betulaceae) produces root clusters (Hurd and Schwintzer, 1996) while two IAA-producing *Bacillus* spp. isolated from its rhizosphere promote overall growth (Gutierrez Mañero et al., 1996). However, any link between the two has not been examined.

Gardner et al. (1982a) obtained many root clusters on *Lupinus albus* in two experiments with non-inoculated, autoclaved sand/vermiculite. This was confounded in a third experiment by the formation of six times fewer clusters with non-inoculated, irradiated sand/vermiculite contaminated with ‘a few microbial species’ compared with the inoculated treatment containing a ‘diverse microflora’. What could have replaced the need for microbes in the first two experiments but not the third? The answer may lie in the way the germinants were treated in the first two experiments: they were placed on trypsin-soybean-agar slopes which so inhibited root growth that the researchers decided to transfer them to tubes of autoclaved vermiculite. Perhaps chemicals absorbed from the agar medium, but normally supplied by rhizosphere bacteria, were already available at suitable levels for initiation of root clusters once the plants were older?

The evidence points to the need for an extra chemical stimulus, normally supplied by bacteria, in root-cluster formation, but there are many unanswered

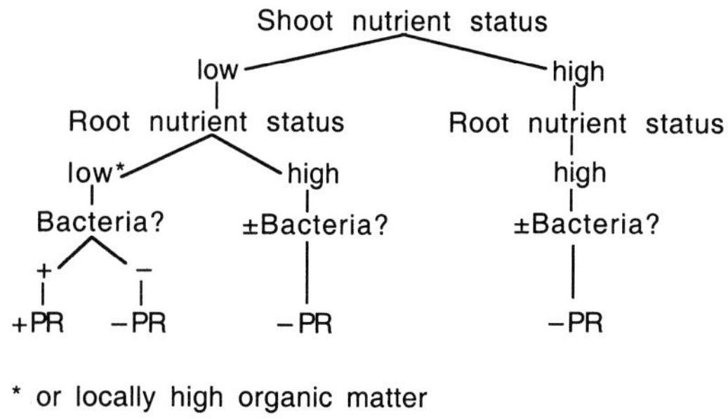

Figure 13. Model of the interaction between shoot nutrient status (a response to overall nutrient status of the soil), root nutrient status (a response to local variations in nutrient status of the soil) and rhizosphere bacterial activity in promoting the formation of proteoid root clusters (PR). Locally high levels of decomposing OM (normally associated with active nutrient release) appear to overcome the usual requirement for a low root nutrient status. + = present or high, – = absent or low. Model based on all material presented in this review. The role of plant hormones is considered in Figure 8.

questions. Is a microbial stimulus universally required? What microbes are involved? Does the high bacterial component explain why proteoid roots are concentrated in the high OM layers of the soil (see Figure 13)? What is the mechanism of stimulation? Is it a growth substance specific for cluster formation or is it a general growth promoter (Gardner et al., 1982a)? Is a precursor secreted by the parent root, or is it solely of bacterial origin? And why do washings of proteoid roots retard rather than promote cluster formation (Lamont and McComb, 1974)? Does this provide an additional explanation to account for the spacing of proteoid roots along the parent root (Figure 1a)? These questions need to be set against the background of any effect of microbes on production of other root types (Bothe et al., 1992). The formation of root clusters is proving to be a response to a complex interaction of internal and external signals to which the process is unusually sensitive compared with other roots.

Acknowledgements

Thanks to Hans Lambers for encouragement, and Corinne de Groot and Simon Kerley for extra information. The three referees provided perceptive comments on the original text.

References

Adams M A, Bell T L and Pate J S 2002 Phosphorus sources and availability modify growth and distribution of root clusters and nodules of native Australian legumes. Plant Cell Environ. 25: (In press).

Barazani O and Friedman J 2001 Allelopathic bacteria and their impact on higher plants. Crit. Rev. Microbiol. 27, 41–51.

Barber A 1995 Soil Nutrient Bioavailability: A Mechanistic Approach. 2nd Edn. John Wiley & Sons, New York.

Borch K, Bouma T J, Lynch J P and Brown K M 1999 Ethylene: a regulator of root architectural responses to soil phosphorus availability. Plant Cell Environ. 22, 425–431.

Bothe H, Körsgen H, Lehmacher T and Hundeshagen B 1992 Differential effects of *Azospirillum*, auxin and combined nitrogen on the growth of the roots of wheat. Symbiosis 13, 167–179.

Braum S M and Helmke P A 1995 White lupin utilizes soil phosphorus that is unavailable to soybean. Plant Soil 176, 95–100.

Brundrett M C and Abbott L K 1991 Roots of jarrah forest plants. I. Mycorrhizal associations of shrubs and herbaceous plants. Aust. J. Bot. 39, 445–457.

Burgess S O, Pate J S, Adams M A and Dawson, T E 1999 Seasonal water acquisition and redistribution in the Australian woody phreatophyte, *Banksia prionotes*. Ann. Bot. 85, 215–224.

Celenza J L, Grisafi P L and Fink G R 1995 A pathway for lateral root formation in *Arabidopsis thaliana*. Genes Dev. 9, 2131–2142.

Crocker L J and Schwintzer C R 1993 Factors affecting formation of cluster roots in *Myrica gale* seedlings in water culture. Plant Soil 152, 287–298.

Crocker L J and Schwintzer C R 1994 Soil conditions affect the occurrence of cluster roots in *Myrica Gale* L. in the field. Soil Biol. Biochem. 26, 615–622.

Dell B, Kuo J and Thomson C J 1980 Development of proteoid roots in *Hakea obliqua* R. Br. (Proteaceae) grown in water culture. Aust. J. Bot. 28, 27–37.

Dinkelaker B, Bengeler C, Neumann G, Eltrop L and Marschner H 1997 Root exudates and mobilization of nutrients. *In* Trees: Contributions to Modern Tree Physiology. Eds. W Eschrich and H Ziegler. pp. 441–451. Backhuys Publishers, Leiden.

Dinkelaker B, Hengeler C and Marschner H 1995 Distribution and function of proteoid roots and other root clusters. Bot. Acta 108, 183–200.

Dinkelaker B, Römheld V and Marschner H 1989 Citric acid excretion and precipitation of calcium citrate in the rhizosphere of white lupin (*Lupinus albus* L.). Plant Cell Environ. 12, 285–292.

Gardner W K, Barber D A and Parbery D G 1982a Effect of microorganisms on the formation and activity of proteoid roots of *Lupinus albus* L. Aust. J. Bot. 30, 303–309.

Gardner W K, Parbery D G and Barber D A 1982b The acquisition of phosphorus by *Lupinus albus* L. I. Some characteristics of the soil/root interface. Plant Soil 68, 19–32.

Gardner W K, Parbery D G and Barber D A 1982c The acquisition of phosphorus by *Lupinus albus* L. II. The effect of varying phosphorus supply and soil type on some characteristics of the soil/root interface. Plant Soil 68, 33–41.

Gardner W K, Parbery D G and Barber D A 1983 The acquisition of phosphorus by *Lupinus albus* L. III. The probable mechanism by which phosphorus movement in the soil/root interface is enhanced. Plant Soil 70, 107–124.

Gerke J, Römer W and Jungk A 1994 The excretion of citric and malic acid by proteoid roots of *Lupinus albus* L.; effects on soil solution concentrations of phosphate, iron, and aluminium in the proteoid rhizosphere in samples of an Oxisol and a Luvisol. Z. Pflanzenernaehr. Bodenk. 155, 339–343.

Gilbert G A, Knight J D, Vance C P and Allan D L 1999 Acid phosphatase activity in phosphorus-deficient white lupin roots. Plant, Cell Environ. 22, 801–810.

Gilbert G A, Knight J D, Vance C P and Allan D L 2000 Proteoid root development of phosphorus deficient lupin is mimicked by auxin and phosphonate. Ann. Bot. 85, 921–928.

Gould S F 1998 Proteoid root mats bind surface materials in Hawkesbury Sandstone biomantles. Aust. J. Soil Res. 36, 1019–1031.

Grierson P F and Attiwill P M 1989 Chemical characteristics of the proteoid root mat of *Banksia integrifolia* L. f. Aust. J. Bot. 37, 137–143.

Grierson P F and Adams M A 2000 Plant species affect acid phosphatase, ergosterol and microbial P in a jarrah (*Eucalyptus marginata* Donn ex Sm.) forest in south-western Australia. Soil Biol. Biochem. 36, 1817–1827.

Gutierrez Mañero F J, Acero N, Lucas J A and Probanza A 1996 The influence of native rhizobacteria on European alder (*Alnus glutinosa* (L.) Gaertn.) growth. Plant Soil 182, 67–74.

Handreck K A 1991 Interactions between iron and phosphorus in the nutrition of *Banksia ericifolia* L. f. var. *ericifolia* (Proteaceae) in soil-less potting media. Aust. J. Bot. 39, 373–384.

Horst W J, Wagner A and Marschner H 1982 Mucilage protects root meristems from aluminium injury. Z. Pflanzenphysiol. 105, 435–444.

Hurd T M and Schwintzer C R 1996 Formation of cluster roots in *Alnus incana* ssp. *rugosa* and other *Alnus* species. Can. J. Bot. 74, 1684–1686.

Jeffrey D W 1967 Phosphate nutrition of Australian heath plants. I. The importance of proteoid roots in *Banksia* (Proteaceae) Aust. J. Bot. 15, 403–411.

Johnson J F, Vance C P and Allan D L 1996 Phosphorus deficiency in *Lupinus albus*: altered lateral root development and enhanced expression of phospho*enol*pyruvate carboxylase. Plant Physiol. 112, 31–41.

Johnson J F, Allan D L and Vance C P 1994 Phosphorus stress-induced proteoid roots show altered metabolism in *Lupinus albus*. Plant Physiol. 104, 657–665.

Keerthisinghe G, Hocking P J, Ryan P R and Delhaize E 1998 Effect of phosphorus supply on the formation and function of proteoid roots of white lupin (*Lupinus albus* L.) Plant Cell Environ. 21, 467–478.

Kerley S J 2000 Chances in root morphology of white lupin (*Lupinus albus* L.) and its adaptation to soils with heterogeneous alkaline/acid profiles. Plant Soil 218, 197–205.

Kerley S J, Leach J E, Swain J L and Huyghe C 2000 Investigations into the exploitation of heterogeneous soils by *Lupinus albus* L. and *L. pilosus* Murr: and the effect upon plant growth. Plant Soil 222, 241–253.

Kuiper D, Kuiper P J C, Lambers H, Schuit J and Staal M 1989 Cytokinin concentration in relation to mineral nutrition and benzyladenine treatment in *Plantago major* ssp. *pleiosperma*. Physiol. Plant. 75, 511–517.

Lambers H, Juniper D, Cawthray GR, Veneklaas EJ and Martinez-Ferri E 2002 The pattern of carboxylate exudation in *Banksia grandis* (Proteaceae) is affected by the form of phosphate added to the soil. Plant Soil 238, 111–122.

Lamont B 1972a 'Proteoid' roots in the legume *Viminaria juncea*. Search 3, 90–91.

Lamont B 1972b The morphology and anatomy of proteoid roots in the genus *Hakea*. Aust. J. Bot. 20, 155–174.

Lamont B 1972c The effect of soil nutrients on the production of proteoid roots by *Hakea* species. Aust. J. Bot. 20, 27–40.

Lamont B 1973 Factors affecting the distribution of proteoid roots within the root systems of two *Hakea* species. Aust. J. Bot. 21, 165–187.

Lamont B 1976 The effects of seasonality and waterlogging on the root systems of a number of *Hakea* species. Aust. J. Bot. 24, 691–702.

Lamont B 1982 Mechanisms for enhancing nutrient uptake in plants, with particular reference to mediterranean South Africa and Western Australia. Bot. Rev. 48, 597–689.

Lamont B 1983a Proteoid roots in the South African Proteaceae. J. S. Afr. Bot. 49, 103–123.

Lamont B 1983b Root hair dimensions and surface/volume/weight ratios of roots with the aid of scanning electron microscopy. Plant Soil 74, 149–152.

Lamont B B 1981 Specialized roots of non-symbiotic origin in heathlands. *In* Heathlands and Related Shrublands of the World B. Analytical Studies. Ed. R L Specht. pp. 183–195. Elsevier Scientific, Amsterdam.

Lamont B B 1986 The significance of proteoid roots in proteas. Acta Hortic. 185, 163–170.

Lamont B B 1993 Why are hairy root clusters so abundant in the most nutrient-impoverished soils of Australia? Plant Soil 156, 269–272.

Lamont B B 1995 Mineral nutrient relations in mediterranean regions of California, Chile and Australia. *In* Ecology and Biogeography of Mediterranean Ecosystems in Chile, California and Australia. Eds. M T Arroya, P H Zedler and M D Fox. pp. 120–132. Springer Verlag, New York.

Lamont B B and Bergl S M 1991 Water relations, shoot and root architecture, and phenology of three co-occurring *Banksia* species: no evidence for niche differentation in the pattern of water use. Oikos 60, 291–298.

Lamont B B and Groom P K 2002 Green cotyledons of two *Hakea* species control seedling mass and morphology by supplying mineral nutrients rather than organic compounds. New Phytol. 153, 101–110.

Lamont B B and McComb A J 1974 Microorganisms and the formation of proteoid roots. Aust. J. Bot. 22, 681–688.

Lamont B, Brown G and Mitchell D T 1984 Structure, environmental effects on their formation and function of proteoid roots in *Leucadendron laureolum* (Proteaceae). New Phytol. 97, 381–390.

Li M, Shinano T and Tadona T 1997 Distribution of exudates of lupin roots in the rhizosphere under phosphorus deficient conditions. Soil Sci. Plant Nutr. 43, 237–245.

Louis I, Racette S and Torrey J G 1990 Occurrence of cluster roots on *Myrica cerifera* L. (Myricaceae) in water culture in relation to phosphorus nutrition. New Phytol. 115, 311–317.

Low A B and Lamont B B 1990 Aerial and below-ground phytomass of *Banksia* scrub-heath at Eneabba, Western Australia. Aust. J. Bot. 38, 351–359.

Lynch J and Brown K M 1997 Ethylene and plant responses to nutritional stress. Physiol. Plant. 100, 613–619.

Malajczuk N and Bowen G D 1974 Proteoid roots are microbially induced. Nature 251, 316–317.

Marschner H, Römheld V and Cakmark I 1987 Root induced changes of nutrient availability in the rhizosphere. J. Plant Nutr. 10, 1175–1184.

McArthur W M 1991 Reference soils of south-western Australia. Department of Agriculture, Western Australia, Perth.

Neumann G, Massonneau A, Langlade N, Dinkelaker B, Hengeler C, Römheld V and Martinoia E 2000 Physiological aspects of cluster root function and development in phosphorus-deficient White Lupin (*Lupinus albus* L.) Ann. Bot. 85, 909–919.

Neumann G, Massonneau A, Martinoia E and Römheld V 1999 Physiological adaptations to phosphorus deficiency during proteoid root development in white lupin. Planta 208, 373–382.

Nichols D G, Jones D L and Beardsell D V 1979 The effect of phosphorus on the growth of *Grevillea* 'Poorinda Firebird' in soil-less potting mixtures. Scient. Hortic. 11, 197–205.

Nichols D G, Jones D L and Beardsell D V 1979 The effect of phosphorus on the growth of Grevillea 'Poorinda Firebird' in soil-less potting mixtures. Scient. Hort. 11, 197–205.

Pate J S and Jeschke W D 1993 Mineral uptake and transport in xylem and phloem of the proteaceous tree *Banksia prionotes*. *In* Plant Nutrition – From Genetic Engineering to Field Practice. Ed. N J Barrow. pp. 313–316. Kluwer Academic Publishers, Dordrecht.

Pate J S, Verboom W H and Galloway P D 2001 Co-occurrence of Proteaceae, laterite and related oligotrophic soils: coincidental associations or causative inter-relationships? Aust. J. Bot. 49, 529–560.

Peiter E, Yan F and Schubert S 2001 Proteoid root formation of *Lupinus albus* L. is triggered by high pH of the root medium. J. Appl. Bot. 75, 50–52.

Purnell H M 1960 Studies of the family Proteaceae. I. Anatomy and morphology of the roots of some Victorian species. Aust. J. Bot. 8, 38–50.

Racette S, Louis I and Torrey J G 1990 Cluster root formation by *Gymnostoma papuanum* (Casuarinaceae) in relation to aeration and mineral nutrient availability in water culture. Can. J. Bot. 68, 2564–2570.

Reddell P, Bowen G D and Robson A D 1986 Nodulation of Casuarinaceae in relation to host species and soil properties. Aust. J. Bot. 34, 435–444.

Reddell P, Yung Y and Shipton W A 1997 Cluster roots and mycorrhizae in *Casuarina cunninghaminana*: their occurrence and formation in relation to phosphorus supply. Aust. J. Bot. 45, 41–51.

Roelofs R F R, Rengel Z, Cawthray G R, Dixon K W and Lambers H 2001 Exudation of carboxylates in Australian Proteaceae: chemical composition. Plant, Cell Environ. 24, 891–903.

Rosenfield C-L, Reed D W and Kent M W 1991 Dependency of iron reduction on development of a unique root morphology in *Ficus benjamina* L. Plant Physiol. 95, 1120–1124.

Salisbury F B and Ross C W 1992 Plant Physiology. 4th edn. Wadsworth Publishing, Belmont.

Schmidt S and Stewart G R 1997 Waterlogging and fire impacts on nitrogen availability and utilization in a subtropical wet heathland (wallum). Plant, Cell Environ. 20, 1231–1241.

Shane M W, De Vos M, De Roock S, Cawthray G R and Lambers H 2003 Effects of external phosphorus supply on internal phosphorus concentrations and the initiation, growth and exudation of cluster roots in *Hakea prostrata* R.Br. Plant Soil 248, 209–219.

Siddiqi M Y and Carolin R C 1976 Studies on the ecology of coastal heath in NSW. II. Proc. Linn. Soc. N.S.W. 101, 38–52.

Skene K R 1998 Cluster roots: some ecological considerations. J. Ecol. 86, 1060–1064.

Skene K R and James W M 2000 A comparison of the effects of auxin on cluster root initiation and development in *Grevillea robusta* Cunn. ex R. Br. (Proteaceae) and in the genus *Lupinus* (Leguminosae). Plant Soil 219, 221–229.

Skene K R, Raven J A and Sprent J I 1998 Aspects of cluster root development in *Grevillea robusta* (Proteaceae). II. Development of the endodermis in a determinate root and in an indeterminate lateral root. New Phytol. 138, 733–742.

Specht R L, Moll E J, Pressinger F and Sommerville J 1983 Moisture regime and nutrient control of seasonal growth in Mediterranean ecosystems. *In* Mediterranean-type Ecosystems: The Role of Nutrients. Ed. F J Kruger, D T Mitchell and J U M Jarvis. pp. 246–273. Springer-Verlag, Berlin.

Subba Rao N S and Rodrígues-Darrucco C 1995 Casuarina. pp. 169–174. Science Publishers, Lebanon.

Sward R J 1978 Studies on vesicular-arbuscular mycorrhizas of some Australian heathland plants. PhD Thesis, Monash University, Victoria, Australia.

Trinick M J 1977 Vesicular-arbuscular infection and soil phosphorus utilization in *Lupinus* spp. New Phytol. 78, 297–304.

van der Werf A and Nagel O W 1996 Carbon allocation to shoots and roots in relation to nitrogen supply is mediated by cytokinins and sucrose: Opinion. Plant Soil 185, 21–32.

Vorster P W and Jooste J H 1986a Potassium and phosphate absorption by excised ordinary and proteoid roots of the Proteaceae. S. Afr. J. Bot. 52, 277–281.

Vorster P W and Jooste J H 1986b Translocation of potassium and phosphate from ordinary and proteoid roots to shoots in the Proteaceae. S. Afr. J. Bot. 52, 282–285.

Waters B M and Blevins D G 2000 Ethylene production, cluster root formation, and localization of iron(III) reducing capacity in Fe deficient squash roots. Plant Soil 225, 21–31.

Watt M and Evans J R 1999 Linking development and determinacy with organic acid efflux from proteoid roots of White Lupin grown with low phosphorus and ambient or elevated atmospheric CO_2 concentration. Plant Physiol. 120, 705–716.

Webb L J 1954 Aluminium accumulation in the Australian – New Guinea flora. Aust. J. Bot. 2, 176–196.

Wenzel C L, Ashford A E and Summerell B A 1994 Phosphate-solubilizing bacteria associated with proteoid roots of seedlings of waratah (*Telopea speciosissima* (Sm.) R.Br. New Phytol. 128, 487–496.

Zaïd H, Arahou M, Diem H G and El Morabet R 2002 Is Fe deficiency rather than P deficiency the cause of cluster root formation in *Casuarina glauca*? Plant Soil 248, 229–235.

Plant and Soil **248:** 21–30, 2003.
© 2003 Kluwer Academic Publishers. Printed in the Netherlands.

The evolution of physiology and development in the cluster root: teaching an old dog new tricks?

Keith R. Skene
Division of Environmental and Applied Biology, Institute of Biological Sciences, University of Dundee, Dundee DD1 4HN, Scotland

Received 16 January 2002. Accepted in revised form 2 April 2002

Key words: cluster roots, high transport capacity (HTC), low transport capacity (LTC), morphogenesis, proteoid roots, pseudo-cluster root, root architecture, transformation theory

Abstract

In this paper we examine the key elements of cluster or proteoid roots, and trace their origins back to regular root properties. By viewing the root system as being composed of two categories of surface, the high transport capacity (HTC) area, just behind the meristem, and the low transport capacity (LTC) area (the rest of the root system), based on export and import capacities, we examine root system architecture in terms of structure–function relationships, and conclude that measuring total root exudation per unit area, volume or mass will not give useful comparative data for root transport properties. Furthermore, the cluster root represents a manipulation of the HTC to LTC root surface area ratio. Increased exudation and P uptake may be no higher in individual rootlets than in other HTC regions of the root system. We also examine the transformation theory (the theory of form resulting from a series of forces, which, when altered, lead to a change, or transformation in form) as an explanation of cluster root evolution, and conclude that the cluster root requires only a change in pericycle response to depleted internal nutrient levels, with the other characteristics representing consequences stemming from the form and constraints of the root system.

Abbreviations: HTC – high transport capacity; LTC – low transport capacity

The cluster root is defined as a densely packed grouping of determinate rootlets. These rootlets develop, grow, cease growth, undergo an exudative burst and absorb phosphate in a synchronous way (Dinkelaker et al., 1995; Neumann et al., 1999, 2000, Skene, 2000; Watt and Evans, 1999). Cluster roots occur in a wide range of families, and in a subset of species within these families (see Skene 1998). Cluster roots are viewed as adaptations for nutrient acquisition, and their development is initiated in plants when internal iron or phosphate levels fall below a critical level (Diem and Arahou, 1996; Marschner et al., 1987). Phylogenetic relationships between species with cluster roots show a wide distribution of the adaptation across many plant families (Skene, 1998). Within a given family, not all genera will display the ability to produce cluster roots, and even within a given genus, not all species are predisposed to forming these structures (Skene, 2000). In this paper we wish to address a number of issues relating to the evolution of these structures.

Firstly, how many changes are needed to produce a cluster root? We can address this by examining, in turn, each of the key characteristics of cluster root structure.

(a) Root initiation

Cluster root morphology is characterized by the massive proliferation of rootlet initiation and concomitant development of rootlets. Through reduction, it

* FAX No: +44-1382-344275. E-mail: k.r.skene@dundee.ac.uk

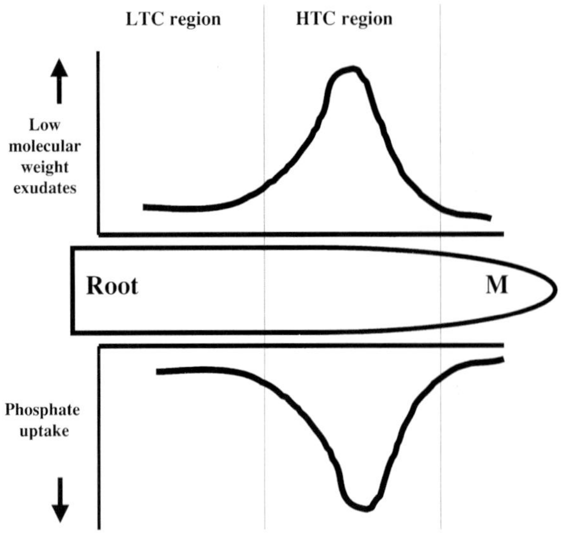

Figure 1. Root exudation and phosphate uptake along a non-mycorrhizal root (schematic presentation based on data from Bowen (1968) and Leisen et al. (1990). Both exudation of low molecular weight exudates and phosphate uptake can be seen to peak at a region just behind the root tip. The root system as a whole can be partitioned into two types: a high transport capacity (HTC) region located behind each root meristem, and measuring typically 2–3 mm, where movement of material in and out of the root is at a maximum, and a low transport capacity (LTC) region, which represents the rest of the root system, where much less material is transported into and out of the root.

becomes obvious that this stems from the pericycle cells, particularly those cells opposite the protoxylem poles (since all species with the capacity to produce cluster roots form lateral roots and cluster rootlets only opposite their protoxylem poles). This is because lateral root initiation begins by cell division in the pericycle (Beeckman et al., 2001). Cluster rootlets are initiated in the same way as, and have a structure identical to lateral roots. In terms of their initiation, the only differences are that there are many more of these rootlets, and they develop opposite *every* protoxylem pole in a radial direction, within the cluster root zone. This is discussed at length in a recent review (Skene, 2001). Thus the cluster root structure does not require any new developmental event, but rather an intensification of an existent programme.

(b) Determinate development

Cluster rootlets are determinate, typically ceasing growth after 3–6 days of growth (Skene et al., 1998). While the mechanism of meristem differentiation is not yet fully understood, it could well relate to the failure or absence of a quiescent centre (Barlow, 1976), the related issue of a sub-optimal meristem size, the starvation of the meristem via plasmodesmata density decrease (Gunning, 1978), reduced expression of a SKP1-like protein (Porat et al, 1998) or increased expression of an ICK1-like protein (Wang et al., 1998). There is not a necessity for a large change in genome to underpin determinate development. Indeed this feature of cluster rootlets may merely be a consequence of the intensification of initiation. As such, determinate development is likely to play a linking role between the development and the physiology, correlated as it is with the onset of the exudative burst.

(c) Exudative burst

As cluster roots develop, there is a sudden, marked increase in exudation of organic acids that lasts for around 2 days (Dinkelaker et al., 1995; Hagstrom et al., 2001; Watt and Evans, 1999). This leads to a large and rapid change in rhizosphere chemistry, resulting in an increase in the available phosphate and iron pool (Dinkelaker et al., 1989). Does the exudative burst require further novel genes and biochemical specialization, or is it a product of constraints inherent in any growing root, later finding a role in cluster root function (known as a first use exaptation (Armbruster, 1997; Arnold, 1994))?

To answer this question, we need to consider the relationship between root development and function and the soil. Nutrients in the soil are partitioned into two pools: the available or extractable nutrient pool and the unavailable nutrient pool. The root can only, directly, absorb nutrients from the available pool, but can increase the size of this pool by recruitment from the unavailable pool.

In terms of nutrient acquisition, roots possess four key properties (aside from symbioses with organisms from other kingdoms):

1. Growth: root growth is an inherent necessity for nutrient acquisition due to the limited size of the available nutrient pool, and the consequences of diffusion. As roots absorb nutrients from a specific point in the soil, the nutrients will become depleted in the soil. Thus, nutrient availability at the rhizoplane will become limited by the diffusion coefficients of the nutrients. One way to improve nutrient availability is to move to another, as yet

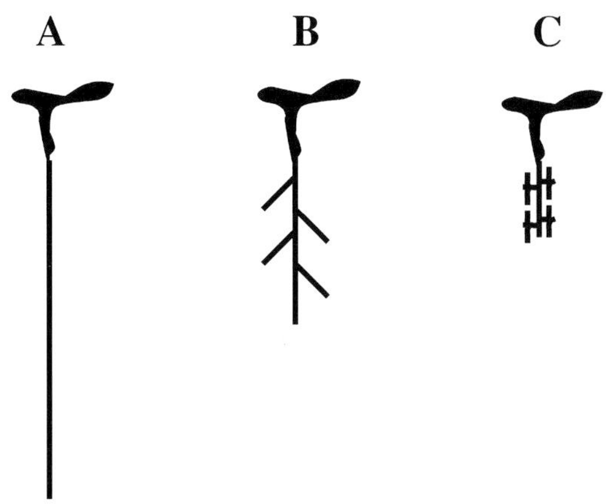

	A	B	C
Root length	**4**	**4**	**4**
Length of primary root	**4**	**2**	**1**
Length of secondary root	**0**	**2**	**1**
Length of tertiary root	**0**	**0**	**2**
Number of meristems	**1**	**5**	**13**
Relative HTC area	**1**	**5**	**13**
Relative LTC area	**15**	**11**	**3**

Figure 2. The relationship between root architecture and transport between root and soil. This schematic diagram shows three model plants with different architecture. Calculations are based on the high transport capacity region measuring 0.25 cm at the end of each root. Several assumptions have been made for simplification: root surface area and density are the same for all orders of roots, the HTC region is the same area in all orders of roots. Length and area units are relative. Relative HTC and LTC area are calculated from the relative lengths of each root type in each architectural model, LTC length being total root length minus HTC length.

un-exploited part of the soil. Roots achieve this by growth.

2. Branching: Indeterminate development of the root system as a whole is an important aspect of nutrient acquisition, meeting the challenge of heterogeneity of nutrients in the soil. This relies, fundamentally on the response of the pericycle to signals as yet poorly defined. The role of internal and external signals is also unclear.
3. Exudation: roots exude many materials, including enzymes, sterols, fatty acids, aromatic acids, aliphatic acids, amino acids, amides and carbohydrates (Bolton et al., 1992). One of the key roles of root exudation is to increase the size of the available nutrient pool, by lowering pH or by chelation (although molybdenum, for example, actually becomes less available at low pH). A second, more controversial role of root exudates is the release of carbon in order to reduce inhibition of photosynthesis by excess shoot carbon (Thomas, 1994).
4. Absorption: the key final stage in nutrient acquisition, allowing nutrient uptake from the available pool.

Accessing the available pool of nutrients in the soil requires only growth and absorption, whereas to access the unavailable pool, exudation, followed by absorption is required.

The root as a two component system

In order to understand, at a holistic level, how these four processes operate at the soil–root interface, we can view the non-mycorrhizal root system (all species that produce cluster roots either are constitutively or environmentally non-mycorrhizal (Skene, 1998)) as being made up of two components: high transport capacity (HTC) root surface (that is, high export and import ability) and low transport capacity (LTC) root surface (low export and import capacity). Rhizodeposition changes quantitatively along the length of the root. The main zones of secretion and exudation are towards the root tip, with lysates being released further back, where outer cells are dying (Bowen and Rovira, 1991; Grayston et al., 1996; Leisen et al., 1990; Marschner, 1996; McDougall and Rovira, 1970) (Figure 1). Interestingly, in non-mycorrhizal roots, P uptake is also mostly limited to a similar region of the root (Bowen, 1968) (Figure 1). Thus the root surface can basically be divided into two categories, areas of high P uptake and exudation (HTC), lying just behind the root tip, and areas of low P uptake and exudation (LTC), lying further back. In Figure 2, we consider three different root architectural layouts. All three plants have the same root surface area, length and volume (exluding the consideration of root hairs, which will increase from A to B to C, magnifying the observed effect). However, it can be seen that the ability to uptake P and exude material differs widely, and is correlated to the number of meristems present. Thus, expressing root exudation or P uptake per unit total root length, total surface area, volume or weight (fresh or dry) is not a useful concept in comparative studies, since the amount of each type of root (LTC or HTC), rather than total root, is the key determinate of both exudation and P uptake. Each plant will perform very differently in terms of total P uptake per unit area of root, not because of branching directly, but because of increased relative HTC root surface area. Then, in Figure 2, all three roots possess the same basic physiological abilities, but differences result from the ratio of LTC to HTC. A root system with more meristems per unit total length of root should have much greater exudation and P uptake than a root system of identical length but different architecture, since the HTC region occurs just behind the meristem. The cluster root would appear to demonstrate this concept, in that it is composed of many root tips, with a concomitant increase in the high P uptake and exudation type of root. By developing many tips, and being determinate, and existing under low internal P or Fe (conditions that lead to greater exudation anyway) these short rootlets are providing maximum high transport root surface, without low transport root surface, at least for the duration of the exudative burst.

How unique is the exudative burst in plant physiology?

In Figure 3, we investigate the change in exudate levels at a particular point in the rhizosphere adjacent to a cluster root over 5 days. Also shown is the effect of a growing lateral root on a point in the soil through which the root elongates, comparing the impact of the exudative burst and of regular root exudation on the soil. It can be seen that both of these effects lead to an exudative burst in terms of localized temporal and spatial events in a given part of the soil. A key difference is related to the fact that all of the exudation occurs at the same location in the rhizosphere of the cluster root, whereas the 'exudative burst' in a regular root moves through the soil profile as the root elongates. This brings us to the second difference in the two systems. In one (the cluster root), the exudative burst occurs upon the onset of meristem differentiation, whereas in the other (the lateral root), exudation occurs at a fixed distance behind the ever-dividing meristem. Thus, in both cases, it is the distance from an active meristem that is the common correlative factor with exudation. In the cluster root this distance results from the complete disappearance of the meristem via differentiation of the meristem initials (Skene et al., 1998), whereas in the lateral root, the meristem moves away, induced by cell expansion behind the meristem.

Age distribution in cluster roots

In the case of cluster roots, the age distribution of cluster roots will have a big impact on the exudation as a whole, particularly since a cluster root that is undergoing the exudative burst will have similar mass and root length as a senescent, non-exuding root. Thus, comparing total plant exudation while ignoring the age distribution (i.e., the proportion of the cluster roots in HTC and LTC mode) of the cluster roots present, again may give rise to misleading conclusions. It appears that cluster roots are produced in 'waves' (Neumann et al., 1999; Skene and James, 2000; Watt and Evans, 1999). Thus, if we compare the exudative profile of two plants that are 2 days out of synchrony in development but otherwise identical in terms of the number and temporal and spatial patterns of development and physiology (Figure 4), it can clearly be seen that apparent differences will occur, depending on the dates when sampling occurs.

(d) Phosphate uptake

As mentioned before, P uptake increases behind the meristem of a non-mycorrhizal root (Bowen, 1968) (Figure 1). Thus the ability of a non-mycorrhizal root to absorb phosphate will be proportional to the number of active meristems present, as was shown for exudation in Figure 2. We would also expect a similar relationship between phosphate and age distribution of cluster roots as was noted for root exudation, as shown in Figure 4. This relationship will be intensified by the paralleled increase of phosphate availability due to the interaction of the soil with the exudates. The cluster root, with a structure that has a large HTC root surface, represents an excellent absorptive strategy.

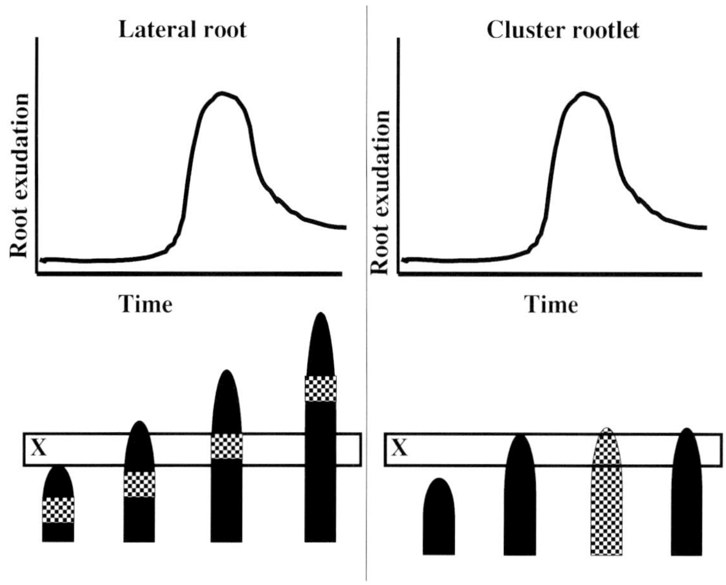

Figure 3. The spatial and temporal exudative profiles of lateral roots and cluster rootlets on the soil. (a) In this figure, the amount of exudate released into a zone of soil, X, is shown in a graph through time, as the lateral root grows through the soil. The high transport capacity (HTC) region of the root is marked by small squares. As the root elongates, there is a transient rise in exudation in zone X. (b) In the case of the cluster root, a similar pattern is observed, with the exudation occurring in zone X during the exudative burst. In this case, the onset and end of exudation is derived from developmental cues, rather than growth-induced movement outwith the zone.

Novelty or transformation?

Thus we can see that the key elements that make up the cluster root are present in regular roots. The differences relate to intensity, and a concomitant manipulation of the HTC to LTC ratio of the root system. There is an increased intensity of rootlet initiation, leading to a much greater concentration of growing rootlets. An exudative burst occurs in each rootlet upon determination of the meristem. The apparently greater rates of exudation and nutrient uptake of the cluster root per unit area of root compared to normal roots, may not require an explanation in terms of increased transport above and beyond any part of the rest of the root system, but rather may be explained by an increased proportion of high transport capacity surface area compared to low transport capacity surface area.

Why are cluster roots different in different species?

This question is often asked, and it highlights important issues in how we categorize cluster roots, returning again to the issue of what is the essence of a cluster root. If we go back to the term adaptation (although the cluster root should, more correctly, be called an aptation or first use exaptation (see Armbuster (1997) and Gould and Vrba (1982) on exaptations, and, more entertainingly, Gould and Lewontin (1979) referring to Voltaire's *Candide),* we can explore the cluster root in light of the definitions formulated by Curio (1973), Tinbergen (1967) and Williams (1966).

The same form or similar forms occur in similar environments in a number of different species, especially in distantly related species

This criterion is easily met in the cluster root. For example, these structures occur in families as diverse as the Proteaceae, Leguminosae, Betulaceae, Myricaceae, Casuarinaceae, Cucurbitaceae, Moraceae and Eleagnaceae, in similar form. *Myrica gale* (Myricaceae, Northern Hemisphere) and *Viminaria juncea* (Leguminosae, Southern Hemisphere) both occupy waterlogged environments. *Casuarina equisetifolia* (Casuarinaceae, Southern Hemisphere) and *Hippophae rhamnoides* (Eleagnaceae, Northern Hemisphere) both occupy pioneer sites, particularly in early succession.

Variant forms of a character in a number of related species accord with differences in the environments of the respective species

Certainly cluster roots in different species show vari-

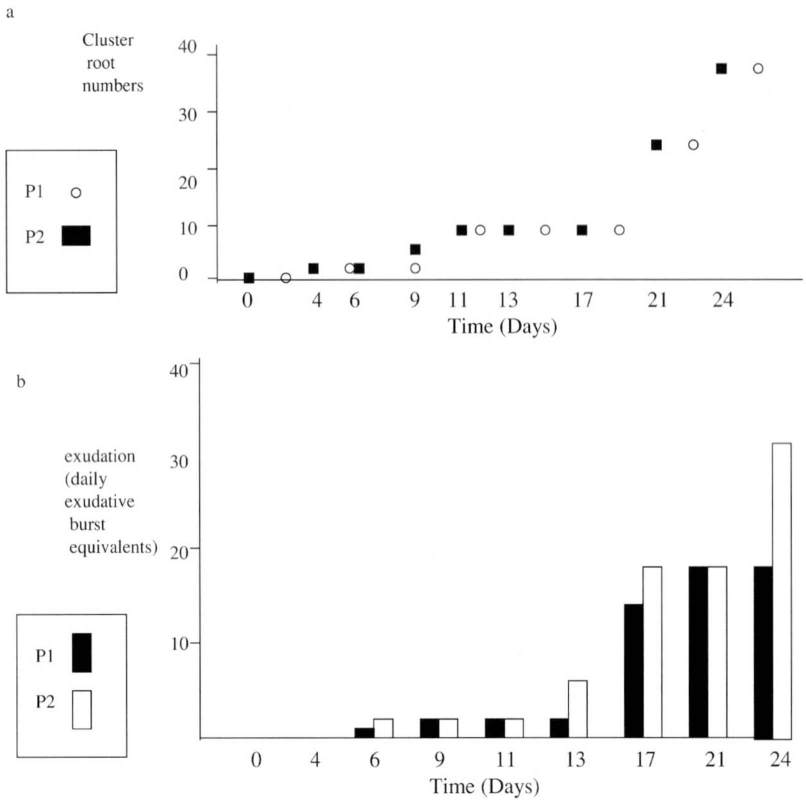

Figure 4. Age distribution and its effect on exudation. (a) In this graph, based on data from Skene and James (2000), the phased production of cluster roots is clearly visible. Two identical plants grown in identical conditions, one (P1) growing 2 days later than the other (P2), produce identical numbers of cluster roots through time. Thus on any given day, P1 will exist in a developmental and physiological state that P2 was in 48 h earlier. (b) In this bar chart we compare the exudation of each plant on particular days. Exudation was modelled as total root exudation, sampled from the hydroponic growth solution. It can be seen that because the exudative burst only occurs on days 4 and 5 of cluster root development, there will be large differences between P1 and P2 at times (e.g. days 13 and 24), and no differences at other times (e.g., days 11 and 24), even though these plants are identical in every way except age. Thus the time of sampling could make a large difference in total exudation measured.

ant physiology. Recent work on *Myrica gale* has shown marked differences to *L. albus* in terms of physiology and structural response to waterlogging (unpublished). Cluster roots of *L. albus* differ in physiology under iron and phosphate stress. It would be most interesting to examine if more closely related species showed differences in cluster root function, in terms of niche separation (Cowling et al., 1994). A fine example would be the Proteaceae of the Fynbos, South Africa.

For complex characters in a particular context, the more their component aspects can be related, point by point, to function in that complex (the 'goodness of design' of Williams)

The cluster root can be seen as an integrated structural and functional unit. Its occurrence when internal P or Fe levels are low, the clustered nature of rootlet initiation, the up-regulation of organic acid production by changes in enzyme activities (both up-regulation and down-regulation) in the TCA cycle, the exudative burst and increase in P or Fe uptake are common to all cluster roots, and each contributes to the successful acquisition of nutrients.

Phylogenetic discontinuity

Another important aspect of the cluster root, is the phylogenetic distribution. While we still do not understand the exact molecular basis of this structure, examination of its occurrence in the genus *Lupinus* points to a very narrow spread within this genus, in closely related members originating within the last 2.5 million years, whereas in the Proteaceae, evidence would suggest that the cluster root in this family

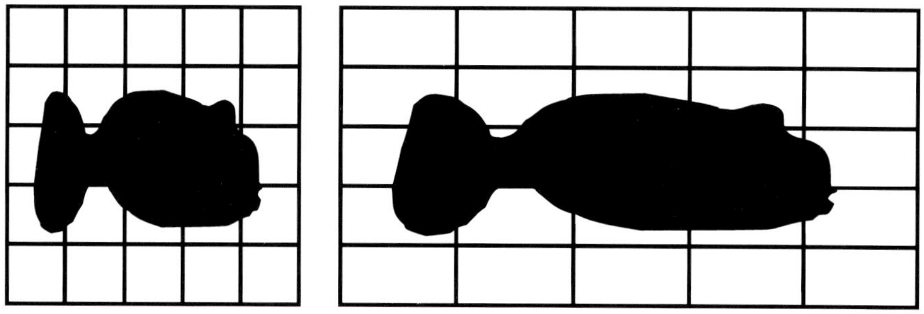

Figure 5. Transformation theory. This figure shows the effect of transforming the co-ordinates of an organism in a simple way ($x = 2y$). Their effect is to produce a different morphology that initially does not appear to be related to the original. Transformation represents the forces that determine final morphology, and combines the influences of genes, material, laws and constraints.

arose 60 million years ago (Skene, 2000) and exists in every member of this ancient family examined thus far, with the exception of the primitive sub-family, Persoonidae. We have shown that there are not many changes needed to produce a cluster root, and that the origin lies in the basic process of rootlet initiation. Thus, its occurrence in a wide range of families, across different geological eras, and with phylogenetic discontinuity at family, genus and species level should not be surprising. Furthermore it points to a simple and basic change, that can occur often, and involves change in a foundational aspect of development.

Transformation theory

Differences in cluster root development and physiology should be expected, due to evolutionary specialization in given habitats. Furthermore, if the molecular change underpinning the cluster root phenotype is a basic one, affecting a core element of development, such as root initiation, then the final appearance will carry the 'personality' of the organism as a whole. Each species will express differently any particular core change. The cluster root most likely represents a transformation of a given set of characteristics, as proposed by Darcy Thompson in his law of transformation. Thompson stated that in examining the relationship between morphology and phylogeny, "a comprehensive 'law of growth' has pervaded the whole structure in its integrity, and ... some more or less simple and recognizable system of forces has been in control" (Thompson, 1917). Thus he understood that there was some underlying relationship between pattern and signal, a concept later developed by other workers (e.g., Goodwin, 1985; Turing, 1952; Wolpert, 1971). By transforming the co-ordinates of a particular structure, he was able to generate another structure. He did this on a wide range of objects, from whole organisms to bones. The concept is explained in Figure 5. A key element in transformation theory is that genes do not work in a vacuum, but, rather, physical laws and constraints play significant roles in the forms produced from gene products. The properties of matter represent the forces controlling form. We can take this a step further in applying transformation theory to physiological evolution as well as morphological evolution. Thus a transformation of a given set of physiological co-ordinates would lead to a change under some *recognizable system of forces* also. An important point, relating to differences in cluster roots between species, is that the transformation may be the same, but the final outcome will depend on the characteristics of the root system upon which the transformation acts (Figure 5). Thus we should expect differences in the final product of transformation, based on the original properties of the object of transformation. If we then look at the events that occur in cluster root formation, we can divide these into two parts, the developmental, and the physiological elements (Figure 6a, b). Here we see that the properties of the cluster root are all related and result from a very few changes. Thus, even a single change in a gene controlling the response of pericycle cells to a low nutrient signal may be enough to produce a cluster root, with the laws of physics and system constraints (the *recognizable system of forces* of Darcy Thompson) imposing the consequences of this. The combination of these developmental and physiological pathways of consequence (Figure 6c) represents an interacting pair of events. Thus the transformation of development and of physiology have repercussions for each other. Hence we are dealing with a multidimensional transformation.

Mn, Cu, Zn, Fe and P, in the detoxification of metals such as Zn and Al, and in the breakdown of wood (Delhaize et al., 1993; Hoffland, 1992; Marschner, 1995; Micales, 1997). Much of the evidence for these purported roles of organic acids comes from *in vitro* experiments performed under conditions which are vastly different from those that exist *in vivo*, and these studies have therefore drawn considerable criticism (Drever and Stillings, 1997; Jones, 1998; Parker and Pedler, 1988).

Since the realization that the root release of organic acids may confer aluminium resistance in plants, there has been great interest in the possibility of manipulating organic acid efflux by altering the expression of organic acid synthesis/efflux genes in plants (Delhaize et al., 2003; Ryan et al., 2001). After release from the root, however, organic acids can suffer a number of fates such as sorption, biodegradation and metal complexation, making their behavior difficult to predict. The aim of this review is to highlight current misconceptions and knowledge gaps surrounding the behavior of organic acids in soil with particular reference to plant nutrition. The review will focus on the main factors controlling the fate of organic acids in soil including their behavior in soil solution, nutrient-mobilization efficiency, sorption to the soil's solid phase, degradation by the soil microbial biomass and role in ecosystem processes. The generic term organic acids is used throughout this review, however, it should be acknowledged that in many soils these compounds are either partially or fully dissociated and as such could be referred to as organic anions.

Organic acid dynamics in soil solution

Across a broad range of ecosystems the concentration of organic acids in soil solution has been shown to be low, typically ranging from 1 to 50 μM (Bazira-makenga et al., 1995; Krzysowska et al., 1996; Strobel, 2001). In comparison with many inorganic solutes in soil solution, such as NO_3^-, Cl^-, K^+, Na^+, Ca^{2+} and Mg^{2+}, these concentrations are an order of magnitude lower, however, they are similar to reports for other low-molecular-weight organic solutes (e.g., amino acids, sugars; Barber, 1995; Kielland, 1995). In the case of amino acids and sugars, significant reservoirs of these compounds exists in a polymeric form, with replenishment of the soil solution pool dependent upon the activity of extracellular enzymes (Stevenson, 1982). In contrast, replenishment of the soil organic acid pool depends almost entirely upon a continued release from either plant or animal cells, and can therefore be expected to be more temporally variable. Based upon this assumption, it might be expected that soil solution organic acid concentrations would vary considerably over both space and time; however, current evidence suggests that this is not the case (Jones, 1998; Strobel, 2001). The question arises as to why organic acid concentrations in soil solutions are always so low and temporally constant? A number of explanations may account for this phenomenon including methods of sampling soil solutions, sorption characteristics and microbial reactions.

Problems with sampling soil solutions for organic acids

The purported role of organic acids in mineral weathering has recently been questioned based upon the premise that the concentration of organic acids in soil solution is too low to have any significant effect on mineral dissolution rates (Drever and Stillings, 1997). Their central argument, however, relies on the accurate quantification of organic acids in soil. To date, there has been relatively little work investigating the impact of soil solution extraction technique on the concentration of organic acids detected, however, based upon other reports this is likely to highly influence the results obtained (Giesler et al., 1996; Grieve, 1996; Jones and Edwards, 1993; Smethurst, 2000; Tyler, 2000). Organic acid analysis, typically involves the bulk removal of water from large volumes of soil by centrifugal drainage, ceramic suction samplers or water extracts. Given the various hypothesized functions of organic acids in soil and the spatial heterogeneity of soil resources such as P, it is likely that the concentration of organic acids may also be highly spatially localized (Figure 1). Although soil solution microsampling techniques have been developed at the millimeter scale (Göttlein et al., 1996), due to the very low diffusion coefficients of most organic acids in soil (e.g., citrate, oxalate; Darrah, 1991), the size of these hot spots may be only a few μm in diameter if release occurs from the tip of a root hair, fungal hyphae or bacterial cell. This is supported by mathematical modeling and experimental simulations of organic acid diffusion in the rhizosphere (Darrah, 1991; Jones et al., 1996a). We hypothesize that current extraction techniques vastly underestimate soil solution organic acid concentrations by maybe up to several orders of magnitude within soil microsites. In addition, soil

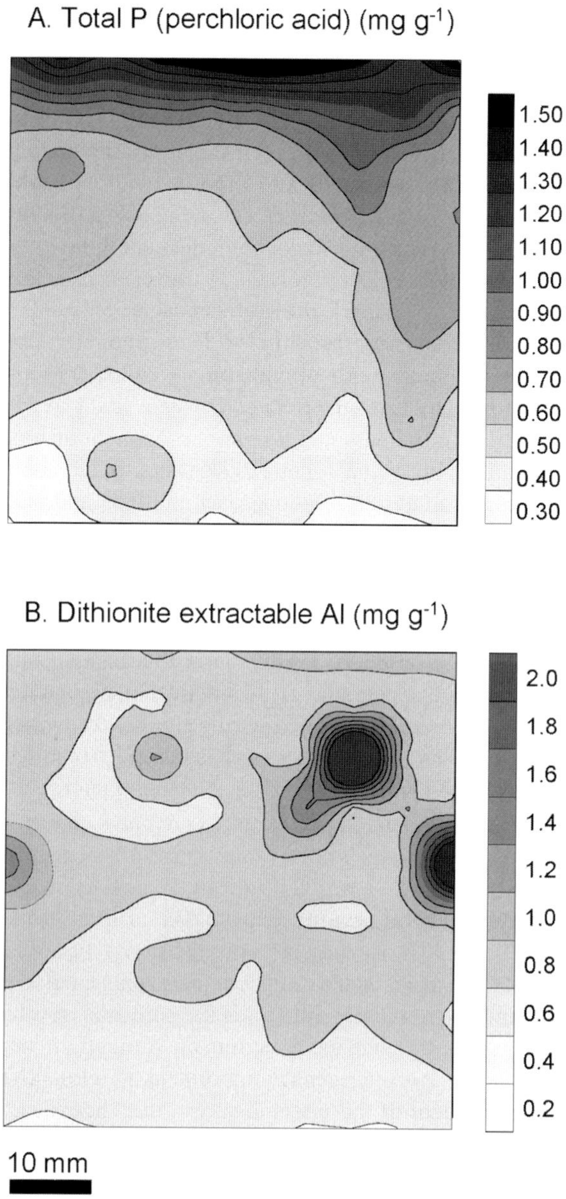

Figure 1. Spatial distribution of phosphorus and aluminium in a 50 mm wide, 50 mm deep block of podzolic soil collected from under an acid grassland in the UK. Soil P and Al analysis was performed at 5 mm intervals and the data interpolated using a computerized minimum curvature routine to obtain a concentration map. The data are representative of other soil blocks analyzed at the same location.

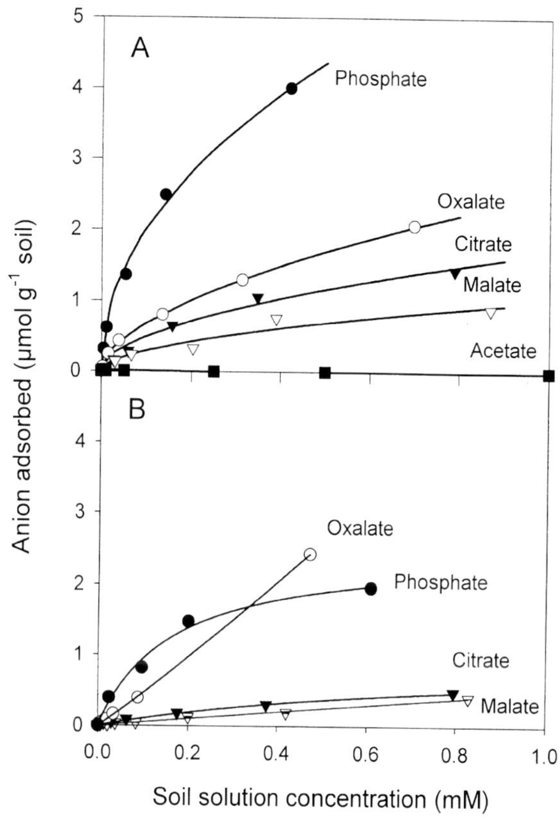

Figure 2. Organic acid and phosphorus sorption on (A) a gleyed subsoil and (B) a calcareous topsoil. Symbols represent experimental data points while lines represent optimized fits to the Langmuir sorption equation. Further details of the soils and experimental procedures can be found in Jones and Brassington (1998) and Ström et al. (2002a,b).

solution sample preparation (e.g., soil storage, etc) and analytical errors may also cause the underestimation of organic acid concentrations in soil (Chapman et al., 1997; Piombo et al., 1996). Unfortunately, due to the lack of specific dyes, electrodes and chemical probes for organic acids, their spatial heterogeneity in soil remains unknown. One solution to this problem could be the development of marker gene bioassays, which involves the tagging of soil bacteria and fungi with reporters based on green fluorescent protein (gfp) or luciferase (lux) that respond to exogenous organic acids (as has been achieved for C, P and amino acids in soil; Killham and Yeomans, 2001; Farrar et al., 2002).

In the case of organic acid-mediated metal detoxification by plant roots, the concentration in the root apoplasm may be more important than the external soil solution concentration (e.g., in malate-induced Al detoxification; Delhaize et al., 1993). To date, there has been no evaluation of apoplastic concentrations, however, this is certainly worthy of further consideration.

Solid phase buffering of soil solution concentrations

Another explanation for the observed near-constant

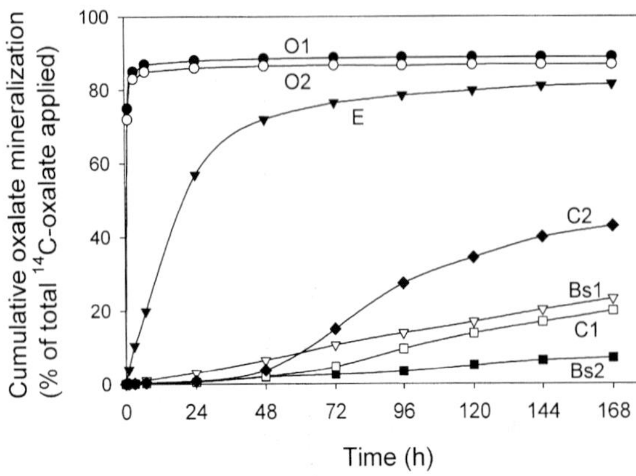

Figure 3. Oxalate mineralization in different horizons of a *Picea abies* forest soil from Sweden. The initial ^{14}C-oxalate concentration was 500 μM. Further details of the soils and experimental procedures can be found in Van Hees et al. (2002).

organic acid concentrations in soil solution may be the high degree of buffering by the soil's solid phase. In contrast to amino acids and sugars, di- and tri-carboxylic acids become strongly associated with the solid phase through the processes of adsorption and precipitation (Jones and Edwards, 1998). In the case of oxalate, if there are significant quantities of free Ca^{2+} in solution, oxalate concentrations will rarely exceed 50 μM, due to the formation of insoluble Ca^{2+}-oxalate (Ström et al., 2002a,b). Additionally, at low organic acid solution concentrations (<50 μM) the soil's solid phase possesses a high affinity and large capacity for organic acid adsorption (e.g., for citrate, malate and aconitate), and these processes will also help maintain low soil solution concentrations (Figure 2).

Microbial regulation of soil solution organic acid concentrations

A third possible explanation for the low organic acid concentrations detected in soil solution is the continual removal by the soil microbial community. In support of this, organic acid mineralization across a wide range of soil types has been shown to be rapid (Figure 3; Jones, 1998). Further, soil solution concentrations reported in the literature are often close to the minimum concentration at which transporters can generate net organic acid uptake (i.e. influx > efflux; 0.1 to 1 μM; Barber, 1995; Jones et al., 1996b). We hypothesize that the soil microbial community plays an active role in scavenging this readily utilizable source of C from the soil solution thereby preventing the accumulation of organic acids in soil.

Where do the organic acids in soil come from?

In many ecosystems root turnover provides the primary source of C entering the soil (Norby and Jackson, 2000). Based upon the typical composition of root sap (40 mM sugars, 10 mM organic acids, 10 mM amino acids; Marschner, 1995; Ryan et al., 2001), organic acids can form a major component of the solutes lost from plant cells upon lysis. However, in comparison with the polymeric reservoirs of sugars and amino acids in root tissue (ca. 20 to 40% of total root dry weight), organic acids provide only a minor fraction of the primary C entering the soil system (ca. 1 to 3% of total root biomass and a minor component of root exudates; Farrar and Jones, 2000). As organic acids possess similar soil solution concentrations and half lives to sugars and amino acids (see later), it implies that the rates of production of all three groups of compounds in soil must be similar. We conclude therefore that most of the organic acid entering the soil solution pool must be produced from a source other than plants (e.g., secondary organic acids produced and excreted by microorganisms formed during the breakdown of other C substrates). Further work is therefore required to elucidate the critical control points of organic acid production in soil.

Measuring total organic acids in soil

Although some methods have been presented for quantifying total organic acids in soil, the results obtained with such methods remain difficult, if not impossible, to interpret (Piombo et al., 1996). The main problem of these techniques is that they measure the sum of all soil pools, many of which may have no direct relationship with the question being asked in the study (e.g., the internal microbial pool in a soil P solubilization study). The main methodological problem that needs to be overcome is finding a way of separating organic acids bound to the solid phase from those held in the microbial biomass. In our studies, we have found that organic acids in organic forest soils can be completely recovered by extraction with KH_2PO_4 (50 mM); however, recovery of organic acids from mineral subsoils remained incomplete even at high KH_2PO_4 concentrations (500 mM; Table 1). Full removal of organic acids from mineral horizons therefore requires extraction with strong acids, which inevitably releases large amounts from soil microorganisms. We estimate that for a typical topsoil the

Table 1. Recovery of ^{14}C-citrate from two *Picea abies* forest soils by KH_2PO_4 (1, 50 and 100 mM) and HCl (10 mM). The soil was $CHCl_3$-fumigated for 2 days, 50 nmol g^{-1} ^{14}C-citrate added (pH 4.5), the soil incubated with a $CHCl_3$ atmosphere for 7 days and then the soil extracted with either KH_2PO_4 (1, 50 or 100 mM) or HCl (10 mM) for 15 min (5:1 v/w solution:soil). Values represent the means of both soils±SEM (*n*=2). Further details of the Swedish forest soils (Heden and Nyänget) can be found in Van Hees et al. (2002)

Horizon	Adsorbed on the solid phase (% of total)[a]	Extractant type for citrate recovery from soil			
		1 mM KH_2PO_4	50 mM KH_2PO_4	100 mM KH_2PO_4	10 mM HCl
		(% of total ^{14}C citrate added)			
O1	19 ± 12	100 ± 3	102 ± 3	102 ± 0	96 ± 2
O2	31 ± 23	95 ± 5	97 ± 3	96 ± 1	96 ± 6
E	80 ± 18	40 ± 20	46 ± 17	52 ± 13	44 ± 16
Bs1	97 ± 2	4 ± 3	16 ± 9	21 ± 7	5 ± 4
Bs2	97 ± 2	6 ± 5	19 ± 10	29 ± 13	5 ± 4
C1	98 ± 1	6 ± 3	23 ± 8	36 ± 11	5 ± 2
C2	98 ± 1	6 ± 1	26 ± 1	39 ± 4	7 ± 2

[a]Estimated from sorption isotherms (Van Hees et al., 2002).

Table 2. Amount of organic acids in the soil solution or associated with the solid phase in the surface horizon of a calcareous Typic rendoll (USDA Soil Taxonomy) soil and in the surface organic and lower mineral horizon of a *Picea abies* forest podzolic soil. Soil solution concentrations were measured directly and the amount associated with the solid phase at this soil solution concentration calculated from sorption isotherms. The values were derived from data presented in Ström et al. (2001, 2002a,b) and Van Hees et al. (2002). n.d. indicates not determined

	Typic rendoll Ah horizon		Podzolic O horizon		Podzolic Bs horizon	
	Soil solution	Adsorbed	Soil solution	Adsorbed	Soil solution	Adsorbed
	pmol mm^{-3}		(pmol mm^{-3})		(pmol mm^{-3})	
Citrate	0.11	0.35	19.5	3.8	0.3	225
Oxalate	0.12	2.76	1.2	30.4	0.9	967
Malate	0.10	0.34	n.d.	n.d.	n.d.	n.d.

amount of organic acids held in the soil microbial biomass is between 0.5 to 100 pmol mm^{-3} soil, given an internal cell concentration of 1 to 10 mM. These concentrations are of a similar magnitude to those held on the soil's solid phase, and therefore extractions should take this factor into account (Ström et al., 2001, 2002a,b; Table 2).

Organic acid-mediated nutrient solubilization and mobilization

One of the main hypothesized functions of organic acids in soil is the solubilization and mobilization of nutrients by plants and microorganisms. Recently, there have been suggestions that organic acid release from plants may provide a practical, green solution for combating declining fertility within resource-poor agroecosystems such as those in parts of South America, Sub-Saharan Africa and Australia (Lopez-Bucio et al., 2000). In the case of nutrients such as P, its solubilization by organic acids such as oxalate and citrate is highly soil dependent (Figure 4). In a screen of 20 contrasting soil types, we found that organic acids mobilized significant amounts of P in 65% of the soils tested ($P<0.05$, in comparison with water) with large amounts of P mobilized in 35% of the soils. The reason for the large differences in the response of the various soil types remains unclear; however, the amount of P mobilized did not correlate well with either total soil P, soil pH, P-sorption capacity or total Fe ($r^2 < 0.5$; data not presented). These differential soil type responses are also supported by the study of Ström et al. (2002a) in which oxalate was capable of increasing plant P acquisition while citrate could not. This result highlights the problems in recommending organic acids as a blanket strategy for enhancing P mo-

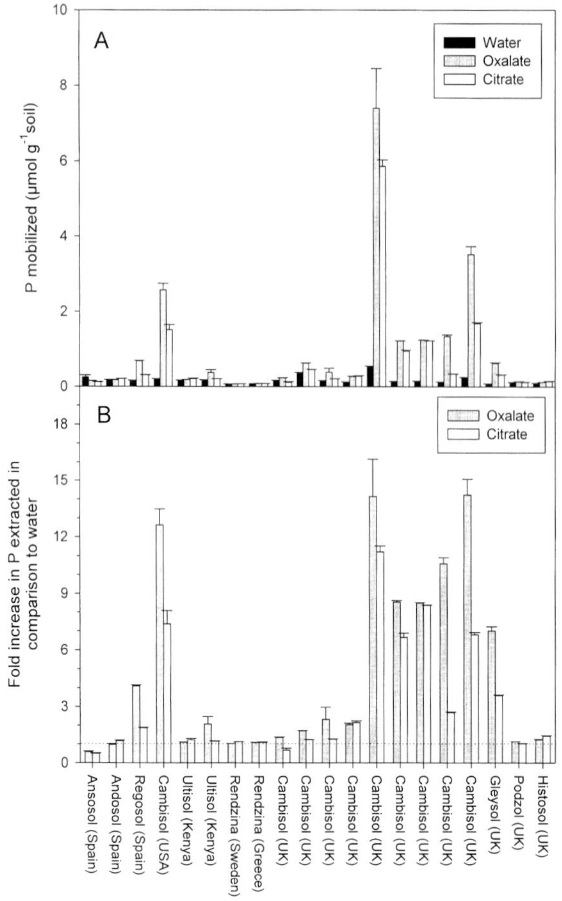

Figure 4. Organic acid-mediated phosphorus solubilization in 20 contrasting soils. Panel A (upper) shows the amount of P solubilized when soils were shaken with either water (control), oxalate (1 mM), or citrate (1 mM) for 30 min at a soil:solution ration of 1:10 (w/v). Panel B (lower) shows the magnitude of the change in P solubilization by oxalate and citrate in comparison to the water extractions. A value of 1 (no enhanced P solubilization) is represented by the dotted line. Further details of the experimental procedures can be found in Jones and Darrah (1994). Values represent means±SEM (*n=3*).

bilization in soil. Similar arguments also exist for other proposed organic acid-mediated nutrient mobilization strategies (e.g., Fe, Zn, Cu) and metal detoxification mechanisms (e.g., Al, Cd).

In the case of Zn and P, a drop in soil pH often enhances the plant availability of these nutrients. The importance of an associated soil pH change on the nutrient mobilization/solubilization efficiency of organic acids, either as a result of the direct release of H^+ with the organic acid (e.g., H^+-citrate) or indirectly from associated H^+-ATPases, also requires further investigation. Similarly, additional work is required to determine the critical concentration of organic acids or anions in the soil that are sufficient to cause a significant increase in plant nutrient uptake. In summary, more research is required to understand the factors that control organic acid-mediated nutrient release in different soil types.

Organic acid biodegradation by soil microorganisms

One of the primary mechanisms by which organic acids are removed from the soil solution is through uptake by soil microorganisms (Jones et al., 1996b; Lundström et al., 1995). Along with other low-molecular-weight compounds such as sugars and amino acids, organic acids represent one of the most labile sources of C in soil and, due to their ubiquity in soil, it can be expected that most soil organisms will be capable of their utilization. A number of studies have shown that the rate of organic acid turnover in soil is extremely rapid with half-lives ranging from 1 to 5 h in organic topsoils, to 5 to 12 h in subsoils (Jones, 1998). This is comparable to the rate of utilization of many other low-molecular-weight solutes in soil (e.g., amino acids, glucose; Coody et al., 1986; Jones, 1999). The finding that organic acid turnover is very fast and that soil solution concentrations are maintained at low concentrations (1 to 50 μM) suggests that the flux through the soil solution is highly dynamic with the free organic acid pool turning over between 500 and 3000 times annually. If the total organic acid pool is considered (i.e. sorbed plus free in solution) the rate of turnover can be estimated to be of the same order of magnitude in organic topsoils where the degree of sorption is low (Table 2). In mineral subsoils, however, where the degree of sorption is much higher and the rate of degradation much lower, the total organic acid pool may turn over at a much slower rate (ca. 0.5 to 10 times annually).

One question that has yet to be adequately resolved is the degree to which soil microorganisms can act as both sources and sinks of organic acids in the soil solution. The excretion of organic acids by soil microorganisms in laboratory liquid cultures is well documented; however, there have been few studies conducted in soil (Dutton and Evans, 1996; Gallmetzer and Burgstaller, 2001; Heyer and Krumbein, 1991; Takao, 1965). It is clear that ectomycorrhizal fungi can release large quantities of organic acids (particularly oxalate and citrate); however, whether they simultaneously act as sources and sinks remains unknown. It could be hypothesized that in these circumstances, distinct spatial regulation of organic acid exudation and uptake occurs in the mycelium in response to localized soil environmental conditions. In the case of silicate weathering by fungal hyphae (Van Breemen et al., 2000), we hypothesize that organic acid exudation occurs primarily at the hyphal tip region with organic acid consumption in basal areas away from the weathering front. This is supported to some extent by laboratory studies with bacteria in which the level of influx is directly regulated by external organic acid supply (Jones et al., 1996b). Hopefully the genome sequencing of various soil bacteria (e.g., *Rhizobium*), combined with the development of molecular tools such as green fluorescent protein marking (Errampalli et al., 1999), will help address issues such as the temporal and spatial dynamics of organic acid transport gene expression in soil microorganisms.

Another consideration in the biodegradation of organic acids is the extent to which organic acids can be accessed when they are either sorbed or fixed to the soil's solid phase or present in a precipitate. Although there have been insufficient studies on this to draw definitive conclusions, the limited information available suggests that organic acids held strongly to the soil's exchange phase (i.e. fixed) are not biodegraded, at least in the short term (Boudot, 1992; Jones and Edwards, 1998). In contrast, organic acids held loosely on soil exchange sites can be rapidly mineralized (Jones and Edwards, 1998). The latter scenario relies on the continual removal of organic acids from solution, which leads to a continual desorption of organic acids off the solid phase. In the case of precipitated organic acids (e.g., Ca-oxalate), indirect studies in calcareous soils have indicated that their rate of biodegradation is extremely slow, although again definitive studies are lacking (Ström et al., 2001).

The extent to which organic acids complexed with metals (e.g., Al-citrate) can be degraded by the soil microbial community remains controversial (Jones, 1998). Studies in liquid culture media have demonstrated that both bacteria and fungi have transporters capable of taking up specific organic acid metal complexes (e.g., Fe-citrate, Ca-citrate) while other reports suggest that these metals are capable of being taken up but not assimilated (Brynhildsen and Allard, 1994; Brynhildsen and Rosswall, 1989; Francis et al., 1992; Madsen and Alexander, 1985). Studies in soil, however, suggest no distinction between the utilization of metal complexed or non-metal complexed organic acids (Brynhildsen and Rosswall, 1997; Jones et al., 2001). This contradiction may be simply due to the degree of adaption in each experimental system. In culture media, it is likely that many 'environment' genes are down-regulated, and therefore when challenged with organic acid-metal complexes not previously exposed to, the cellular machinery is not present. Alternatively, C is provided by another component of the media, making utilization of the organic acids unnecessary. If these complexes are common in their natural environment (e.g., Al-citrate in a podzolic soil) it can be expected that the microbes will possess the transporters capable of using this labile substrate (given that organic acids do not accumulate in soil). Again definitive studies are required to clarify this.

Organic acid interactions with the solid phase

Individual organic acids interact differently with the soil's solid phase. Generally, the degree of association is directly related to the amount of charge the organic acid carries and the solution pH (Figure 2; Jones and Brassington, 1998; Karltun, 1998; Ström et al., 2001; Van Hees et al., 2002). Monovalent organic acids such as acetate, lactate and propionate are weakly adsorbed to the soil's solid phase, whereas divalent organic acids such as malate, fumarate and succinate are adsorbed to a greater degree (Figure 2). The greatest adsorption occurs for trivalent organic acids such as citrate, isocitrate and aconitate, with the degree of sorption often being similar to that for phosphate (Figure 2). A major problem that remains to be resolved, however, exists in the interpretation of conventional sorption isotherm experiments. This mainly centers around the issue of whether these experiments actually measure an adsorption process in its strictest sense. Typically, sorption isotherms for organic acids have been constructed using the standard methodology applied to phosphate which involves shaking soil with known concentrations of organic acids at a high soil:solution ratio (e.g., 1:10 w/v; Kuo, 1996). This methodology may not be appropriate for organic acids or P for a number of reasons (Li and Stanforth, 2000). Firstly, unless steps are taken to sterilize the soil, significant microbial degradation of the organic acids can occur during the shaking process, thereby overestimating the degree of sorption. Degradation of organic acids by the soil microbial community can be suppressed through the addition of high concentrations of chemical sterilants (e.g., Na-azide or $HgCl_2$; de Maagd et al., 1998), but many of these also interact with the soil's solid phase, possibly reducing organic acid sorption. Moreover, they can cause the lysis and release of competing anionic substances from microbial cells. Alternatives, including chloroform fumigation, may also release significant quantities of potentially competing substances into the soil solution from ruptured microbial cells, while autoclaving significantly affects the structure and chemistry of Fe- and Al-hydroxide sorption surfaces (Otabbong and Barbolina, 1998; Xie and Mackenzie, 1991). To reduce organic acid microbial degradation, short shaking times have been advocated (5 to 10 min; Jones and Brassington, 1998). Whilst this will reduce microbial degradation, it probably also greatly underestimates the amount of sorption. Evidence for this comes from studies on phosphate where sorption is a multiphasic process involving a fast sorption phase which typically reaches equilibrium within 1 h, and a second slower sorption phase, which may not reach equilibrium for many days (Hope and Syers, 1976). Given the short residence time of organic acids in the rhizosphere, it could be argued that for plant nutrient uptake only the first sorption phase is important, however, for pedogenic processes the second phase becomes more important. Other experimental factors such as the soil:solution ratio, the background ionic strength, the control of solution pH during the shaking period, the organic acid balancing cation (Na^+, H^+, Ca^{2+}, etc.) are also major considerations which may affect the final result (Courchesne, 1991; Rupa et al., 2001). As yet there have been no studies to address the importance of these issues.

Of greater concern is the extent to which these experiments are actually measuring adsorption *per se* or other chemical events such as precipitation. Here we define sorption as the reversible binding of organic acids to the soil's solid phase. In contrast, we define fixation as the process by which organic acids become bound to soil particles in an almost irreversible manner unless extreme chemical conditions, which rarely occur in the soil, are employed to remove them (e.g., 1 *M* HCl). While sorption and fixation are reactions occurring at the surface of minerals, precipitation can occur purely in the solution phase. Using the current experimental procedures for quantifying sorption, precipitates formed in the soil solution will be accounted for as being adsorbed (e.g., Ca-oxalate). The differences in sorption between free organic acids (e.g., H_3-citrate^0, citrate^{3-}) and organometallic complexes (e.g., Fe-citrate^0, Al-citrate^0) also remains unknown but deserves further study.

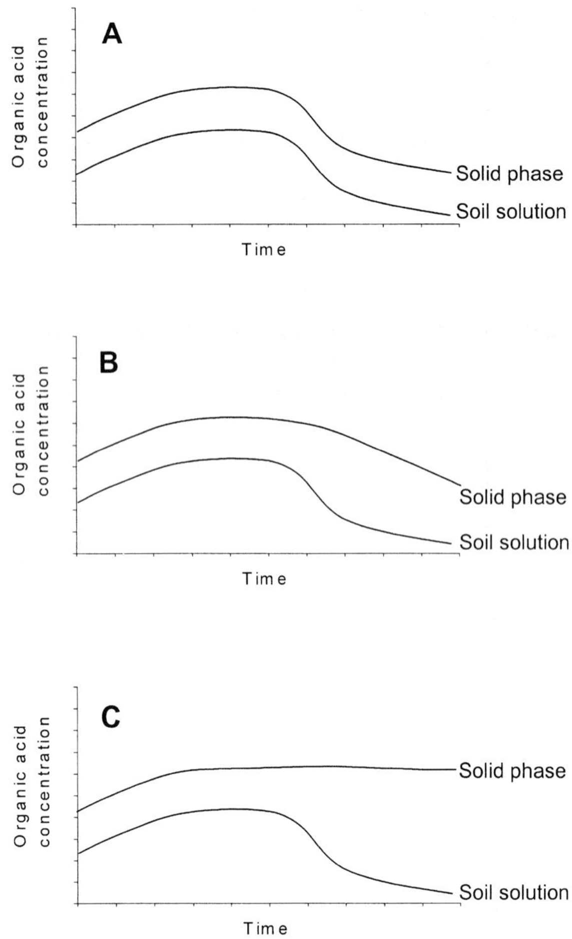

Figure 5. Schematic representations of the sorption and desorption reactions of organic acids in soil. Panel A (upper) illustrates the fast equilibrium reaction between organic acids held in solution and on the solid phase. Panel B (middle) shows a slow rate of desorption compared to the rate of sorption. Panel C (lower) shows almost no desorption when soil solution concentrations fall. See text for further details.

Another major issue that is rarely considered, and which would help address the issues raised above, is the rate of desorption of organic acids from the solid phase. Studies on the fate of organic acids in soil have mainly focused on the removal of organic acids from the soil solution as this directly affects the efficiency of rhizosphere-based nutrient acquisition or metal detoxification reactions (Ryan et al., 2001). However, desorption is also important, particularly in the computer modelling of organic acid dynamics where it is often assumed that the rate of sorption and desorption from the solid phase are identical (Barber, 1995). Typically this reaction is described by a constant, the buffer power (b), which is defined as

$$b = (OA_{tot}/OA_{sol}), \quad (1)$$

where OA_{tot} is the total amount of organic acid in the soil (sorbed plus solution) and OA_{sol} is the amount of organic acid in solution. The extent to which this relationship holds, however, remains unknown. Three different scenarios are shown in Figure 5 to illustrate the point. At low soil solution concentrations the strength of organic acid binding to the solid phase is typically high but at higher soil solution concentrations organic acid binding to the solid phase becomes weak (Figure 2). As most reported soil solution concentrations are <100 μM, we will only consider the lower part of the isotherm where Equation 1 is satisfied (i.e. a linear relationship between OA_{sol} and OA_{tot}). Figure 5A shows the assumed situation in Equation 1 where the concentration of organic acids in solution is always in equilibrium with the soil solid phase. Figure 5B shows the situation where the rate of desorption is not as fast as the rate of sorption, while Figure 5C illustrates very slow desorption or fixation/precipitation. Which of these scenarios occurs in a particular soil remains unknown. From the available evidence, we hypothesize that monovalent and some divalent organic acids (e.g., malate, oxaloacetate etc.) will conform to the pattern observed in Figure 5A at all solution concentrations in both mineral and organic horizons. Divalent organic acids like oxalate, which undergo strong sorption and precipitation at mineral surfaces will probably conform to pattern Figure 5C in mineral soils. In the case of trivalent TCA cycle organic acids such as citrate and aconitate, our results indicate that in highly organic soil horizons sorption follows the pattern shown in Figure 5A (Table 2). In contrast, in mineral soil horizons (e.g., Bs, Ah horizons), at high solution concentrations where sorption is weak, they will conform to the pattern of Figure 5A but at low solution concentrations where sorption is strong they will follow Figure 5B.

The preceding discussion on organic acid solid phase reactions highlights our poor understanding of how organic acids behave in soil. Without this information, mathematical modeling will only remain purely speculative, and our understanding of rhizosphere and pedogenic processes involving organic acids rudimentary.

Concluding remarks

Organic acids have been hypothesized to be important in many soil processes; however, consideration of the importance of these processes at the ecosystem level has rarely been explored. This is partly due to a lack of fundamental knowledge about their behavior in soil and the short-term nature of many experiments. More effort is therefore required to unravel the complexities of organic acids in soil with specific emphasis on gaining greater spatial and temporal resolution in measurements and also in validating the procedures used in these studies. Further work is also required to clarify the sequence of events and importance of the various fates of organic acids in the rhizosphere.

Acknowledgement

Thanks are due to Hans Lambers at the University of Western Australia for providing the incentive to write this review and to Phillipe Hinsinger for critical comments. This project was funded jointly by the Biotechnology and Biological Sciences Research Council and the Natural Environment Research Council, UK.

References

Barber S A 1995 Soil Nutrient Bioavailability: A Mechanistic Approach. John Wiley & Sons, New York. 426 pp.

Baziramakenga R, Simard R R and Leroux G D 1995 Determination of organic acids in soil extracts by ion chromatography. Soil Biol. Biochem. 27, 349–356.

Barbour W M, Hattermann D R and Stacey G 1991 Chemotaxis of *Bradyrhizobium japonicum* to soybean exudates. Appl. Environ. Microbiol. 57, 2635–2639.

Beyer L and Bolter M 2000 Chemical and biological properties, formation, occurrence and classification of Spodic Cryosols in a terrestrial ecosystem of East Antarctica (Wilkes Land). Catena 39, 95–119.

Boudot J P 1992 Relative efficiency of complexed aluminum, noncrystalline Al hydroxide, allophane and imogolite in retarding the biodegradation of citric acid. Geoderma 52, 29–39.

Brynhildsen L and Allard B 1994 Influence of metal complexation on the metabolism of citrate by *Klebsiella-oxytoca*. Biometals 7, 163–169.

Brynhildsen L and Rosswall T 1989 Effects of cadmium, copper, magnesium, and zinc on the decomposition of citrate by a *Klebsiella* sp. Appl. Environ. Microbiol. 55, 1375–1379.

Brynhildsen L and Rosswall T 1997 Effects of metals on the microbial mineralization of organic acids. Water Air Soil Poll. 94, 45–57.

Chapman P J, Shand C A, Edwards A C and Smith S 1997 Effect of storage and sieving on the phosphorus composition of soil solution. Soil Sci. Soc. Am. J. 61, 315–321.

Coody P N, Sommers L E and Nelson D W 1986 Kinetics of glucose uptake by soil microorganisms. Soil Biol. Biochem. 18, 283–289.

Courchesne F 1991 Electrolyte concentration and composition effects on sulfate sorption by 2 spodosols. Soil Sci. Soc. Am. J. 55, 1576–1581.

Darrah P R 1991 Measuring the diffusion-coefficient of rhizosphere exudates in 2. The diffusion of sorbing compounds. J. Soil Sci. 42, 421–434.

Delhaize E, Ryan P R and Randall P J 1993 Aluminum tolerance in wheat (*Triticum aestivum* L.). 2. Aluminum stimulated excretion of malic acid from root apices. Plant Physiol. 103, 695–702.

Delhaize E and Ryan P R 2003 Effects of altered citrate synthase and isocitrate dehydrogenase expression on internal citrate concentrations and citrate efflux from tobacco (*Nicotiana tabacum* L.) roots. Plant Soil 248, 137–144.

de Maagd P G J, Sinnige T L, Schrap S M, Opperhuizen A and Sijm D T H M 1998 Sorption coefficients of polycyclic aromatic hydrocarbons for two lake sediments: Influence of the bactericide sodium azide. Environ. Toxicol. Chem. 17, 1899–1907.

Drever J I and Stillings L L 1997 The role of organic acids in mineral weathering. Colloids Surfaces A: Physicochem. Eng. Aspects 120, 167–181.

Dutton M V and Evans C S 1996 Oxalate production by fungi – its role in pathogenicity and ecology in the soil environment. Can. J. Microbiol. 42, 881–895.

Erampalli D, Leung K, Cassidy M B, Kostrzynska M, Blears M, Lee H and Trevors J T 1999 Applications of the green fluorescent protein as a molecular marker in environmental microorganisms. J. Microbiol. Meth. 35, 187–199.

Farrar J F and Jones D L 2000 The control of carbon acquisition by roots. New Phytol. 147, 43–53.

Farrar J F, Hawes M, Jones D L and Lindow S 2002 How roots control the flux of carbon to the rhizosphere. Ecology (In press).

Francis A J, Dodge C J and Gillow J B 1992 Biodegradation of metal citrate complexes and implications for toxic-metal mobility. Nature 356, 140–142.

Gallmetzer M and Burgstaller W Citrate efflux in glucose-limited and glucose-sufficient chemostat culture of *Penicillium simplicissium*. Anton. Leeuw. Int. J. Gen. Mol. Microbiol. 79, 81–87.

Giesler R, Lundström US and Grip H 1996 Comparison of soil solution chemistry assessment using zero-tension lysimeters or centrifugation. Eur. J. Soil Sci. 47, 395–405.

Göttlein A, Hell U and Blasek R A 1996 System for microscale tensiometry and lysimetry. Geoderma 69, 147–156.

Grierson P F 1992 Organic-acids in the rhizosphere of *Banksia integrifolia* L.. Plant Soil 144, 259–265.

Grieve I C 1996 Effects of the centrifuge drainage method on total organic carbon concentrations in soil solutions from peaty soils. Goederma 74, 115–123.

Heyer H and Krumbein W E 1991 Excretion of fermentation products in dark and anaerobically incubated cyanobacteria. Arch. Microbiol. 155, 284–287.

Hoffland E 1992 Quantitative evaluation of the role of organic-acid exudation in the mobilization of rock phosphate by rape. Plant Soil 140, 279–289.

Hope G D and Syers J K 1976 Effects of solution:soil ratio on phosphate sorption by soils. J. Soil. Sci. 27, 301–306.

Jones D L 1998 Organic acids in the rhizosphere – a critical review. Plant Soil 205, 25–44.

Jones D L 1999 Amino acid biodegradation and its potential effects on organic nitrogen capture by plants. Soil Biol. Biochem. 31, 613–622.

Jones D L and Brassington D S 1998 Sorption of organic acids in acid soils and its implications in the rhizosphere. Eur. J. Soil Sci. 49, 447–455.

Jones D L and Edwards A C 1998 Influence of sorption on the biological utilization of two simple carbon substrates. Soil Biol. Biochem. 30, 1895–1902.

Jones D L, Darrah P R and Kochian L V 1996a Critical-evaluation of organic-acid mediated iron dissolution in the rhizosphere and its potential role in root iron uptake. Plant Soil 180, 57–66.

Jones D L, Prabowo A M and Kochian L V 1996b Kinetics of malate transport and decomposition in acid soils and isolated bacterial-populations – the effect of microorganisms on root exudation of malate under Al stress. Plant Soil 182, 239–247.

Jones D L, Eldhuset T, de Wit H A and Swensen B 2001 Aluminium effects on organic acid mineralization in a Norway spruce forest soil. Soil Biol. Biochem. 33, 1259–1267.

Karltun E 1998 Modelling $SO_4{}^{2-}$ surface complexation on variable charge minerals. II. Competition between $SO_4{}^{2-}$, oxalate and fulvate. Eur. J. Soil Sci. 49, 113–120.

Kelly E F, Chadwick O A and Hilinski T E 1998 The effect of plants on mineral weathering. Biogeochemistry 42, 21–53.

Kielland K 1995 Landscape pattern of free amino acids in arctic tundra soils. Biogeochemistry 31, 85–98.

Killham K and Yeomans C 2001 Rhizosphere carbon flow measurement and implications: from isotopes to reporter genes. Plant Soil 232, 91–96.

Kuo S 1996 Phosphorus. *In* Methods of Soil Analysis. Part 3. Chemical Methods. Ed. D L Sparks. pp. 869–919. American Society of Agronomy and Soil Science Society of America, Madison, WI.

Krzyszowska A J, Blaylock M J, Vance G F and David M B 1996 Ion-chromatographic analysis of low molecular weight organic acids in spodosol forest floor solutions. Soil Sci. Soc. Am. J. 60, 1565–1571.

Li L and Stanforth R 2000 Distinguishing adsorption and surface precipitation of phosphate on goethite (α-FeOOH). J. Colloid Interf. Sci. 230, 12–21.

Lopez-Bucio J, de la Vega O M, Guevara-Garcia A and Herrera-Estrella L 2000 Enhanced phosphorus uptake in transgenic tobacco plants that overproduce citrate. Nature Biotechnol. 18, 450–453.

Lundström U S, Van Breemen N and Jongmans A G 1995 Evidence for microbial decomposition of organic acids during podzolization. Eur. J. Soil Sci. 46, 489–496.

Madsen E L and Alexander M 1985 Effects of chemical speciation on the mineralization of organic compounds by microorganisms. Appl. Environ. Microbiol. 50, 342–349.

Marschner H 1995 Mineral Nutrition of Higher Plants. Academic Press, London.

Micales J A 1997 Localization and induction of oxalate decarboxylase in the brown-rot wood decay fungus *Postia placenta*. Int. Biodeterioration Biodegrad. 39, 125–132.

Millet M, Wortham H, Sanusi A and Mirabel P 1997 Low molecular weight organic acids in fogwater in an urban area: Strasbourg (France). Sci. Tot. Env. 206, 57–65.

Norby R J and Jackson R B 2000 Root dynamics and global change: seeking an ecosystem perspective. New Phytol. 147, 3–12.

Otabbong E and Barbolina I 1998 Changes in solubility of various phosphorus fractions promoted by autoclaving soil and sewage sludge. Swed. J. Agr. Res. 28, 129–135.

Parker D R and Pedler J F 1998 Probing the 'malate hypothesis' of differential aluminium tolerance in wheat by using other rhizotoxic ions as proxies for Al. Planta 205, 389–396.

Piombo G, Babre D, Fallavier P, Cazevieille P, Arvieu J C and Callot G 1996 Oxalate extraction and determination by ionic chromato-

graphy in calcareous soils and in mycorrhized roots environment. Comm. Soil Sci. Plant Anal. 27, 1663–1677.
Pohlman A A and McColl J G 1986 Kinetics of metal dissolution from forest soils by soluble organic acids. J. Environ. Qual. 15, 86–92.
Rupa T R, Tomar K P, Rao C S and Rao A S 2001 Kinetics of phosphate sorption-desorption as influenced by soil pH and electrolyte. Agrochimica 45, 124–133.
Ryan P R, Delhaize E and Randall P J 1995 Characterization of Al-stimulated efflux of malate from the apices of Al-tolerant wheat roots. Planta 196, 103–110.
Ryan P R, Delhaize E and Jones D L 2001 Function and mechanism of organic anion exudation from plant roots. Annu. Rev. Plant Physiol. Plant Mol. Biol. 52, 527–560.
Smethurst P J 2000 Soil solution and other soil analyses as indicators of nutrient supply: a review. For. Ecol. Manag. 138, 397–411.
Stevenson F J 1982 Nitrogen in Agricultural Soils. ASA-CSSA-SSSA, Madison WI. 940 pp.
Strobel B W 2001 Influence of vegetation on low-molecular-weight carboxylic acids in soil solution – a review. Geoderma 99, 169–198.
Ström L, Owen A G, Godbold D L and Jones D L 2001 Organic acid behaviour in a calcareous soil: sorption reactions and biodegradation rates. Soil. Biol. Biochem. 2125–2133.
Ström L, Owen A G, Godbold D L and Jones D L 2002a Organic acid mediated P mobilization in the rhizosphere and uptake by maize roots. Soil Biol. Biochem. (In press).
Ström L, Owen A G, Godbold D L and Jones D L 2002b Organic acid behaviour in a calcareous soil: Implications for rhizosphere nutrient cycling. Soil Biol. Biochem. (In press).
Takao S 1965 Organic acid production by basidiomycetes I. Screening of organic acid producing strains. Appl. Microbiol. 13, 732–737.
Tyler G 2000 Effects of sample pretreatment and sequential fractionation by centrifuge drainage on concentrations of minerals in a calcareous soil solution. Geoderma 94, 59–70.
Van Breemen N, Finlay R, Lundström U S, Jongmans A G, Giesler R and Melkerud P A 2000 Mycorrhizal weathering: A true case of mineral plant nutrition? Biogeochemistry 49, 53–67.
Van Hees P A W, Jones D L and Godbold D L 2002 Biodegradation of low molecular weight organic acids in coniferous forest podzolic soils. Soil Biol. Biochem. (In press).
Xia J H and Roberts J K M 1994 Improved cytoplasmic pH regulation, increased lactate efflux, and reduced cytoplasmic lactate levels are biochemical traits expressed in root-tips of whole maize seedlings acclimated to a low-oxygen environment. Plant Physiol. 105, 651–657.
Xie R J and Mackenzie A F 1991 Effects of autoclaving on surface-properties and sorption of phosphate and zinc in phosphate-treated soils. Soil Sci. 152, 250–258.

Plant and Soil **248:** 43–59, 2003.
© 2003 *Kluwer Academic Publishers. Printed in the Netherlands.*

Origins of root-mediated pH changes in the rhizosphere and their responses to environmental constraints: A review

Philippe Hinsinger[1], Claude Plassard[1], Caixian Tang[2] & Benoît Jaillard[1]
[1]*INRA UMR Sol & Environnement, Place Pierre Viala, 34060 Montpellier Cedex 1, France.* [2]*Soil Science and Plant Nutrition, University of Western Australia, 35 Stirling Hwy, Crawley WA 6009, Australia*

Received 4 February 2002. Accepted in revised form 22 April 2002.

Key words: acidification, alkalinisation, proton, rhizosphere, root, soil pH

Abstract

The aim of the present review is to define the various origins of root-mediated changes of pH in the rhizosphere, i.e., the volume of soil around roots that is influenced by root activities. Root-mediated pH changes are of major relevance in an ecological perspective as soil pH is a critical parameter that influences the bioavailability of many nutrients and toxic elements and the physiology of the roots and rhizosphere microorganisms. A major process that contributes root-induced pH changes in the rhizosphere is the release of charges carried by H^+ or OH^- to compensate for an unbalanced cation–anion uptake at the soil–root interface. In addition to the ions taken up by the plant, all the ions crossing the plasma membrane of root cells (e.g., organic anions exuded by plant roots) should be taken into account, since they all need to be balanced by an exchange of charges, i.e., by a release of either H^+ or OH^-. Although poorly documented, root exudation and respiration can contribute some proportion of rhizosphere pH decrease as a result of a build-up of the CO_2 concentration. This will form carbonic acid in the rhizosphere that may dissociate in neutral to alkaline soils, and result in some pH decrease. Ultimately, plant roots and associated microorganisms can also alter rhizosphere pH via redox-coupled reactions. These various processes involved in root-mediated pH changes in the rhizosphere also depend on environmental constraints, especially nutritional constraints to which plants can respond. This is briefly addressed, with a special emphasis on the response of plant roots to deficiencies of P and Fe and to Al toxicity. Finally, soil pH itself and pH buffering capacity also have a dramatic influence on root-mediated pH changes.

Introduction

Changes of pH in the rhizosphere, i.e., the volume of soil around roots that is influenced by the activity of living plant roots (definition derived from Hiltner, 1904, by Darrah, 1993; Hinsinger, 1998), are by far the best documented chemical interaction occurring at the soil–root interface. As early as a century ago, Dehérain (1902) reported in his Treatise of Agricultural Chemistry on an experiment by Sachs, which demonstrated, by growing roots of beans over the surface of a polished marble plate, that roots secreted an acid strong enough to dissolve calcium carbonate, thereby leaving clearly visible imprints in the rock (see also Trolldenier, 1987). Such imprints of root systems can also be observed *in situ* for plant roots growing in cracks between dense limestone plates (Jaillard and Hinsinger, 1993) (Figure 1). In the first half of the 20th Century, such acidic root secretions were attributed to carbonic and organic acids produced by rhizosphere microflora and roots through respiration and exudation. Since the late 1960s, evidence has accumulated that roots can substantially change their rhizosphere pH by releasing H^+ or OH^- to compensate for an unbalanced cation–anion uptake at the soil–root interface (Riley and Barber, 1969, 1971, and many other authors since then, see below). The relationship of such H^+ release with root cell growth (acid growth theory) will not be discussed here as it has been discussed elsewhere (Jaillard et al., 2002). The aim of the present review is to characterise the various origins of such pH changes, namely: (i) cation–anion exchange

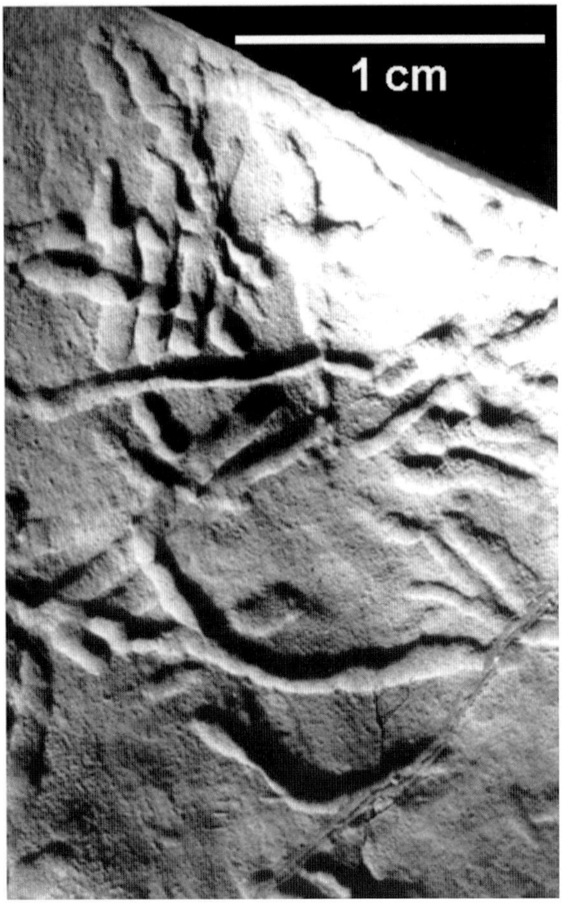

Figure 1. Imprints of roots growing in a crack between limestone plates, sampled *in situ* in the 'garrigues' bushland of St Gély du Fesc, Southern France (taken from Jaillard and Hinsinger, 1993, with kind permission from Ibis Press).

balance, (ii) organic anion release, (iii) root exudation and respiration and (iv) redox-coupled processes. In the recent years, numerous authors have shown that these various processes that contribute some proportion of root-mediated pH changes in the rhizosphere also depend on environmental constraints, especially nutritional constraints to which plant can respond. These will be addressed briefly in a fifth section of this review, with special emphasis on Al toxicity and Fe and P deficiencies. Furthermore, effects of soil pH *per se* and pH buffering capacity on and the ecological significance of root-mediated pH changes will be discussed in the final section.

Cation–anion exchange balance

A major source of H^+ fluxes in the rhizosphere is related to the differential uptake of cations and anions by plant roots (Haynes, 1990; Hinsinger, 1998; Jaillard et al., 2002; Marschner, 1995; Nye, 1981, 1986; Tang and Rengel, 2002). This arises from the necessary compensation of electrical charges and regulation of cellular pH that are required in the root cells: the cytosolic pH is maintained within a narrow range of values around 7.3 via an efficient pH-stat system. For more details about the physiological mechanisms of the cation–anion balance, pH-stat in the root cells and the involvement of the ATP-driven H^+-pump, see Haynes (1990) or Marschner (1995). The pH-stat system comprises both biophysical (H^+ exchange) and biochemical components. The latter involves the production and consumption of H^+ as a result of carboxylation and decarboxylation of organic acids within the root cell (Haynes, 1990; Marschner, 1995). These internal buffers play a key role in the efficiency of the pH-stat system as recently emphasised by Gerendás and Schurr (1999): ATPases cannot control the pH of either the apoplasm or the cytosol on their own, and should therefore be considered to mainly act through energising ion transport across the membrane that itself result in significant pH changes. ATPases provide an adequate model for understanding cation uptake; however, there is still a need for a better understanding of the mechanisms of charge regulation involved in the uptake of anions. In that respect, the possible involvement of plasma membrane-bound reductases that are involved in the transfer of negative charges (as electrons) from the cytosol into the apoplasm has been questioned (see Haynes, 1990).

Whatever the exact mechanisms involved in the cation–anion balance and charge-pH regulation in root cells, when more cations than anions are taken up, H^+ will be released into the apoplasm to compensate for the excess of positive charges entering the cell, resulting in increased pH in the cytosol. This is typically what would occur when supplying a plant with a K_2SO_4 solution, which leads to a larger uptake of K^+ than the poorly permeating SO_4^{2-} (Haynes, 1990; Hiatt, 1967; Marschner, 1995). Conversely, when more anions are taken up than cations, OH^- (or HCO_3^- resulting from OH^- carbonatation) will be released into the apoplasm (or H^+ will be taken up from the apoplasm) to compensate for the excess of negative charges entering the cell, resulting in decreased pH in the cytosol. This is typically what would occur when supplying a plant with a $CaCl_2$ solution, which leads to less uptake of Ca^{2+} than the highly permeating Cl^- (Haynes, 1990; Hiatt, 1967; Marschner, 1995). This

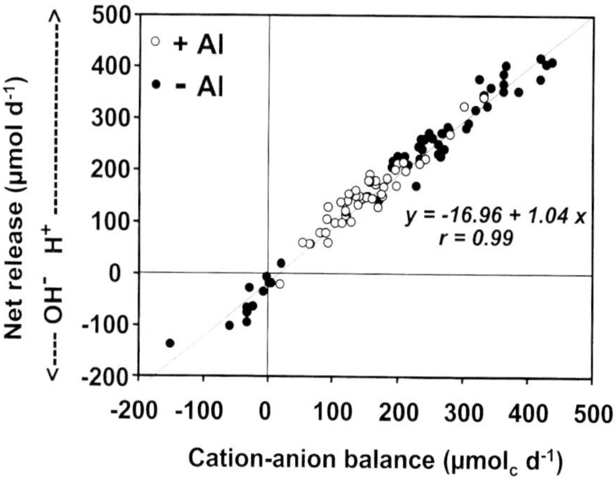

Figure 2. Relationship between the amount of H^+ released and the balance of cations over anions taken up by seedlings of various genotypes of banana (*Musa* spp.) grown in solution culture in a complete nutrient solution including both ammonium and nitrate and with or without 78.5 μM Al (adapted from Rufyikiri et al. (2001), with kind permission from EDP Sciences S.A.).

Table 1. Uptake of cations and anions by chickpea (*Cicer arietinum* L.) plants grown in culture solution for 24 days and supplied with either NO_3^- or NH_4^+ (initial pH 6.0) All values are expressed as mmol of negative/positive charge per plant (adapted from Le Bot et al., 1990, with kind permission from Marcel Dekker Inc.)

	NO_3^-	NH_4^+
NH_4^+	–	7.82
K^+	3.10	1.38
Na^+	0.02	0.02
Ca^{2+}	2.00	0.63
Mg^{2+}	0.92	0.66
ΣC (Sum of cations)	6.04	10.51
NO_3^-	7.39	–
$H_2PO_4^{2-}$	0.71	0.63
SO_4^{2-}	0.79	0.82
Cl^-	0.05	0.12
ΣA (Sum of anions)	8.94	1.57
$\Sigma C - \Sigma A$	−2.90	8.94
H^+/OH^- release*	−3.03	6.84

*As measured by titrating the nutrient solution back to its initial pH (negative and positive values stand for net H^+ and OH^- release, respectively).

results in a strong relationship between H^+ release and cation–anion balance, as illustrated by the work of Rufyikiri et al. (2001) on banana seedlings (*Musa* spp.). Figure 2 shows for a range of hydroponically grown genotypes of banana that this relation is indeed fairly close to $y = x$.

Nitrogen plays a prominent role in the cation–anion balance, because it is the nutrient that is taken up at the highest rate by most plant species (Marschner, 1995; Mengel et al., 2001) (Table 1). In addition, it can be taken up as a cation (ammonium, NH_4^+) or as an anion (nitrate, NO_3^-), or even as an uncharged species (gaseous N_2°) in the case of N_2-fixing plants such as legumes living in symbiotic association with N_2-fixing bacteria. In addition, plants can directly use a significant proportion of N as amino acids that can either be positively charged, neutral or negatively charged (Jones and Darrah, 1994a). There are numerous papers showing that (i) plants supplied with NO_3^- will counterbalance the corresponding excess of negative charges by releasing equivalent amounts of OH^- or HCO_3^- into the rhizosphere, thereby increase rhizosphere pH, and (ii) plants receiving NH_4^+ will counterbalance the corresponding excess of positive charges by releasing equivalent amounts of H^+ in the rhizosphere, thereby decrease rhizosphere pH (Gahoonia et al., 1992; Imas et al., 1997a; Jarvis and Robson, 1983; Kirkby and Mengel, 1967; Marschner and Römheld, 1983; Raven, 1986; Riley and Barber, 1971; Römheld, 1986; Weinberger and Yee, 1984) (Table 1).

In a similar manner, plants relying on atmospheric N_2, such as legumes, take up more cations than anions, and hence release the excess of positive charges as H^+ and acidify their rhizosphere (Jarvis and Robson, 1983; McLay et al., 1997; Römheld, 1986; Tang et al., 1997, 2001a). Tang et al. (1997) have shown a strong linear relationship between the amounts of H^+ released by the roots and the excess of cations over anions taken up by the plants ($\Sigma C - \Sigma A$) for 12 species of pasture legumes (H^+ release=0.24+1.06×($\Sigma C - \Sigma A$) when expressed in mmol of charge per plant, with r=0.97): for all of them, more cations than anions were taken up, resulting in a net release of an equivalent amount of H^+.

The cation–anion balance does not only include the ions that are taken up by plant roots, but should include all the charges/ions that cross the root cell plasma membrane via either influx or efflux. In that respect, the release of organic anions can be a major component of cation–anion balance and thereby influence the net release of H^+ or OH^- equivalents (e.g., Dinkelaker et al., 1989). This will be further discussed in the following section ('Organic anion release'). Besides, amino acids that have been already referred to as a possible source of N for the plants (influx) can also be released as root exudates (efflux), and therefore both processes and/or the net balance of their influx/efflux should be accounted for (Jones and Darrah, 1994a). In addition, these various organic ions and also inorganic ions such as phosphate ions can bear various charges, as dependent on the ambient pH and the different p*K* values of the corresponding acids. For phosphoric acid the three p*K* values are 2.15, 7.20 and 12.34, which means that the dominant ion is $H_2PO_4^-$ for soil pH values in the acidic range (below 7.2) and that the dominant ion is HPO_4^{2-} for soil pH values in the alkaline range (above 7.2). In calcareous environments, roots may have to take up HPO_4^{2-} ions as this is by far the dominant species (amounting to a molar fraction of 0.94 at pH 8.4). It is well known, however, that monovalent phosphate ions, $H_2PO_4^-$, can much more easily permeate through the plasma membrane than divalent phosphate ions, HPO_4^{2-}, and are therefore expected to be preferentially taken up by plant roots. In addition, root apoplastic pH is normally less than 6 (Yu et al., 2000), and is not markedly affected by external pH (e.g., increased by 0.3 as external pH increased from 5.2 to 7.5 in *Lupinus angustifolius* L.; Yu et al., 2001). Therefore, the major form of P taken up (across the membrane) by plants in high-pH soils would still be $H_2PO_4^-$. Besides, because of the proximity of their pK and the cytosolic pH, the $H_2PO_4^-/HPO_4^{2-}$ system plays a prominent role in the buffering of cytosolic pH, in addition to organic acid/anions buffers that have been previously mentioned (Gerendás and Schurr, 1999).

Another important feature of root-induced changes of pH in the rhizosphere is their spatial variation along the root axes. Although some authors have shown that the whole root system may sometimes behave evenly, exhibiting a generalised alkalinisation under excess anion supply (in particular NO_3^-), or a generalised acidification under excess cation supply, as would occur with NH_4^+ supply (Marschner et al., 1982; Römheld, 1986), many others have reported that the different portions of a single root or of a whole root system may behave differently. For instance, some portions of the roots, and most often that portion that is just behind the root apex, may release H^+ while basal parts of the roots release OH^- in the rhizosphere (e.g., Häussling et al., 1985; Jaillard et al., 2002; Marschner and Römheld, 1983; Plassard et al. 1999; Römheld, 1986). This would suggest that the cation–anion balance varies accordingly along the root. This is likely related to cations and anions being not evenly taken up along the root, although there is little experimental evidence supporting this hypothesis, probably because of methodological difficulties to locate such fluxes. To fill this knowledge gap, further research providing combined measurements of fluxes of H^+ and major nutrients along root axes is thus clearly needed.

Organic anion release

Organic acids have often been referred to as a possible source of rhizosphere acidification (e.g., Hoffland et al., 1989). However, very different levels of exudation of organic acids can be found in the literature, depending on the species and on environmental constraints (see the section 'Effect of environmental and nutritional constraints'), e.g., P deficiency or Al toxicity (Jones, 1998; Jones and Darrah, 1994b; Jones et al., 2002; Neumann and Römheld, 1999; Neumann et al., 1999, 2000). In maize for instance, the contribution of exuded organic acids to acidification was rather negligible, not exceeding 0.2–0.3% (Petersen and Böttger, 1991). In white lupin in contrast, cluster roots can release such large amounts of organic anions that they can precipitate and accumulate in the rhizosphere and strongly acidify it (Dinkelaker et al., 1989). In addition, these organic acids comprise a wide variety of compounds that vary with plant species, plant age and environmental constraints (e.g., Dakora and Phillips, 2002; Jones, 1998; Neumann and Römheld, 1999).

Of the organic acids, citric, oxalic and malic acids are the most frequently referred to for their potential effect on the rhizosphere (e.g., Jones and Brassington, 1998; Jones et al., 2002), but the extent to which they are involved in changing rhizosphere pH is still a question for debate (see below). It is noteworthy that these acids are present in large concentrations in root cells where they play an important role in the buffering of cytosolic pH (see above, Marschner, 1995). In many plant species, these will be stored in the vacuoles of root cells, while a substantial proportion may be exuded in the rhizosphere only in some species (Jones, 1998; Neumann and Römheld, 1999). However, as already pointed out by several authors (Haynes, 1990; Hedley et al., 1982; Hinsinger, 1998, 2001a; Jones and Darrah, 1994b; Nye, 1986), it is clear that these organic acids are dissociated in the cytosol because of their low p*K* values compared with the neutral pH of the cytosol. They are thus not expected to be released as acids but as their conjugate base, i.e., as organic anions: at pH around 7.3, citrate is predominantly present as citrate^{3-} (p*K* for citrate^{2-}/citrate^{3-}

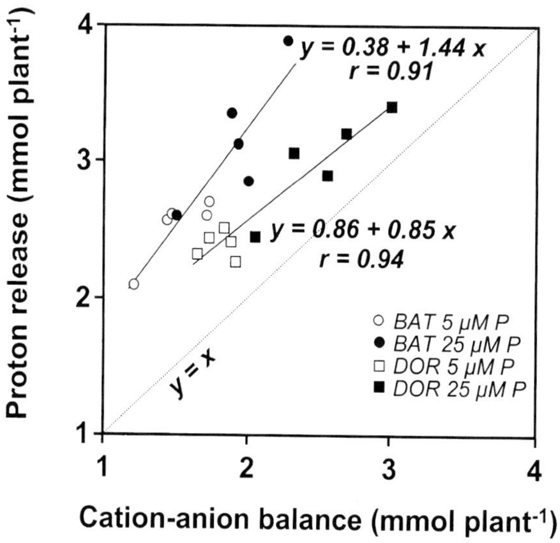

Figure 3. Relationship between total H^+ release and the uptake of excess cations over anions (i.e., excess of positive charges taken up) by two genotypes of bean (*Phaseolus vulgaris*) grown at 5 (deficient) and 25 μM P (sufficient) for 42 days (adapted from Tang et al., 2001b, with kind permission from EDP Sciences S.A.).

is 6.40 and p*K* for citrate$^-$/citrate^{2-} is 4.76), malate as malate^{2-} (p*K* for malate$^-$/malate^{2-} is 5.11) and oxalate as oxalate^{2-} (p*K* for oxalate$^-$/oxalate^{2-} is 4.19) (Jones and Brassington, 1998). Therefore, the release of such organic anions is a process that is not different, from a purely chemical point of view, from the uptake of inorganic ions: the charges carried by organic anions should thus be counterbalanced by an equivalent influx of OH^- or efflux of H^+. That is, the release of these organic anions by the root should be taken into account in the overall balance of cations and anions that are crossing the plasma membrane. Hence, differences between measured H^+ efflux and independently determined cation–anion balance can be attributed to organic anion release, whenever the other sources of H^+ (such as root respiration) can be discounted (Durand et al., 2001). Figure 3 shows a larger discrepancy between H^+ release and excess cation uptake in one bean genotype (BAT477) than in another one (DOR364), which could be explained by a genotypic difference in carboxylate exudation (Tang et al., 2001b), irrespective of whether P supply was deficient or sufficient. However, just like phosphate (second p*K*=7.20) or bicarbonate (first p*K*=6.36) anions, carboxylate ions can react with free H^+ in solution and thus buffer the pH changes induced by root activity. Organic anions might result in an increase in rhizosphere pH if they are protonated because of a lower external pH than cytosolic pH (Jones and Darrah, 1994b) or if they are decomposed (Yan et al., 1996). Citrate for instance has its third p*K* (citrate^{2-}/citrate^{3-}) at 6.40, and is thus able to react with H^+ at pH values commonly found in the rhizosphere. In addition, if organic anions are released with an accompanying cation such as K^+, the net flux of protons released par roots is then nil. This nonetheless shows that the release of organic anions might affect the rhizosphere pH as much as the exchange of inorganic nutrient ions and should thus be accounted for in the overall balance of cations and anions.

Release of large quantities of citrate has been reported to be associated with strong rhizosphere acidification in cluster roots of white lupin (*Lupinus albus* L.) (e.g., Dinkelaker et al., 1989; Gardner et al., 1982, 1983; Neumann and Römheld, 1999). In this particular case, Dinkelaker et al. (1989) showed that the efflux of citrate was far larger than the measured excess of cations over inorganic anions taken up by the roots of white lupin. They suggested that in that case, H^+ released to accompany the efflux of citrate was a major component of the observed acidification of the rhizosphere around cluster roots. This will be further discussed in the 'Response to environmental and nutritional constraints' section. However, for most other plant species than cluster-rooted species, the rates of release of organic anions are minute (two or three orders of magnitude lower) compared with the rates of H^+ release that arise from an excess of cations over anions taken up, as shown for chickpea (*Cicer arietinum* L.) (Ohwaki and Sugahara, 1997), maize (*Zea mays* L.) (Petersen and Böttger, 1991) and tomato (*Lycopersicon esculentum* L.) (Imas et al., 1997a).

The synthesis of carboxylic anions and their ultimate exudation in the rhizosphere also depend upon the cation–anion balance. Indeed, the cytosolic pH-stat in root cells largely relies on carboxylation–decarboxylation of organic acids/anions (Haynes, 1990; Marschner, 1995). This means that carboxylation will occur when a net excess of cations over anions enters the root cell, ultimately resulting in the synthesis of malic acid that will dissociate and counterbalance the corresponding elevation of cytosolic pH (buffer). Concentrations of organic anions such as malate will thus build up in the vacuole as a consequence of excess cation uptake, and the reverse will take place when more anions than cations are taken up (Haynes, 1990; Hiatt, 1967; Marschner, 1995). However, when considering the case of nitrogen, this simple picture becomes more complicated because of the differential effect of NO_3^- and NH_4^+ assimilation (incorporation into amino acids) on intracellular OH^- and H^+ production and, thereafter, on organic acid synthesis. Larger concentrations of carboxylates are thus generally found in root cells of plants fed with NO_3^-, compared with plants fed with NH_4^+ (Haynes, 1990; Marschner, 1995). Accordingly, Imas et al. (1997a) found that larger exudation of carboxylates such as citrate, malate and fumarate occurred for tomato plants relying on NO_3^- relative to those relying on NH_4^+.

Root exudation and respiration

Plant roots provide a significant proportion of CO_2 input into the soil. This can originate both from root respiration and root exudation of organic carbon that will then be degraded by rhizosphere microorganisms. Indeed, one to two thirds of the carbohydrates translocated from the shoots to the roots are ultimately respired and this commonly represents 10–50% of the photosynthates produced daily (Lambers et al., 1996). Several ^{14}C labelling studies have shown that, for cereals such as maize (*Zea mays* L.) or wheat (*Triticum aestivum* L.), a total of 30–50% is allocated to the below-ground parts of the plants, 10–30% accumulates in the roots, 10–20% is respired in the rhizosphere and 1–5% accumulates as organic material and microbial cells in the rhizosphere (Merckx et al., 1986; Nguyen et al., 1999). Helal and Sauerbeck (1989) estimated that roughly 20% of photosynthates are exuded by maize roots into the rhizosphere and that three quarters of these exudates are ulimately transformed into CO_2 as a consequence of microbial respiration. Although published figures show large variations that may partly result of discrepancies between the various methods used to estimate the fluxes of exudation and CO_2 production (Meharg, 1994), there is little doubt that such fluxes are a considerable source of C into the rhizosphere.

Reported fluxes of CO_2 produced by roots are in the order of 50–200 nmol g^{-1} DW s^{-1} in *Cucumis sativus*, *Lycopersicon esculentum* and *Holcus lanatus* (Lambers et al., 1996, and references therein; Van der Westhuizen and Cramer, 1998). There are few studies that enable a direct comparison with the values obtained for other acidic root exudates and their relation with rhizosphere acidification. Fluxes of released CO_2 of 100–200 nmol $plant^{-1}$ s^{-1} have been reported in maize by Durand and Bellon (1993), while in another experiment with maize seedlings of a similar age, H^+ effluxes of about 5–10 nmol H^+ $plant^{-1}$ s^{-1} were found by Durand et al. (2001). Yan et al. (1992) simultaneously measured fluxes of about 3 nmol O_2 and 0.2 nmol H^+ g^{-1} FW s^1 in maize. These various figures thus suggest that CO_2 is released by roots at fluxes of an order of magnitude larger than H^+ released to balance an excess of cations over anions. In addition, rhizosphere microorganisms can contribute an additional release of CO_2 due to their own respiration being boosted by the supply of C-compounds from the roots (exudates).

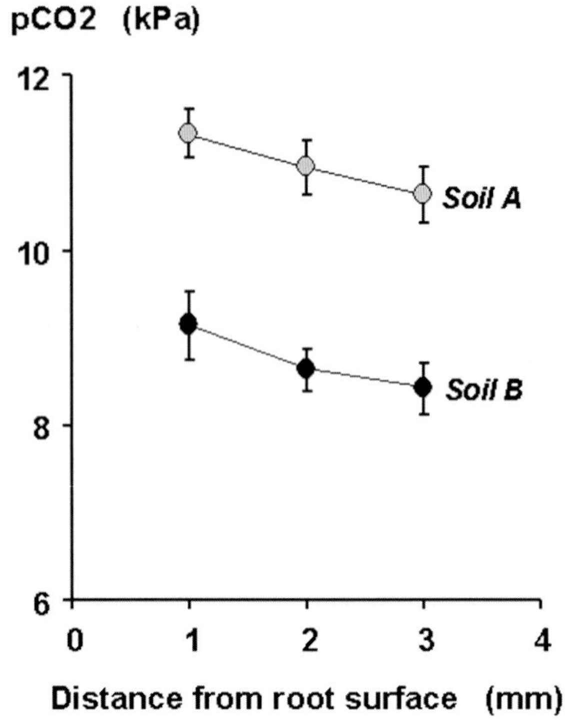

Figure 4. Profiles of CO_2 partial pressure as a function of the distance from roots in the rhizosphere of Sordan (Sorghum×sudangrass) grown in two calcareous soils (adapted from Gollany et al., 1993, with kind permission from Elsevier Science).

Root and rhizosphere microbial respiration is thus expected to potentially contribute some significant proportion of changes in CO_2 concentration in soil atmosphere, and thereby in soil pH. There is, however, a limited number of papers to provide quantitative insight into these processes, and more so into how they relate to one another. In fact, there are very few figures showing elevated CO_2 concentrations in the rhizosphere as a consequence of the respiration of soil-borne organisms including plant roots. Most of the available data refer to bulk soil atmosphere. For instance, CO_2 concentrations ranging from 11 to 32 mmol mol^{-1}, i.e., 30–100-fold greater than atmospheric CO_2 concentrations (0.36 mmol mol^{-1}) were reported in the soil beneath an alfalfa (*Medicago sativa* L.) stand (Norstadt and Porter, 1984). As gaseous CO_2 diffuses much faster than HCO_3^- ions in solution, it can be expected to have a measurable effect only locally, unless the physical properties of the soil severely restrict its diffusion (Anoua et al., 1997). Soil compaction for instance can reduce its ability to diffuse and result in larger increases in bulk soil CO_2 concentration as reported for soil columns planted with *Phaseolus vulgaris* (Asady and Smucker, 1989). As these are bulk soil values, one might expect even greater CO_2 concentrations in the rhizosphere, especially in those soils that exhibit a poor permeability to gases, such as compacted or water-saturated soils. However, because of the technical difficulties for sampling gases at this scale, to our knowledge there is only one report of measured rhizosphere CO_2 concentrations: Gollany et al. (1993) measured the pCO_2 in the rhizosphere of two genotypes of sordan (*Sorghum bicolor/Sorghum sudanese* hybrid) as a function of the distance from root surface and found fairly large values and a smooth gradient indicating a diffusion away from the roots (Figure 4). There are, however, several experiments using small pots or rhizoboxes that provide additional, relevant data. For instance for maize seedlings, Durand and Bellon (1993) measured 5- to 10-fold increase in CO_2 concentrations on average, and sometimes up to 100-fold in the rhizosphere of maize seedlings. Similar increases were also reported for western wheatgrass (*Agropyron smithii* Rydb.), with the largest values obtained for VA-mycorrhizal plants (Knight et al., 1989). The development of new techniques for measuring local changes in gas concentrations may provide more precise data about their variation in the rhizosphere. For instance, by using O_2-microelectrodes Bidel et al. (2000) were able to detect spatial variations in root respiration along the root system of peach (*Prunus persica* L.) seedlings grown in agar plates. This confirmed earlier works (Lambers et al., 1996) showing that root respiration peaked near the root tip, in the meristematic zone (beginning of elongation zone).

There are, unfortunately, no data providing an estimate of the quantitative contribution of changes in CO_2 concentrations to changes in rhizosphere pH. This is of concern only for neutral to alkaline soils because H_2CO_3 with its first pK of 6.36, remains undissociated at acidic pH values (Lindsay, 1979). Therefore, in acid soils the contribution of root respiration and exudation, and consequent microbial respiration, to changes in rhizosphere pH can be neglected (Nye, 1981). Conversely, in calcareous soils, the soil pH can be considerably decreased with increasing CO_2concentration (Gras, 1974). Figure 5 shows that the pH of a calcareous soil, which is typically equal to 8.3 at ambient atmospheric pCO_2, can decrease by 0.67 pH units for every 10-fold increase in pCO_2, and can thereby reach values as low as 6.7 at a pCO_2 of 0.1 mol mol^{-1}. For instance the pCO_2 values ranging from 0.08 to 0.11 mol mol^{-1} reported by Gollany et al. (1993) would thus yield pH values between 6.70 and 6.78 in the rhizosphere. Clearly, root and microbial respiration can have a dramatic influence on rhizosphere pH in calcareous soils. More quantitative data are needed on the potential effect of pCO_2 on rhizosphere pH in neutral and mildly acidic soils.

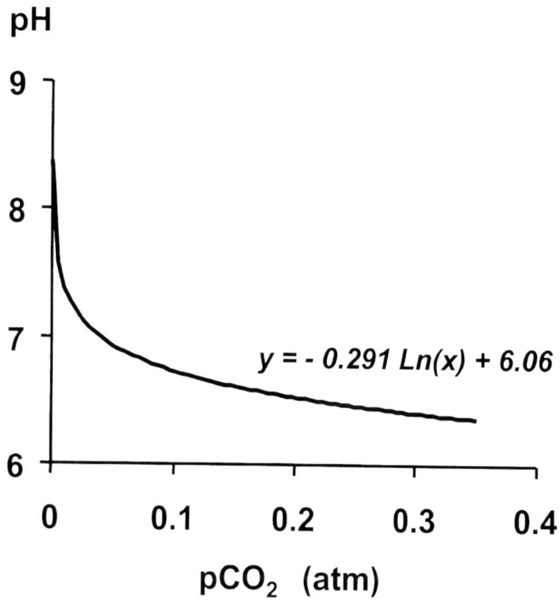

Figure 5. Relationship between pH and pCO_2 (partial pressure) in calcareous soils (derived from the equation supplied by Gras, 1974).

Redox-coupled processes

The major chemical reactions involving changes in the oxidation state of Fe, Mn and N also imply the consumption or production of H^+, and thereby a coupling of redox potential and pH. This is illustrated here for the case of iron transformation, which is quantitatively the most important and visible redox phenomenon in most soils (Van Breemen, 1987). If reduction of Fe occurs, that is if electrons (e^-) are supplied, the dissolution of Fe^{III}-bearing minerals such as goethite, FeOOH, and combined reduction of Fe^{III} into Fe^{II} can be described by the following equation (Lindsay, 1979):

$$FeOOH + 3H_3O^+ + e^- \Leftrightarrow Fe^{2+} + 5H_2O \qquad (1)$$

Most often, electrons are supplied by the oxidation of organic matter (including microbial or root respiration). The following equation equation given by Sposito (1989) is an example which applies to formate, CHO_2^-, but could be generalized to any other organic compound:

$$CHO_2^- + H_2O \Leftrightarrow CO_2 + H_3O^+ + 2e^- \qquad (2)$$

Combining these two redox half reactions (1) and (2) so that electrons do not appear explicitly (Sposito, 1989) gives the overall redox reaction that actually occurs in the soil (Sposito, 1989; Van Breemen, 1987):

$$2FeOOH + CHO_2^- + 5H_3O^+ \Leftrightarrow 2Fe^{2+} + CO_2 + 9H_2O \qquad (3)$$

This clearly shows that the reduction of Fe^{III} contained in Fe oxides should be accompanied by the oxidation of other compounds such as organic matter and by a consumption of H^+ (H_3O^+) that will ultimately result in an increase in soil pH. Conversely, the oxidation of Fe^{II} will be accompanied by a decrease in soil pH. These soil alkalinisation–acidification processes are well known to take place in seasonally flooded soils (Van Breemen, 1987).

In the rhizosphere (see the 'Root respiration and exudation' section), O_2 concentration decreases (Bidel et al., 2000) while CO_2 concentration increases (Durand and Bellon, 1993) as a result of root and microbial respiration, which means that electrons will be produced during this process of oxidation of organic matter (cf. Equation (2)). Besides, reduction processes can also occur at the root surface as related to enzymatic systems. These reductase systems are bound to the plasma membrane of root cells (Bienfait et al., 1983), with NADPH serving as an electron donor in the root cells (Marschner and Römheld, 1994; Sijmons et al., 1984). Such a reduction process that is stimulated by Fe deficiency in all strategy I plant species (Marschner and Römheld, 1994; Römheld and Marschner, 1986) has been demonstrated to occur behind root apex by using redox dye indicators (Brown, 1978; Dinkelaker et al., 1993; Marschner et al., 1986) and Pt microelectrodes (Fischer et al., 1989). From a physicochemical point of view, according to Equation (3), this reduction process should be accompanied by a consumption of H^+ and contribute *per se* to an increase in rhizosphere pH. However, as this process is combined at a physiological point of view with a release of H^+ by plant roots (Haynes, 1990; Marschner and Römheld, 1994), the zones of reduction around roots are rather expected to be located where acidification takes place (Fischer et al., 1989).

In the case of plants that occur under reducing conditions such as wetland plants and lowland rice (*Oryza sativa* L.), oxidation of Fe^{II} occurs in the rhizosphere, as a consequence of O_2 release (leakage) from the roots, O_2 being transferred from the shoots through aerenchyma (Ando et al., 1983; Armstrong, 1967). This root-induced oxidation of Fe was demonstrated in various ways. Firstly, an iron plaque made of precipitates of Fe oxides often forms at the surface of the roots and in root cell walls of rice (Chen et al., 1980; Green and Etherington, 1977). Secondly, the concentration of Fe^{II} increases while that of Fe^{III} decreases with the distance from the surface of rice roots (Begg et al., 1994). Thirdly, redox potential increases close to root surface of rice plants, as shown by using redox indicators (Trolldenier, 1988) or redox electrodes (Flessa and Fischer, 1992). Interestingly, Begg et al. (1994)

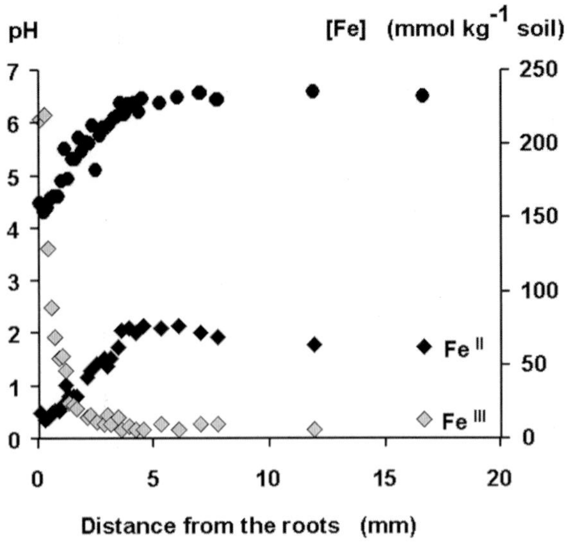

Figure 6. Profiles of contents of Fe^{II}, Fe^{III} (diamonds) and pH (circles) as a function of the distance from root surface in the rhizosphere of lowland rice (*Oryza sativa*) grown in reduced soil conditions (adapted from Begg et al., 1994, with kind permission from Blackwell Science).

have detected a strong acidification that accompanied the oxidation of the rhizosphere (Figure 6). This is in agreement with what might be expected because of the coupling of these two processes (see Equation (3)). However, part of the reported acidification of rice rhizosphere was due to an excess of cations over anions taken up, resulting from NH_4^+ uptake in such reducing, ambient conditions (Begg et al., 1994). Using the following equation that relates H^+ production with Fe^{II} oxidation in the presence of O_2 (Ahmad and Nye, 1990), Begg et al. (1994) and Kirk and Le Van Du (1997) estimated that this oxidation process contributed a significant or even major proportion of the measured rhizosphere acidification:

$$4Fe^{2+} + O_2 + 14H_2O \Leftrightarrow 4FeOOH + 8H_3O^+ \quad (4)$$

Redox processes in the rhizosphere, such as those described for Fe, are thus intimately coupled with pH changes and thus need to be taken into account for understanding all the mechanisms underlying root-induced pH changes. The same rule would apply for other important redox processes, such as those involved in the dynamics of elements like Mn, but also N and S (Sposito, 1989; Van Breemen, 1987).

Figure 7. Acidification at the root tip of an Fe-deficient tobacco (*Nicotiana tabaccum*) seedling grown in an agarose gel containing a pH dye indicator (bromocresol purple) that turns yellow at acidic pH (from Vansuyt G, Souche G, Straczek A, Briat J F and Jaillard B, unpublished).

Effect of environmental and nutritional constraints

Localised release of H^+ in some portions of the roots, most often in the subapical zone, has frequently been reported to occur as a response to various types of environmental stresses, including shortage of Fe or P and toxicity of Al and possibly other metals (Haynes, 1990; Hinsinger, 1998, 2001a, b; Marschner, 1995). The case of Fe-deficiency is by far the most documented. Most plant species (all but the Graminaceae) belong to the strategy I group of plant species, which means that they respond to Fe-deficiency with an enhanced capacity of their roots to acidify the rhizosphere (via H^+ release) and to reduce Fe^{III} (via a reductase system) (e.g., Marschner and Römheld, 1994; Marschner et al., 1986; Römheld and Marschner, 1986). For instance, a strong acidification of the rhizosphere occurs locally, behind the root tips of Fe-deficient sunflower (*Helianthus annuus* L.) plants (Marschner et al., 1982; Römheld and Marschner, 1981; Römheld et al., 1984). This increased capacity of localised portions of the roots to release H^+ has been attributed to rhizodermal transfer cells, which are induced by Fe-deficiency and have a strong H^+-ATPase activity (Haynes, 1990; Römheld and Marschner, 1986; Römheld et al., 1984). Figure 7 provides an illustration of this process in Fe-deficient roots of tobacco.

Enhanced H^+ release also occurs as a response to P-shortage (Bertrand et al., 1999; Grinsted et al., 1982; Imas et al., 1997b; Moorby et al., 1988; Neumann and Römheld, 1999; Ruiz, 1992). Such a response is also rather localised behind the root tips (Gregory and Hinsinger, 1999; Hinsinger P, Jaillard B, Souche G and Rengel Z, unpublished; Ruiz, 1992), as previously described for Fe deficiency. For instance, Ruiz (1992) measured an efflux of 18 μmol H^+ h^{-1} g^{-1} of fresh roots behind the root tip of a P-deficient rape (*Brassica napus* L.) seedling and an efflux of 9 μmol OH^- h^{-1} g^{-1} for basal parts of the same root. In comparison, no acidification occurred for P-sufficient rape and an average efflux of about 12 μmol OH^- h^{-1} g^{-1} fresh roots was found along the root. Figure 8 shows that decreased alkalinisation/increased acidification occurred in roots of bean (*Phaseolus vulgaris*) supplied with 1 μm P (deficient level) compared with 25 μm P (suffi-

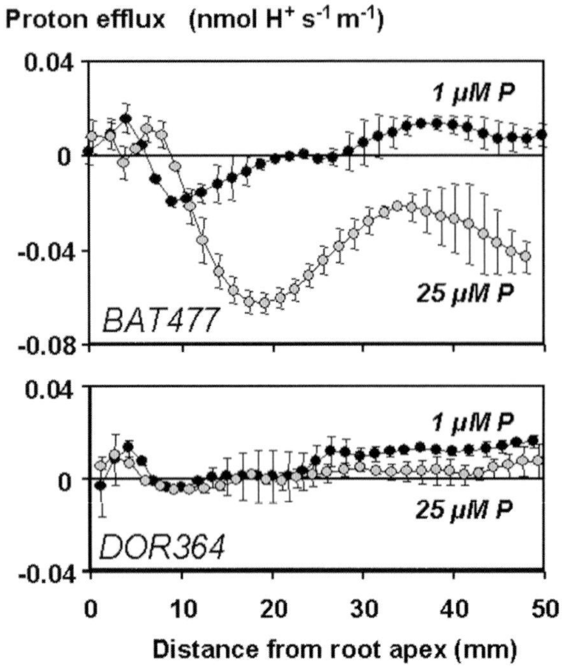

Figure 8. Profiles of H^+ efflux along single roots of two genotypes of nitrate-fed bean (*Phaseolus vulgaris*) grown at 1 (deficient) and 25 μM P (sufficient) for 29 days. These profiles were obtained by the videodensitometry of dye indicator technique developed by Jaillard et al. (1996): two roots of 29-day-old plants were embedded for each P treatment and the points are the average value of the two (from Tang C, Drevon J J, Jaillard B, Souche G and Hinsinger P, unpublished).

cient level). Such a response was found in nitrate-fed bean, but not in bean plants relying on N_2 fixation (Tang C, Drevon J J, Jaillard B, Souche G and Hinsinger P, unpublished). This enhanced acidification of the rhizosphere in nitrate-fed plants might be related to an inhibition of NO_3^- uptake, and to the consequent increase in the excess of cation over anion uptake, in response to P deficiency, as suggested by several authors (Kirk and Le Ven Du, 1997; Le Bot et al., 1990; Neumann and Römheld, 1999; Neumann et al., 1999, 2000). By contrast, only enhanced acidification occurred in medic (*Medicago truncatula*) relying exclusively on N_2 fixation (Tang et al., 2001a), as shown in Figure 9. In white lupin, although a significant increase in citrate exudation occurred under P deficiency (see below) which might contribute some of the observed acidification of the hydroponic solution in P-deficient relative to P-sufficient plants (Neumann and Römheld, 1999; Neumann et al., 2000), the major contribution to the acidification was the decreased uptake of nitrate (Neumann et al., 1999, 2000). This probably played a prominent role in the reported acidification that was induced by P-deficient plants.

Similarly, it has been shown that Al toxicity induces a larger decrease in anion uptake, and particularly NO_3^- uptake (Rufty et al., 1995), than cation uptake by plant roots and thereby caused an additional acidification or decreased alkalinisation of the rhizosphere (Calba and Jaillard, 1997). These authors and others (see Haynes, 1990) showed that such a response to Al toxicity was typical of Al-sensitive genotypes, whereas Al-resistant genotypes were capable of better maintaining their capacity to take up anions and thereby alkalinising their environment, a possible way of ultimately alleviating Al toxicity in acid soils (e.g., Calba and Jaillard, 1997; Degenhardt et al., 1998; Foy, 1988).

It is noteworthy that the exudation of organic anions is often enhanced in response to the same environmental stresses as above-mentioned for the release of H^+, namely Fe deficiency (Ohwaki and Sugahara, 1997), P deficiency (Hoffland et al., 1989; Imas et al., 1997b; Keerthisinghe et al., 1998; Neumann and Römheld, 1999) and Al toxicity (Delhaize et al., 1993). The physiological mechanisms involved in the regulation of such a response is still a question for debate and definitely needs further research, as stressed by Jones (1998). In the case of Al toxicity, reported effluxes are fairly small, and the question of the efficiency of the protection provided by such low concentrations of malate or citrate is still debated (Parker and Pedler, 1998). However, this is possibly efficient at a localised level within the apoplasm of the root tip. In comparison, considerable amounts of carboxylates are released in response to P deficiency, especially in cluster-rooted plants. In the case of white lupin, a low P supply clearly favours the development of cluster roots (Johnson et al., 1994, 1996; Keerthisinghe et al., 1998; Neumann et al., 1999), which exhibit considerably faster rates of citrate and malate exudation than other roots (Keerthisinghe et al., 1998; Neumann and Römheld, 1999). Such exudation of citrate and malate is confined to portions of the root system (cluster roots) and to stages of development of these cluster roots: citrate efflux is high in mature cluster roots and low in juvenile and senescent clusters (Dinkelaker et al., 1995; Neumann et al., 1999). Interestingly, strong acidification as evidenced by using dye indicators seems to be confined to the same regions of the root system and stages of development (mature root clusters, Neumann et al., 1999). This suggests that the exudation of citrate and malate is balanced or, at least

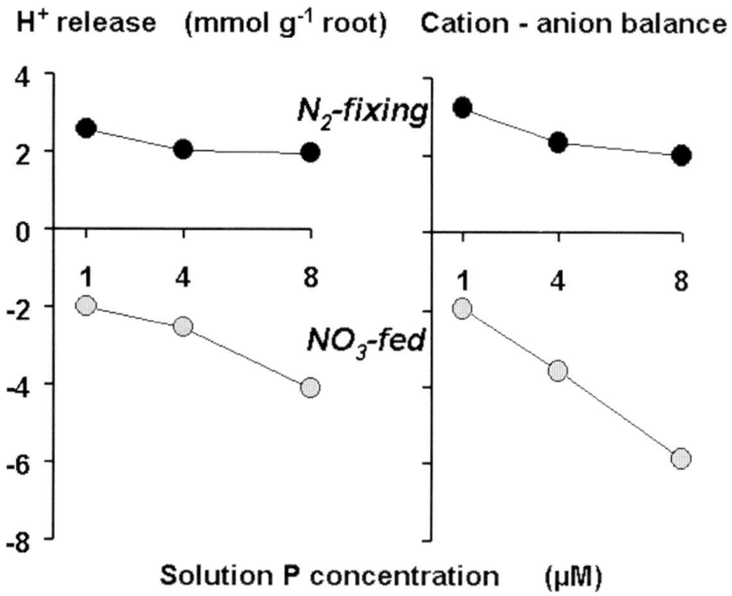

Figure 9. H^+ efflux and corresponding uptake of excess cations over anions in *Medicago truncatula* grown in solution culture for 35 days at various concentrations of P, from 1 (deficient) to 8 μM (sufficient). Plants were either relying on N_2 fixation only (adapted from Tang et al., 2001a, with kind permission from Oxford University Press) or on nitrate only (Tang C, Drevon J J, Jaillard B and Hinsinger P, unpublished).

accompanied by some H^+ release in mature cluster roots. To draw a definitive conclusion in that respect, there is, however, a lack of quantitative data about the efflux of H^+ achieved locally in these portions of the root system of white lupin. At the level of the whole root system, the acidification that occurs under P deficiency (Neumann et al., 1999, 2000) is driven by the considerably reduced uptake of nitrate, rather than by the increased citrate efflux (Figure 10), when considering their respective effect on charge balance. In N_2-fixing white lupin, the acidification is closely related to excess uptake of cations over inorganic anions but the ratio of H^+ release to excess cation uptake increased with decreasing P supply (Sas et al., 2001).

In other cluster-rooted plants such as Proteaceae, similar increased formation of cluster ('proteoid') roots has been reported to occur and to be responsible for very fast rates of exudation of carboxylates in response to P deficiency (Dinkelaker et al., 1995; Roelofs et al., 2001; Shane et al. 2002; Skene et al., 1996). In *Banksia integrifolia*, larger concentrations of citrate and much lower pH values in water extracts were found in rhizosphere soil sampled around cluster roots of plant grown *in situ* than in the bulk soil (Grierson, 1992; Grierson and Attiwill, 1989). This would suggest that cluster roots of this species release both citrate and H^+. In contrast, for a range of other Proteaceae species that were grown in hydroponic conditions, Roelofs et al. (2001) showed that the considerable rates of carboxylate released by cluster roots (with citrate being a major component) were not resulting in any substantial pH change. Instead, K concentrations increased in the collection solution, suggesting that K^+ efflux accompanied the efflux of negatively charged exudates (carboxylates). Similarly, Al toxicity-induced release of malate in the apex of wheat seems to be accompanied by K^+ efflux rather than H^+ efflux (Ryan et al., 1995). Enhanced release of organic anions in response to either P deficiency or Al toxicity is therefore not always responsible for significant pH change in the rhizosphere.

Effect of bulk soil pH and pH buffering capacity: Ecological significance of root-mediated pH changes

The effect of bulk soil pH on root-induced changes in rhizosphere pH has rarely been reported in the literature. However, it can be rather dramatic, as shown by Youssef and Chino (1989) for barley and soybean grown in a soil at three pH values obtained by liming (Figure 11). These authors showed indeed that alkalinisation occurred in the rhizosphere at an initially low pH (pH 4.8), while acidification occurred at higher initial pH values (pH 7.1). Such changes amounted up

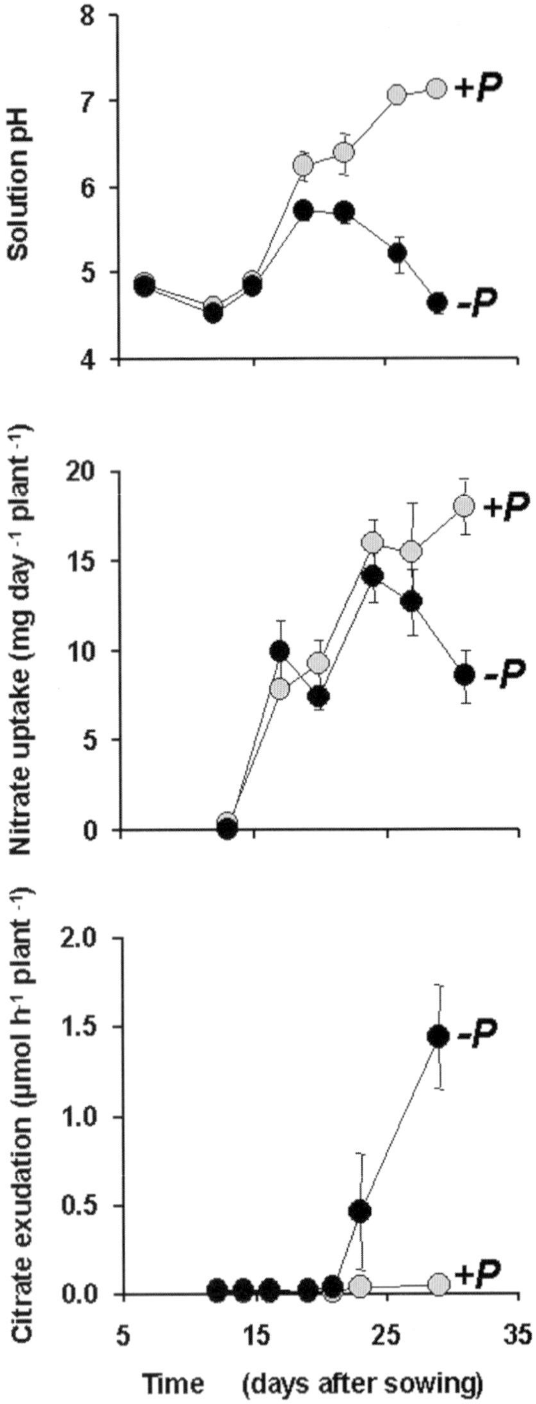

Figure 10. Temporal evolution of solution pH, nitrate uptake and citrate exudation in white lupin (*Lupinus albus*) grown with adequate supply of P (+P) or no supply of P (−P) in solution culture (adapted from Neumann et al., 1999, with kind permission from Springer).

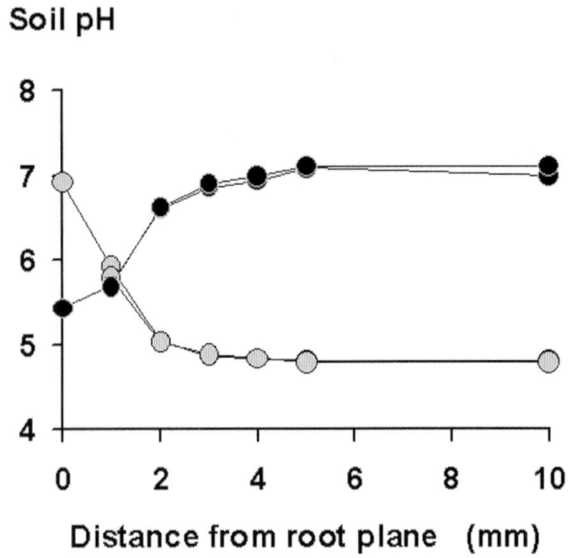

Figure 11. Profiles of soil pH as a function of the distance from the root surface barley (*Hordeum vulgare*) grown in a clay loam either prior (initial bulk pH 4.8) or after liming to yield an initial bulk pH 7.1 (adapted from Youssef and Chino, 1989, with kind permission from the Japanese Society of Soil Science and Plant Nutrition).

to a maximum of about 2 pH units and were detected up to 2–3 mm from root surface. Similarly, alkalinisation in an acidic soil and acidification in a calcareous soil was reported to occur in the rhizosphere of oilseed rape and tomato (Chaignon et al., 2002). A recent experiment by Quesnoit M, Chaignon V and Hinsinger P (unpublished) confirmed such results over a larger range of pH values achieved by liming an initially acidic soil (pH 3.7 in $CaCl_2$ background electrolyte). The rhizosphere pH was measured on the bulk of a 2-mm thick slice of soil on top of which rape plants had been grown for 7 days with the root-mat technique designed by Guivarch et al. (1999). Rape consistently produced a significant alkalinisation at a pH below 4.8 and, conversely, a significant acidification above such a pH. Schubert et al. (1990) also showed for *Vicia faba* grown in a simple solution of $CaSO_4$ that net H^+ release occurred at a pH above 6 while alkalinisation occurred for pH 4 or 5. They suggested that ATP-ase activity was probably altered at low pH, resulting in decreased ability of the H^+ pump to release H^+, and ultimately in a net influx of H^+ at elevated concentrations of H^+ in the ambient solution. In the pH range of 3.7–6.0 studied by Quesnoit M, Chaignon V and Hinsinger P (unpublished), the contribution of root and microbial respiration was probably negligible. The observed pH-dependency of root-mediated pH changes was thus probably the consequence of

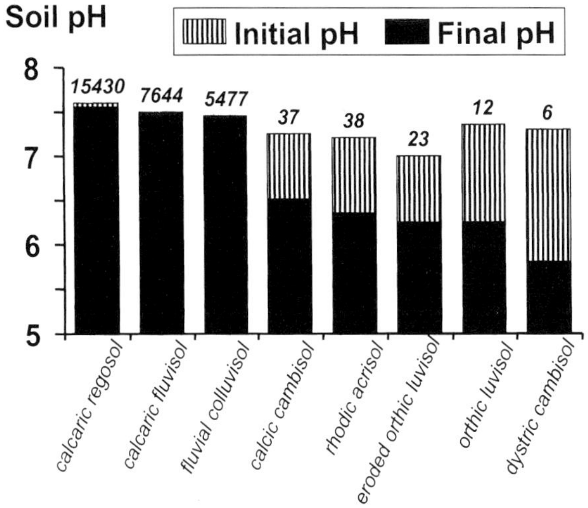

Figure 12. Change of soil pH and computed H^+ release during the growth of N_2-fixing faba bean (*Vicia faba*) grown in soils of different pH buffer capacity (adapted from Schubert et al., 1990, with kind permission from Kluwer Academic Publishers).

a different cation–anion balance at various soil pH values. Youssef and Chino (1989) concluded that the effect of the initial soil pH was much larger than that of N nutrition, although the form of N is a major component of the overall balance of cation–anion taken up by plants. Whatever the origin of such pH changes, it is remarkable that those various plant species behaved as if their roots were avoiding extreme (either too low or too high) pH values, and thereby alleviating the associated constraints such as, for instance, Al toxicity at acidic pH or Fe deficiency at alkaline pH. Further studies are clearly needed to elucidate the origins of such plant responses, which point to 'adaptive' reactions. Such a plasticity of a given plant species to extreme bulk soil pH values by either reducing or increasing rhizosphere pH, i.e., a typical avoidance strategy, is likely to be of great ecological significance. Root-mediated pH changes would certainly help a species to colonise different environments. In addition, when considering the relationships between root respiration and soil pH, it should also be stressed that interactions can occur as pH and other related environmental factors such as CO_2 concentration in the soil atmosphere and Al concentration in acidic soil solution can affect root respiration one way or the other (Lambers et al., 1996; Van der Westhuizen and Cramer, 1998).

Next to soil pH itself, the pH buffering capacity is a chemical parameter that plays a key role in root-induced pH changes in the rhizosphere, as stressed by Schaller (1987) and Schubert et al. (1990). The pH buffering capacity of soils is generally high, because of the ability of numerous soil constituents to react with H^+ via (i) cation exchange reactions, (ii) protonation–deprotonation of weak acid groups, and (iii) dissolution or precipitation of soil minerals (reactions that consume or produce H^+, respectively). It can, however, considerably vary with soil composition and is expected to be minimal in sandy soils, which contain little organic matter and maximal in calcareous soils (because of the consumption of H^+ when Ca carbonates dissolve). Schaller (1987) showed that the extension of root-mediated pH changes in the rhizosphere of peanut was larger in a poorly buffered soil with an initial pH of 5.5 (2.8 mm) than in more buffered alkaline and acidic soils (1.4 mm), in good agreement with Nye's calculations (1981). Schubert et al. (1990) showed that a given plant species induced fairly different changes in soil pH according to the pH buffering capacity of the soil (Figure 12): the greater the pH buffering capacity, the smaller the plant-induced pH change; for the two calcareous soils that exhibited by far the largest pH buffering capacity, no significant pH change was found. Remarkably, Schubert et al. (1990) found that the initial soil pH had no effect on the H^+ efflux involved in such pH changes, as calculated from the pH buffering capacity and the pH change, whenever it was significant (Figure 12). No such calculation was done for the two calcareous soils as they did not yield a significant pH change. Durand et al. (2001) also showed that fluxes of H^+ released by maize when grown in various conditions of pH and pH buffering capacities were constant, depending only on the water uptake and consequent biomass production. In contrast, the pH changes that they observed in the culture solution varied greatly with growth conditions. In addition, in alkaline conditions the variations in free H^+ in solution corresponded to less than one per thousand of the whole flux of charges released by roots, which means that 99.9% of the released H^+ had been consumed in chemical reactions in solution (Durand et al., 2001). These data clearly highlight that we should be very careful when interpreting pH changes in the rhizosphere, without taking account of the pH buffering capacity of the solution, and more so that of the soil. In particular, it is suggested that no significant pH change does not necessarily mean that the roots (and associated microbes in the rhizosphere) do not release any H^+ or OH^-. These might have been consumed in the many chemical reactions that contribute to the buffering capacity of the soil. Hinsinger and Gilkes

(1996, 1997) calculated that much larger pH changes than those measured in the rhizosphere of plants supplied with rock phosphate would have been found if the dissolution of rock phosphate had not consumed the majority of the H^+ released by plants roots. More than the observed pH change, it is the actual amount of H^+ exchanged by the root that is ecologically relevant. Indeed, the involvement of those H^+ ions in chemical reactions occurring in the rhizosphere is of prime significance, due to their implications in the acquisition of many mineral nutrients such as P and Fe (e.g., Bertrand et al., 1999; Gahoonia et al., 1992; Grinsted et al., 1982; Hinsinger and Gilkes, 1996; Hinsinger et al., 2001; Riley and Barber, 1971; and reviews by Hinsinger, 1998, 2001a; Marschner, 1995; Römheld and Marschner, 1986) but also in the mobilisation/immobilisation of potentially toxic metals such as Al (e.g., Bakker et al., 1999; Calba et al., 1999; Göttlein et al., 1999; Nietfeld, 2001) and heavy metals (e.g., Chaignon et al., 2002; and reviews by Hinsinger, 2001b; McLaughlin et al., 1998).

Conclusions

The various origins of H^+ released by roots in their immediate vicinity, i.e., the rhizosphere, and the underlying physiological mechanisms implied are now well elucidated. However, the contribution of some of these processes such as respiration, exudation of organic anions and redox-coupled processes in the rhizosphere deserves being further clarified. In contrast, the implication of the balance of inorganic and organic ions taken up or excreted by roots is well understood. In higher plants, those H^+ play a major role in compensating the charges crossing root cells (charge-balanced H^+ exchanges) and exchanged with the root environment (Raven, 1991). The resulting H^+ fluxes depend upon a range of internal, regulatory processes and external constraints: cell elongation, mineral nutrition, respiration and exudation, redox activity, environmental stresses such as nutrient deficiencies and metal toxicities. Those fluxes vary considerably, both in direction and intensity, according to the physiological status of the plant and the location along the root. The complex combination of the many underlying processes and determining factors involved in root-mediated pH changes is such that the response of plant roots to environmental constraints in terms of H^+ fluxes is not always appropriate for overcoming adverse growth conditions. In some cases such as the response of strategy I plant species to Fe deficiency, H^+ release is an effective way to alleviate the constraint. It can therefore be considered as an adaptive response. However, in other cases such as the decreased nitrate uptake that occurs in response to Al toxicity, the resulting increase in H^+ release probably enhances the constraint instead of reducing it. In any case, those actions of the root and the related responses of the plant to environmental conditions are of prime ecological significance because of the direct implication of H^+ in most physico-chemical processes that occur in the soil. In addition, these root-mediated pH changes will not only interact directly with plant roots because soil pH is a major factor influencing the dynamic of growth and composition of rhizosphere microflora.

Acknowledgements

Philippe Hinsinger thanks Prof Hans Lambers for kindly inviting him to present this review at the Cluster Root Workshop in Perth and for his support to write up this paper. Prof Zed Rengel is also acknowledged for helpful discussions on an earlier version of the manuscript.

References

Ahmad A R and Nye P H 1990 Coupled diffusion and oxidation of ferrous iron in soils. I. Kinetics of oxygenation of ferrous iron in soil suspension. J. Soil Sci. 41, 395–409.

Ando T, Yoshida S and Nishiyama I 1983 Nature of oxidizing power of rice roots. Plant Soil 72, 57–71.

Anoua M, Jaillard B, Ruiz T, Bénet J C 1997 Couplages entre transferts de matière et réactions chimiques dans un sol. II. Application au transfert de matière dans la rhizosphère. Entropie 207, 13–24.

Armstrong W 1967 The use of polarography in the assay of oxygen diffusing from roots in anaerobic media. Physiol. Plant. 20, 540–553.

Asady G H and Smucker A J M 1989 Compaction and root modifications of soil aeration. Soil Sci. Soc. Am. J. 53, 251–254.

Bakker M R, Dieffenbach A and Ranger J 1999 Soil solution chemistry in the rhizosphere of roots of sessile oak (*Quercus petraea*) as influenced by lime. Plant Soil 209, 209–216.

Begg C B M, Kirk G J D, MacKenzie A F and Neue H-U 1994 Root-induced iron oxidation and pH changes in the lowland rice rhizosphere. New Phytol. 128, 469–477.

Bertrand I, Hinsinger P, Jaillard B and Arvieu J C 1999 Dynamics of phosphorus in the rhizosphere of maize and rape grown on synthetic, phosphated calcite and goethite. Plant Soil 211, 111–119.

Bienfait H F, Bino R J, van der Bliek A M, Duivenvoorden J F and Fontaine J M 1983 Characterization of ferric reducing activity in roots of Fe-deficient *Phaseolus vulgaris*. Physiol. Plant. 59, 196–202.

Bidel L P R, Renault P, Pages L and Riviere L M 2000 Mapping meristem respiration of *Prunus persica* (L.) Batsch seedlings: potential respiration of the meristems, O_2 diffusional constraints and combined effects on root growth. J. Exp. Bot. 51, 755–768.

Brown JC 1978 Mechanism of iron uptake by plants. Plant Cell Environ. 1, 249–257.

Calba H and Jaillard B 1997 Effect of aluminium on ion uptake and H^+ release by maize. New Phytol. 137, 607–616.

Calba H, Cazevieille P and Jaillard B 1999 Modelling of the dynamics of Al and protons in the rhizosphere of maize cultivated in acid substrate. Plant Soil 209, 57–69.

Chaignon V, Bedin F and Hinsinger P 2002 Copper bioavailability and rhizosphere pH changes as affected by nitrogen supply for tomato and oilseed rape cropped on an acidic and a calcareous soil. Plant Soil 243, 219–228

Chen C C, Dixon J B and Turner F T 1980 Iron coatings on rice roots: mineralogy and quantity influencing factors. Soil Sci. Soc. Am. J. 44, 635–639.

Dakora F D and Phillips D A 2002 Root exudates as mediators of mineral acquisition in low-nutrient environments. Plant Soil 245, 35–47.

Darrah P R 1993 The rhizosphere and plant nutrition: a quantitative approach. Plant Soil 155/156, 1–20.

Degenhardt J, Larsen P B, Howell S H and Kochian L V 1998 Aluminum resistance in the Arabidopsis mutant alr-104 is caused by an aluminum-induced increase in rhizosphere pH. Plant Physiol. 117, 19–27.

Dehérain P P 1902 Traité de Chimie Agricole. Masson, Paris, 969 pp.

Delhaize E, Ryan P R and Randall P J 1993 Aluminium tolerance in wheat (*Triticum aestivum* L.): II. Aluminium-stimulated excretion of malic acid from root apices. Plant Physiol. 103, 695–702.

Dinkelaker B, Römheld V and Marschner H 1989 Citric acid excretion and precipitation of calcium citrate in the rhizosphere of white lupin (*Lupinus albus* L.). Plant Cell Environ. 12, 285–292.

Dinkelaker B, Hahn G and Marschner H 1993 Non-destructive methods for demonstrating chemical changes in the rhizosphere. II. Application of methods. Plant Soil 155/156, 73–76.

Dinkelaker B, Hengeler C and Marschner H 1995 Distribution and function of proteoid roots and other root clusters. Bot. Acta 108, 183–200.

Durand R and Bellon N 1993 Comparaison d'une méthode de diffusion et d'une méthode géochimique pour estimer la production racinaire de CO_2. Sci. Sol 31, 93–107.

Durand R, Bellon N and Jaillard B 2001 Determining the net flux of charge released by maize roots by directly measuring variations of the alkalinity in the nutrient solution. Plant Soil 229, 305–318.

Fischer W R, Flessa H and Schaller G 1989 pH values and redox potentials in microsites of the rhizosphere Z. Pflanzenernaehr. Bodenkd. 152, 191–195.

Flessa H and Fischer W R 1992 Plant-induced changes in the redox potentials of rice rhizospheres. Plant Soil 143, 55–60.

Foy C D 1988 Plant adaptation to acid, aluminium toxic soils. Commun. Soil Sci. Plant Anal. 19, 959–987.

Gahoonia T S, Claassen N and Jungk A 1992 Mobilization of phosphate in different soils by ryegrass supplied with ammonium or nitrate. Plant Soil 140, 241–248.

Gardner W K, Parberry D G and Barber D A 1982 The acquisition of phosphorus by *Lupinus albus* L. I. Some characteristics of the soil/root interface. Plant Soil 68, 19–32.

Gardner W K, Barber D A and Parberry D G 1983 The acquisition of phosphorus by *Lupinus albus* L. III.The probable mechanism by which phosphorus movement in the soil/root interface is enhanced. Plant Soil 70, 107–124.

Gerendás J and Schurr U 1999 Physicochemical aspects of ion relations and pH regulation in plants — a quantitative approach. J. Exp. Bot. 50, 1101–1114.

Gollany H T, Schumacher T E, Rue R R and Liu S-Y 1993 A carbon dioxide microelectrode for *in situ* pCO_2 measurement. Microchem. J. 48, 42–49.

Göttlein A, Heim A and Matzner E 1999 Mobilization of aluminium in the rhizosphere soil solution of growing tree roots in an acidic soil. Plant Soil 211, 41–49.

Gras F 1974 Une méthode expérimentale pour la détermination du pH et de la solubilité du carbonate de calcium en sols calcaires. Bull. Assoc. Fr. Etude Sol 1, 33–48.

Green M S and Etherington J R 1977 Oxidation of ferrous iron by rice (*Oryza sativa* L.) roots: a mechanism for waterlogging tolerance? J. Exp. Bot. 28, 678–690.

Gregory P J and Hinsinger P 1999 New approaches to studying chemical and physical changes in the rhizosphere: an overview. Plant Soil 211, 1–9.

Grierson P F 1992 Organic acids in the rhizosphere of *Banksia integrifolia* L. f. Plant Soil 144, 259–265.

Grierson P F and Attiwill P M 1989 Chemical characteristics of the proteoid root mat of *Banksia integrifolia* L. Austr. J. Bot. 37, 137–143.

Grinsted M J, Hedley M J, White R E and Nye P H 1982 Plant-induced changes in the rhizosphere of rape (*Brassica napus* var. Emerald). I. pH changes and the increase in P concentration in the soil solution. New Phytol. 91, 19–29.

Guivarch A, Hinsinger P and Staunton S 1999 Root uptake and distribution of radiocaesium from contaminated soils and the enhancement of Cs adsorption in the rhizosphere. Plant Soil 211, 131–138.

Häussling M, Leisen E, Marschner H and Römheld V 1985 An improved method for non-destructive measurements of the pH at the root–soil interface (rhizosphere). J. Plant Physiol. 117, 371–375.

Haynes R J 1990 Active ion uptake and maintenance of cation–anion balance: A critical examination of their role in regulating rhizosphere pH. Plant Soil 126, 247–264.

Hedley M J, Nye P H and White R E 1982 Plant-induced changes in the rhizosphere of rape (*Brassica napus* var. Emerald) seedlings. II. Origin of the pH change. New Phytol. 91, 31–44.

Helal H M and Sauerbeck D 1989 Carbon turnover in the rhizosphere. Z. Pflanzenernaehr. Bodenkd. 152, 211–216.

Hiatt A J 1967 Relationship of cell pH to organic acid change during ion uptake. Plant Physiol. 42, 294–298.

Hiltner L 1904 Über neuere Ehrfahrungen und Problem auf dem Gebiet der Bodenbakteriologie unter besonderer Berücksichtigung der Grundüngung und Brache. Arb. Dtsch. Landwirt. Ges. 98, 59–78.

Hinsinger P 1998 How do plant roots acquire mineral nutrients? Chemical processes involved in the rhizosphere. Adv. Agron. 64, 225–265.

Hinsinger P 2001a Bioavailability of soil inorganic P in the rhizosphere as affected by root-induced chemical changes: a review. Plant Soil 237, 173–195.

Hinsinger P 2001b Bioavailability of trace elements as related to root-induced chemical changes in the rhizosphere. *In* Trace Elements in the Rhizosphere. Eds G R Gobran, W W Wenzel and E Lombi. CRC Press LCC, Boca Raton, FL, pp. 25–41.

Hinsinger P and Gilkes R J 1996 Mobilization of phosphate from phosphate rock and alumina-sorbed phosphate by the roots of ryegrass and clover as related to rhizosphere pH. Eur. J. Soil Sci. 47, 533–544.

Hinsinger P and Gilkes R J 1997 Dissolution of phosphate rock in the rhizosphere of five plant species grown in an acid, P-fixing mineral substrate. Geoderma 75, 231–249.

Hinsinger P, Fernandes Barros O N, Benedetti M F, Noack Y and Callot G 2001 Plant-induced weathering of a basaltic rock: experimental evidence. Geochim. Cosmochim. Acta 65, 137–152.

Hoffland E, Findenegg G R and Nelemans J A 1989 Solubilization of rock phosphate by rape. II. Local root exudation of organic acids as a response to P-starvation. Plant Soil 113, 161–165.

Imas P, Bar-Yosef B, Kafkafi U and Ganmore-Neumann R 1997a Release of carboxylic anions and protons by tomato roots in response to ammonium nitrate ratio and pH in nutrient solution. Plant Soil 191, 27–34.

Imas P, Bar-Yosef B, Kafkafi U and Ganmore-Neumann R 1997b Phosphate induced carboxylate and proton release by tomato roots. Plant Soil 191, 35–39.

Jaillard B and Hinsinger P 1993 Alimentation minérale des végétaux dans le sol. *In* Techniques Agricoles, n. 87 (6-93), fasc. 1210, pp. 1–13. Editions Techniques, Paris.

Jaillard B, Plassard C and Hinsinger P 2002 Measurements of H^+ fluxes and concentrations in the rhizosphere. *In* Handbook of Soil Acidity. Ed. Z Rengel. Marcel Dekker, New York (in press).

Jarvis S C, Robson A D 1983. The effects of nitrogen nutrition of plants on the development of acidity in Western Australian soils. I. Effects with subterranean clover grown under leaching conditions. Aust. J. Agric. Res. 34, 341–353.

Johnson J F, Allan D L and Vance C P 1994 Phosphorus stress induced proteoid roots show altered metabolism in *Lupinus albus*. Plant Physiol. 104, 657–665.

Johnson J F, Allan D L, Vance C P and Weiblen G 1996 Root carbon dioxide fixation by phosphorus-deficient *Lupinus albus*: Contribution to organic acid exudation by proteoid roots. Plant Physiol. 112, 19–30.

Jones D L 1998 Organic acids in the rhizosphere — A critical review. Plant Soil 205, 25–44.

Jones D L and Brassington D S 1998 Sorption of organic acids in acid soils and its implications in the rhizosphere. Eur. J. Soil Sci. 49, 447–455.

Jones D L and Darrah P R 1994a Amino-acid influx and efflux at the soil-root interface of *Zea mays* L. and its implications in the rhizosphere. Plant Soil 163, 1–12.

Jones D L and Darrah P R 1994b Role of root derived organic acids in the mobilization of nutrients from the rhizosphere. Plant Soil 166, 247–257.

Jones D L, Dennis P G, Owen A G and van Hees P A W 2002 Organic acid behaviour in soils — Misconceptions and knowledge gaps. Plant Soil 248, 31–41.

Keerthisinghe G, Hocking J P, Ryan P R and Delhaize E 1998 Effect of phosphorus supply on the formation and function of proteoid roots of white lupin (*Lupinus albus* L.). Plant Cell Environ. 21, 467–478.

Kirk G J D and Le Van Du 1997 Changes in rice root architecture, porosity, and oxygen and proton release under phosphorus deficiency. New Phytol. 135, 191–200.

Kirkby E A and Mengel K 1967 Ionic balance in different tissues of the tomato plants in relation to nitrate, urea and ammonium nutrition. Plant Physiol. 42, 6–14.

Knight W G, Allen M F, Jurinak J J and Dudley L M 1989 Elevated carbon dioxide and solution phosphorus in soil with vesicular-arbuscular mycorrhizal western wheatgrass. Soil Sci. Soc. Am. J. 53, 1075–1082.

Lambers H, Scheurwater I and Atkin O K 1996 Respiratory patterns in roots in relation to their functioning. *In* Plant Roots. The Hidden Half. 2nd ed. Eds Y Waisel, A Eshel and U Kafkafi. pp. 529–556. Marcel Dekker, New York.

Le Bot J, Alloush G A, Kirkby E A and Sanders F E 1990 Mineral nutrition of chickpea plants supplied with NO_3 or NH_4-N. II. Ionic balance in relation to phosphorus stress. J. Plant Nutr. 13, 1591–1605.

Lindsay W L 1979 Chemical Equilibria in Soils. John Wiley & Sons, New York. 449 pp.

Marschner H 1995 Mineral Nutrition of Higher Plants. 2nd ed. Academic Press, London. 889 pp.

Marschner H and Römheld V 1983 *In vivo* measurement of root-induced pH changes at the soil–root interface: effect of plant species and nitrogen source. Z. Pflanzenernaehr. Bodenkd. 111, 241–251.

Marschner H and Römheld V 1994 Strategies of plants for acquisition of iron. Plant Soil 165, 261–274.

Marschner H, Römheld V and Ossenberg-Neuhaus H 1982 Rapid method for measuring changes in pH and reducing processes along roots of intact plants. Z. Pflanzenphysiol. 105, 407–416.

Marschner H, Römheld V, Horst W J and Martin P 1986 Root-induced changes in the rhizosphere: importance for the mineral nutrition of plants. Z. Pflanzenernaehr. Bodenkd. 149, 441–456.

McLaughlin M J, Smolders E and Merckx R 1998 Soil-root interface: Physicochemical processes. *In* Soil Chemistry and Ecosystem Health, Special Publication no 52. pp. 233–277. Soil Science Society of America, Madison, WI.

McLay C D A, Barton L and Tang C 1997 Acidification potential of ten grain legume species grown in nutrient solution. Aust. J. Agric. Res. 48, 1025–1032.

Meharg A A 1994 A critical review of labelling techniques used to quantify rhizosphere carbon-flow. Plant Soil 166, 55–62.

Mengel K, Kirkby E A, Kosegarten H and Appel T 2001 Principles of Plant Nutrition, 5th Ed. Kluwer Academic Publishers, Dordrecht. 864 pp.

Merckx R, van Ginkel J H, Sinnaeve J and Cremers A 1986 Plant-induced changes in the rhizosphere of maize and wheat. I. Production and turnover of root-derived material in the rhizosphere of maize and wheat. Plant Soil 96, 85–93.

Moorby H, White R E and Nye P H 1988 The influence of phosphate nutrition on H ion efflux from the roots of young rape plants. Plant Soil 105, 247–256.

Nietfeld H W F 2001 Modeling the dynamics of the rhizosphere chemistry in acid forest soils. *In* Trace Elements in the Rhizosphere. Eds G R Gobran, W W Wenzel and E Lombi. pp. 25–41. CRC Press LCC, Boca Raton, FL.

Neumann G and Römheld V 1999 Root excretion of carboxylic acids and protons in phosphorus-deficient plants. Plant Soil 211, 121–130.

Neumann G, Massonneau A, Martinoia E and Römheld V 1999 Physiological adaptations to phosphorus deficiency during proteoid root development in white lupin. Planta 208, 373–382.

Neumann G, Massonneau A, Langlade N, Dinkelaker B, Hengeler C, Römheld V and Martinoia E 2000 Physiological aspects of cluster root function and development in phosphorus-deficient white lupin (*Lupinus albus* L.). Ann. Bot. 85, 909–919.

Nguyen C, Todorovic C, Robin C, Christophe A and Guckert A 1999 Continuous monitoring of rhizosphere respiration after labelling of plant shoot with $^{14}CO_2$. Plant Soil 212, 191–201.

Norstadt F A and Porter L K 1984 Soil gases and temperatures: a beef cattle feedlot compared to alfalfa. Soil Sci. Soc. Am. J. 48, 783–789.

Nye P H 1981 Changes of pH across the rhizosphere induced by roots. Plant Soil 61, 7–26.

Nye P H 1986 Acid–base changes in the rhizosphere. Adv. Plant Nutr. 2, 129–153.

Ohwaki Y and Sugahara P 1997 Active extrusion of protons and exudation of carboxylic acids in response to iron deficiency by roots of chickpea (*Cicer arietinum* L.). Plant Soil 189, 49–55.

Parker D R and Pedler J F 1998 Probing the 'malate hypothesis' of differential aluminium tolerance in wheat by using rhizotoxic ions as proxies for Al. Planta 205, 389–396.

Petersen W and Böttger M 1991 Contribution of organic acids to the acidification of the rhizosphere of maize seedlings. Plant Soil 132, 159–163.

Plassard C, Meslem M, Souche G and Jaillard B 1999 Localization and quantification of net fluxes of H^+ along roots of maize by combined use of videodensitometry of indicator dye and ion-selective microelectrodes. Plant Soil 211, 29–39.

Raven J A 1986 Biochemical disposal of excess H^+ in growing plants? New Phytol. 104, 175–206.

Raven J A 1991 Terrestrial rhizophytes and H^+ currents circulating over at least a millimetre: an obligate relationship ? New Phytol 117, 177–185.

Riley D and Barber S A 1969 Bicarbonate accumulation and pH changes at the soybean (*Glycine max* (L.) Merr.) root–soil interface. Soil Sci. Soc. Am. Proc. 33, 905–908.

Riley D and Barber S A 1971 Effect of ammonium and nitrate fertilization on phosphorus uptake as related to root-induced pH changes at the root–soil interface. Soil Sci. Soc. Am. Proc. 35, 301–306.

Roelofs R F R, Rengel Z, Cawthray G R, Dixon K W and Lambers H 2001 Exudation of carboxylates in Australian Proteaceae: Chemical composition. Plant Cell Environ. 24, 891–904.

Römheld V 1986 pH-Veränderungen in der Rhizosphäre verschiedener Kulturpflanzenarten in Abhängigkeit vom Nährstoffangebot. Potash Rev. 55, 1–8.

Römheld V and Kramer D 1983 Relationship between proton efflux and rhizodermal transfer cells induced by iron deficiency. Z. Pflanzenphysiol. 113, 73–83.

Römheld V and Marschner H 1981 Iron deficiency stress induced morphological and physiological changes in root tips on sunflower. Physiol. Plant. 53, 354–360.

Römheld V and Marschner H 1986 Mobilization of iron in the rhizosphere of different plant species. Adv. Plant Nutr. 2, 155–204.

Römheld V, Müller Ch and Marschner H 1984 Localization and capacity of proton pumps in roots of intact sunflower plants. Plant Physiol. 76, 603–606.

Ruiz L 1992 Mobilisation du Phosphore des Apatites dans la Rhizosphère. Rôle de l'Excrétion de Protons par les Racines. PhD Thesis, Montpellier II University, 125 pp.

Rufyikiri G, Dufey J E, Nootens D and Delvaux B 2001 Effect of aluminium on bananas (*Musa* spp.) cultivated in acid solutions. II. Water and nutrient uptake. Fruits 56, 5–16.

Rufty T W, MacKown C T, Lazof D B, Carter T E 1995 Effects of aluminium on nitrate uptake and assimilation. Plant Cell Environ. 18, 1325–1331.

Ryan P R, Delhaize E and Randall P J 1995 Characterisation of Al-stimulated efflux of malate from the apices of Al-tolerant wheat roots. Planta 196, 103–110.

Sas L, Rengel Z and Tang C 2001 Root morphology, excess cation uptake, and extrusion of proton and organic acid anions in *Lupinus albus* L. under phosphorus deficiency. Plant Sci. 160, 1191–1198.

Schaller G 1987 pH changes in the rhizosphere in relation to the pH-buffering of soils. Plant Soil 97, 444–449.

Schubert S, Schubert E and, Mengel K 1990 Effect of low pH of the root medium on proton release, growth, and nutrient uptake of field beans (*Vicia faba*). Plant Soil 124, 239–244.

Shane M W, De Roock S, De Vos M, Cawthray G R and Lambers H 2002 Internal phosphorus concentration modifies the initiation, growth and functioning of cluster roots in *Hakea prostrata*. Plant Soil 248, 209–219.

Sijmons P C, Briel W van den and Bienfait H F 1984 Cytosolic NADPH is the electron donor for extracellular Fe^{III} reduction in iron-deficient bean roots. Plant Physiol. 75, 219–221.

Skene K R, Kierans M, Sprent J I and Raven J A 1996 Structural aspects of cluster root development and their possible significance for nutrient acquisition in *Grevillea robusta* (Proteaceae). Ann. Bot. 77, 443–451.

Sposito G 1989 The Chemistry of Soils. Oxford University Press, New York. 277 p.

Tang C and Rengel Z 2002 Role of plant cation/anion uptake ratio in soil acidification. *In* Handbook of Soil Acidity. Ed. Z Rengel. Marcel Dekker, New York (in press).

Tang C, McLay C D A and Barton L 1997 A comparison of proton excretion of twelve pasture legumes grown in nutrient solution. Aust. J. Exp. Agric. 37, 563–570.

Tang C, Hinsinger P, Drevon J J and Jaillard B 2001a Phosphorus deficiency impairs early nodule functioning and enhances proton release in roots of *Medicago truncatula* L. Ann. Bot. 88, 131–138.

Tang C, Hinsinger P, Jaillard B, Rengel Z and Drevon J J 2001b Effect of phosphorus deficiency on the growth, symbiotic N_2 fixation and proton release by two bean (*Phaseolus vulgaris*) genotypes. Agronomie 21, 683–689.

Trolldenier G 1987 Rhizosphere organisms — Potassium interactions with emphasis on methodology. *In* Methodology in Soil-K Research. Proceedings of the 20th Colloquium of the International Potash Institute (I.P.I.), Baden bei Wien, Austria. Ed. I.P.I. pp. 283–297. I.P.I., Bern.

Trolldenier G 1988 Visualization of oxidizing power of rice roots and of possible participation of bacteria in iron deposition Z. Pflanzenernaehr. Bodenkd. 151, 117–121.

Van Breemen N 1987 Effects of redox processes on soil acidity. Neth. J. Agric. Sci. 35, 271–279.

Westhuizen M M van der and Cramer M D 1998 The influence of elevated rhizosphere dissolved inorganic carbon concentrations on respiratory O_2 and CO_2 flux in tomato plants. J. Exp. Bot. 49,1977–1985.

Weinberger P and Yee D 1984 The influence of nitrogen sources on root-mediated changes in substrate pH. Can. J. Bot. 62, 161–162.

Yan F, Schubert S and Mengel K 1992 Effect of low root medium pH on net proton release, root respiration, and root growth of corn (*Zea mays* L.) and broad bean (*Vicia faba* L.). Plant Physiol. 99, 415–421.

Yan F, Schubert S and Mengel K 1996 Soil pH increase due to biological decarboxylation of organic anions. Soil Biol. Biochem. 28, 17–24.

Youssef R A and Chino M 1989 Root-induced changes in the rhizosphere of plants. I. pH changes in relation to the bulk soil. Soil Sci. Plant Nutr. 35, 461–468.

Yu Q, Tang C and Kuo J 2000 A critical review on methods to measure apoplastic pH in plants. Plant Soil 219, 29–40.

Yu Q, Tang C and Kuo J 2001 Apoplastic pH change in roots of *Lupinus angustifolius* L. in response to pH>6. *In* Plant Nutrition: Food Security and Sustainability of Agro-ecosystems through Basic and Applied Research. Eds. W J Horst et al. pp. 242–243. Kluwer Academic Publishers, Dordrecht.

metabolism, is important in the control of efflux (Ryan et al., 2001; Watt and Evans, 1999).

Regardless of these issues, it is clear that the transport of these organic anions across the root-cell membranes is a critical step in the exudation process. Since the pH of the cell cytosol is near-neutral, organic acids will most likely leave the plant cells as anions and reports using the patch-clamp technique have now demonstrated that Al can activate anion-permeable channels in the root cells of wheat and maize (Kollmeier et al., 2001; Piñeros and Kochian, 2000; Ryan et al., 1997; Zhang et al., 2001). These channels allow the thermodynamically passive flow of malate and citrate anions from the cytosolic side of the membrane to the apoplasm. While the physiology of this process is being revealed, the transport proteins involved, or the genes that encode them, are still unknown. In this paper we describe several strategies used in our laboratory and elsewhere, aimed at identifying the genes and proteins involved in regulating the release of organic anions across the plasma membrane of root cells.

Choosing a strategy: knowing some biology helps

Here we refer to the 'target genes' as those that encode proteins that facilitate the efflux of organic anions such as malate, citrate and oxalate. The strategies chosen to identify these genes will depend on the biology of the plant species used. For instance, it is helpful to know whether organic anion efflux from a particular species requires the induction of specific genes because this can influence the experimental approach. The rapid activation by Al of malate release from wheat roots suggests that gene induction is not required (Delhaize et al., 1993; Ryan et al., 1995). By contrast, in rye, triticale and *Cassia tora* a lag of several hours occurs between the application of Al and the time at which maximum citrate and malate efflux occurs (Li et al., 2000; Ma et al., 2000) suggesting that the induction of specific genes is a prerequisite for efflux (Ma et al., 2001). Therefore, a strategy that identifies genes that are induced by Al treatment would be more sensible for rye than for wheat. It is also useful to know whether organic anion release is localized to certain zones of the root. For instance, malate efflux from Al-resistant wheat and citrate efflux from P-deficient *Brassica napus* are restricted to regions at or near the root apicies (Hoffland et al., 1989; Ryan et al., 1995). A strategy that focuses on these specific tissues will increase the likelihood of success.

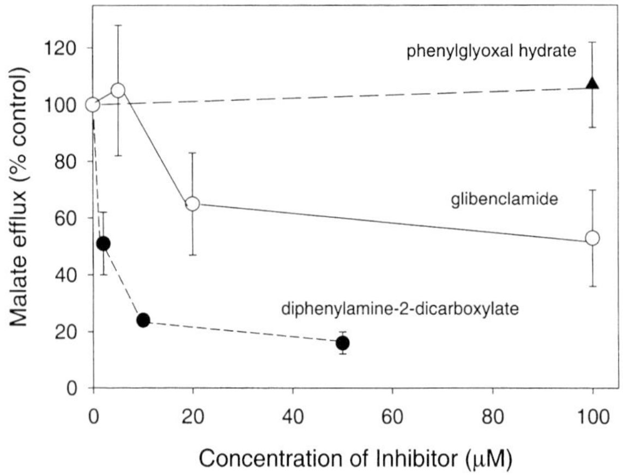

Figure 1. Inhibition of the Al-activated malate efflux from wheat roots by three different antagonists. Root apices from the wheat line ET8 were collected as described previously and treated with 100 μM $AlCl_3$ in 0.2 mM $CaCl_2$ (pH 4.2) in the presence or absence of a antagonist (Ryan et al., 1995). Stock solutions for the antagonists were prepared as 20 mM in deionised water except for glibenclamide which was dissolved in DMSO.

Physiological studies can also provide clues to the identity of target genes by revealing similarities with other systems that are better characterised. For example, evidence that anion channels were involved in the release of malate from Al-treated wheat roots arose from the use of anion-channel antagonists (Ryan et al., 1995). We show here that malate efflux is inhibited by additional antagonists including diphenylamine-2-dicarboxylic acid (DPC) and glibenclamide but not by phenylglyoxal hydrate (Figure 1). These results parallel the pharmacology described for the slow anion channel (S-type) in guard cells (Leonhardt et al., 1999; Schroeder, 1995; Schroeder et al., 1993). Although the identity of the S-type channel remains unknown, Leonhardt et al (1999) have proposed that it is either a member of the ATP-binding cassette (ABC) family of proteins or its activity is regulated by an ABC protein. Both the S-type channel and the Al-activated anion channel in wheat are inhibited by niflumic acid, NPPB, IAA-94, DPC, A-9-C, and glibenclamide but are unaffected by the stilbene derivative DIDS. The activity of the S-type anion channel is increased by abscisic acid and by a rise in cytosolic Ca^{2+} concentration. Malate efflux from wheat roots, by contrast, is not activated by either ABA or the Ca^{2+} ionophore A23187 (data not shown). Therefore, although these channels appear to be activated differently, the similarities in aspects of their pharmacology raise the possibility that they are members of the same protein

Table 1. Diversity of proteins that are known to function as anion channels. Amino acid sequences were compared by aligning each against the *Arabidopsis* chloride channel protein CLC-a with the Genestream program (http://www2.igh.cnrs.fr/bin/align-guess.cg i). Tubulin is included as a control sequence since it is a structural protein and not involved in ion transport

Protein	Length (aa)	% similarity to CLC-a	Accession
CLC-a (*Arabidopsis*)	775	100	AAC05742
CLC-d (*Arabidopsis*)	780	88	AAL32596
Cl^- channel (bacterial)	425	16	AAD53928
CFTR (human)	1480	16	P13569
VDAC (yeast)	281	11	P40478
AQP6 (rat)	276	12	AFO83879/AAD29856
NCC27 (human)	241	10	O00299
Tubulin (human)	444	14	XP_004389

family. Organic anion efflux in other species is also sensitive to a variety of inhibitors and these results can provide clues to the identity of the proteins involved in those systems as well. For example, A-9-C and ethacrynic acid inhibit citrate efflux from the cluster roots of white lupin (Neumann and Römheld, 1999) while DIDS and niflumic acid inhibit the Al-dependent efflux of organic anions from maize roots (Kollmeier et al., 2001). By contrast the Al-activated efflux of oxalate in buckwheat was stimulated by niflumic acid but inhibited by phenylglyoxal hydrate, a general antagonist of anion transport (Zheng et al., 1998). However, many of these inhibitors are not as specific as first believed and we now know that anion channel antagonists such as NPPB and niflumic acid can also inhibit cation channels (Garrill et al., 1996). Therefore definitive conclusions about transport mechanisms are difficult when inhibitors provide the primary source of information.

While these types of studies are useful they cannot reveal the identity of the target transporter genes with any certainty. Even the strong indication that anion channels are involved in malate and citrate efflux from wheat and maize is not necessarily helpful in finding the genes involved because these types of proteins show considerable variation in their sequence and structure. To emphasise this, Table 1 compares the alignment of a chloride channel from *Arabidopsis*, CLC-a, with several other proteins that are also known to function as anion channels. We see that CLC-a is equally similar to tubulin, a structural protein, as it is to many of the other anion channel proteins (Barbier-Brygoo et al., 2000; Dulhunty et al., 2001; Ryan et al., 2001; Yasui et al., 1999). Because of this diversity, simply looking for sequence homologies to known anion channels is unlikely to be a successful strategy. It is also possible that channel-mediated transport is only one of several routes contributing to organic anion efflux. Therefore different types of analyses or screens are necessary to find the target genes. Genes selected in these screens become candidates if they show homology to known transport proteins or if they encode proteins with trans-membrane domains. Several Web sites are available that predict trans-membrane regions in proteins

(eg. http://www.enzim.hu/hmmtop/; http://sosui.proteome.bio.tuat.ac.jp/sosuiframe0.html http://www.ch.embnet.org/software/TMPRED_form.html; http://www.cbs.dtu.dk/services/TMHMM-2.0/).

Yet even defining membrane spanning regions is not straightforward because the algorithms used in these programs are imperfect. For example, three of the four Web sites provided above predict an anion channel protein from the CLIC family, NCC27, to be a soluble protein (Dulhunty et al., 2001). As many different algorithms as possible should be tried and the predictions from all treated with some caution.

Expression cloning in yeast

The use of baker's yeast, *Saccharomyces cerevisiae*, as a heterologous expression system has proved to be a successful strategy for isolating genes that encode transporters. Yeast is well suited for the functional ex-

Table 2. Plant genes that confer Al tolerance when expressed in yeast. cDNA libraries were prepared from plant tissues and cloned in to the pYES2 expression vector as described previously (Delhaize et al., 1999). *Saccharomyces cerevisiae* strain INVSc2 (Invitrogen, MAT**a** *his3-Δ1 ura3-52*) was transformed with the library and spread on agar plates containing concentrations of Al sufficient to prevent growth of control cells containing an empty vector. Plant cDNAs conferring Al tolerance were selected and likely functions were obtained from sequence similarity using the BLASTX algorithm. The table shows a selection of these cDNAs with their corresponding GeneBank numbers

Source of cDNA Library	Clones conferring Al tolerance to yeast	GeneBank number
Triticum aestivum (roots)	blue copper-binding	AF031195
	expansin	U91981
	phosphatidylserine synthase	U91983
	lipid desaturase	AF031194
	contains a FYVE Zn-finger domain	AF004816
	glycerophosphoryl diester phosphodiesterase	AF031231
Stylosanthes hamata (roots)	2 × EREB transcription factors	U91982, U91857
	GTP dissociation inhibitor	
	7 × no homology	

pression of plant transporter genes because, like plant cells, yeast is eukaryotic, it contains vacuoles and a cell wall, and it has a large negative electrical potential difference across the plasma membrane. Furthermore, many thousands of yeast cells can be screened quickly and easily on a single agar plate. Indeed, this approach was used to identify the first K^+ channels from plants almost 10 years ago (Anderson et al., 1992; Sentenac et al., 1992). In those studies, a yeast strain was generated by mutating all of the endogenous K^+ transporter genes. The mutant strain was unable to take up K^+ and therefore incapable of growing on low concentrations of K^+. These cells were then transformed with cDNA libraries in a yeast-expression vector prepared from plant tissue (wheat and *Arabidopsis*) and spread on agar plates containing low concentrations of K^+. Each yeast cell contained a plasmid with a single plant cDNA. The expression of any plant genes, which allowed the yeast to acquire K^+ from the media (ie to complement the mutant) grew to form a colony. These colonies were picked off the plate and the plant gene providing the K^+ transport capacity was sequenced and identified by homology with other genes.

This scheme is straightforward when cloning transporters involved in the uptake of essential nutrients as long as the appropriate yeast mutants are available or can be generated. However, this strategy is not easily applied to the cloning of transporters involved in the efflux of nutrients or metabolites. The major challenge in this situation is to develop an appropriate screen. In our case, the screen should ideally select for an increase in organic anion efflux from yeast cells. We have used expression cloning for several years in an attempt to identify genes that confer Al tolerance to yeast and in particular genes that encode organic anion transporters. We were unable to use yeast mutants for a complementation approach (described above) because mutants defective in citrate or malate efflux have not yet been identified. Instead we transformed wild-type yeast with cDNA libraries from plant tissues and spread them on agar plates containing Al at concentrations which prevented the growth of wild-type cells. This screen was chosen because the efflux of organic anions from roots has been implicated in the Al tolerance of many plants (see above) and organic anions can also protect yeast from Al toxicity (MacDiarmid and Gardner, 1996). Using this approach we have identified over 20 plant genes that confer Al tolerance to yeast. A selection of these cDNAs from two libraries is shown in Table 2. Approximately half of the putative proteins encoded by these genes are similar to proteins with known functions. One gene encoding a phosphatidylserine synthase has been expressed in plants and characterized in detail (Delhaize et al., 1999). To date none of the genes isolated in these screens have been confirmed to be organic anion channels or transporters and none appear to increase malate or citrate release from yeast. This indicates that they have increased the Al tolerance in the yeast cells by other pathways. Using a similar approach MacDiarmid and Gardner (1998) screened a yeast genomic library and isolated two genes that increased the Al tolerance of yeast. Characterisation of these genes indicated that they encode transporters that are permeable to magnesium and other divalent cations. Although organic anion transporters were not identified, this study does demonstrate that the heterologous expression of plant cDNAs in yeast, combined with a screen based on Al tolerance, can identify novel membrane proteins.

Heterologous expression systems have some disadvantages that limit their usefulness. For instance, target genes are unlikely to be selected from a screen if they need to interact with other plant proteins to function normally or if the target protein is comprised of different subunits. Furthermore, if the target cDNA is large (> 3 kb) then it might not be represented in the library. This risk, at least, can be minimized by fractionating the cDNA to enrich the library with larger cDNAs.

Expression cloning with Xenopus *oocytes*

Oocytes of the African clawed frog, *Xenopus laevis*, represent another heterologous expression system that has been successfully used to clone novel transporters. Typically, mRNA is isolated from tissue expressing the transporter of interest and injected into the oocytes. After 2–4 days a novel transport activity is measured on the oocytes using an assay that is specifically developed to identify the function of the target protein. The assay might use radio-nucleotides to estimate uptake of a substrate or it could use the voltage-clamp technique to measure net current across the oocyte membrane. Once the transport activity is detected the mRNA can be size-fractionated and the fraction conferring the greatest activity is used to prepare a cDNA library. The cDNA library is divided into pools and used to prepare cRNA. Each pool of cRNA is injected into additional oocytes to determine which ones induce the novel transport activity. The selected pools of cDNA are sub-divided further and analysed in the same way. This continues with progressively smaller fractions of cDNA until the clone causing the transport activity can be isolated. The gene can then be sequenced and characterised. This approach has been used to isolate many animal membrane receptors and ion channels including voltage-dependent anion channels (Jentsch et al., 1990) and ligand-gated anion channels (Pena et al., 1992). However, to date no plant membrane transporters have been isolated with this approach.

Transcript analysis

Another strategy to identify the target genes is to analyse the genes expressed in plant tissues that are actively releasing organic anions. One option is to prepare a cDNA library from a suitable tissue and to sequence random clones. However, there are likely to be thousands of genes expressed in normal active tissues and trying to isolate the target sequence against such a high background would be difficult, especially when the identity of the target is unknown. The effectiveness of this method can be improved by focusing only on those genes whose expression is increased during organic anion release. This can be achieved by several techniques including differential display or the microarray technology which analyses thousands of transcripts simultaneously. Massonneau et al. (2001) recently applied a technique based on an amplified fragment length polymorphism (AFLP) method to identify genes specifically expressed in cluster roots of white lupin. Several genes induced by Al treatment have also been isolated by Gardner and colleagues using differential hybridisation in tobacco, wheat and *Arabidopsis* (Ezaki et al., 1995, 1996; Richards et al., 1998; Snowden and Gardner, 1993; Snowden et al., 1995). These studies were aimed at isolating any genes that were induced by Al treatment and not just those that encode transporters. We now realize that this strategy is unlikely to isolate the target transport genes in wheat and tobacco because increased gene transcription is unlikely to be required for the Al-dependent efflux of organic anion anions in these species.

Subtraction techniques can also be used to enrich a library for cDNAs associated with organic anion efflux. In this technique, two cDNA populations are prepared from tissues which, ideally, are identical except that organic anion efflux occurs from one and not the other. We assume that the transcript for the target genes is specific to, or more highly expressed in, the tissue exuding organic anions. Subsequent procedures allow all cDNAs common to both pools to be subtracted away leaving only the cDNAs that are involved with efflux. This PCR-based procedure has the added advantage of increasing the representation

Table 3. Identity of some of the clones selected from a cDNA subtration library prepared from cluster roots of P-deficient white lupin. The library was prepared with the Clontech PCR-Select^{TM} cDNA subtraction kit using the tissues discussed in the text. Randomly selected clones were sequenced and analysed with BLASTX. The Table shows the frequency with which these clones appeared in the first 250 sequences

Frequency	Possible identities of genes identified from BLASTX analysis
60	Sequences with no similarities to known proteins (BLASTX <50)
40	Similar to *Arabidopsis* proteins of unknown function
31	Several different acid phosphatases (secretory, purple etc)
11	Monogalactosyldiacylglycerol synthase
7	PEP carboxylase
7	Glyceraldehyde 3-P dehydrogenase
6	Malate synthase (glyoxysomal)
4	B12D protein
3	Catalase
3	Chitinase+endochitinase
2	Acyl-CoA thioesterase
2	Ca^{2+}-transporting plasmalemma ATPase
2	K^+ transporter (high affinity)
2	Enolase
2	Cysteine protinease
2	Hypothetical 12.1 kDa protein
2	MAP kinase
2	Carbonic anhydrase
2	Basic blue copper protein

of transcripts that are expressed at very low levels. The choice of tissues used for the subtraction is critical because subtractions that ignore the biology of the system can remove the target genes from the final library.

We prepared a cDNA subtraction library from the cluster roots of white lupin. cDNA was prepared from cluster roots collected at various developmental stages mindful that the expression of the target genes may precede organic anion release. cDNA was also prepared from root apices of P-replete lupin plants. cDNA common to both pools were removed and those that remained were ligated into a cloning vector. There were two reasons for this choice of tissues: root apices were chosen as a means of removing genes involved with root growth from both pools and P-replete plants were chosen as a precautionary measure because it is not known whether increases in organic anion release also occur from the apices of lateral roots during P-deficiency. The putative identity of some of the clones in the resulting library is shown in Table 3. Most genes show no homology to known proteins or were similar to proteins with unknown function. Many of the other cDNAs encode proteins that are consistent with the symptoms of P-deficiency such as phosphatases or enzymes involved in glycolysis and organic anion metabolism. As yet, none have been found that encode for anion channels or transporters.

Protein-based strategies

Another strategy to identify the target genes is to isolate the proteins that they encode. This approach is valuable when expression of the target genes change through time or show differences between tissues or genotypes. Proteins extracted from tissues that release organic anions are separated on the basis of size and charge with two-dimensional polyacrylamide gels and compared with the proteins present in similar tissues that do not release organic anions. For instance, protein profiles can be compared in plant tissues before and after they are treated to induce organic anion release (e.g., Al treatment of rye roots, or P deficiency in white lupin) or they can be compared in different genotypes where one shows organic anion release and the other does not. Once again, the assumption is that proteins which are specific to, or appear more abundantly in, the actively exuding tissue are involved with that process. Candidate proteins can then be sequenced and investigated further.

Several attempts have been made to identify proteins associated with Al tolerance in wheat (Basu et al., 1994; Delhaize et al., 1991; Putterill et al., 1991). For instance, Delhaize (1991) isolated proteins from the root apices of Al-sensitive and Al-tolerant wheat cultivars and separated them with two-dimensional polyacrylamide gels. Eleven proteins were found to be specific to the tolerant cultivar but none were subsequently found to co-segregate with Al tolerance in a set of closely related wheat lines derived from the parental lines. In another set of studies using different wheat genotypes, two 51-kDa hydrophobic proteins were shown to be induced by Al in an Al-tolerant cultivar but not in a sensitive cultivar (Taylor et al., 1997). These proteins were localized to the tonoplast and their accumulation co-segregated with Al tolerance in a population arising from a cross between an Al-sensitive and an Al-tolerant parent. Sequence analysis revealed that they are homologous to the B subunit of the vacuolar H^+-ATPase (V-ATPase) and the α and β subunits of the mitochondrial ATPase (Hamilton et al., 2001). Although these proteins are not involved in organic anion transport they demonstrate the potential of proteomic-type approaches to identify membrane proteins. A disadvantage of this approach is that proteins involved in transport are often expressed at low levels and it may be difficult to observe them against the background of the more abundant soluble proteins. Furthermore many membrane-bound polypeptides are difficult to solubilise and most extractions are unlikely to represent all the very hydrophobic proteins. New techniques are being developed to overcome these problems and several groups are constructing publically-available databases of the membrane bound proteins from model species (http://sphinx.rug.ac.be:8080/ppmdb/index.html).

Martinoia et al. (1991) and Steiger et al. (1997) attempted to purify the proteins associated with malate transport across the tonoplast of barley and *Kalanchöe diagremontiana* cells. Tonoplast proteins were purified by affinity chromatography and fractions of the eluate were tested for the presence of transport proteins by reconstituting the proteins into liposomes and measuring the uptake of labeled malate. Steiger et al (1997) reported that a 32-kDa protein appeared to facilitate a high malate transport capacity when reconstituted into liposomes. Subsequent analysis has revealed that this protein was not responsible for malate transport suggesting that the transport activity was caused by another protein in the preparation (personal communication). Photoaffinity labeling has also been used to isolate transport and receptor proteins in plants and bacteria (Lahjouji et al., 1996). This technique relies on the availability of a transport analogue that can be photolysed when exposed to ultraviolet light. Once a radio-labelled analogue interacts with the target protein it is lysed with ultraviolet light to form a highly reactive compound that reacts covalently with its immediate environment. The radiaolabelled proteins are then purified from the tissues by standard techniques and the putative transport proteins analysed. Lahjouji, et al. (1996) used azido-salicylyl aspartic acid as an analogue for malate to isolate a 37-kDa protein from *Catharanthus roseus* and provided some physiological evidence to support the involvement of this protein in the transport of malate across the tonoplast. However, to date there are no reports that the gene encoding this protein has been isolated.

When target genes have been isolated: What then?

When candidate cDNAs have been isolated they need to be tested to determine whether they do indeed encode organic anion transporters. There are a number of possible procedures and a few of these are mentioned here. The cDNAs can be expressed in yeast or a model plant to determine whether organic anion efflux or Al tolerance is increased. The transgenic cells can then be analysed with the patch-clamp technique to detect novel anion-dependent currents. Alternatively the cRNAs can be injected into other heterologous expression systems such as *Xenopus* oocytes where the induction of novel ion currents can also be measured with the voltage-clamp technique. Localisation to tissues and membranes can be assessed by transforming plants or yeast with the cDNAs fused in-frame to the gene encoding the green fluorescent protein. In addition, the function of the candidate cDNAs can be clarified by knocking out the genes in the source species or homologous genes in a model species (e.g., *Arabidopsis*, tobacco), and examining the mutant phenotypes. Once these genes are identified they will provide new opportunities for increasing plant production on hostile soils. The genes could be used to transform important crop and pasture species to increase their tolerance to Al stress and also to improve their capacity to acquire P from the soil.

References

Anderson J A, Huprikar S S, Kochian L V, Lucas W J and Gaber R F 1992 Functional expression of a probable *Arabidopsis thaliana* potassium channel in *Saccharomyces cerevisiae*. Proc. Natl. Acad. Sci. USA 89, 3736–3740.

Barbier-Brygoo H, Vinauger M, Colcombet J, Ephritikhine G, Frachisse J-M and Maurel C 2000 Anion channels in higher plants: functional characterization, molecular structure and physiological role. Biochim. Biophys. Acta 1465, 199–218.

Basu A, Basu U and Taylor G J 1994 Induction of microsomal membrane proteins in roots of an aluminum-resistant cultivar of *Triticum aestivum* L. under conditions of aluminum stress. Plant Physiol. 104, 1007–1013.

Delhaize E, Higgins T J and Randall P J. (1991). Aluminium tolerance in wheat: analysis of polypeptides in the root apices of tolerant and sensitive genotypes. *In* Plant-soil Interactions at Low pH. Eds. R J Wright, V C Baligar and R P Murrmann. pp. 1071–1079. Kluwer Academic Publishers, Dordrecht.

Delhaize E, Ryan P R and Randall P J 1993 Aluminum tolerance in wheat (*Triticum aestivum* L.) II. Aluminum-stimulated excretion of malic acid from root apices. Plant Physiol. 103, 695–702.

Delhaize E, Hebb D M, Richards K D, Lin J-M, Ryan P R and Gardner R C 1999 Cloning and expression of a wheat (*Triticum aestivum* L.) phosphatidylserine synthase cDNA: overexpression in plants alters the composition of phospholipids. J. Biol. Chem. 274, 7082–7088.

Dinkelaker B, Hengeler C and Marschner H 1995 Distribution and function of proteoid roots and other root clusters. Bot. Acta 108, 183–200.

Dulhunty A, Gage P, Curtis S, Chelvanayagam G and Board P 2001 A glutathione transferase structural family includes a nuclear chloride channel and a ryanodine receptor calcium release channel modulator. J. Biol. Chem. 276, 3319–3323.

Ezaki B, Yamamoto Y and Matsumoto H 1995 Cloning and sequencing of the cDNAs induced by aluminum treatment and Pi starvation in tobacco cultured cells. Physiol. Plant. 93, 11–18.

Ezaki B, Tsugita S and Matsumoto H 1996 Expression of a moderately anionic peroxidase is induced by aluminum treatment in tobacco cells: possible involvement of peroxidase isozymes in aluminum ion stress. Physiol. Plant. 96, 21–28.

Garrill A, Tyerman S D, Findlay g P and Ryan P R 1996 Effects of NPPB and niflumic acid on outward K^+ and Cl^- currents across the plasma membrane of wheat root protoplasts. Aust. J. Plant Physiol. 23, 527–534.

Hamilton C A, Good A G and Taylor G J 2001 Induction of vacuolar ATPase and mitochondrial ATP synthase by aluminum in an aluminum-resistant cultivar of wheat. Plant Physiol. 125, 2068–2077.

Hoffland E, Findenegg G R and Nelemans J A 1989 Solubilization of rock phosphate by rape.1. Evaluation of the role of the nutrient uptake pattern. Plant Soil 113, 155–160.

Jentsch T J, Steinmeyer K and Schwarz G 1990 Primary structure of a *Torpedo marmorata* chloride channel isolated by expression cloning in *Xenopus* oocytes. Nature 348, 510–514.

Johnson J F, Allan D L and Vance C P 1994 Phosphorus stress-induced proteoid roots show altered metabolism in *Lupinus albus*. Plant Physiol. 104, 657–665.

Jones D L 1998 Organic acids in the rhizosphere – a critical review. Plant Soil 205, 25–44.

Keerthisinghe G, Hocking P J, Ryan P R and Delhaize E 1998 Effect of phosphorus supply on the formation and function of proteoid roots of white lupin (*Lupinus albus* L.). Plant Cell Environ. 21, 467–478.

Kollmeier M, Dietrich P, Bauer C S, Horst W J and Hedrich R 2001 Aluminum activates a citrate-permeable anion channel in the aluminum-sensitive zone of the maize root apex. A comparison between an aluminum-sensitive and an aluminum-resistant cultivar. Plant Physiol. 126, 397–410.

Lahjouji K, Carrasco A, Bouyssou H, Cazaux L, Marigo G and Canut H 1996 Identification with a photoaffinity reagent of a tonoplast protein involved in vacuolar malate transport of *Catharanthus roseus*. Plant J. 9, 799–808.

Leonhardt N, Vavasseur A and Forestier C 1999 ATP binding cassette modulators control abscisic acid-regulated slow anion channels in guard cells. Plant Cell 11, 1141–1151.

Li X F, Ma J F and Matsumoto H 2000 Pattern of Al-induced secretion of organic acids differs between rye and wheat. Plant Physiol. 123, 1537–1543.

Ma J F, Taketa S and Yang Z M 2000 Aluminum tolerance genes on the short arm of chromosome 3R are linked to organic acid release in Triticale. Plant Physiol. 122, 687–694.

Ma J F, Ryan P R and Delhaize E 2001 Aluminium tolerance in plants and the complexing role of organic acids. Trends Plant Sci. 6, 273–278.

MacDiarmid C W and Gardner R C 1996 Al toxicity in yeast: a role for Mg? Plant Physiol. 112, 1101–1109.

MacDiarmid C W and Gardner R C 1998 Overexpression of the *Saccharomyces cerevisiae* magnesium transport system confers resistance to aluminum ion. J. Biol. Chem. 273, 1727–1732.

Martinoia E, Vogt E, Rentsch D and Amrhein N 1991 Functional reconstitution of the malate carrier of barley mesophyll vacuoles in liposomes. Biochim. Biophys. Acta 1062, 271–278.

Massonneau A, Langlade N, Leon S, Smutny J, Vogt E, Neumann G and Martinoia E 2001 Metabolic changes associated with cluster root development in white lupin (*Lupinus albus* L.): relationship between organic acid excretion, sucrose metabolism and energy status. Planta 213, 534–542.

Neumann G, Massonneau A, Martinoia E and Römheld V 1999 Physiological adaptations to phosphorus deficiency during proteoid root development in white lupin. Planta 208, 373–382.

Neumann G and Römheld V. (2000). The release of root exudates as affected by the plant's physiological status. *In* The Rhizosphere: Biochemistry and organic substances in the soil-plant interface. Eds. R Pinton, Z Varanini and Z Nannipieri. pp. 41–93. Marcel Dekker, New York/Basel.

Neumann G and Römheld V 1999 Root excretion of carboxylic acids and protons in phosphorus-deficient plants. Plant Soil 211, 121–130.

Pena P D L, Delgado L M, Camino D D, Barros F, Pena P D L and Camino D D 1992 Cloning and expression of the thyrotropin-releasing hormone receptor from GH3 rat anterior pituitary cells. Biochem. J. 284, 891–899.

Piñeros M A and Kochian L V 2000 A patch clamp study on the physiology of aluminium toxicity and aluminium tolerance in *Zea mays*: Identification and characterisation of Al^{3+}-induced anion channels. Plant Physiol. 125, 292–305.

Putterill J J, Richards K D, Boyd L, Konlgstorfer A, Richardson T E and Gardner R C 1991 Molecular approaches to aluminum tolerance in plants. Curr. Top. Plant Biochem. Physiol. 10, 142–147.

Richards K D, Schott E J S, Y.K., Davis K R and Gardner R C 1998 Aluminum induces oxidative stress genes in *Arabidopsis thaliana*. Plant Physiol. 116, 409–418.

Ryan P R, Delhaize E and Randall P J 1995 Characterisation of Al-stimulated efflux of malate from the apices of Al-tolerant wheat roots. Planta 196, 103–110.

Ryan P R, Skerrett M, Findlay G P, Delhaize E and Tyerman S D 1997 Aluminum activates an anion channel in the apical cells of wheat roots. Proc. Natl. Acad. Sci. USA 94, 6547–6552.

Ryan P R, Delhaize E and Jones D L 2001 Function and mechanism of organic anion exudation from plant roots. Annu. Rev. Plant Physiol. Plant Mol. Biol. 52, 527–560.

Schroeder J I 1995 Anion channels as central mechanisms for signal transduction in guard cells and putative functions in roots for plant-soil interactions. Plant Mol. Biol. 28, 353–361.

Schroeder J I, Schmidt C and Sheaffer J 1993 Identification of high-affinity slow anion channel blockers and evidence for stomatal regulation by slow anion channels in guard cells. Plant Cell 5, 1831–1841.

Sentenac H, Bonneaud N, Minet M, Lacroute F, Salmon J, Gaymard F and Grignon C 1992 Cloning and expression in yeast of a plant potassium ion transport system. Science 256, 663–665.

Snowden K C and Gardner R C 1993 Five genes induced by aluminum in wheat (*Triticum aestivum* L.) roots. Plant Physiol. 103, 855–861.

Snowden K C, Richards K D and Gardner R C 1995 Aluminum-induced genes. Induction by toxic metals, low calcium, and wounding and pattern of expression in root tips. Plant Physiol. 107, 341–348.

Steiger S, Pfeifer T, Ratajczak R, Martinoia E and Lüttge U 1997 The vacuolar malate transporter of *Kalanchoë daigremontiana*: a 32 kDa polypeptide? J. Plant Physiol. 151, 137–141.

Takita E, Koyama H and Hara T 1999 Organic acid metabolism in aluminum-phosphate utilizing cells of carrot (*Daucus carota L.*). Plant Cell Physiol. 40, 489–495.

Taylor G J, Basu A, Basu U, Slaski J J, Zhang G-C, Good A and Zhang G-C 1997 Al-induced, 51-kilodalton, membrane-bound proteins are assocaited with resistance to Al in a segregating population of wheat. Plant Physiol. 114, 363–372.

Watt M and Evans J R 1999 Linking development and determinacy with organic acid efflux from proteoid roots of white lupin grown with low phosphorus and ambient or elevated atmospheric CO_2 concentration. Plant Physiol. 120, 705–716.

Yasui M, Hazama A, Kwon T-H, Nielsen S, Guggino W B and Agre P 1999 Rapid gating and anion permeability of an intracellular aquaporin. Nature 402, 184–187.

Zhang W-H, Ryan P R and Tyerman S D 2001 Malate-permeable channels and cation channels activated by aluminum in the apical cells of wheat roots. Plant Physiol. 125, 1459–1472.

Zheng S J, Ma J F and Matsumoto H 1998 High aluminum resistance in buckwheat. 1. Al-induced specific secretion of oxalic acid from root tips. Plant Physiol. 117, 745–751.

Plant and Soil **248:** 71–83, 2003.
© 2003 *Kluwer Academic Publishers. Printed in the Netherlands.*

Phosphate transport in plants

Frank W. Smith[1], Stephen R. Mudge, Anne L. Rae & Donna Glassop
CSIRO Plant Industry, Long Pocket Laboratories, 120 Meiers Rd, Indooroopilly, Qld 4068 Australia
[1]*Corresponding author**

Received 25 March 2002. Accepted in revised form 11 June 2002.

Key words: cotransporters, ion transport, membrane proteins, phosphate transport, phosphate transporters, phosphate uptake

Abstract

Transport of inorganic phosphate (P_i) through plant membranes is mediated by a number of families of transporter proteins. Studies on the topology, function, regulation and sites of expression of the genes that encode the members of these transporter families are enabling roles to be ascribed to each of them. The Pht1 family, of which there are nine members in the *Arabidopsis* genome, includes proteins involved in the uptake of P_i from the soil solution and the redistribution of P_i within the plant. Members of this family are $H_2PO_4^-/H^+$ symporters. Most of the genes of the Pht1 family that are expressed in roots are up-regulated in P-stressed plants. Two members of the Pht1 family have been isolated from the cluster roots of white lupin. These same genes are expressed in non-cluster roots. The evidence available to date suggests that there are no major differences between the types of transport systems that cluster roots and non-cluster roots use to acquire P_i. Differences in uptake rates between cluster and non-cluster roots can be ascribed to more high-affinity P_i transporters in the plasma membranes of cluster roots, rather than any difference in the characteristics of the transporters. The efficient acquisition of P_i by cluster roots arises primarily from their capacity to increase the availability of soil P_i immediately adjacent to the rootlets by excretion of carboxylates, protons and phosphatases within the cluster. This paper reviews P_i transport processes, concentrating on those mediated by the Pht1 family of transporters, and attempts to relate those processes involved in P_i acquisition to likely P_i transport processes in cluster roots.

Introduction

Most plant processes require phosphorus. It is present both as a constituent of such compounds as nucleic acids, phospholipids and ATP, and as a metabolite involved in energy transfer, the activation of proteins and the regulation of metabolic processes (Marschner, 1995). The primary source of phosphorus for plants is inorganic phosphate (P_i). Plant growth is limited by the availability of P_i in most natural ecosystems. This has resulted in the evolution in plants of a range of adaptive morphological, biochemical and symbiotic strategies that increase the acquisition of P_i and/or improve the efficiency of internal utilization of P_i (Smith et al., 2001). Amongst these adaptations is the formation of cluster roots by some species. These combine a high density of determinate lateral roots in a localised soil volume for gathering P_i, with a synchronised release of organic acids and acid phosphatases that increase the availability of P_i in the immediate vicinity of these roots (Dinkelaker et al., 1995; Neumann et al., 2000).

Specific transport systems are essential for the uptake of P_i and for its internal redistribution within plants. Considerable progress on the elucidation of these transport processes has been made in recent years through the adoption of molecular techniques. Model systems such as *Arabidopsis*, *Medicago truncatula* and rice are proving to be particularly useful. Whilst none of these species form cluster roots, they are enabling the basic molecular mechanisms underlying P_i transport to be worked out and the genes

* FAX No: +61-7-3214-2288. E-mail: frank.smith@csiro.au

that encode and regulate the proteins involved in these processes to be identified and characterised. Both the physiological and molecular evidence that is emerging highlight the fact that plant growth and development requires the coordinated application of many different P_i transport processes in plants. This paper reviews these processes and then attempts to relate those processes involved in P_i acquisition to likely P_i transport processes in cluster roots.

Phosphate concentration in soil solution

The P_i that plant roots acquire from the soil solution is in equilibrium with phosphate sorbed onto soil minerals and colloids in the soil solid phase. The sorption reactions, defined by P sorption isotherms (Fox and Kamprath, 1970), buffer the amount of P_i in solution. Because of the high sorptive capacity and reactivity for P of many components of the soil solid phase, only a small fraction of the total soil P is in solution, and solution concentrations are usually low. Organic P in the soil is not directly available to plants and must first be mineralised to P_i, which then enters into the equilibrium reactions defined by the P-sorption isotherm. The result is that, even in the most fertile soils, P_i concentrations in soil solution rarely exceed 8 μM (Barber et al., 1962). Solution P_i concentrations in highly weathered soils, sandy soils and alkaline soils are commonly less than 1 μM (Reisenauer, 1966).

Phosphate ions in solution move to the sites of uptake on the root by the relatively slow process of diffusion. This results in depletion of P_i in solution around plant roots growing in P-deficient soils. Whilst root tips will be provided with external P_i concentrations similar to those in the bulk soil through extension of roots into undepleted zones, further back along the root axis the volume of soil explored for phosphate is considerably increased by the presence of root hairs. Still further back the soil volume explored by some species growing in low-phosphate soils may be enhanced by the presence of hyphae of mycorrhizal fungi, which can extend several centimetres from the root surface. A cone of soil in which the concentration of P_i in solution is depleted thus develops back from the root tip. Plants have adapted a number of morphological and biochemical strategies to reduce this depletion zone and increase the concentration of P_i in solution immediately adjacent to the sites of uptake in the roots. Some species develop more highly branched root systems with more root apices that are efficient in acquiring P. Studies on *Phaseolus* have shown that P-acquisition efficient genotypes have an altered geotropic response which enables their lateral roots to grow out at angles that better explore the upper layers of the soil (Lynch, 1995; Lynch and Beebe, 1995). The surface area of root in contact with the soil may also be increased in some species by an increase in the diameter of roots when the plants are P-stressed (Ma et al., 2001), and both the density and length of root hairs may increase (Bates and Lynch, 1996, 2000; Foehse and Jungk, 1983). Lateral root development may be enhanced in localised zones of the soil that are rich in P_i (Drew and Saker, 1978; Jackson et al., 1990). Some species develop the specialised cluster or proteoid roots (Dinkelaker et al., 1995) that were the particular subject of a workshop on response to P-stress (Lambers and Poot, 2002). Excretion of carboxylates and protons by plant roots aids in the release of P_i from iron and aluminium compounds from the soil through both desorption of P from sesquioxide surfaces and chelation of iron and aluminium from Fe-phosphates and Al-phosphates (Gerke, 1992). There can be dramatic increases in this excretion from the roots of some species when subjected to P stress (Dinkelaker et al., 1989, 1995; Hoffland et al., 1989; Ohwaki and Hirata, 1992). This provides a very valuable mechanism for acquiring more P by these species, particularly when combined with changes in root morphology as occur with cluster roots.

P uptake: apoplastic and symplastic transport in plant roots

The apoplasm comprises the walls of root epidermal and cortical cells and their associated intercellular spaces. These walls are composed of interlaced fibres that form an open latticework in young roots (Peterson and Cholewa, 1998). Soil solution is able to move radially towards the central stelar region of the root through the intercellular spaces and the pores in this latticework of the cell walls until it meets the suberised Casparian band around the transverse and radial walls of the endodermal cells that surround the central stele. This band also restricts nutrients in apoplastic spaces within the stele from leaking out into the apoplasm surrounding cortical and epidermal cells. Although suberised, the selectivity of the endodermis can change during root development and result in variations in the passage of water and various solutes (Clarkson, 1993; Steudle and Frensch, 1996). Older areas of the roots of some species have another layer of suberised cells in the outer layers of cortical cells that form an exodermis. This layer further restricts apoplastic movement of external soil solution in these regions of the root. Recently it has been demonstrated that the permeation properties of the exodermis can also vary, thereby providing plants with another adaptive mechanism for regulating the uptake or loss of water and solutes (Hose et al., 2001). The exodermis may be formed closer to the tip in slower-growing roots, such as those on plants subjected to stress (Perumalla and Peterson, 1986). Movement of solutes through the apoplasm also appears to be restricted near the meristematic region close to the root tip where the microfibrils of the cell walls appear densely packed (Peterson and Cholewa, 1998).

The mucilages that surround many roots and are excreted into cell walls contain negatively charged hydroxyl groups that can modify the flow of anions. These mucilages, and other root excretions, also provide substrates for rhizosphere microorganisms that influence nutrient concentrations close to the root surface. The interlacing fibres of cell walls in the apoplasm increase the path length over which ions must diffuse to the underlying uptake sites on the plasma membrane. A net negative charge is generated by carboxyl groups associated with the pectic polysaccharides of the cell wall fibres. Anions such as P_i are repelled by this charge and restricted to the larger pores within the apoplasm. The net effect is that movement of P_i may be impeded within the apoplast, further lowering the P_i concentration at the outer surface of the plasma membrane.

The boundary between the apoplasm and the symplasm of roots involved in nutrient uptake is the plasma membrane of root epidermal and cortical cells. Once inside the symplasm, nutrient ions in the cytosol can move radially to the stele through plasmodesmata connections without encountering further membrane barriers (Clarkson, 1993). Thus, movement of ions across the plasma membrane of root epidermal and cortical cells is a crucial step in the transport of nutrients into the plant. This semipermeable membrane functions to both mediate and regulate nutrient uptake. Although the physiological processes by which nutrients are transported across the plasma membrane have been understood for a long time (Epstein, 1953; Epstein and Hagen, 1952), it has only been in recent years that the specific proteins involved in this transport have been characterised and the genes encoding them identified. Uptake of phosphate into the root symplasm involves transport from the apoplasm where P_i concentrations are commonly less than 2 μM, across the plasma membrane to the cytosol where phosphate concentrations of the order of 5–17 mM are maintained (Mimura, 1999; Mimura et al., 1996). This large difference between external and internal concentrations, together with the net negative charge on the inside of the plasma membrane, necessitates that strong electrochemical gradients must be overcome for transport of phosphate anions into root cells. Transport of P_i from the external soil solution into plant roots therefore requires a high-affinity, energy-driven transport mechanism. This role is facilitated by a number of proteins belonging to the Pht1 family of plant phosphate transporters. Nomenclature for members of the Pht1 family has been suggested by the Commission for Plant Gene Nomenclature and is defined in the review of Bucher et al. (2001). The application of this nomenclature to the nine members of the Pht1 family in *Arabidopsis* is provided in Table 1 of Mudge et al. (2002).

The internal utilization of P_i within the plant requires transport through membranes wherever symplastic connections are lost. Thus, membrane transport systems are also required for the distribution and remobilization of P_i throughout shoot tissues. A number of members of the Pht1 family of plant phosphate transporters fulfil this role (Mudge et al., 2002), together with at least one member of another family of P_i transporters, the Pht2 family (Daram et al., 1999). The Pht2 family has similarities to some mammalian Na^+/phosphate cotransporters, but functions as a $H^+/H_2PO_4^-$ cotransporter in plants. The Pht2;1 transporter, which possesses a long N-terminal tail, is primarily expressed in shoot tissues where it is postulated that it plays a role in the internal cycling of P_i.

Intra-cellular phosphate transport

The physiological functioning of cells requires that P_i concentrations in the cytosol be maintained within a narrow range. As indicated above, cytosolic P_i concentrations in the range of 5–17 mM have been measured (Mimura et al., 1996). P_i stored in plant vacuoles provides a readily available buffer to help maintain cytosolic P_i concentrations and P_i homeostasis. Cells of plants well supplied with P may have a vacuolar P_i concentration as high as 120 mM (Mimura et al., 1990). This store of P_i can be drawn upon under P-stress conditions until it is virtually exhausted. Transport between the vacuole and cytosol occurs through the tonoplast membrane and involves both influx and efflux. Stimulation by ATP (Mimura, 1995) suggests that an active transport system may be involved in influx. It has also been demonstrated that the permeability of the tonoplast increases considerably under P-deficient conditions (Mimura, 1995). Unfortunately, the molecular mechanisms that mediate and regulate P_i transport across the tonoplast have not yet been elucidated nor have the genes involved in these processes been identified.

Transport of P_i between the cytosol and the stroma of the chloroplast is mediated by triose phosphate/P_i translocators (Flugge, 1999; Flugge et al., 1989). These transporters are localised in the inner membrane of the chloroplast envelope. They are encoded by nuclear genes and, when synthesised, contain chloroplast-targeting sequences, which are subsequently removed upon transfer to the chloroplast (Fischer et al., 1994). Transporters that mediate exchange of glucose-6-phosphate and P_i have also been isolated from plant plastids, as have phosphoenolpyruvate/P_i exchangers (Fischer et al., 1997; Kammerer et al., 1998). P_i is also essential for the metabolic functioning of mitochondria. Movement of P_i into mitochondria is facilitated by mitochondrial phosphate transporters that are similar to those in yeast and mammals (Takabatake et al., 1999). All of these proteins that mediate transport of P_i between the cytosol and organelles belong to distinct classes of proteins that are quite different to the Pht1 family of transporters involved in the initial acquisition of P_i from the external soil solution.

The Pht1 family of phosphate transporters

The Pht1 family of P_i transporters are of particular interest for research on cluster roots because of their important role in P_i acquisition (Raghothama, 1999). Pht1 type phosphate transporters were first isolated from yeast (Bun-ya et al., 1991), filamentous fungi (Versaw, 1995) and mycorrhizal fungi (Harrison and van Buuren, 1995). The appearance in sequence databases of an expressed sequence tag from *Arabidopsis* that contained similarities to these yeast and fungal sequences resulted in the cloning of the first members of the Pht1 family from plants (Muchhal et al., 1996: Smith et al., 1997a). Subsequently, additional members of the family have been isolated from potato (Leggewie et al., 1997), *Arabidopsis* (Mitsukawa et al., 1997a), tomato (Daram et al., 1998; Liu et al., 1998a), *Catharanthus* (Kai et al., 1997), *Medicago* (Liu et al., 1998b), barley (Smith et al., 1999) and *Lupinus albus* (Liu et al., 2001). Eight different members of this family of phosphate transporters have been isolated from the barley genome to date (Smith et al., 1999), and sequencing of the *Arabidopsis* genome reveals there are nine closely related members of the family in *Arabidopsis* (Mudge et al., 2002).

Members of the Pht1 family of transporter proteins are approximately 58 kDa in size and contain 520–550 amino acids. They contain a number of highly hydrophobic membrane-spanning domains (MSDs), each of which is usually composed of 17–25 amino acids arranged in a helix that crosses the membrane. The Pht1 family is predicted to have 12 MSDs arranged in a well-defined 6+6 configuration that contains a long central loop between MSD6 and MSD7. The MSDs are separated by extracellular and intracellular loops of more highly charged hydrophilic amino acids. It is predicted that the N-terminal, C-terminal and the long central hydrophilic loop are oriented towards the inner surface of the membrane. This configuration is a characteristic of the major facilitator superfamily of proteins (Pao et al., 1998; Saier and Reizer, 1991) to which the Pht1 family belongs under the designation of the PHS (P_i:H^+ symporter) family. Within the major facilitator superfamily there are membrane transporters with a similar configuration that are involved in the transport of organic acids, amino acids, sugars and inorganic ions. This configuration suggests a model for P_i transport by members of the Pht1 family in which the 12 MSD helices form a doughnut-shaped structure in the membrane. $H_2PO_4^-$ ions and protons move from the outside of the membrane to the inside through the inner pore of this structure. An H^+-ATPase 'pumps' protons to the outer surface of the membrane, thereby maintaining an electrochemical potential gradient and energizing the process. The outer hydrophilic loops and possibly the inner surfaces of the pore formed by the MSDs probably contain critical domains that define the specificity of the transporter for $H_2PO_4^-$ ions, the 'binding capacity', and hence the affinity and K_m of the process for phosphate. Little is known about such structure/function relationships, but site-directed mutagenesis aimed at altering potentially important domains within the protein should enable these relationships to be explored.

The functional analysis of plant phosphate transporters

Many plant transporters have been functionally characterised by heterologous expression in the yeast, *Saccharomyces cerevisiae*. This system, which relies upon complementation of a mutation in the yeast by expression of plant cDNA, has been used successfully with plant sulfate and potassium transporters. In addition to verifying the function of the transporter, yeast complementation has been used to determine transport kinetics and assess the effect of pH and inhibitors of transport (Smith et al., 1995, 1997b). Unfortunately, complementation of yeast mutants has proved less effective as a means of characterising plant phosphate transporters. A yeast mutant with a defect in a gene encoding a phosphate transporter (PHO84) has been available (Bunya et al., 1991), but plant genes encoding putative phosphate transporters do not appear to fully complement this mutation. Michaelis constant (K_m) values calculated for plant transporters expressed in this mutant have generally been much higher than expected, making it difficult to obtain reliable kinetic data. A yeast strain, PAM2, with mutations in two genes encoding phosphate transporters has been constructed (Martinez and Persson, 1998) and offers improved prospects for analysis of plant transporters. This strain has defects in both the gene encoding the PHO84 phosphate transporter and the gene encoding the Na^+-coupled PHO89 phosphate transporter. Using this double mutant, apparent K_m values for phosphate have been estimated at 31 μM for the tomato LePT1 phosphate transporter (Daram et al., 1998), and 64 μM for the StPT3 transporter from potato (Rausch et al., 2001). These values may still be higher than expected for a high-affinity phosphate transporter, but they are an order of magnitude lower than the results from complementation of the single *PHO84* mutant.

The most accurate functional analysis of a plant phosphate transporter was achieved following biolistic transformation of tobacco suspension-cultured cells with a construct containing an *Arabidopsis* cDNA under the control of a strong constitutively expressed promoter. Using this system, Mitsukawa et al. (1997b) demonstrated that the PHT1 phosphate transporter from *Arabidopsis* (= Pht1;1) was a high-affinity transporter with an apparent K_m for phosphate of 3.1 μM. The pH dependence of the PHT1 transporter and the effect of a range of inhibitors could also be assessed in the transgenic tobacco cells.

All of the plant phosphate transporters from the Pht1 family that have been analysed to date appear to have a high affinity for phosphate. This is consistent with a role in primary uptake of P_i from the low P_i concentrations in soil solution. The sequences and expression patterns of some of the transporters that have been isolated from *Arabidopsis* (Smith et al., 1997a) and barley (Smith et al., 1999) are almost identical, and might suggest that these transporters have similar roles and functions. Indeed it seems likely that some redundancy of genes encoding phosphate transporters has developed to ensure a critical function. However, it would be premature to conclude that all transporters in this family have similar kinetic properties. Amongst the plant sulfate transporters, closely related sequences have been shown to encode both high- and low-affinity transporters with diverse expression patterns (Smith et al., 1995). A more detailed examination of the kinetic properties of further transporters from the Pht1 family may reveal a similar diversity. Interestingly, the Pht2;1 phosphate transporter from *Arabidopsis*, which is predominantly expressed in leaves, has been shown to have a low affinity for phosphate (Daram et al., 1999). This is consistent with a role in internal redistribution or intracellular transport, where phosphate concentrations are typically in the millimolar range.

The sensitivity of plant phosphate transporters to proton concentration suggests that they operate as $H^+/H_2PO_4^-$ symporters. The rates of phosphate uptake by the high-affinity transporters from potato and *Arabidopsis* were reduced when the pH was increased over the range 4.5–7.5 (Leggewie et al., 1997; Mitsukawa et al., 1997b). Uptake by the tomato phosphate transporter LePT1 was also dependent upon the proton gradient across the membrane (Daram et al., 1998). Physiological measurements indicate that the concentration of protons in the cytosol increases during phosphate uptake (Mimura et al., 1992; Ulrich and Novacky, 1992) and that agents that depolarise the membrane inhibit phosphate transport (Mitsukawa et al., 1997b). Transport of phosphate therefore requires the activity of a proton pump to maintain a proton gradient across the membrane.

Regulation of plant phosphate transporters

P_i concentrations within the cytosol are maintained within a narrow range by close regulation of both influx and efflux processes (Mimura, 1999). Although little is known of the molecular mechanisms of phosphate efflux, physiological measurements have shown that phosphate-deprived plants rapidly increase their capacity for phosphate uptake (Clarkson and Luettge, 1991; Clarkson and Scattergood, 1982; Cogliatti and Clarkson, 1983; Lefebvre and Glass, 1982). The genes encoding phosphate transporters have provided useful tools to unravel the control processes behind phosphate influx. The rate of transport of phosphate across plasma membranes will be partially determined by the number of transporters in the membranes. This is dependent upon the rates of transcription of the genes encoding the transporters, the rates of mRNA decay and the rates of protein turnover. In addition to the number of transporters in the membrane, the rate of phosphate uptake will be modulated by allosteric regulation of the transporter proteins.

Transcriptional regulation appears to be an important control point for many of the plant P_i transporters. The expression of mRNA corresponding to the *Arabidopsis* P_i transporters Pht1;1 and Pht1;2 increased greatly when roots were deprived of P_i, and decreased rapidly upon P_i resupply to these plants (Smith et al., 1997a). Similar regulation of gene transcription by negative feed-back has been demonstrated for P_i transporters isolated from potato (Leggewie et al., 1997), tomato (Daram et al., 1998; Liu et al., 1998a), *Medicago* (Liu et al., 1998b), barley (Smith et al., 1999) and *Lupinus albus* (Liu et al. 2001). By analogy with the regulation of plant sulfate transporters, positive feed-back may also play a role in the regulation of transcription of P_i transporters. The expression of genes encoding sulfate transporters increases in response to decreased sulfate supply. However, an increase in the concentration of *O*-acetyl serine, a non-sulfated precursor in the sulfate assimilation pathway, has also been shown to cause an increase in transcription of a sulfate transporter gene (Smith, 1997b).

The overall nutritional status of the plant may also influence simple feed-back regulation of the transcription of P_i transporters. It appears that when growth is limited by deficiency of a second nutrient, such as sulfur or nitrogen, P_i transporters fail to be induced by P_i deprivation (Smith et al., 1999), thereby restricting uptake of P_i that cannot be effectively utilised. Whilst this phenomenon probably occurs with a number of nutrients, zinc appears to have a more specific effect on the expression of some genes encoding Pht1 transporters in plant roots. Accumulation of very high levels of P in zinc-deficient plants has been known for many years (Welch et al., 1982).

Experiments with barley have indicated that the expression of genes encoding some of the Pht1 high-affinity phosphate transporters in roots is up-regulated in zinc-deficient plants, even when those plants are grown at a high external P_i concentration (Huang et al., 2000). This effect of zinc could not be replaced by a similar divalent cation, manganese. These interesting results suggest that zinc may have a specific role in the signal-transduction pathway that regulates the expression of these Pht1 genes in roots.

For transcriptional control to be effective, the rate of protein turnover in the membrane must be rapid. Antibodies specific for phosphate transporters have been used to investigate the localisation and concentration of transporter proteins (Chiou et al., 2001; Muchhal and Raghothama, 1999). An antibody raised to a peptide from the MtPT1 transporter confirmed that the protein was located in the plasma membrane, and indicated that protein levels increase and decrease concomitantly with mRNA levels in response to phosphate supply (Chiou et al., 2001). These experiments support the hypothesis that control of transcription is a primary mechanism for transporter regulation.

Split-root studies have provided some insights into the nature of the signal or signals that mediate transcription of the transporter genes. In portions of the barley root system deprived of external P_i, the ability to take up P_i did not increase when the P_i supply to the remainder of the root system was adequate (Cogliatti and Clarkson, 1983). These results suggest that the activity of the transporters primarily responds to the P status of the whole plant rather than the local concentration in or around the root. Recently we have demonstrated that this regulation occurs at the transcriptional level through split-root experiments with transgenic *Arabidopsis* plants that express the green-fluorescent protein (GFP) reporter gene under the control of the *Pht1;1* and *Pht1;2* promoters. These experiments demonstrated that the promoters of these genes were not induced in the portion of the root system deprived of external P_i when P_i was supplied to the other portion of the root system. This implies the existence of a mechanism for assessing the P status of the whole plant and for transmitting a systemic signal via the shoot to all parts of the root system. Generation of mutant *Arabidopsis* plants that are unable to perceive this signal may be an effective approach to identifying the components of this signalling pathway.

A complex interplay of regulatory sequences in the promoters of genes encoding P_i transporters must exist to control expression in response to metabolite levels, developmental and tissue-specific signals. Some progress has already been made in identification of transcription factor binding sites by DNA mobility-shift assays (Mukatira et al., 2001; Raghothama, 2000). A *cis*-acting element with a core sequence of GNATATNC has been identified in the promoter regions of some phosphate-responsive genes, including some P_i transporters (Rubio et al., 2001). PHR1, A MYB transcription factor that binds to this element was isolated from *Arabidopsis* and shown to be involved in modulating the transcription of genes in response to phosphate starvation. The gene encoding a similar transcription factor, PSR1, has also been isolated from the unicellular algae, *Chlamydomonas reinhardtii*. The availability of full promoter sequences from the *Arabidopsis* genome project is providing a valuable resource for dissection of the control elements in these promoters.

Post-translational modification is another potential control point for P_i transporter activity. Most transporters in the Pht1 family share highly conserved sequences that may be targets for phosphorylation or glycosylation. There is also some evidence that interactions between the transporters and other proteins may be important in phosphate uptake. In yeast, the *PHO* regulon that controls P_i uptake is composed of a number of proteins (Ogawa et al., 2000; Oshima et al., 1996) and this may also be true in plants. Poor complementation of the yeast PHO84 phosphate transporter by plant proteins may indicate that other interacting proteins are required for a fully functional uptake system.

Expression of transporter sequences in transgenic plants may help to determine the relative contributions of transcriptional and post-transcriptional control of phosphate uptake. Over-expression of an *Arabidopsis* transporter was clearly able to increase the capacity of tobacco suspension-cultured cells to take up P_i (Mitsukawa et al., 1997). It will be interesting to determine whether simply inserting more copies of the transporter protein into the membrane results in increased rates of uptake in whole plants. The results will have implications for designing molecular strategies to increase phosphate uptake by crop plants.

The roles of different phosphate transporters in plants

Information on the localisation of plant P_i transporters provides a key to assigning possible roles to each of them. Northern analysis has shown that almost all of the Pht1 genes studied to date are expressed in roots, and are up-regulated under conditions of P_i starvation. This suggests a role for them in uptake of P_i from the soil solution. *In situ* hybridisation studies have indicated that the high-affinity LePT1 transporter from tomato is expressed most strongly in the root epidermis and root hairs, and in the root cap, with some expression also in the outer layers of the cortex (Daram et al., 1998; Liu et al., 1998). Similarly, in roots of *Medicago truncatula*, the MtPT1transporter is expressed in the epidermis and root hairs using both *in situ* hybridisation and immunolocalisation (Chiou et al., 2001). The proteins encoded by these two transporters have also been localised to the plasma membrane of these cells (Chiou et al., 2001; Muchhal and Raghothama, 1999). The StPT3 transporter from potato has recently been localised to cells containing arbuscules within the roots of plants colonised by endomycorrhizal fungi (Rausch et al., 2001). These findings further support the idea that these transporters are involved in the primary uptake of P_i, since these cell types are either in direct contact with the soil solution or with P_i delivered to the root cell/arbuscule interface. The observed expression in root hair cells (trichoblasts) is of particular interest, as root hairs have recently been shown to be the major site of P_i uptake in rye (Gahoonia and Nielsen, 1998). As already noted, both root hair length and root hair frequency in *Arabidopsis* increase in response to P_i starvation (Ma et al., 2001), greatly increasing the surface area of the root in contact with the soil solution.

Once taken up by these epidermal cells, P_i can move symplastically through the root towards the stele. A relatively high density of plasmodesmata has been observed at the base of trichoblasts compared to atrichoblasts (Vakhmistrov et al., 1981), and a recent study of plasmodesmatal frequencies in *Allium cepa* roots has demonstrated, for the first time, that a continuous symplastic pathway exists from the epidermis to the stele (Ma and Peterson, 2001). Although some degree of apoplastic movement into the root can also occur, the presence of the suberised exodermis in the roots of most angiosperm species restricts this pathway (Hose et al., 2001), so transport of P_i across the cortex and into the stele is likely to be symplastic. In

trical potential across the membrane of root cells to derive kinetic constants for P_i uptake by cluster and non-cluster roots of lupin deprived on P_i. They found that the V_{max} for cluster roots was approximately twice that of non-cluster roots, but there was no significant difference between the K_m values. Using ^{33}P, Neumann et al. (2000) found rates of P_i uptake by juvenile and mature root clusters were twice those measured in the apices of non-cluster roots. They measured a K_m of 30.7 μM for P_i uptake by non-cluster roots of P-sufficient plants, but approximately 8.5 μM for both non-cluster and juvenile root clusters of P-starved plants. These data suggest that higher-affinity uptake systems are induced in both cluster roots and lateral roots of P-starved plants. The fact that neither of these studies detected differences in K_m between cluster and non-cluster roots suggests that there are no major differences between the high-affinity transport systems that operate in both types of roots. However, higher V_{max} values for cluster roots point to considerably more of these high-affinity transporters in the plasma membranes of cluster roots.

Two genes that encode P_i transporters have been isolated from the cluster roots of white lupin (Liu et al., 2001). The LaPT1 and LaPT2 transporters encoded by these genes are members of the Pht1 family. *LaPT1* is expressed in both non-cluster roots and cluster roots, and is strongly up-regulated in P-stressed plants. Its expression in cluster roots increases as those roots develop up to 14 days following their emergence. This pattern corresponds to the developmental pattern of cluster roots and would ensure a high density of transporters was present when the soil P_i level was being increased by organic acid and phosphatase excretion from these roots. *LaPT2* is also expressed in both non-cluster and cluster roots, but is only slightly up-regulated in P-stressed plants. Unlike *LaPT1*, *LaPT2* is expressed at all stages of cluster root formation. It appears to behave, therefore, as a constitutive P_i transporter in both cluster roots and non-cluster roots. Although searches for other P_i transporters need to be conducted and additional work defining the roles of LaPT1 needs to be done, LaPT1 emerges as a candidate for the transporter responsible for the higher rates of P_i uptake by cluster roots. The expression of this same gene in non-cluster roots is consistent with the results from kinetic experiments and adds weight to the view that there are no major differences between the transport systems that cluster roots and non-cluster roots use to acquire their P_i.

Finally, it should be emphasised that P_i transport in plants relies upon many different transporters, each functioning wherever P_i needs to be moved across a membrane. These operate in shoot tissues as well as roots, and within individual cells at the membrane boundaries around vacuoles and organelles. Growth and development requires the coordinated operation of all of these processes in a manner that ensures a supply of P_i is maintained to support those metabolic reactions and structural components required for each particular stage of development. The capacity of most plants to employ a number of quite different strategies to acclimate to phosphate stress (Smith, 2001) adds to the complexity of the regulatory circuits through which this coordination is achieved. As we have pointed out, molecular approaches are contributing to our knowledge of the regulation of individual P_i transport systems. How all these individual systems are coordinated in a way that supports plant growth and the orderly progression of developmental cycles during the life of a plant are important and exciting challenges for the future. Happily, excellent progress is being made on developing the large-scale molecular screening and analysis tools and the genetic information systems that will enable these challenges to be addressed.

Acknowledgements

Research at Commonwealth Scientific and Industrial Research Organization (CSIRO) Plant Industry is supported in part by the Australian Grains Research and Development Corporation.

References

Barber S A, Walker J M and Vasey E H 1962 Principles of ion movement through the soil to the plant root. Trans. Joint Meeting Commission IV & V Internat. Soil Sci., New Zealand 1962. pp 121–124.

Bates T R and Lynch J P 1996 Stimulation of root hair elongation in *Arabidopsis thaliana* by low phosphorus availability. Plant Cell Environ. 19, 529–538.

Bates T R and Lynch J P 2000 The efficiency of *Arabidopsis* root hairs in phosphorus acquisition. Am. J. Bot. 87, 964–970.

Bun-ya M., Nishimura M, Harashima S and Oshima Y 1991 The *PHO84* gene of *Saccharomyces cerevisiae* encodes an inorganic phosphate transporter. Mol. Cell. Biol. 11, 3229–3238.

Chiou T-J, Liu H and Harrison M J 2001 The spatial expression patterns of a phosphate transporter (MtPt1) from *Medicago truncatula* indicate a role in phosphate transport at the root/soil interface. Plant J. 25, 281–293.

Clarkson DT 1993 Roots and the delivery of solutes to the xylem. Phil. Trans. R. Soc. London B 341, 5–17.

Clarkson D T and Lüttge U 1991 Mineral nutrition: Inducible and repressible nutrient transport systems. Progr. Bot. 52, 61–83.

Clarkson D T and Scattergood C B 1982 Growth and phosphate transport in barley and tomato plants during development of and recovery from phosphate stress. J. Exp. Bot. 33, 865–875.

Cogliatti D H and Clarkson D T 1983 Physiological changes in phosphate uptake by potato plants during development of and recovery from phosphate deficiency. Physiol. Plant. 58, 287–294.

Daram P, Brunner S, Persson B L, Amrhein N and Bucher M 1998 Functional analysis and cell-specific expression of a phosphate transporter from tomato. Planta 206, 225–233.

Daram P, Brunner S, Rausch C, Steiner C, Amrhein N and Bucher M 1999 *Pht2;1* encodes a low-affinity phosphate transporter from *Arabidopsis*. Plant Cell 11, 2153–2166.

Delhaize E and Randall P 1995 Characterisation of a phosphate-accumulator mutant of *Arabidopsis thaliana*. Plant Physiol. 107, 207–213.

Dinkelaker B, Romheld V and Marschner H 1989 Citric acid excretion and precipitation of calcium citrate in the rhizosphere of white lupin (*Lupinus albus* L.) Plant Cell Environ. 12, 285–292.

Dinkelaker B, Hengeler C and Marschner H 1995 Distribution and function of proteoid roots and other root clusters. Bot. Acta 108, 183–200.

Drew M C and Saker L R 1978 Nutrient supply and growth of the seminal root system in barley. III. Compensatory changes in growth of lateral roots and in rates of phosphate uptake in response to a localised supply of phosphate. J. Exp. Bot. 29, 435–451.

Epstein E 1953 Mechanism of ion absorption by roots. Nature 171, 83–84.

Epstein E and Hagen C E 1952 A kinetic study of the absorption of alkali cations by barley roots. Plant Physiol. 27, 457–474.

Fischer K, Arbinger B, Kammerer B, Busch C, Brink S, Wallmeier H, Sauer N, Eckerskorn C and Flugge U I 1994 Cloning and in vivo expression of functional triose phosphate/phosphate translocators from C_3- and C_4-plants: Evidence for the putative participation of specific amino acid residues in the recognition of phospho*enol*pyruvate. Plant J. 5, 215–226

Fischer K, Kammerer B, Gutensohn M, Arbinger B, Weber A, Hausler R E and Flugge U I 1997 A new class of plastidic phosphate translocators: a putative link between primary and secondary metabolism by the phosphoenolpyruvate/phosphate antiporter. Plant Cell 9, 453–462.

Fluegge U I 1999 Phosphate translocators in plastids. Annu. Rev. Plant Physiol. Plant Mol. Biol. 50, 27–45.

Flugge U I, Fischer K, Gross A, Sebald W, Lottspeich F and Ackerskorn C 1989 The triose-3-phosphoglycerate-phosphate translocator from spinach chloroplasts: nucleotide sequence of a full-length cDNA clone and import of the *in vitro* synthesized precursor protein into chloroplasts. EMBO J. 8, 39–46.

Foehse D and Jungk A 1983 Influence of phosphate and nitrate supply on root hair formation of rape, spinach and tomato plants. Plant Soil 74, 359–368.

Fox R L and Kamprath E J 1970 Phosphate sorption isotherms for evaluating the phosphorus requirement of soils. Soil Sci. Soc. Am. Proc. 34, 902–907.

Gahoonia T S and Nielsen N E 1998 Direct evidence on participation of root hairs in phosphorus (^{32}P) uptake from soil. Plant Soil 198, 147–162.

Gaymard F, Pilot G, Lacombe B, Bouchez D, Bruneau D, Boucherez J Michaux-Ferriere N, Thibaud J-B and Sentenac H 1998 Identification and disruption of a plant Shaker-like outward channel involved in K^+ release into the xylem sap. Cell 94, 647–655.

Gerke J 1992 Phosphate, aluminium and iron in the soil solution of three different soils in relation to varying concentrations of citric acid. Z. Pflanzenernähr. Bodenk. 155, 339–343.

Gottwald J R, Krysan P J, Young J C, Evert R F and Sussman M R 2000 Genetic evidence for the *in planta* role of phloem-specific plasma membrane sucrose transporters. Proc. Natl. Acad. Sci. USA 97, 13979–13984.

Hamburger D, Rezzonico E, Petetot J M-C, Somerville C and Poirier Y 2002 Identification and characterization of the *Arabidopsis PHO1* gene involved in phosphate loading to the xylem. Plant Cell 14, 889–902.

Harrison M J 1999 Molecular and cellular aspects of the mycorrhizal symbiosis. Annu. Rev. Plant Physiol. Plant Mol. Biol. 50, 361–389.

Harrison M J and van Buuren M L1995 A phosphate transporter from the mycorrhizal fungus *Glomus versiforme*. Nature 378, 626–629.

Himelblau E and Amabino R M 2001 Nutrients mobilized from leaves of *Arabidopsis thaliana* during leaf senescence. J. Plant Physiol. 158, 1317–1323.

Hoffland E, Findenegg G R and Nelemans J A 1989 Solubilization of rockphosphate by rape. 2. Local root exudation of organic acids as a response to P starvation. Plant Soil 113, 161–165.

Hose E., Clarkson D T, Steudle E, Schreiber l and Hartung W 2001 The exodermis: a variable apoplastic barrier. J. Exp. Bot. 52, 2245–2264.

Huang C, Barker S J, Langridge, P Smith F W and Graham R D 2000 Zinc deficiency up-regulates expression of high-affinity transporter genes in both phosphate-sufficient and -deficient barley roots. Plant Physiol. 124, 415–422.

Jackson R B, Manwaring J H and Caldwell M M 1990 Rapid physiological adjustment of roots to localised soil enrichment. Nature 344, 58–60.

Kai M, Masuda Y, Kikuchi Y, Osaka M and Tadano T 1997 Isolation and characterization of a cDNA from *Catharanthus roseus* which is highly homologous with phosphate transporter. Soil Sci. Plant Nutr. 83, 227–235.

Kammerer B, Fischer K, Hilpert B, Schubert S, Gutensohn M, Weber A and Flugge U I 1998 Molecular characterization of a carbon transporter in plastids from heterotrophic tissues: the glucose 6-phosphate/phosphate antiporter. Plant Cell 10, 105–107

Karley A J, Leigh R A and Sanders D 2000 Where do all the ions go? The cellular basis of differential ion accumulation in leaf cells. Trends Plant Sci. 5, 465–470

Keerthisinghe D, Hocking P J, Ryan P R and Delhaize E 1998 Effect of phosphorus supply on the formation and function of proteoid roots of white lupin (*Lupinus albus* L.). Plant Cell Environ. 27, 467–478.

Lambers H and Poot P 2002 Structure and functioning of cluster roots and plant responses to phosphate deficiency. Plant Soil (in press).

Lefebvre D D and Glass A D M 1982 Regulation of phosphate influx in barley roots: Effects of phosphate deprivation and reduction in influx with provision of orthophosphate. Physiol. Plant. 54, 199–206.

Leggewie G, Willmitzer L and Reismeier J W 1997 Two cDNAs from potato are able to complement a phosphate uptake-deficient yeast mutant: Identification of phosphate transporters from higher plants. Plant Cell 9, 381–392.

Liu C, Muchhal U S, Uthappa M, Kononowicz A K and Raghothama K G. 1998a Tomato phosphate transporter genes are differentially regulated in plant tissues by phosphorus. Plant Physiol. 116, 91–99.

Liu H, Trieu A T, Blaycock L A and Harrison M J.1998b Cloning and characterization of two phosphate transporters from *Medicago truncatula* roots: regulation in response to phosphate and to colonization by arbuscular mycorhizal (AM) fungi. Mol. Plant-Microbe Interact. 11, 14–22.

Liu J, Uhde-Stone C, Li A, Vance C and Allan D 2001 A phosphate transporter with enhanced expression on proteoid roots of white lupin (*Lupinus albus* L.). Plant Soil 237, 257–266.

Lynch J P 1995 Root architecture and plant productivity. Plant Physiol. 109, 7–13.

Lynch J P and Beebe S E 1995 Adaptation of beans (*Phaseolus vulgaris* L.) to low phosphorus availablility. HortScience 30, 1165–1171.

Ma F and Peterson C A 2001 Frequencies of plasmodesmata in *Allium cepa* L. roots: implications for solute transport pathways. J. Exp. Bot. 52, 1050–1061.

Ma Z, Bielenberg D G, Brown K M and Lynch J P 2001 Regulation of root hair density by phosphorus availability in *Arabidopsis thaliana*. Plant Cell Environ. 24, 459–467.

Maldonado-Mendoza I E, Dewbre G R, and Harrison M J.2001 A phosphate transporter gene from the extra-radical mycelium of an arbuscular mycorrhizal fungus *Glomus intraradices* is regulated in response to phosphate in the environment. Mol. Plant-Microbe Interact. 14, 1140–1148.

Marschner H 1995 Mineral Nutrition of Higher Plants. Academic Press, London. 889 pp.

Martinez P and Persson B L 1998. Identification, cloning and characterization of a derepressible Na^+-coupled phosphate transporter in *Saccharomyces cerevisiae*. Mol. Gen. Genet. 258, 628–638.

Mimura T 1995 Homeostasis and transport of inorganic phosphate transport in plants. Plant Cell Physiol. 36, 1–7.

Mimura T 1999 Regulation of phosphate transport and homeostasis in plant cells. Int. Rev. Cytol. 191, 149–200.

Mimura T, Dietz K-J, Kaiser W Schramm M J, Kaiser G and Heber U 1990 Phosphate transport across biomembranes and cytosolic phosphate homeostasis in barley leaves. Planta 180, 139–146.

Mimura T, Yin Z-H, Wirth E and Dietz K-J 1992 Phosphate transport and apoplastic phosphate homeostasis in barley leaves. Plant Cell Physiol. 33, 563–568.

Mimura T, Sakano K and Shimmen T 1996 Studies on distribution, re-translocation and homeostasis of inorganic phosphate in barley leaves. Plant Cell Environ. 19, 311–320.

Mitsukawa N, Okumura S and Shibata D 1997a High-affinity phosphate transporter genes of *Arabidopsis thaliana*. *In* Plant nutrition – for sustainable food production and environment. Proceedings of the XIII International Plant Nutrition Colloquium, 13–19 September 1997, Tokyo, Japan. Development in Plant and Soil Sciences 78. Eds. T Ando, K Fulita, T Mae, H Matsumoto, S Mori and J Sekiya. pp. 187–190. Kluwer Academic Publishers, Dordrecht.

Mitsukawa N, Okumura S, Shirano Y, Sato S, Kato T, Harashima S and Shibata D 1997b Overexpression of an *Arabidopsis thaliana* high-affinity phosphate transporter gene in tobacco cultured cells enhances cell growth under phosphate-limited conditions. Proc. Natl. Acad Sci. USA 94, 7098–7102.

Muchhal U S and Raghothama K G 1999 Transcriptional regulation of plant phosphate transporters. Proc. Natl. Acad. Sci. USA 96, 5868–5872.

Muchhal U S, Pardo J M and Raghothama K G 1996 Phosphate transporters from the higher plant *Arabidopsis thaliana*. Proc. Natl. Acad. Sci. USA 93, 10519–10523.

Mudge S R, Rae A L, Diatloff E and Smith F W 2002 Expression analysis suggests novel roles for members of the Pht1 family of phosphate transporters in *Arabidopsis*. Plant J. 31, 341–353.

Mukatira U T, Liu C, Varadarajan D K and Raghothama K G 2001 Negative regulation of phosphate starvation-induced genes. Plant Physiol. 127, 1854–1862.

Neumann G and Romheld V 1999 Root excretion of carboxylic acids and protons in phosphorus-deficient plants. Plant Soil 211, 121–130.

Neumann G, Massonneau A, Martinoia E and Romheld V 1999 Physiological adaptations to phosphorus deficiency during proteoid root development. Planta 208, 373–382.

Neumann G, Massonneau A, Langlade N, Dinkelaker B, Hengeler, C Romheld V and Martinoia E 2000 Physiological aspects of cluster root function and development in phosphorus-deficient white lupins (*Lupinus albus* L.). Ann. Bot. 85, 909–919.

Ogawa N, DeRisi J and Brown P O 2000 New components of a system for phosphate accumulation and polyphosphate metabolism in *Saccharomyces cerevisiae* revealed by genomic expression analysis. Mol. Biol. Cell 11, 4309–432

Ohwaki Y and Hirata H 1992 Differences in carboxylic acid exudation among P-starved leguminous crops in relation to carboxylic acid contents in plant tissues and phospholipid levels in roots. Soil Sci. Plant Nutr. (Tokyo) 38, 235–243.

Oshima Y, Ogawa N and Harashima S 1996 Regulation of phosphatase synthesis in *Saccharomyces cerevisiae* – a review. Gene 179, 171–177.

Pao S S, Paulsen I T and Saier M H 1998 Major facilitator superfamily. Microbiol. Mol. Biol. Rev. 62, 1–34.

Perumalla C J and Peterson C A 1986 Deposition of Casparian bands and suberin lamellae in the exodermis and endodermis of young corn and onion roots. Can. J. Bot. 64, 1873–1878.

Peterson C A and Cholewa E 1998 Structural modifications of the apoplast and their potential impact on ion uptake. Z. Pflanzenernähr. Bodenk. 161, 521–531.

Poirier Y, Thomas S, Somerville C and Scheifelbein J 1991 A mutant of *Arabidopsis* deficient in xylem loading of phosphate. Plant Physiol. 97, 1087–11093

Raghothama K G 1999 Phosphate acquisition. Annu. Rev. Plant Physiol. Plant Mol. Biol. 50, 665–693.

Raghothama K G 2000 Phosphate transport and signalling. Curr. Opin. Plant Biol. 3, 182–187.

Rausch C, Daram P, Brunner S, Jansa J, Laloi M, Leggewie G, Amrheln N and Bucher M 2001 A phosphate transporter expressed in arbuscule-containing cells in potato. Nature 414, 462–466.

Reisenauer H M 1966 Mineral nutrients in soil solution. In Environmental Biology. Eds P L Altman and D S Dittmer. pp 507–508. Fed. Am. Soc. Exp. Biol., Bethesda.

Rosewarne G M, Barker S J, Smith S E, Smith F A and Schachtman D P 1999 A *Lycopersicon esculentum* phosphate transporter (LePT1) involved in phosphorus uptake from a vesicular-arbuscular mycorrhizal fungus. New Phytol. 144, 507–516.

Rubio V, Linhares F, Solano R, Martin A C, Iglesias J, Leyva A and Paz-Ares J 2001 A conserved MYB transcription factor involved in phosphate starvation signalling both in vascular plants and in unicellular algae. Genes Dev. 15, 2122–2133.

Saier M H and Reizer J 1991 Families and superfamilies of transport proteins common to prokaryotes and eukaryotes. Curr. Opin. Struct. Biol. 1, 362–368.

Skene K R 2000 Pattern formation in cluster roots: Some developmental and evolutionary considerations. Ann. Bot. 85, 901–908.

Skene K R 2002 The evolution of physiology and development in the cluster root: teaching an old dog new tricks? Plant Soil (in press).

Skene K R, Sutherland J M, Raven J A and Sprent J I 1998 Cluster root development in *Grevillea robusta* (Proteaceae). II. The development of the endodermis in a determinate root and in an indeterminate, lateral root. New Phytol. 138, 733–742.

Smith F W 2001 Plant responses to nutritional stress. *In* Molecular Analysis of Plant Adaptation to the Environment, Ed. M J Hawkesford and P Buchner. pp. 249–269. Kluwer Academic Publishers, Dordrecht.

Smith F W, Ealing P M, Hawkesford M J and Clarkson D T 1995 Plant members of a family of sulfate transporters reveal functional subtypes. Proc. Natl. Acad. Sci USA 92, 9373–9377.

Smith F W, Ealing P M, Dong B and Delhaize E 1997a The cloning of two *Arabidopsis* genes belonging to a phosphate transporter family. Plant J. 11, 83–92.

Smith F W, Hawkesford M J, Ealing P M, Clarkson D T, Vanden Berg P J. Belcher A R and Warrilow A G S 1997b Regulation of expression of a cDNA from barley roots encoding a high affinity sulphate transporter. Plant J. 12, 875–884.

Smith F W, Cybinski D and Rae A L 1999 Regulation of expression of genes encoding phosphate transporters in barley roots. *In* Plant Nutrition – Molecular Biology and Genetics. Ed. G Gissel-Nielsen and A Jensen. pp. 145–150. Kluwer Academic Publishers, Dordrecht.

Smith S E and Read D J 1997 Mycorrhizal Symbiosis. Academic Press, San Diego.

Steudle E and Frensch J 1996 Water transport in plants: role of the apoplast. Plant Soil 198, 67–79.

Takabatake R, Hata S, Taniguchi M, Kouchi H, Sugiyama T and Izui K 1999 Isolation and characterization of cDNAs encoding mitochondrial phosphate transporters in soybean, maize, rice and *Arabidopsis*. Plant Mol. Biol. 40, 479–486.

Ulrich C I and Novacky A J 1992 Extra and intracellular pH and membrane potential changes induced by K^+, Cl^- $H_2PO_4^-$ and NO_3^- uptake and fusicoccon in root hairs of *Limnobium stoloniferum*. Plant Physiol. 94, 1561–1567.

Vakhmistrov D B, Kurkova E B and Zlotnikova I F 1981 Symplastic connections and intracellular activity of K^+ in the rhizodermis of *Raphanus sativus*. Soviet J. Plant Physiol. 28, 1142–1150 (English translation).

Versaw W K 1995 A phosphate-repressible, high-affinity phosphate permease is encoded by the *pho-5*$^+$ gene of *Neurospora crassa*. Gene 153, 135–139.

Welch R W, Webb M J and Loneragan J F 1982 Zinc in membrane function and its role in phosphorus toxicity. *In* Plant Nutrition 1982, Proc 9th Int. Plant Nutrition Colloq., Warwick, UK. Ed. A Scaife. pp. 710–715. CAB International, Farnham Royal, UK.

Plant and Soil **248:** 85–96, 2003.
© 2003 *Kluwer Academic Publishers. Printed in the Netherlands.*

Development of the adhesive pad on climbing fig (*Ficus pumila*) stems from clusters of adventitious roots

Edwin P. Groot[1], Eileen J. Sweeney & Thomas L. Rost[2]
*Section of Plant Biology, University of California, Davis, CA 95616 USA; [1]Current address: Institute of Systematic Botany, University of Mainz, Bentzelweg 9a, D-55099, Mainz, Germany. [2]Corresponding author**

Received 8 November 2001. Accepted in revised form 21 March 2002

Key words: adhesive pad, adventitious roots, climbing fig, cluster roots, *Ficus pumila*, vines

Abstract

Vines have different climbing strategies; one type, called clinging vines, has developed a specialized structure, the adhesive pad that secretes a sticky substance that adheres to almost any substrate. In this study on the climbing fig, *Ficus pumila* L., we report on the developmental anatomy of clusters of adventitious roots that become transformed into this unique structure. Clustered adventitious roots in *F. pumila* are initiated in pairs on either side of a vascular bundle at the 2nd to 3rd internodes of young stems. After emergence through the cortex and epidermis, root hairs form, which secrete a substance that stains positively for polysaccharide and protein. The adventitious roots and root hairs stick together forming the adhesive pad. Pads stick to almost any substrate. If the adventitious roots do not touch a substrate they usually dry up, if they touch moist soil they tend to branch and change to a terrestrial form. We discuss experiments on the shoot developmental behavior leading to adhesive pad formation and the possible roles of auxin.

Introduction

Naturally occurring clusters of roots occur in all 1800 species of the family Proteaceae and in members of seven other families, including one member of the Moraceae, *Ficus benjamina* (Dinkelaker et al., 1995; Lamont, 1972a, b; Skene, 1998). This close spacing of lateral roots can also be induced in *Lupinus albus* and other species by treatments with low phosphorus, or iron (Dinkelaker et al., 1995; Hagström et al., 2001). Clustered lateral roots can be experimentally induced by treatment with mitotic inhibitors; e.g., *Cucumis sativus* (Segawa and Kondo, 1980), or by root tip excision and treatment with auxin (Gilbert et al., 2000) and coupled with auxin-transport inhibitors; e.g., *Pisum sativum* (Hinchee and Rost, 1992a). Close spacing of lateral root primordia is due to maximizing cell division initiation sites in the pericycle thereby reducing the spacing intervals between lateral root primordia (Hinchee and Rost, 1992b). In this study we are reporting on another example of clustered roots, the clustered adventitious roots of the stems of the climbing fig, *Ficus pumila*. The cluster of roots in *F. pumila* become further modified into an adhesive pad responsible for adhesion of this vine to many different substrates.

Vines fall into three different types depending on how the shoot climbs on its substrate: clinging, twining or winding (Black, 2001; Musgrove and Williams, 2001). Twining vines wrap around branches, winding vines support themselves by tendrils, and clinging vines (e.g., English ivy, *Hedera helix*; and climbing fig, *Ficus pumila*) climb by means of clusters of adventitious roots that emerge from internodes and sometimes form into an aggregate structure called an adhesive disk or pad.

Ficus pumila (syn. *F. repens*) is a climbing fig commonly used as an ornamental. It is a very aggressive vine that rapidly grows on almost any surface. In the present study we trace the development of the clustered adventitious roots as they develop into an adhesive pad. We will also describe some experiments

* FAX No.: +530-752-5410; E-mail: tlrost@ucdavis.edu

done to examine how these processes might be controlled. Interestingly, the first known published report of *Ficus repens* (now called *F. pumila*) adhesive pads was by Darwin (1875) in his book on climbing plants. Our study reconnects these old observations and adds new approaches to the descriptions and analysis of the structure of the adhesive pad.

Materials and methods

Branches of juvenile *F. pumila* L. were collected from north-facing privacy walls in Davis, CA, USA between the months of April and September. They were stripped of their leaves and cut into about 5 mm lengths that included the adhesive pad. Segments were then sorted into nine groups based on visual determination of developmental stage, and fixed separately under gentle agitation overnight in a solution of 2.5% glutaraldehyde (w/v), 0.3% paraformaldehyde (w/v) and 0.025 *M* PIPES (piperazine-*N*, *N*′-bis [2-ethanesulfonic acid]) buffer (pH 7.2). The specimens were aspirated, rinsed with 0.025 *M* PIPES and agitated three times for a total of 60 min. Following the final rinse each group was divided into two halves, with one half to be used for fresh sectioning and the other half to be infiltrated and embedded.

Some unfixed samples were sectioned using double-edged razor blades with a Vibratome 1000 (Technical Products International, Inc., St. Louis, MO USA). Both cross and longitudinal sections were made at 60 μM. Different sections from each sample were each stained in one of the following: 1.64 m*M* toluidine blue$_{aq}$(0.05%), or filtered saturated Sudan IV in 70% (v/v) ethanol. Incubation time was approximately 10–20 min before rinsing in ddH_2O or 70% (v/v) ethanol (with the latter being used only with the sections stained with Sudan IV). Sections were mounted in fresh rinsing media on glass slides. Of these stains, toluidine blue worked best as a general stain. Images were recorded with the use of an Olympus Vanox - AHBT photomicroscope and Scion Image software.

Tissues that were to be embedded were dehydrated in an ethanol series increasing by 10% (v/v) with each successive step up to 90% (v/v), followed by 95% (v/v) and then three times in absolute ethanol (20 min each step). This was followed by infiltration with 3:1,1:1 and 1:3 ethanol: Histocryl (Electron Microscope Sciences, Fort Washington, PA) solutions for no less than 2 h each. Infiltration continued with three changes of 5% DMSO (v/v) in Histocryl for at least 1 day each change. The pieces were embedded in fresh Histocryl under a vacuum in aluminum weigh boats on a chilled plate until fully hardened. The stem pieces were cut out of the disc with a jeweler saw and mounted on wood dowels. A Riechert-Jung 2050 microtome outfitted with a dry glass knife was used to make 3 μm thick sections. The sections were placed individually onto water drops on gelatin-coated glass slides and dried on a slide warmer set at 60 °C. The slides were mounted in 2.05 *M* sucrose (70%) and observed microscopically.

To observe the effect of auxin on adhesive pad development, 70 juvenile *F. pumila* branches were selected from five plants at the original site. Each was numbered and ten branches were randomly assigned to each of the following treatments:

1. Control, no treatment.
2. Removal of apex and young leaves.
3. Removal of apex and young leaves, application of lanolin paste to wounds.
4. Removal of apex and young leaves, application of 5.4 m*M* NAA (0.1%) in lanolin to wounds. That concentration was selected based on its use in other studies (Cline, 2000; Jacobs, 1998).
5. Abrasion of exposed side of stem on internode lacking pad.
6. Abrasion of exposed side of stem on internode lacking pad, application of lanolin to wound.
7. Abrasion of exposed side of stem on internode lacking pad, application of 5.4 m*M* NAA lanolin paste to wounds. Rubbing Celite 545 with a gloved finger effected the abrasion treatments.

Rooted cuttings of *F. pumila* were raised in 10 cm pots in a growth chamber for the purposes of investigating adhesion of the pads. The chamber conditions were set to 20 °C and 16 h per day of fluorescent lamp illumination. Microscope slides were taped to the stem tips, and the slides removed at various intervals for staining and observation. Slides were stained with either periodic acid-Schiff's reagent (O'Brien and McCully, 1981), filtered saturated Sudan IV in 70% (v/v) ethanol, or 8.1 m*M* aniline blue black (0.5%) in 7% (v/v) acetic acid (O'Brien and McCully, 1981). Stained slides were observed and photographed using a compound microscope.

To investigate the tropism response of *F. pumila* stems, 15 horizontally growing branches were selected from the study site. A 15 mm high and 6 mm wide stick of balsa wood was fixed to the wall under each branch tip. The youngest internode with developing pads rested on the stick so that the internodes lacking

Table 1. Chronology of adhesive pad stages. Stage designations are as described in Figure 1

Stage	Time (d)
A to B	1.12 ($\pm$ 0.23[a])
B to C	0.76 ($\pm$ 0.08)
C to D	0.78 ($\pm$ 0.09)
D to E	1.63 ($\pm$ 0.54)
Total	4.29 ($\pm$ 0.58)

[a] Standard error of the mean.

pads were not in contact with the wall. At various intervals after the treatment, internodes that had not been in contact with the wall were harvested, prepared and observed as previously described.

Results

One adhesive pad developed per internode in *F. pumila* on the side of the stem facing the climbing substrate, and just basal to a node. The pad consisted of a dense cluster of short adventitious roots arranged in several rows along the length of the stem (Figure 1e). There were 2–12 adventitious roots per row, 2–6 rows, and 30 – 50 roots per pad.

Adhesive pad developmental stages

We divided adhesive pad development into five stages, which will be referred to as stages A–E (Figure 1a–e, Table 1). Initially, the young internode lacked external signs of adventitious roots (Figure 1a). Root clusters were first apparent at the second or third internode from the shoot tip, appearing as several longitudinal ridges the same color of the stem (Figure 1b), this

⟶

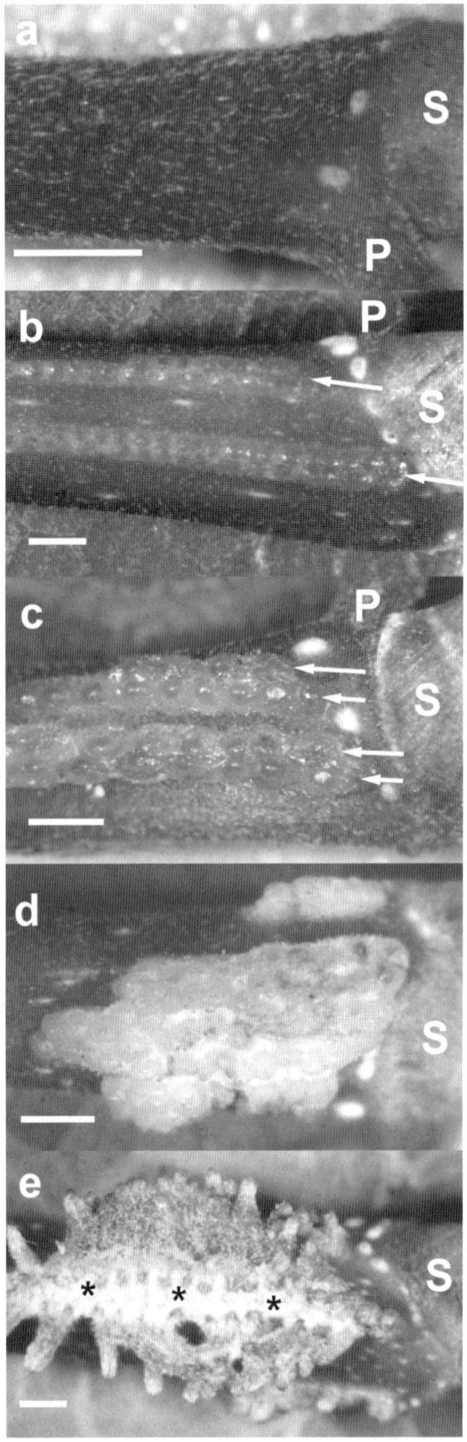

Figure 1. Developmental stages of the adhesive pad in *F. pumila*. Each stage is photographed just basal to a node, showing the stipule (S) and the petiole (P), where visible. (a) Young internode at stage A lacks adhesive pads. (b) The roots that will form an adhesive pad are first discernable at stage B as longitudinal ridges (arrows) on the stem. (c) Roots of stage C are noticeably lighter in color than the stem. Each ridge of the pad is clearly comprised of two rows (arrows) of roots. (d) Stage D pads have much longer roots arrayed in a compact mass. Signs of root hair formation are visible as a slight pubescence. (e) A stage E adhesive pad adhered to a glass substrate. Medial roots (asterisks) are still arrayed compactly, but exmedial roots splay from the median. Root hairs are obvious as a white or tan pubescence. All scale bars 1 mm.

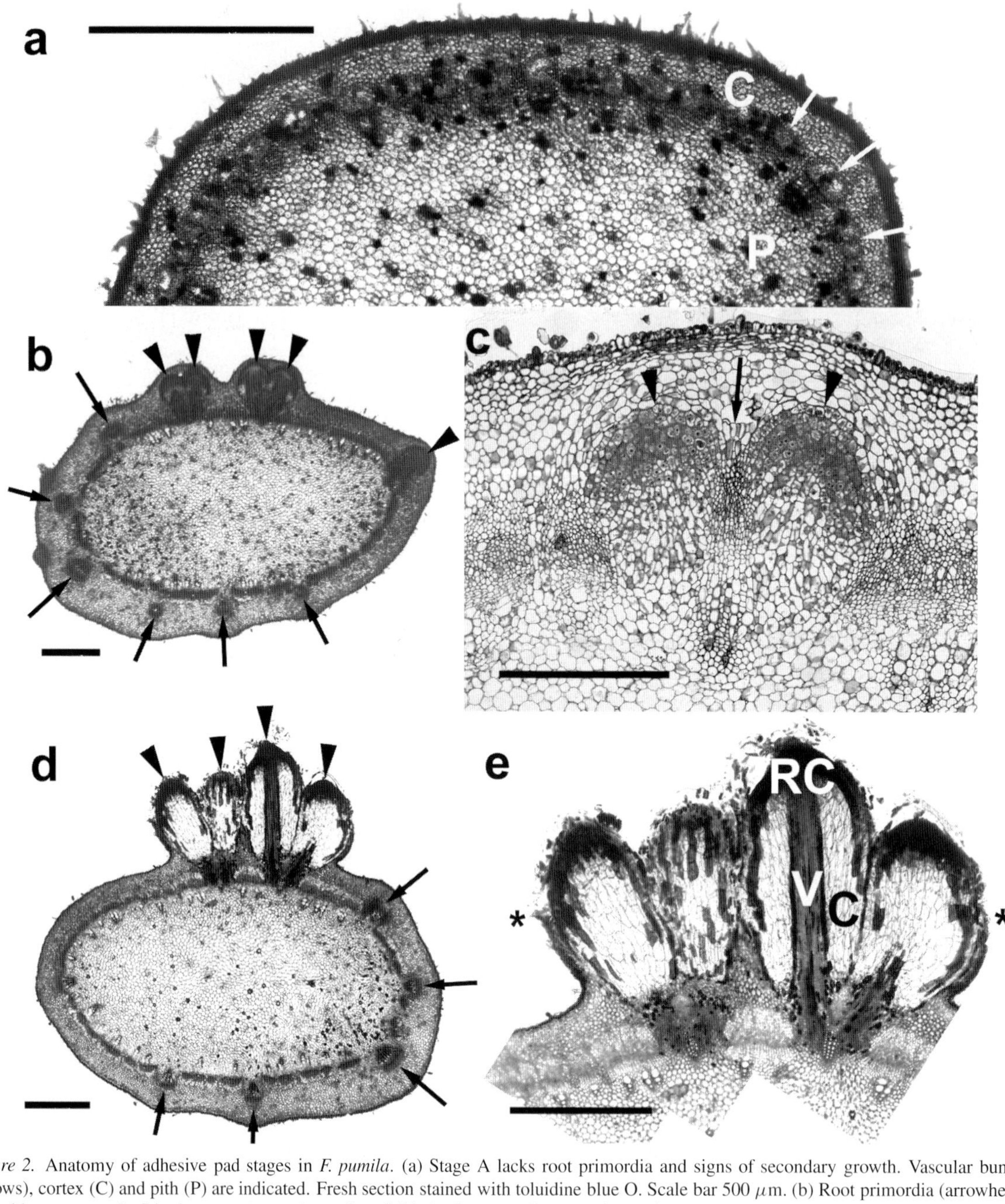

Figure 2. Anatomy of adhesive pad stages in *F. pumila*. (a) Stage A lacks root primordia and signs of secondary growth. Vascular bundles (arrows), cortex (C) and pith (P) are indicated. Fresh section stained with toluidine blue O. Scale bar 500 μm. (b) Root primordia (arrowheads) are evident at stage B. Stipule traces (arrows) are also visible in the cortex because the section was taken near the node. Fresh section stained with toluidine blue O. Scale bar 500 μm. (c) Thin section (3 μm) of embedded stage B material stained with toluidine blue O. Root primordia (arrowheads) initiate between vascular bundles in the zone that will become the interfascicular cambium. Sister roots have one intervening vascular bundle (arrow). Scale bar 200 μm. (d) Stage C roots (arrowheads) have just erupted through the stem. Stipule traces (arrows) are also visible in the cortex. Fresh section stained with toluidine blue O. Scale bar 500 μm. (e) Root tissues of stage C have differentiated into root cap (RC), vascular cylinder (V) and cortex (C). Stem tissue covers the base (asterisks) of the roots. Secondary growth has commenced. Enlargement of (d). Scale bar 500 μm.

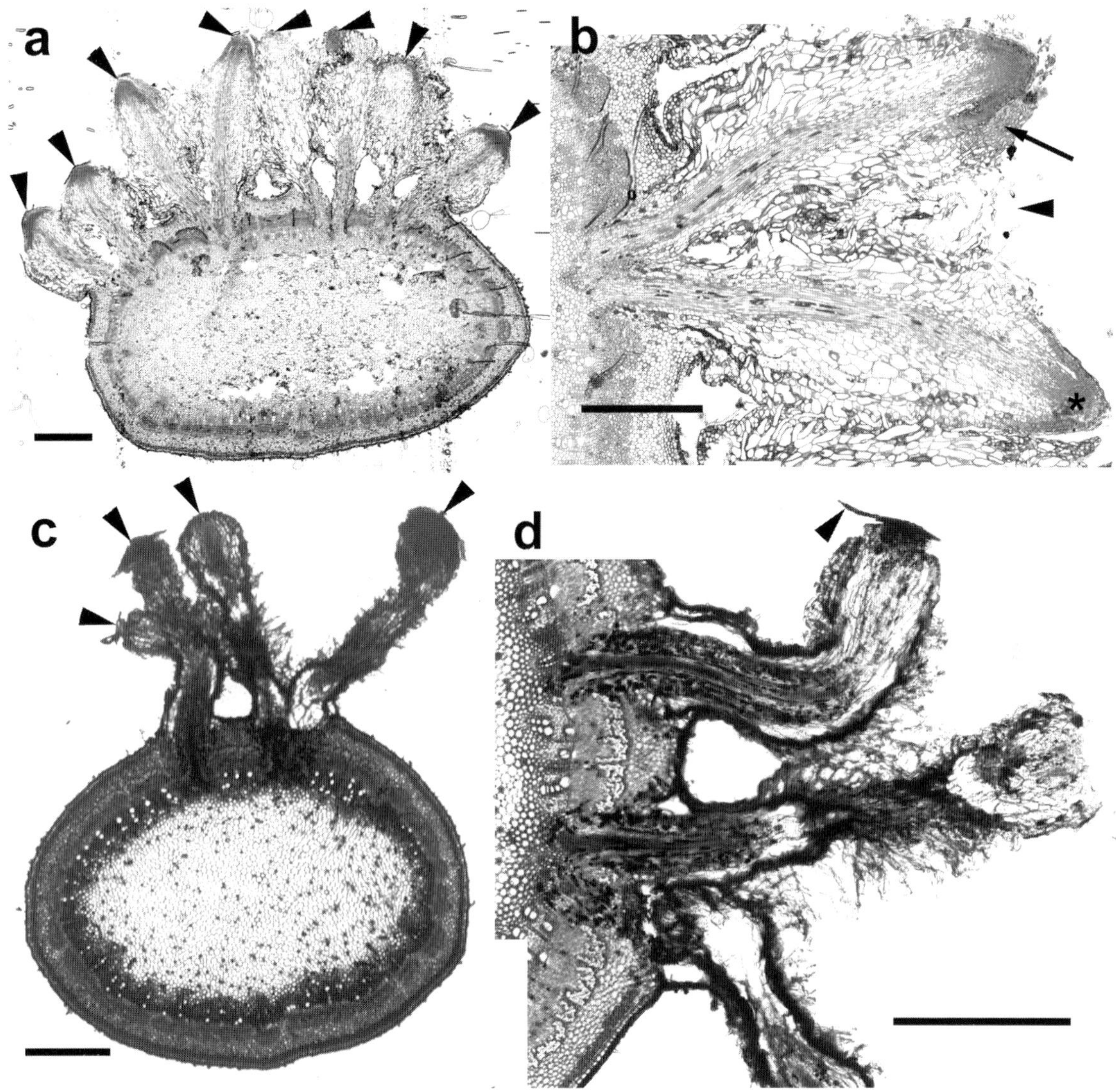

Figure 3. Anatomy of adhesive pad stages in *F. pumila*. (a) Stage D roots (arrowheads) are longer, but remain closely associated. Thin section (3 μm) of embedded material stained with toluidine blue O. (b) At higher magnification the root apical meristem (asterisk) is short and broad. Root hairs (arrowhead) are evident between roots. Root cap cells in a mucilagenous matrix (arrow) are found near the tip of some roots. Enlargement of (c) Stage E roots (arrowheads) are curved and the cortex tissue is collapsing at the root bases. Fresh section stained with toluidine blue O. (d) Roots have a dense mat of root hairs. Dry flakes (arrowhead) of root cap tissue appear to be shedding from the root tip. Enlargement of (c). All scale bars 500 μm.

stage took just over 1 day (Table 1). The stem color at this stage varied from red in warm, sunny and dry locations to green in shadier, cooler and moister locations. The ridges progressively became more distinct as tightly packed individual adventitious roots (Figure 1c). Each ridge consisted of two rows of adventitious roots. The root clusters elongated, became a cream color and formed root hairs (Figure 1d); both stages taking just under one day (Table 1). At that time the root cluster (we'll now call it an adhesive pad) were able to adhere to stucco walls at our study site, this was stage D. After the pads adhered to a substrate the roots became darker (Figure 1e). This final stage lasted almost two days (Table 1). It was difficult to remove intact stage E pads from a substrate without tearing them from the stem. The pads adhered to many substrates, including: wood, concrete, brick, glass, cardboard, and even other stems of *F. pumila*. The roots were determinate and grew no longer than 5 mm on 1-year-old stems. The total time from the first

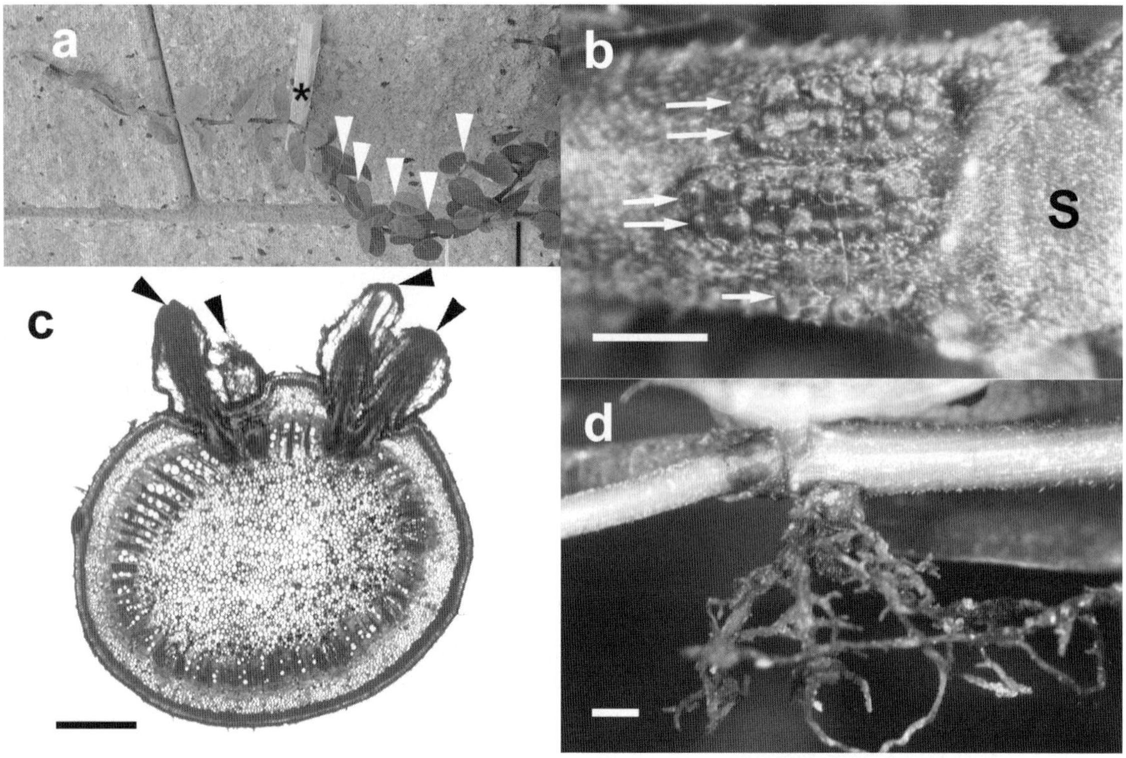

Figure 4. (a) Wooden block experiment 3.5 months after setup. The portion of the stem to the left of the block (asterisk) has grown to the wall. The portion of the stem to the right of the block is raised from the wall and branches (arrowheads) are growing towards the wall. (b) Pad that has developed from the stem not in contact with the wall has rows of short, dry, brown roots (arrows). The stipule (S) is persistent. Scale bar 1 mm. (c) Roots (arrowheads) of such pads lack root hairs characteristic of pads that have adhered to the wall. Fresh section stained with toluidine blue O. Scale bar 500 μm. (d) Adhesive pad that contacted soil of an adjacent pot. A highly branched root system has developed from the adhesive pad. Scale bar 1 mm.

visible sign of a cluster of adventitious roots to full pad development was just over 4 days (Table 1).

Developmental anatomy of an adhesive pad

At stage A there are no visible root primordia (Figure 2a). Cross sections of developing pads showed that the adventitious roots were initiated in pairs at stage B (Figure 2b). The stem cross sections were elliptical, with the minor axis perpendicular to site of pad projection. The adventitious roots originated at the interfascicular region, between vascular bundles in the stem (Figure 2c). The two adventitious roots making a pair formed on either side of one vascular bundle. Pairs of adventitious roots were separated by more than one vascular bundle. By stage C the adventitious roots had erupted through the stem cortex and epidermis (Figure 2d). The root tissues were distinct consisting of a densely staining vascular cylinder, a cortex consisting of enlarged, vacuolate cells, and a root cap (Figure 2e). As the roots elongated they maintained intimate contact with each other except at their tips (Figure 3a). Their root apical meristems were short and broad (Figure 3b). Root hairs were apparent at stage D, when pads begin adhering to a substrate (Figure 3b). In a mature adhesive pad (stage E) each root was covered with a dense mat of root hairs (Figure 3c); the cortex was collapsed, and root cap tissue flaked off the root tips (Figure 3d). After this stage, root anatomy appeared unchanged, and the roots continued elongating to their final determinate length.

It appeared that every internode produced an adventitious root cluster, but not all clusters became a mature adhesive pad. There were two alternate pathways for pad development.

It was observed that horizontally and vertically growing stems tended to closely follow the contours of a wall or other substrate. They grew around corners and in-and-out of cracks. Experiments were done to determine the growth response of stems lifted off the wall substrate. At the commencement of the experi-

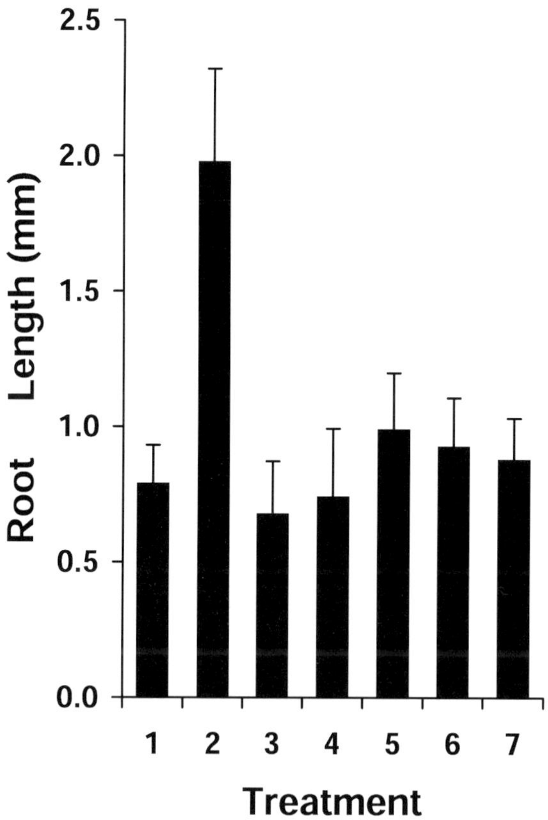

Figure 5. Length of roots of treated internodes 8 days after treatment. Sample size ranges from 18 to 20 for each treatment. Error bars represent standard error of the mean. Treatments: 1. Control, no treatment; 2. Removal of apex and young leaves; 3. Removal of apex and young leaves, application of lanolin paste to wounds; 4. Removal of apex and young leaves, application of 5.4 m*M* NAA (0.1%) in lanolin to wounds; 5. Abrasion of exposed side of stem on internode lacking pad: 6. Abrasion of exposed side of stem on internode lacking pad, application of lanolin to wound. 7. Abrasion of exposed side of stem on internode lacking pad, application of 5.4 m*M* NAA (0.1%) in lanolin to wound.

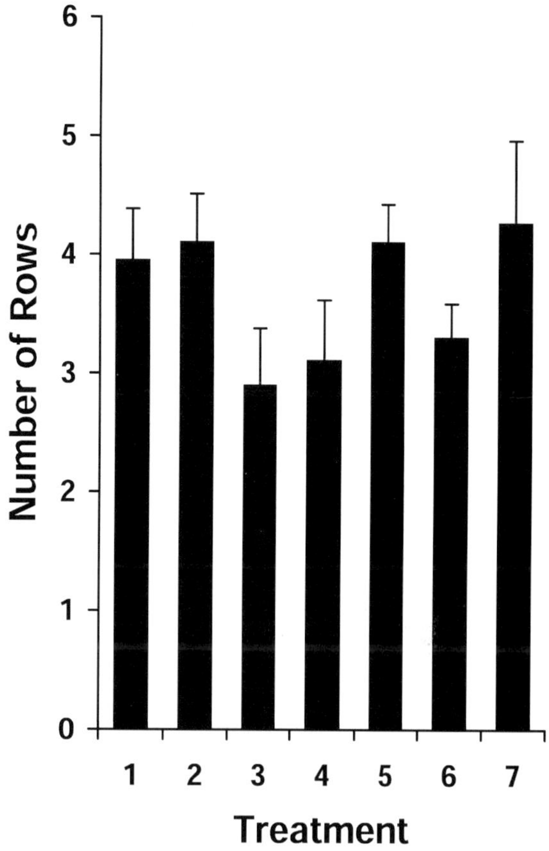

Figure 6. Number of root rows in treated internodes 8 days after treatment. Sample size ranges from 18 to 20 for each treatment. Error bars represent standard error of the mean. Treatment numbers are as described in Figure 5.

ment, the internode between the 2nd and 3rd arrowheads in Figure 4a was rested on a small wood block. About two internodes were present to the left of the block, no adhesive pads had developed, and those internodes had not contacted the wall yet. Within 1 week the stem had curved strongly toward the wall, and made contact. The steep angle of the shoot toward the wall combined with the force of the elongating stem proximal to the tip caused the stem to bow and to be lifted off the block. This bowing of the stem caused the internode (that was resting on the block at the start) to be displaced away from the wall surface. As the stem tip grew further to the left of the block the angle of growth became less steep so that more of the stem just proximal to the tip was able to be out of the bowed portion and to rest on the wall surface. This appearance in Figure 4a was typical of the other branches in the experiment. The adventitious root clusters that formed on the bowed part of the stem tended to turn brown, not form root hairs (compare Figures 1e and 4b), and not form into an adhesive pad. Another interesting observation was that the new stem turned back toward the substrate, not up or down in reaction to light or gravity, and then continued to grow along its horizontal track (Figure 4a).

Another growth strategy was for adventitious root clusters to form an apparently absorptive root system. In this case, if adhesive pad adventitious roots were in contact with moist soil, they branched and formed a terrestrial type root system (Figure 4d).

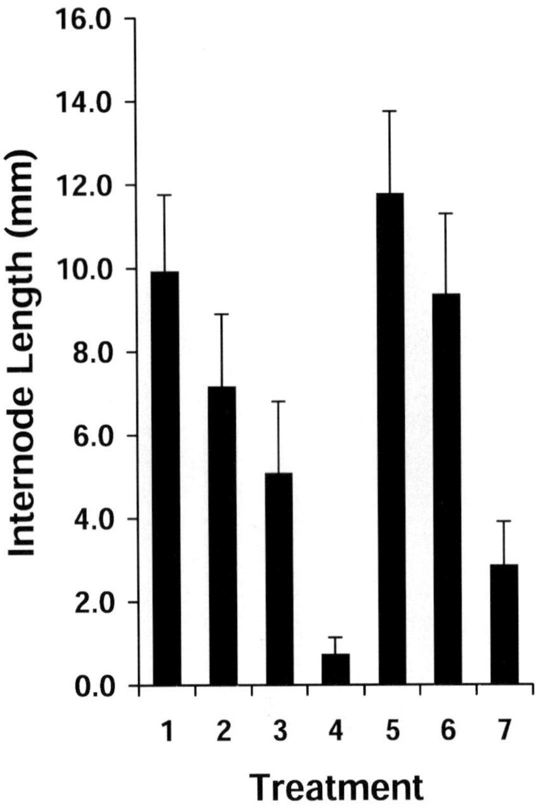

Figure 7. Eight-day growth increment of treated internodes. Sample size ranges from 7 to 10 for each treatment. Error bars represent standard error of the mean. Treatment numbers are as described in Figure 5.

A possible role for auxin

Measurements of the length of the roots and numbers of rows of roots within each pad on treated nodes were taken and compared between treatment groups of the auxin experiment (Figures 5 and 6). Root length became twice as long as the control in the decapitation treatment (Figure 5). None of the other decapitation treatments, or the abrasion treatments had any significant effect on root length. Decapitation had no significant effect on number of root rows (Figure 6), but this treatment in addition to either lanolin or NAA in lanolin resulted in about one fewer root row than in the control group. However, the most apparent quantitative difference between the treatments was the difference of average growth of the treated internodes (Figure 7). Both auxin treatments (abrasion and decapitation) resulted in a dramatic (about 3-fold and 10-fold, respectively) decrease in internode elongation (Figure 7). The other two decapitation treatments both experienced a modest decrease in growth. The stem abrasion treatments (except for the auxin application) resulted in no significant difference to internode elongation of the control (Figure 7). Auxin treatment visibly reduced the vigor of the branch portion between the treated nodes and the tip during the five weeks of this experiment. Our data suggests the application of auxin does not have a large effect on the size of the normal adhesive pad or the roots within it, but auxin clearly affects internode elongation.

In the abrasion and auxin treatment, auxin in lanolin paste was applied to the youngest 2–3 internodes on the side of the stem opposite the climbing substrate after abrasion with Celite 545 powder. Within 1 week, the youngest nodes of several branches in this treatment showed clear curvature away from the climbing substrate (Figure 8a). Concurrently, the site of NAA application resulted in lighter stem color and callus like outgrowths (Figure 8a) for all branches in this treatment. One week later, adventitious roots were visible at the locations of treatment in addition to the presence of a normal adhesive pad (Figure 8b). The roots stimulated by NAA were of similar length as those in the adhesive pad clusters, but they were not as closely initiated, or arranged in uninterrupted rows. The NAA-induced roots lacked root hairs, tended to desiccate, and were sometimes saw-tooth-shaped rather than radially symmetrical. Cross sections of stems revealed that NAA stimulated root initiation singly, rather than in pairs (Figure 8c). None of the other five treatments resulted in the formation of adventitious roots in locations other than the adhesive pad or curvature away from the growth substrate. Decapitation plus NAA did not elicit the curvature away from the wall or additional root formation. Similarly, lanolin paste alone did not trigger a change in growth direction or stimulate adventitious root formation whether applied to sites of shoot tip and leaf removal, or when applied to the abraded stem. Thus, these alterations in growth are attributed to the application of auxin to abraded stem tissue.

The adhesive substance

Stems were taped to glass slides to maintain contact with the glass. At different intervals the stems were lifted from the slides. In each case, after even 5 min, a mark was left on the slide from the adhesive pad. The roots seemed to have collapsed where they contacted the glass (Figure 9a). The marks on the glass were strongly PAS and aniline blue black positive,

Figure 8. (a) Stem one week after treatment with auxin paste. One treated internode has bent away from the wall, and callus-like growth is present (arrowhead). The pad at the older, untreated internode has adhered to the wall (oval). (b) Stem two weeks after treatment with auxin paste. Auxin induced roots (arrowheads) are present on the treated side of the stem. They lack root hairs and many appear dried and shrunken. The untreated side of the stem has a normally developed pad (oval) with root hairs. The petiole (P) and the stipule (S) are visible. Scale bar 1 mm. (c) Cross section of a treated stem with a normal pad on the upper side, and auxin-induced roots on the lower side. The normal roots tend to be in pairs, but the auxin-induced roots are borne singly. Scale bar 500 μm.

indicating polysaccharide and protein as their major components. The material did not stain with Sudan IV, except for some weak staining of the root cap cells. Under magnification, the pattern of root arrangement was evident (Figure 9b). Sheets of flattened cells, actually detached root caps, were also found on the glass slides (Figure 9c). Impressions of root hairs were also apparent on the slides (Figure 9d).

Discussion

Vines have developed three different growth strategies – twining, winding and clinging (Black, 2001; Musgrove and Williams, 2001). Twining vines (e.g., Wisteria, *Wisteria sinensis*) encircle poles and branches, and winding vines (e.g., Clematis, *Clematis hybrida*) support themselves by tendrils. Clinging vines (e.g., climbing fig, *Ficus pumila*, English ivy, *Hedera helix*; Virginia creeper, *Parthenocissus quinquifolia*) climb by means of clusters of adventitious roots that emerge from internodes and form into an aggregate structure called an adhesive disk or adhesive pad.

Ficus pumila (syn. *F. repens*) is a climbing fig that is commonly used as an ornamental to cover walls and fences. It is very aggressive and sprouts plagiotropic shoots that rapidly cover almost any surface including glass. The plant is dimorphic with a juvenile stage consisting of a vine habit with small heart-shaped leaves. This stage produces adventitious root clusters that become adhesive pads, and cuttings can be easily induced to form terrestrial type adventitious roots

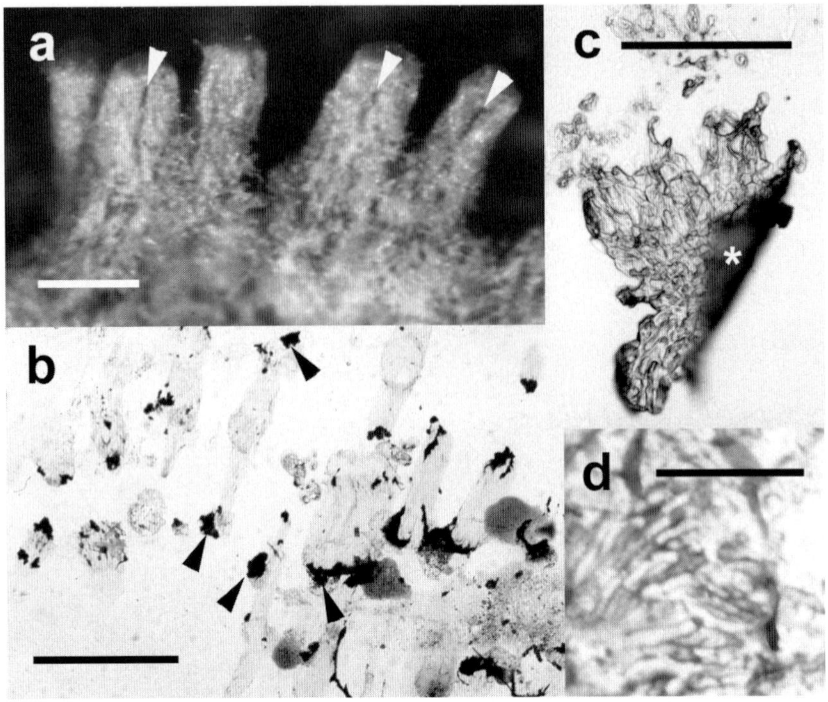

Figure 9. (a) Roots in the process of adhering to glass. The contact zones are collapsing, leaving a longitudinal groove on the roots (arrowheads). Scale 500 μm. (b) Low magnification of smudge left on glass by an adhering pad, stained with periodic acid-Shiff's reagent (PAS). The dark spots (arrowheads) are root cap tissue. Scale bar 500 μm. (c) Root cap flake stained with Sudan IV. The cap is weakly Sudan-positive. One side of the cap has folded over (asterisk). Scale bar 100 μm. (d) Higher magnification of smudge stained with PAS reveals impression of root hairs in a PAS-positive matrix. Scale bar 100 μm.

without any supplemental treatments (Davies et al., 1982). Horizontal shoots growing on a solid substrate, such as a concrete wall, will follow the contour of the wall, growing into and out of cracks. We observed them grow over obstructions placed in their path, but they will always end up growing toward the wall, without the characteristic nutation of twining vines. This is an interesting observation because it suggests a different type of tropism – toward a substrate. This kind of growth behavior is also seen in English ivy and is thought to be negative phototropism (Hershey, 1999), an instance where the stem grows away from the light. Mohr and Schopfer (1995) exemplify *Monstera gigantea* in a strategy that involves both negative and positive phototropism. They claim that seedlings of *M. gigantea* are negatively phototropic at high light intensity so that they can find the nearest dark host tree. At low light intensity they are positively phototropic so that they can climb to locations optimal for their photosynthesis. In *F. pumila*, this strategy is supported by the observation that the juvenile stage stems grow in and then out of cracks. The apical few mm of a juvenile shoot tip growing along a wall actually points away from the wall. This apical portion is perhaps positively phototropic, but the remainder of the stem is negatively phototropic so that it remains appressed to the wall. This strategy may protect the shoot apex from abrasion as the stem grows along the wall.

Adult plants of *F. pumila* have larger leathery oval-shaped leaves, a more upright habit, and they produce inedible, fig-like fruit. Cuttings of adult shoots may be induced to produce adventitious roots, but only with auxin treatments (Davies et al., 1982).

Perhaps the first report of the clinging vine mechanism in *F. pumila* (he called it *F. repens*) was by Charles Darwin (1875) in his book, 'The Movements and Habits of Climbing Plants.' In his report he notes that roots on the stems secrete a 'viscid fluid' possibly 'india-rubber' that adheres to any surface. Darwin conducted simple experiments by pressing roots against glass plates for different periods from 1 week to 23 days. He noted that a small droplet of clear yellowish fluid was deposited on the plate after a week, and that when these slides were stored the droplets did not evaporate. After roots were on a plate for 2 weeks a larger droplet formed and the material could be drawn into threads, and after 23 days the droplet dried and the stem was cemented to the glass. We repeated Darwin's experiments and never observed a fluid droplet. In our hands the root clusters did leave a smudge on the microscope slides within minutes of pressing on a slide, and the material left behind did react to carbohydrate and protein dyes, but weakly to lipophilic dyes. Pads of stage D left a smudge most readily on the slide than any other stage of pad development. Definitive, chemical analysis of the material left on the slide will remain for future study.

Model for development of the adhesive pad

The adhesive pad in *F. pumila* develops from a cluster of adventitious roots that are initiated just basal to a node. The developmental steps are similar to those reported for adventitious roots in many other plants (Barlow, 1986; DeKlerk et al., 1999) with the exceptional difference that the primordia are initiated in a clustered pattern with minimal distance between them. Most authors who have reported on adventitious root development agree on the basic steps as follows:

1. Dedifferentiation of cells usually between vascular bundles, from phloem parenchyma or the vascular cambium (Barlow, 1986; DeKlerk et al., 1999);
2. Induction of the dedifferentiated cells into an organized primordium;
3. Elongation of the primordium through the stem cortex and epidermis.

In *F. pumila* we didn't actually observe the initial dedifferentiation step, but all of these steps were basically consistent. Once emerged from the stem, at stage D, masses of root hairs emerge from the adventitious roots. These small roots usually grow to about 3–5 mm, but sometimes as long as 10 mm. In all cases they are determinate and stop elongating. Coupled with the determination step the root apical meristem is diminished and the root cap loosens at the tip. At this time the root hairs secrete some kind of sticky substance that stains positively for protein and carbohydrate in our hands, and could be the India rubber suggested by Darwin (1875), however we did not conduct biochemical analyses to verify this. The next step is that the secreted substance acts to stick the root hairs and clustered roots together and to the substrate forming the adhesive pad. Root tips in contact with the substrate partially collapse, enabling greater contact area between the pad and the substrate. The pad eventually dries into a hard structure and it is very difficult to pull from the substrate. Development of an adhesive pad took just over 4 days under our conditions.

We conducted several different experiments to determine the effects of raising a horizontal stem from a substrate, forcing the emerged root clusters onto a substrate and applying auxin. If after emerging the cluster is not allowed to touch a substrate the adhesive pad fails to fully form and it dries and hardens. Our hypothesis is that touch is a necessary stimulus for the complete development and adhesion of a pad. On the other hand, if the root cluster touches moist soil the roots elongate and branch and change their form to a terrestrial type. This difference in growth form between the aerial form of an adventitious root and the soil form has also been observed in *Monstera deliciosa* (Hinchee, 1981).

Many studies have examined the effects of auxin concentrations and types on adventitious root development from cuttings (Davies, 1984; Davis and Haissig, 1990; DeKlerk et al., 1999). Auxin is known from these studies to stimulate adventitious root initiation. In *F. pumila* we applied auxin in lanolin paste to the surface of horizontally growing stems using a concentration (5.4 m*M*) of NAA previously used in other studies and known to replace the endogenous source of auxin after decapitation. This application stimulated the initiation of adventitious roots, but the new roots did not initiate in dense clusters and they did not appear in pairs. Since each pair of cluster roots leading to pad formation formed on either side of a vascular bundle, we assume that the signal (possibly auxin) for adventitious root initiation is a diffusible substance probably originated from the bundle. These observations suggest that auxin may be important for adventitious root initiation, and that locations other than that of the normal pad are competent in forming aerial roots, but there are obviously other steps and other factors that control the organization of the cluster itself and the later stages of pad formation. Notice that the adhesive pad always formed toward the substrate. The observation of root clusters forming only on one stem surface has been made in many other instances, for example rhizomes of grasses grow horizontally along a surface and at each node sprout a cluster of roots on the substrate side (Barlow, 1994).

Is the formation of clusters of adventitious roots that transform into an adhesive pad in any way similar to cluster (proteoid) roots that form in the soil in several plants (Dell et al., 1980; Lamont, 1972b, Skene, 1998)? There are some similarities – both consist of clustered roots, both are determinate, in both the root apical meristem seems to differentiate, both produce a proliferation of root hairs, and both have a secretory function associated with an early development stage (Dinkelaker et al., 1995, Skene, 1998). There are some obvious differences – cluster roots are lateral roots growing in soil; climbing fig adhesive pad roots are aerial adventitious roots. Cluster roots grow around a soil-borne root and act as absorbing organs, climbing fig roots secrete a sticky substance creating a structure designed to support the stem to a substrate.

Acknowledgements

Keri Chapman, Cyril A. Dalmon, Julie L. Howard, and Allen Hahus are acknowledged for their assistance in conducting experiments and preparing materials during this study.

References

Barlow P W 1986 Adventitious roots of whole plants: their forms, functions, and evolution. *In* New Root Formations in Plants and Cuttings. Ed. M B Jackson. pp. 68–110. Martinus Nijhoff Publishers, Dordrecht.

Barlow P W 1994 The origin, diversity and biology or shoot-borne roots. *In* Biology of Adventitious Root Formation. Eds. T D Davis and B E Haissig. pp. 1–23. Plenum Press, New York.

Black R J 2001 Vines for Florida. Univ. of Florida Cooperative Extension Service. Circular 860.

Cline M G 2000 Execution of the auxin replacement apical dominance experiment in temperate woody species. Amer. J. Bot. 87, 182–190.

Darwin C 1875 The Movements and Habits of Climbing Plants. John Murray, London. pp. 185–188.

Davies F T, Jr 1984 Shoot RNA, cambial activity and indolebutyric acid effectivity in seasonal rooting of juvenile and mature *Ficus pumila* cuttings. Physiol. Plant. 62, 571–575.

Davies F T, Jr, Lazarte J E and Joiner J N 1982 Initiation and development of roots in juvenile and mature leaf bud cuttings of *Ficus pumila* L. Amer. J. Bot. 69, 804–811.

Davis T D and Haissig B E 1990 Chemical control of adventitious root formation in cuttings. PGRSA Quart. 18, 1–17.

De Klerk G, Van Der Krieken W and De Jong J C 1999 The formation of adventitious roots: New concepts, new possibilities. In Vitro Cell. and Dev. Biol. Plant. 35, 189–199.

Dell B, Kuo J and Thomson G J 1980 Development of proteoid roots in *Hakea obliqua* R. Br. (Proteaceae) grown in water culture. Aust. J. Bot. 28, 27–37.

Dinkelaker B, Hengeler C and Marschner H 1995 Distribution and function of proteoid roots and other root clusters. Bot. Acta 108, 183–200.

Gilbert GA Knight JD Vance CP and Allan DL 2000 Proteoid root development of phosphorus deficient lupin is mimicked by auxin and phosphonate. Ann. Bot. 85, 921–928.

Hagström J, James W M and Skene K R 2001 A comparison of structure, development and function in cluster roots of *Lupinus albus* L. under phosphate and iron stress. Plant and Soil 232, 81–90.

Hershey, D 1999 How do vines grow? http://www.madsci.org/posts/archives/dec99/ 945534660.Bt.r.html [accessed September 2001].

Hinchee M A W 1981 Morphogenesis of aerial and subterranean roots of *Monstera deliciosa*. Bot. Gaz. 142, 347–359.

Hinchee M A W and Rost T L 1992a The control of lateral root development in cultured pea seedlings. II. Root fasciation induced by auxin inhibitors. Bot. Acta 105, 121–126.

Hinchee M A W and Rost T L 1992b The control of lateral root development in cultured pea seedlings. III. Spacing intervals. Bot. Acta 105, 127–131.

Jacobs W P 1998 Hormone transport and action in the green shoot: long-term studies of a clonal stock of *Coleus blumei* (Labiatae). Amer. J. Bot. 85, 1353–1363.

Lamont B 1972a The morphology and anatomy of proteoid roots in the genus *Hakea*. Aust. J. Bot. 20, 155–172.

Lamont B 1972b 'Proteoid' roots in the legume *Viminaria juncea*. Search 3, 90–91.

O'Brien TP and McCully ME 1981 The Study of Plant Structure: Principles and Selected Methods. Termarcarphi Pty., Melbourne, Australia.

Mohr H and Schopfer P 1995 Plant Physiology, 4th edition, *translated by* G Lawlor and D W Lawlor. Springer Verlag, Berlin.

Musgrove M B and Williams J D 2001 Vines for Alabama Landscapes. http://www.aces.edu [accessed September 2001].

Segawa M and Kondo K 1980 Root fasciation in *Cucumis sativus* and *Vicia faba* treated with vinblastine. Environ. Exp. Bot. 20, 201–206.

Skene K R 1998 Cluster roots: some ecological considerations. J. Ecol. 86, 1060–1064.

Pathways of carbon metabolism related to enhanced exudation of organic acids and acid phosphatase

Plant and Soil **248:** 99–116, 2003.
© 2003 Kluwer Academic Publishers. Printed in the Netherlands.

Acclimation of white lupin to phosphorus deficiency involves enhanced expression of genes related to organic acid metabolism

Claudia Uhde-Stone[1,3], Glena Gilbert[1,4], Jane M-F Johnson[1,5], Ralph Litjens[3,6], Kelly E. Zinn[1,3], Stephen J. Temple[3,7], Carroll P. Vance[2,3] & Deborah L. Allan[1,8]

[1]*Department of Soil, Water and Climate, University of Minnesota, 1991 Upper Buford Circle, St. Paul, MN 55108, USA.* [2]*USDA-ARS, Plant Science Research Unit, 1991 Upper Buford Circle, St. Paul, MN 55108, USA.* [3]*Department of Agronomy and Plant Genetics, University of Minnesota, St. Paul, MN 55108, USA.* [4]*Viterbo University, Biology Dept., 815 Ninth Street South, LaCrosse, WI 54601, USA.* [5]*USDA-ARS-MWA, 803 Iowa Ave, Morris, MN 56267, USA.* [6]*Higher Institute for Professional Education Venlo, Division of Laboratory Science, Department of Technical Microbiology, 5822 BD Venlo, The Netherlands.* [7] *Forage Genetics International, N5292 South Gills Coulee Rd, West Salem, WI 54669, USA.* [8]*Corresponding author**

Received 15 December 2001. Accepted in revised form 22 March 2002

Key words: cluster root, *Lupinus albus* L., malate dehydrogenase, phosphorus deficiency, phospho*enol*pyruvate carboxylase, proteoid root

Abstract

White lupin (*Lupinus albus* L.) acclimates to phosphorus deficiency (−P) by the development of short, densely clustered lateral roots called proteoid (or cluster) roots. These specialized plant organs display increased exudation of citric and malic acid. The enhanced exudation of organic acids from P stressed white lupin roots is accompanied by increased *in vitro* phospho*enol*pyruvate carboxylase (PEPC) and malate dehydrogenase (MDH) activity. Here we report the cloning of full-length white lupin PEPC and MDH cDNAs. RNA blot analysis indicates enhanced expression of these genes in −P proteoid roots, placing higher gene expression at the site of organic acid exudation. Correspondingly, macroarray analysis of about 1250 ESTs (expressed sequence tags) revealed induced expression of genes involved in organic acid metabolism in −P proteoid roots. *In situ* hybridization revealed that PEPC and MDH were both expressed in the cortex of emerging and mature proteoid rootlets. A C_3 PEPC protein was partially purified from proteoid roots of P deficient white lupin. Native and subunit Mr were determined to be 440 kD and 110 kD, respectively. Citrate and malate were effective inhibitors of *in vitro* PEPC activity at pH 7. Addition of ATP partially relieved inhibition of PEPC by malate but had little effect on citrate inhibition. Taken together, the results presented here suggest that acclimation of white lupin to low P involves modified expression of plant genes involved in carbon metabolism.

Introduction

Phosphorus (P) is an essential macronutrient that plays a key role in many plant processes such as energy metabolism and the synthesis of nucleic acids and membranes (Raghothama, 1999). Low availability of P is a limiting factor in crop cultivation in many soils (Marschner, 1995). In response to low P availability, higher plants have evolved highly specialized physiological and biochemical mechanisms, including formation of mycorrhizae and proteoid or cluster roots (Marschner, 1995; Raghothama, 1999). White lupin (*Lupinus albus* L.), a legume capable of hosting nitrogen fixing bacteria, is a plant with extreme tolerance for low P availability and one of few agronomically important species that develop proteoid roots in response to P deficiency (Avio et al., 1990; Gardner et al., 1982; Trinick, 1977). Proteoid roots excrete large amounts of the organic acids citrate and malate (Johnson et al., 1996a; Marschner, 1986, 1987;

* FAX No: 612-625-2208. E-mail: dallan@soils.umn.edu

Neumann et al., 1999), which help increase the availability of mineral bound phosphates (Dinkelaker et al., 1989; Gardner et al., 1983) and the release of phosphates from humic substances (Braum and Helmke, 1995). Acid phosphatases that may aid in the release of organic P from soil (Tadano and Sakai, 1991) are excreted coincident with the exudation of organic acids from proteoid roots (Gilbert et al., 1999; Miller et al., 2001; Neumann et al., 1999).

Root and shoot ^{14}C-labeling studies showed that a significant portion of the C excreted as malate and citrate was fixed by PEPC in roots (Johnson et al., 1996b). Excretion of organic acids from proteoid roots is accompanied by increased activity of malate dehydrogenase (MDH) and phosphoenolpyruvate carboxylase (PEPC) (Johnson et al., 1996a). Phosphoenolpyruvate carboxylase plays an important role in photosynthetic reactions in C_4 and CAM species. In C_3 species, PEPC has an anaplerotic function providing additional carbon for the formation of citric acid cycle intermediates (Melzer and O'Leary, 1987). The amount of C supplied via PEPC can be significant. PEPC can provide up to 25% of the carbon required for assimilation of nitrogen derived by symbiosis (Maxwell, 1982; Vance et al., 1983) and provides 25–34% of the C excreted as citrate and malate by P deficient proteoid roots of white lupin (Johnson et al., 1996b). Plant PEPC expression is controlled by environment (light, salt) and internal stimuli (hormonal, developmental) and has been shown to be under both positive (sugar phosphates) and negative (malate) allosteric control (Lepiniec et al., 1994).

Many of the physiological functions of PEPC in C_3 and C_4 plants are coupled to malate dehydrogenase (Lance and Rustin, 1984; Latzko and Kelly, 1983). MDH catalyzes the readily reversible conversion of carbon between oxaloacetic acid and malate (Gietl, 1992). MDH is ubiquitous in plants, playing key roles in the citric acid cycle, glyoxylate bypass, amino acid synthesis, gluconeogenesis and facilitation of the exchange of metabolites between cytoplasm and subcellular organelles (Musrati et al., 1998). Plant mitochondrial, cytosolic, glyoxysomal, chloroplast, and nodule-enhanced forms of the enzyme have been cloned and characterized (Cretin et al., 1988; Gietl, 1992; Guex et al, 1995; Kim and Smith, 1994; Metzler et al., 1989; Miller et al., 1998). While MDH can use either NADH or NADPH as reductant, only the chloroplast form of MDH utilizes NADPH (Gietl, 1992). Plant mitochondria use the concerted action of MDH, NAD-dependent malic enzyme (ME), and pyruvate dehydrogenase to provide citrate in the anaplerotic function of the TCA cycle (Gietl, 1992). Carbon can enter the TCA cycle in the form of cytosolic oxaloacetic acid or malate, the latter produced by the successive operation of PEPC and MDH in the cytosol (Gietl, 1992).

Objectives of this research were to analyze expression of genes related to organic acid metabolism in −P proteoid roots, with special focus on PEPC and MDH, and to examine factors that might affect PEPC activity in proteoid roots. To achieve this goal we wanted to isolate and characterize cDNAs encoding PEPC and MDH from white lupin, examine their expression over time, in specific tissues and in response to P deficiency. Our strategy for isolating PEPC and MDH cDNA was to use heterologous PEPC and MDH cDNAs from alfalfa to screen a cDNA library prepared from −P proteoid roots of white lupin. One full-length PEPC and two full-length MDH cDNA clones were isolated and characterized. The three cDNAs were sequenced and the deduced amino acid sequences were compared to other reported PEPC and MDH sequences. Transcript abundance was evaluated in different plant tissues and at different stages of proteoid root development in RNA gel blots. *In situ* hybridizations were performed to assess the cellular localization of PEPC and MDH transcripts in just emerged and developed proteoid rootlets. Macroarrays with ESTs from −P proteoid roots were performed to analyze expression patterns of other genes related to organic acid metabolism. A PEPC isoform predominant in P deficient proteoid roots was partially purified and characterized. The activity of phosphorylated and dephosphorylated PEPC was determined in response to several possible effectors at different pH in order to analyze regulation of PEPC activity.

Materials and methods

Plant material

White lupin (*Lupinus albus* L. var Ultra) plants were grown in the growth chamber in sand culture under growth conditions as previously described (Johnson et al., 1996a). P sufficiency or deficiency was defined by the presence or absence of 0.5 m*M* $Ca(H_2PO_4)_2$ in the nutrient solutions, which were replenished every 2 d (Gilbert et al., 2000). To maintain appropriate Ca concentrations, the nutrient solutions for the −P plants contained 0.5 m*M* $CaSO_4$.

Preparation and screening of a cDNA library

Proteoid roots of white lupin grown under P deficient conditions were harvested 7 and 10 DAE (days after emergence), and poly(A)$^+$ RNA was isolated. The poly(A)$^+$ RNA obtained from plants 7 and 10 DAE was combined in a 1:1 ratio and 7 μg total RNA was used for construction of a proteoid root cDNA library in the phage ZAPII vector according to the manufacturer's instructions (Stratagene, La Jolla, CA, USA). The original library represented approximately 2.6×10^6 transformants.

A 2 kb *Eco*RI fragment containing the 5′ portion of a *Medicago sativa* PEPC cDNA (Pathirana et al., 1992) and 2 *Sal*I fragments, which together contained the full-length cDNA encoding for a cytosolic MDH of *Medicago sativa* (Miller et al., 1998) were ^{32}P labeled and used for screening the cDNA library of white lupin proteoid roots (−P) harvested at 7 and 10 DAE.

RNA gel blots

Fifteen μg total RNA from different tissues were separated electrophoretically on 1.6% denaturing agarose gels and transferred to Hybond N+ nylon membrane following standard capillary blotting procedures (Sambrook et al., 1989). Blots were hybridized to a ^{32}P-labeled 937bp *Bgl*IIB fragment of LaPEPC1 cDNA and to the full-length LaMDH1 cDNA clone (1290 bp), respectively. Both probes were prepared by random primer-labeling. Hybridizations were performed in 0.5 *M* Na_2HPO_4, 7% SDS, and 10 m*M* EDTA at 65 °C. Blots were washed with three subsequent washes (2× SSC, 0.5% SDS; 1×SSC, 0.5% SDS; and 0.5×SSC, 0.5% SDS) before autoradiography. Equivalent loading of each lane was assessed by probing blots with ribosomal RNA and determining counts in the ribosomal bands. Loading variability was no greater than 20%.

Generation of ESTs

Proteoid roots of white lupin grown under P-deficient conditions were harvested 12 and 14 DAE. A proteoid root cDNA library was constructed in the phage λZAPII vector according to the manufacturers instructions (Stratagene). 5′ single strand sequences were edited and compared with the non-redundant database at the National Center for Biotechnology (NCBI) using the BlastX program.

Sequencing and computational analyses

Sequencing was carried out at the Advanced Genetic Analysis Center (St. Paul, MN, USA). Sequences were processed with the UW-GCG program (Devereux et al., 1984).

Macroarray assays

A 0.5 μl (about 0.2 μg aliquot) of PCR amplified cDNA was spotted in parallel onto Gene Screen Plus membranes (NEN^{TM} Life Science Products, Boston, MA, USA). ^{32}P-labeled first strand cDNA probe was synthesized by reverse transcription of 30 μg total RNA using SuperScriptII reverse transcriptase according to manufacturers instructions (Stratagene). The RNA was isolated from −P proteoid roots, +P and −P normal roots and +P and −P leaves at 14 DAE. Hybridizations were performed in 50% formamide, 0.5 *M* Na_2HPO_4, 0.25 *M* NaCl, 7% SDS, and 1 m*M* EDTA at 42 °C. Blots were washed with three subsequent washes (1× SSC, 0.1% SDS; 0.5×SSC, 0.1% SDS; and 0.1×SSC, 0.1% SDS). Radioactivity of each spot was quantified using a Phosphor Screen imaging system (Molecular Dynamics, Sunnyvale, CA, USA).

In situ *hybridization*

A 937bp *Bgl*IIB fragment of LaPEPC1 and the 1290 bp LaMDH1 cDNA, both subcloned into pBluescript, were used to obtain RNA probes for *in situ* hybridization. Linearized constructs were transcribed *in vitro* with T3 or T7 RNA polymerases (Stratagene) using [^{35}S]UTP according to the manufacturer's instructions. The probes were degraded partially to a length of about 150 nucleotides by heating at 60 °C in a 0.06 *M* Na_2CO_3/0.04 *M* $NaHCO_3$ solution.

In situ hybridization was carried out as described by Fleming et al. (1993). Tissues were fixed under mild vacuum at 4 °C in 4% paraformaldehyde and 0.25% glutaraldehyde in 50 m*M* phosphate buffer, pH 7.2. The tissues were then rinsed three times in deionized water, and dehydrated in a degraded ethanol series. After the absolute ethanol was replaced with tertiary butanol, tissues were embedded in Paraplast (Oxford Labware, St. Louis, MO). The embedded tissues were sectioned at a thickness of 10 μM and affixed onto poly-L-Lysin-coated slides. Hybridizations were performed in mineral oil and the final wash conditions were 0.1 × SSC including 1 m*M* DTT at 50 °C for 20 min. The slides were exposed to photo emulsion for 10–14 d. After development, the sections were stained with 0.05% toluidine blue O, dehydrated and mounted with Permount (Fisher Scientific). Sections were viewed and photographed with a Labophot microscope (Nikon) equipped with dark- and bright-field optics.

Enzyme purification

Proteoid roots were collected, quick frozen in liquid N_2 and stored at −80 °C until used for PEPC extraction. All subsequent steps were conducted at 4 °C. Frozen tissue was pulverized in liquid N_2 in a pre-chilled mortar and pestle. Phosphoenolpyruvate carboxylase was extracted from the pulverized tissue (2.5 mL buffer g^{-1} tissue) with potassium phosphate (KP_i) buffer, pH 7.3 (100 m*M* KP*i*, 10 m*M* $MgCl_2$, 10 m*M* malate, 1 m*M* EDTA, 0.3 m*M* sucrose, 20% v/v glycerol, 4% PEG-2000, 15 m*M* 2-mercaptoethoanol, 2 m*M* PMSF (phenylmethylsulfonyl fluoride), 1 mg L^{-1} trans epoxysuccinyl-*L*-leucylamindo (4-guaninino)-butane, 10 mg L^{-1} chrymostatin, 4 mg L^{-1} 3,4, dichloroisocoumarin, 0.5 mg L^{-1} leupeptin) based on Zhang et al. (1995) in a Waring blender for 1 min. The 2-mercaptoethanol, PMSF and protease inhibitors were added fresh on the day of use to prevent oxidation. Homogenate was filtered through several layers of miracloth before centrifuging at 10 000 ×*g* for 30 min. The supernatant containing the crude PEPC extract was fractionated with 50% polyethylene glycol-8000 in 100 m*M* KP*i*, 2 m*M* EDTA (pH 7.3) to 24% (Schuller, et al., 1990). The 24% fraction was separated by centrifugation at 10 000 ×*g* for 30 min. The resulting pellet was suspended in buffer A (20 m*M* KP*i*, 5 m*M* malate, 10% glycerol, 1 m*M* DTT (dithiothreitol) fresh, 1 m*M* PMSF (fresh) and 5 mg L^{-1} chrymostatin (pH 7.8)) and loaded onto a DEAE-sepharose (1.5 × 7.5 cm). The sample was eluted with a 120-mL linear gradient of 0–40% v/v of buffer A with 1 *M* KCl at a flow rate of 1.0 mL min^{-1}. Samples were collected in 1.5 mL aliquots and assayed for PEPC activity. Peak activity fractions were pooled and loaded on a hydroxylapatite column (BioRad Econopac 1 × 5 HT). The sample was eluted with a 40-mL linear gradient 0–100% from buffer A to buffer C (400 m*M* KP*i*, 5 m*M* malate, 10% glycerol, 1 m*M* fresh DTT) followed by 10 mL isocratic buffer C, at a flow rate of 0.5 mL min^{-1}. Ten mL aliquots were collected to assay for activity. The active peaks were desalted and concentrated with a 30 K filter microseparation device into buffer A, unless sample was to be used for malate kinetics, in which case it was desalted into buffer A without malate. The partially purified and concentrated PEPC could be stored for several weeks at −80 °C without loss of activity.

Protein electrophoresis and immunoblotting

Total soluble proteins were separated by SDS–PAGE or Native–PAGE using PhastGel 10–15% gradient acrylamide gels and transferred to Immobilon P (Millipore, Bedford, MA) (Braun and Abraham, 1989). Rabbit polyclonal antibodies to alfalfa nodule PEPC were used to detect PEPC on protein immunoblots. SDS-PAGE gels with FPLC-purified PEPC were silver-stained to check for protein purity (Heukeshovewn and Dernick, 1988). Phosphoenolpyruvate carboxylase was detected on Native-Page with fast violet B (Scrutton and Fatebene, 1975).

PEPC assay

The activity of PEPC was determined spectrophotometrically by monitoring the disappearance of NADH at A340 for two min in a MDH (EC 1.1.1.37) coupled reaction (Vance et al., 1983). Standard assay conditions were 100 m*M* bicine at pH 8, 5 m*M* $MgCl_2$, 10 m*M* H_2CO_3, 2 m*M* PEP (phosphoenolpyruvate), 0.16 m*M* NADH and 5 units MDH. The final volume was 1 mL; the reaction was initiated with the addition of the extract or partially purified PEPC. Total protein concentration was determined by Bio-Rad protein assay reagent (Bio-Rad, Hercules, CA, USA) for micro amounts of protein, using bovine serum albumin standards.

PEPC kinetic studies and metabolite regulation

Partially purified PEPC that had been stored at −80 °C in 5 m*M* malate and 10% glycerol was used for kinetic studies. The reaction was initiated with the addition of 10 μl partially purified PEPC enzyme. The K_m and V_{max} were determined for PEP using a range of 0.05 to 30 m*M* and Mg^{2+} using a range of 0.1–10 m*M* $MgCl_2$, with the remaining reagents maintained at standard assay conditions and pH 8.0. Kinetics were repeated four times using different PEPC preparations.

Test for regulatory properties of PEPC

The regulatory effects of several metabolites on PEPC isolated from P-deficient proteoid roots were assessed in the presence and absence of ATP, at pH 7 and at pH 8. The standard assay conditions were maintained with the exception of addition of metabolites, ATP and pH. ATP activation of PEPC was adapted from Schuller et al. (1993). Briefly, 5 g of roots were ground in 15 mL buffer (25 m*M* Tris-HCl, 0.5 *M* sucrose, 0.01 *M* DTT, pH 7.0). Extract was filtered through 4 layers of gauze and 1 layer of miracloth. The tubes were centrifuged at 26 000 $\times g$ for 10 min. Ten mL of supernatant fraction was desalted over a 10 cm Sephadex G-25 column (1×8 cm) and equilibrated in the grinding buffer. Fifty μL of crude extract were used for each assay. ATP was added at room temperature and samples were taken every 3 min and assayed for PEPC activity using standard assay conditions described above. Assays were repeated at least three times for each component and condition.

Results

Identification of a PEPC and two MDH cDNAs

In order to identify a cDNA encoding for PEPC in white lupin, a PEPC cDNA of *Medicago sativa* (Pathirana et al., 1992) was used as a probe for screening a cDNA library prepared from −P proteoid roots of white lupin 7 and 10 DAE. A cDNA-containing clone was identified and designated pLaPEPC1. Double-stranded sequencing confirmed a full-length PEPC cDNA.

To identify cDNA encoding for MDH in white lupin, the full-length cDNA of a cytosolic MDH from *Medicago sativa* (Miller et al., 1998) was used to screen the white lupin proteoid root cDNA library. Two cDNA containing clones were identified, designated pLaMDH1 and pLaMDH2. Double-stranded sequencing of the identified clones revealed two full-length MDH cDNAs.

The 3.3 kb cDNA inserted in pLaPEPC1 contains a 2904 bp ORF, designated *LaPEPC1*, which encodes a deduced protein of 967 amino acids with a Mr of 104 kD and a pI of 5.99. The deduced amino acid sequences of LaPEPC1 display the phosphoenolpyruvate carboxylase active sites, PEPCASE_1 and PEPCASE_2 (Figure 1A) characteristic for PEPCs. A putative phosphorylation site typical for plant PEPCs (Huber et al., 1994; Lepiniec et al., 1994; Zhang et al., 1995) is present in the N-terminus (Figure1A).

Both cDNA inserted in pLaMDH1 and pLaMDH2, respectively, contain 999 bp ORFs, designated *LaMDH1* and *LaMDH2*, which encode deduced proteins of 332 amino acids. Both proteins display a Mr of 35.6 and similar pI (6.1 for LaMDH1, 6.0 for LaMDH2). The deduced amino acid sequences of LaMDH1 and LaMDH2 are 95% identical. Both sequences contain an MDH active site signature (Figure 1B). There are no significant variations in the amino acid sequence between the two clones. Even the 3′ untranslated regions of *LaMDH1* and *LaMDH2* display 78% homology. The nucleotide sequences for *LaPEPC1, LaMDH1* and *LaMDH2* can be accessed through GenBank accession numbers AF459644, AF459645, and AF459646, respectively.

Table 1. Percentage of amino acid sequence identity between LaPEPC1 from white lupin and PEPCs from other organisms

	GenBank Accession #	% identity to *Lupinus albus* LaPEPC1
Glycine max, housekeeping form	Q02909	91
G. max, nodule enhanced	BAA23419	89
Medicago sativa, nodule enhanced	AAB46618	88
Pisum sativum, nodule enhanced	BAA10902	86
Zea mays, root form	CAA33663	81
Z. mays, C_4-form	QYZM	75
E. coli	P00864	42

Homology to other known PEPC and MDH

Protein sequence alignments of LaPEPC1 revealed 91% sequence homology to a C_3 isoform of PEPC from *Glycine max* and over 80% to root or nodule enhanced isoforms of *Glycine max, Medicago sativa, Solanum tuberosum* and *Zea mays* but only 75% homology to a C_4 form of *Zea mays* (Table 1). There was only 42% sequence homology with the bacterial PEPC form from *E. coli*.

Both putative MDH proteins, LaMDH1 and LaMDH2, display about 90% homology to other NAD-dependent MDH proteins (EC 1.1.1.37), e.g. from *M. sativa*, *N. tabacum*, common iceplant (*Mesembryanthemum crystallinum*), *Z. mays* and *A. thaliana* (Table 2). Both forms are only about 40% homologous to an NADP-dependent form of MDH from *A. thaliana* chloroplasts.

Expression of PEPC and MDH is induced under P deficiency

To assess the expression of PEPC and MDH in various tissues of white lupin, RNA gel blots were performed

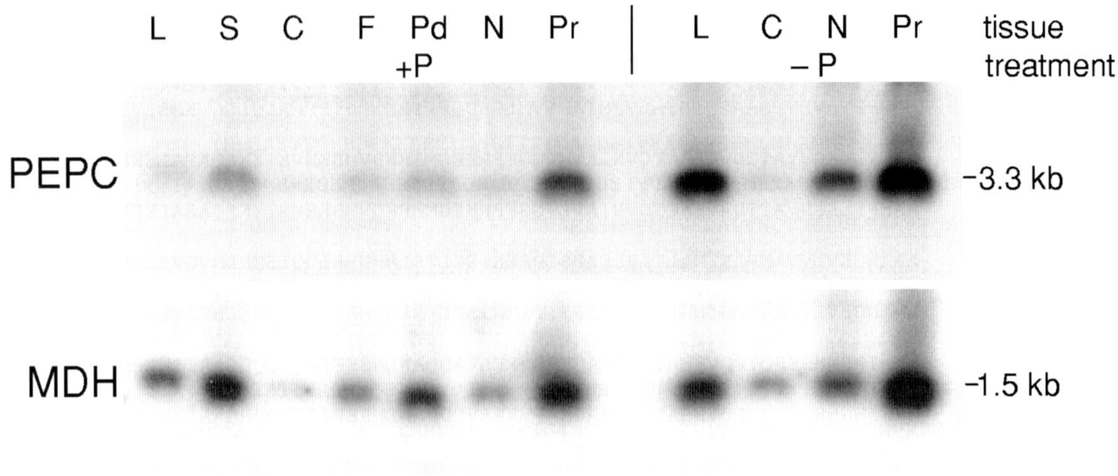

Figure 2. Gel blot analysis of RNA isolated from different tissues of +P and −P treated white lupin. LaPEPC1 cDNA was used as a probe in the upper panel while in the lower panel LaMDH1 was used as a probe. Plants grown in presence of sufficient P are designated by (+) and those grown under P stress are designated by (–). Letters above each lane represent: L = leaves; S = stems; C= cotyledons; F = flowers; Pd= pods; N= normal roots; Pr = proteoid roots. Flowers and pods did not form on −P plants. Each lane contains 15 μg of total RNA.

under −P conditions by 14 DAE (Figure 3). A similar expression pattern was found for MDH. Induction of MDH in −P normal and proteoid root tissue was not apparent at 10 DAE but MDH was strongly expressed by 14 DAE (Figure 3).

In situ *hybridization*

Single stranded sense and antisense RNA, derived from LaPEPC1 and LaMDH1 cDNAs, were used as probes for *in situ* hybridization of just emerging (9 DAE) and developed (12 DAE) proteoid roots as well as of +P normal roots (12 DAE) to ascertain the cellular localization of the corresponding transcripts. *In situ* hybridization with the LaPEPC1 antisense probe, seen in the dark field as bright silver grain, revealed expression in the just emerging rootlet (Figure 4A) but stronger expression throughout the proteoid rootlet 12 DAE (Figure 4C). Transverse sections of the rootlet displayed no detectable transcript at the root cap (Figure 4E), but high transcript abundance in the cortex of a cross section about 500 μM distant to the tip (Figure 4G). Normal roots showed high transcript abundance in the apical meristem and to a lesser degree in the more distal cortex (Figure 4I).

After hybridization with the LaMDH1 antisense probe, dark-field microscopy revealed high transcript abundance throughout an emerging rootlet 9 DAE (Figure 5A). Proteoid rootlets 12 DAE display strong expression over a region of several hundred μM along the rootlet but less so at the tip or base (Figure 5C). Accordingly, no hybridization was detected in a cross section from the root tip (Figure 5E). Dark field photography of a transverse section approximately 500 μM distant to the proteoid rootlet tip revealed high transcript abundance that seems most pronounced in the cortex (Figure 5G). Normal roots have lower expression overall but a similar expression pattern, displaying highest transcript abundance in the cortex at some distance from the root tip (Figure 5I). No significant hybridization was detected when normal roots were probed with LaPEPC1 (Figure 4K) and LaMDH1 (Figure 5K) sense probes, respectively.

Macroarray analysis

To assess C-metabolism related gene expression in proteoid roots on a wider scale, we performed macroarray analyses of about 1250 ESTs derived from −P proteoid roots of white lupin. A total of 55 ESTs (4.4%), displayed significant homology (E-value $\leq 10^{-10}$) to genes involved in the glycolytic pathway. Two ESTs each displayed homology to PEPC and MDH. Alignments of the corresponding ESTs with the complete PEPC and MDH cDNAs revealed sequence identity to LaPEPC1 and LaMDH2, respectively.

Gene expression was compared in −P proteoid roots, +P normal roots, −P normal roots, +P leaves and −P leaves at 14 DAE. Macroarray analyses revealed at least 2-fold induction in −P proteoid roots,

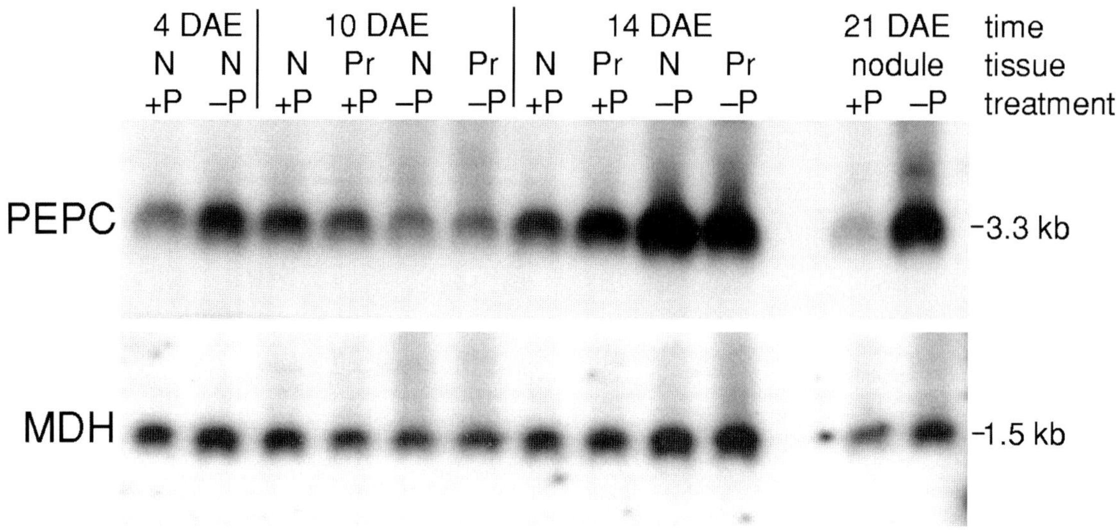

Figure 3. Gel blot analysis of RNA isolated from +P and −P normal (N) and proteoid (Pr) roots at 4, 10, and 14 DAE and nodules (Nd) at 21 DAE. LaPEPC1 cDNA (upper panel) and LaMDH1 cDNA (lower panel) served as a probe. Each lane contains 15 μg of total RNA.

compared to +P normal roots, of the majority of glycolytic pathway-related genes (Table 3, Figure 6). In contrast, the majority of these genes did not show induced expression in −P leaves compared to +P leaves, nor in −P normal roots compared to +P normal roots (Table 3). RNA gel blots performed with selected ESTs confirmed the results obtained by macroarray analysis. Figure 7 shows induced expression of a glyceraldehyde-3P dehydrogenase in −P proteoid roots but not in +P or −P normal roots or leaves.

Partial purification and characterization of a proteoid root enhanced PEPC isoform

Phospho*enol*pyruvate carboxylase enzyme protein was purified about 70 fold from −P proteoid roots (Table 4). After desalting and concentrating, the specific activity was 18.3 units. A unit of PEPC specific activity is defined as the conversion of 1 μmol NADH to NAD^+ min^{-1} mg^{-1} protein.

The purified PEPC from −P proteoid roots displayed a K_m PEP of 0.055 m*M* (±0.01) and a K_m Mg^{2+} of 0.24 m*M* (± 0.095) with V_{max} of 0.07 (±0.028) and 0.046 (±0.019) units, respectively. These measurements were done at standard assay conditions at pH 8 using PEPC stored at −80 °C in the presence of 10% glycerol and 5 m*M* malate. Only 10 μL of PEPC was used in each assay; therefore the concentration of malate and glycerol in the assay were negligible.

Native protein immunoblots and native PAGE activity stainings indicated a native molecular weight of about 440 kD for the partially purified PEPC (Figure 8). SDS-PAGE as well as Superose 6 analysis indicated a subunit size of 110 kD (data not shown).

Metabolite regulation of partially purified proteoid root-enhanced PEPC

Several metabolites were analyzed in regard to inhibition or activation of PEPC at pH 7 and pH 8, with and without preincubation with ATP (Table 5) to assess possible effects of endogenous phosphorylation (Schuller and Warner, 1993). Hexose phosphates showed strongest activation in absence of ATP at pH 7. Citrate and 2-oxoglutarate were strong inhibitors independent of pH and independent of the presence of ATP. Malate was the strongest inhibitor at pH 7; inhibition was reduced by preincubation with ATP or by increasing the pH to pH 8. The metabolites 3-phosphoglycerate and 2-phosphoglycerate displayed 20–40% inhibition after preincubation with ATP and at pH 8, but showed no effect at pH 7 in the absence of ATP. DHAP increased activity 7% at pH 7 and 15% at pH 8 when assayed without preincubation, but had no effect when PEPC was preincubated with ATP. The amino acids Asp and Gln caused less than a 15% decrease in PEPC activity at pH 7 and had marginal effects at pH 8. Addition of P_i in concentrations ranging from 10 to 30 m*M* showed some activation at pH 7 in

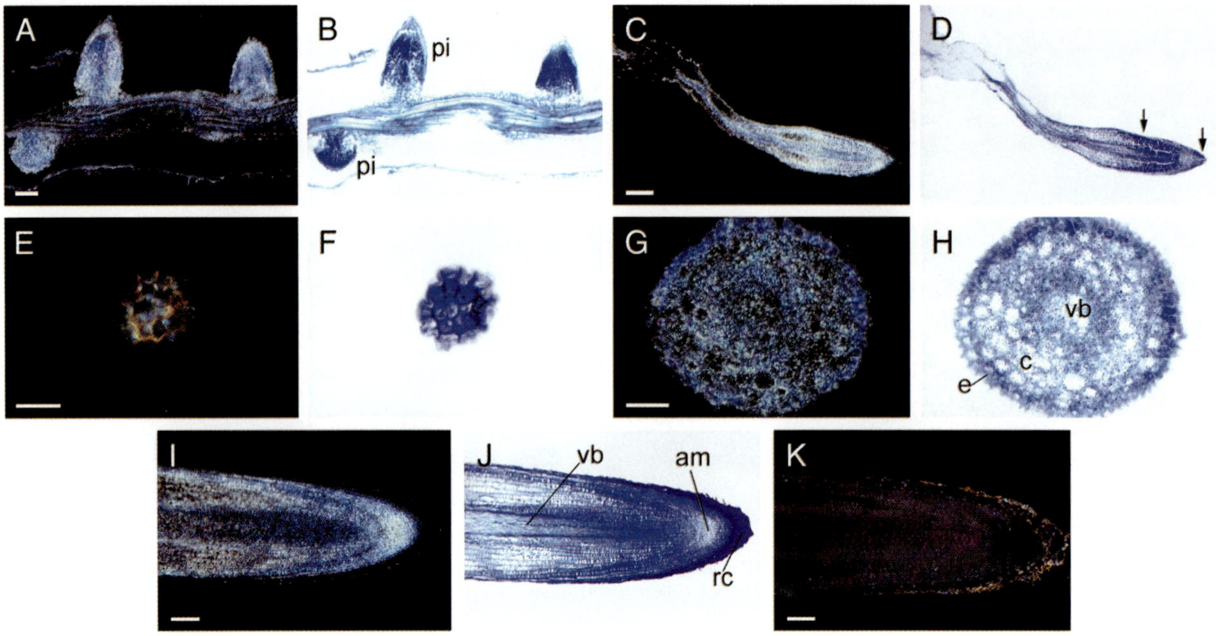

Figure 4.

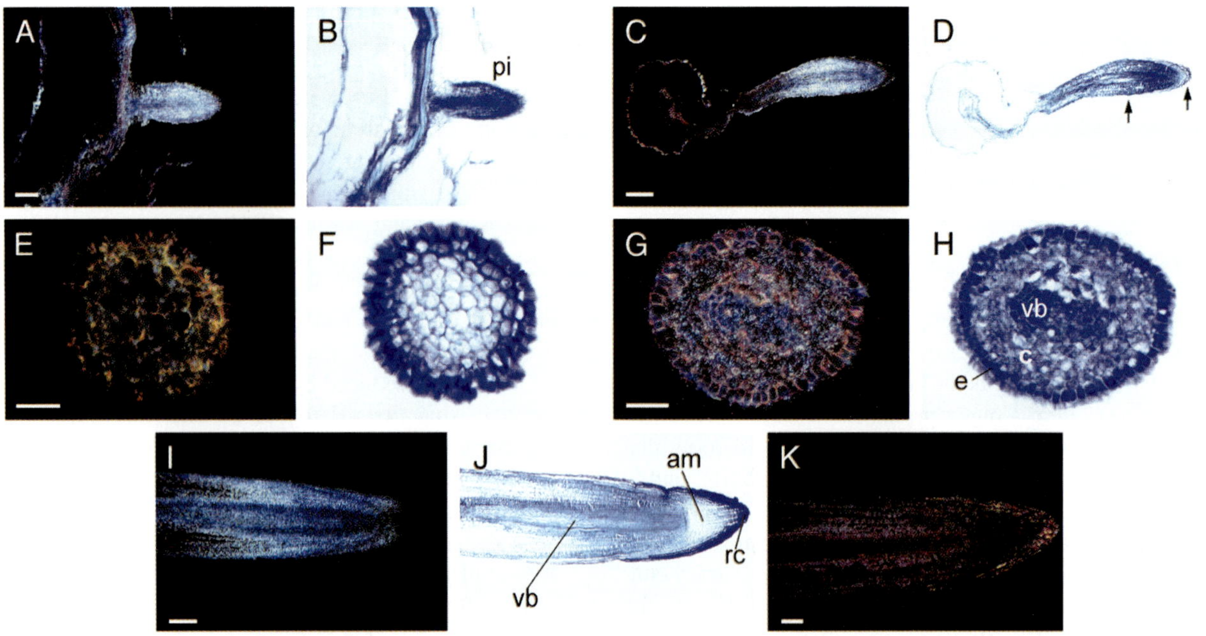

Figure 5.

Table 3. Macroarray analysis revealed −P proteoid root-induced expression of genes with putative function in the glycolytic pathway. Expression in leaves and normal roots was not induced under P deficiency

Annotation	Number of ESTs[b]	Ratio of gene expression[a]		
		−P proteoid/ +P normal root	−P leaf/ +P leaf	−P normal root/ +P normal root
Sucrose synthase	6	3	1	0.8
Fructokinase	1	2.2	0.5	0.8
Phosphoglucomutase	2	2.1	0.7	0.9
Glucose-6P isomerase	1	2.1	1.3	0.8
PPi-dependent phosphofructokinase	5	7.3	0.9	1.9
Fructose-1,6-bisP aldolase	12	4.2	0.8	1
Triose-phosphate isomerase	3	3.2	0.9	1.4
Glyceraldehyde-3P dehydrogenase	12	3	0.8	1.3
Phosphoglycerate kinase	4	3	1.7	1
Enolase	4	2.5	0.9	1.1
Phospho*enol*pyruvate carboxylase	2	6	0.5	1.4
Malate dehydrogenase	2	4.9	0.6	1.4
Malic enzyme	1	1.2	2	1.6

[a]Ratios are the average of corresponding ESTs from two independent macroarray experiments.
[b] Number of ESTs with corresponding BLASTX annotation.

Table 4. Partial purification of PEPC from 94 g of −P proteoid root tissue collected from plants 14 DAE

Sample	mg protein ml^{-1}	Total protein mg[a]	PEPC specific activity	Protein total U[b] mg^{-1}	Purification	PEPC yield	Protein yield %
Crude extract	0.478	119.55	0.27	32.75	1.00	100.00	100.00
4–24% PEG fraction	0.919	27.56	0.70	19.38	2.57	59.18	23.06
DEAE	0.142	2.35	7.97	18.69	29.09	57.08	1.96
HT	0.093	0.56	13.06	7.28	47.67	22.22	0.47
Desalted and concentrated	0.123	0.32	18.28	5.86	66.72	17.89	0.27

[a]Protein determined with Bio-Rad assay for micro amount of protein based on Bradford (1976).
[b]One unit of PEPC specific activity is the conversion of 1 μmol NADH to NAD^+ min^{-1} mg^{-1} protein at A_{340} in an MDH coupled assay.

←
Figure 4. In situ localization of PEPC in −P proteoid roots of white lupin. Sections A–J were hybridized with ^{35}S-labeled antisense RNA derived from LaPEPC1 cDNA, whereas section K was hybridized with the corresponding sense RNA probe. The dark field (A, C, E, G, I, K) and bright field (B, D, F, J) images were longitudinal and transverse sections through proteoid rootlets 9 DAE (A, B) and 12 DAE (C-H) as well as through 12 DAE +P normal roots (I, J, K). Arrows in panel D indicate the approximate location of the two transverse sections shown in panel E–H. Bar sizes represent 50 μM (E, G), 100 μM (A), and 200 μM (C, I, K), respectively. Letters represent: pi = proteoid initials, vb = vascular bundle, c = cortex, e = epidermis, am = apical meristem, rc = rootcap.
Figure 5. Localization of MDH in −P proteoid roots of white lupin by *in situ* hybridization. Sections A–J were hybridized with ^{35}S-labeled antisense RNA, derived from LaMDH1 cDNA, section K was hybridized with the corresponding sense RNA probe. The dark field (A, C, E, G, I, K) and bright field (B, D, F, J) images were longitudinal and transverse sections through proteoid rootlets 9 DAE (A, B) and 12 DAE (C–H) as well as through 12 DAE +P normal roots (I, J, K). Arrows in panel D indicate the approximate location of the two transverse sections shown in panel E–H. Bar sizes represent 50 μM (E, G), 100 μM (A, K), and 200 μM (C, I), respectively. Letters represent: pi = proteoid initials, vb = vascular bundle, c = cortex, e = epidermis, am = apical meristem, rc = rootcap.

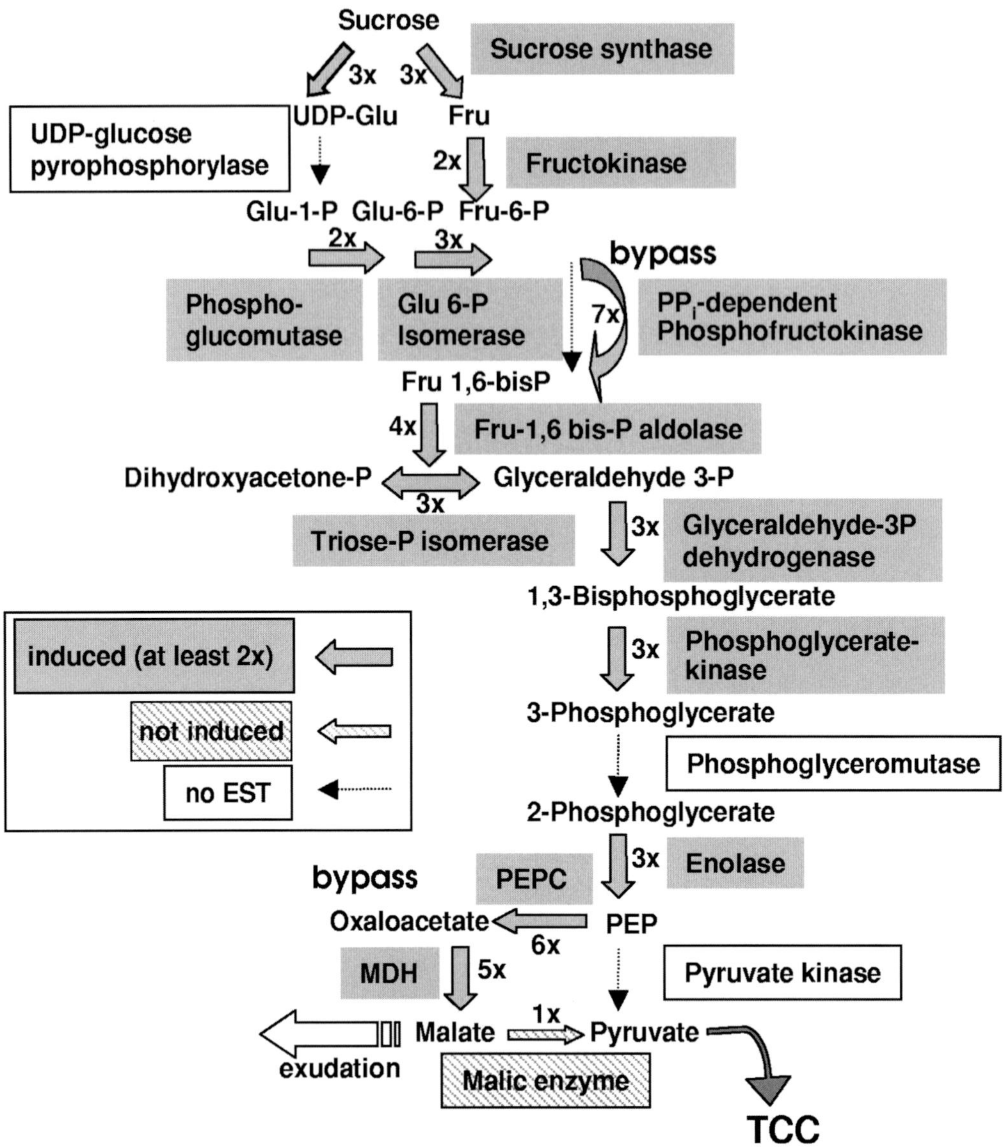

Figure 6. Schematic representation of the glycolytic pathway. ESTs with induced expression in −P proteoid roots, compared to +P normal roots, are represented in gray boxes. The average of gene induction as determined by two independent macroarrays is indicated at the corresponding arrows (e.g. 2x). Enzymes of the glycolytic pathway that were not found in the collection of 1250 ESTs are shown in white boxes and are represented by dotted arrows. The majority of ESTs with possible function in the glycolytic pathway displayed increased expression in −P proteoid roots, compared to +P normal roots.

the absence of ATP but inhibition to various degrees under all other conditions tested.

Discussion

In this report, we have advanced the understanding of how enzymes involved in organic acid synthesis, particularly MDH and PEPC, participate in the process of acclimation to P deficiency in white lupin. This was done by (1) isolating and sequencing cDNAs with high homology to PEPC and MDH; (2) demonstrating induced expression of PEPC and MDH under P deficiency in different tissues but most pronounced in proteoid roots; (3) revealing different expression patterns of PEPC and MDH in just emerged and developed proteoid rootlets by *in situ* hybridization; (4) show-

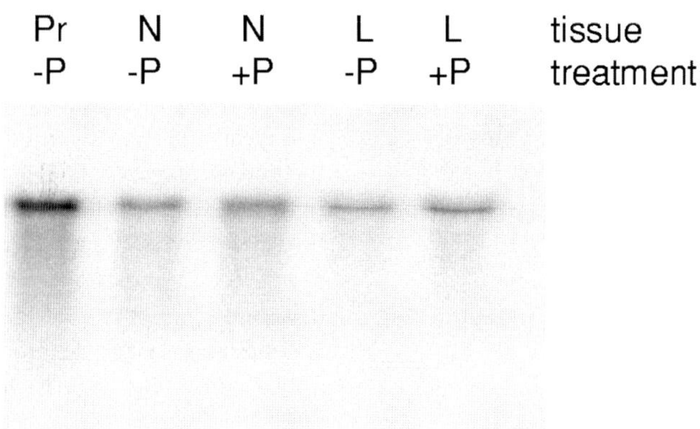

Figure 7. RNA gel blot analysis to confirm the differential expression pattern indicated by macroarray analysis. RNA was isolated from −P proteoid roots (Pr), +P and −P normal roots (N) and +P and −P leaves (L) at 14 DAE. Each lane contains 15 μg of total RNA. An EST with homology to NAD-dependent glyceraldehyde-3P dehydrogenase was used as a probe.

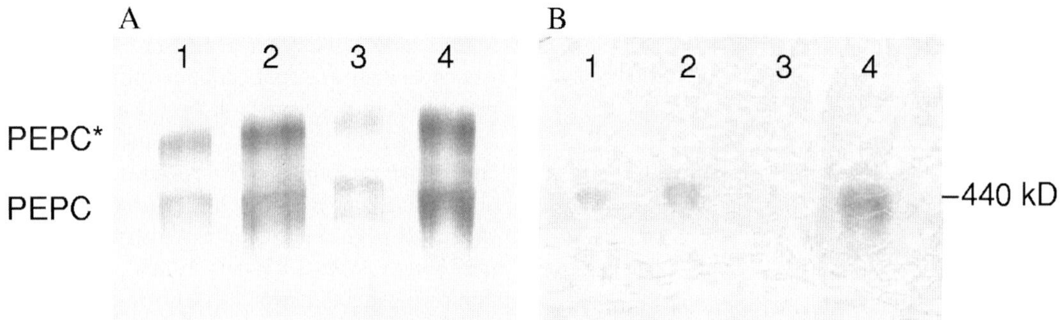

Figure 8. Native protein immunoblot (A) and native PAGE activity staining (B) of various samples obtained during the partial purification of PEPC from proteoid roots of white lupin. The band detected by activity staining corresponds to the lower band of the protein gel blot and represents the 440 kD tetramer of PEPC. The upper band visible in the protein gel blot was not detected in activity stainings and might reflect a different phosphorylation status of PEPC. Lane 1 = crude extract, 482 ng total protein/lane; Lane 2 = PEG cut, 729 ng total protein/lane; Lane 3 = DEAE cut, 72 ng total protein/lane, Lane 4 = HT column concentrated, 131 ng/lane.

ing differential expression of genes with homology to enzymes of the glycolytic pathway by macroarray analysis of proteoid root derived ESTs; and (5) partially purifying and characterizing a proteoid root-enhanced PEPC isoform.

Both PEPC and MDH exist in small multigene families. Each member encodes a distinct isoform associated with its respective metabolic pathway. The different isoforms play different physiological roles depending on location and function in the plant (Ting and Osmond, 1973). Root forms of PEPC are known to have various anaplerotic functions that include providing the carbon skeletons for nitrogen assimilation, pH maintenance, and osmolarity regulation (Latzko and Kelly, 1983). Like PEPC, MDH is a widely distributed enzyme playing a crucial role in many metabolic pathways.

To identify cDNAs of PEPC and MDH isoforms that are expressed in proteoid roots, we screened a proteoid root cDNA library of white lupin with heterologous PEPC and MDH cDNAs from *M. sativa*. With this approach we isolated a putative PEPC cDNA and two putative MDH cDNAs. Protein sequence alignments of LaPEPC1 revealed higher sequence homology (80–90%) to root or nodule enhanced C_3 isoforms than to a C_4 form of *Z. mays* (75%) (Table 1). LaMDH1 and LaMDH2 both displayed high homology (about 90%) to cytosolic MDH forms of *M. sativa*, *Z. mays* and *A. thaliana* but only relatively low homology (40%) to a NADP-dependent form from *A. thaliana* chloroplasts (Table 2).

RNA gel blot experiments revealed induced expression of PEPC in −P root and leaf tissues with highest expression in −P proteoid roots. The high expression in −P proteoid roots is consistent with results

Table 5. Comparison of regulatory properties of PEPC in −P proteoid roots in absence and presence of 5 m*M* ATP

Metabolite	Relative activity in %[a]			
	pH 7		pH 8	
	− ATP	+ ATP	− ATP	+ ATP
4 m*M* Glu-6-P	135	127	113	123
4 m*M* Glu-1-P	120	105	110	108
4 m*M* Fru-6-P	134	112	104	106
4 m*M* DHAP	107	99	115	101
5 m*M* 3-PGA	102	72	65	58
5 m*M* 2-PGA	121	81	69	78
1.25 m*M* Malate	26	54	82	86
1.25 m*M* Asp	88	86	97	95
1.25 m*M* Glutamine	87	94	104	106
2.5 m*M* Citrate	18	24	30	45
5 m*M* 2-OG	55	58	67	61
10 m*M* P_i	121	92	83	81
20 m*M* P_i	133	82	67	62
30 m*M* P_i	117	72	80	60

[a]Enzyme activity is expressed relative to the respective control set at 100%. Each value represents the average of at least three separate measurements. Standard errors did not exceed 15%.

from previous work showing increased in vitro activity and enzyme protein level in P deficient proteoid roots (Johnson, 1996b; Neumann et al., 1999, 2000). Expression of MDH follows a similar pattern with some induction under P deficiency in leaves and normal roots but highest expression in −P proteoid roots. Induction of PEPC and, to a lesser degree MDH, under P deficiency has also been reported in tomato (Pilbeam et al., 1993) and *Sesbania rostrata* (Aono et al., 2001).

Time course experiments revealed greater expression of PEPC and MDH in mature proteoid roots (14 DAE) than in just emerged proteoid roots (10 DAE). This timing corresponds to the timing of organic acid exudation and increased PEPC activity previously reported for white lupin (Johnson et al., 1996b). Watt and Evans (1999a), however, reported maximal PEPC activity 1 day before the onset of citrate efflux and 2 days before its peak, followed by a 75% decrease of PEPC activity over a period of 3 days. The observation that maximal PEPC activity precedes citrate efflux might be explained by the increase of internal citrate concentrations prior to excretion (Neumann et al., 1999). The early decrease in PEPC activity is not reflected in the transcript abundance and might be directly related to maximum inhibition by increased concentrations of internal citrate. The time discrepancy between maximum *in vitro* PEPC activity and citrate efflux led to the authors' postulation that another factor, possibly an organic acid transporter, might be responsible for the timing of organic acid efflux (Watt and Evans, 1999a). Maximum acid phosphatase activity and strong expression of a putative high affinity phosphate transporter have also been observed in mature proteoid roots at 14 DAE (Gilbert et al., 1999; Liu et al., 2001), indicating a precise coordination of the different components of the proteoid root response to P deficiency. In this regard, it is noteworthy that plant uptake of P hydrolyzed by acid phosphatase is thought to be improved by the presence of citrate, a chelate that effectively binds metals that would otherwise compete for released P (Braum and Helmke, 1995; Watt and Evans, 1999b).

To reveal the spatial and temporal distribution of PEPC and MDH expression in proteoid roots, *in situ* hybridizations were performed on −P proteoid rootlets and +P normal roots. Mature proteoid rootlets display a differentiated apex, a loosely anchored root cap, an endodermis with a Casparian band, a diarch stele, and numerous root hairs (for reviews see Dinkelaker et al., 1995; Skene et al., 1998b). Citrate efflux has been shown to begin shortly after the rootlets reach their final length (Watt and Evans, 1999a). *In situ* hybridization revealed expression of PEPC in the just emerging proteoid rootlets but as the proteoid rootlet matured the transcript abundance increased (Figure 4). These findings are consistent with the RNA gel blot time course experiments reported here that displayed low PEPC expression for juvenile (10 DAE) but high expression in mature (14 DAE) proteoid roots. As can be seen in Figure 4, PEPC expression is localized primarily in the cortex and in the meristem. This first report of the site of organic acid synthesis in proteoid roots is consistent with the likely routes of organic acid excretion including secretion into the cortical apoplast or cell to cell transport and excretion from the epidermis.

In situ hybridization of MDH also revealed expression in the cortex but not in the meristem of mature rootlets. Malate dehydrogenase displayed expression throughout the just emerging proteoid rootlet; however, in mature proteoid rootlets expression was restricted to a central zone along the rootlet with no detectable expression at the tip and base (Figure 5). This finding might indicate that part of the oxaloacetate that is produced by PEPC is not converted into malate by the subsequent action of MDH. An alternative fate of oxaloacetate is its conversion into citrate by citrate synthase in the mitochondria. Thus, in mature proteoid root tissues that display PEPC but not MDH, a flux of oxaloacetate might be directed to citrate rather than to malate. These findings correspond to reports that show a significant increase of citrate and decrease of malate exudation during the development of proteoid roots (Neumann et al., 1999).

To gain insight into expression patterns of other genes related to organic acid metabolism, we performed macroarray analyses with about 1250 ESTs from −P proteoid roots. The majority of ESTs with homology to enzymes of the glycolytic pathway, including ESTs with homology to LaPEPC1 and LaMDH2, were significantly induced in P deficient proteoid roots (Table 2; Figure 6). These results correspond well to findings from Massonneau et al. (2001) who reported increased expression and enzyme activity of sucrose synthase, fructokinase and phosphoglucomutase in P deficient proteoid roots of white lupin.

An interesting aspect of organic acid metabolism under P deprivation in plants is the induction of alternative glycolytic pathways (reviewed by Theodorou and Plaxton, 1993). In P-stressed proteoid roots, PEPC not only provides C for the synthesis of organic acids but may also recycle P_i. Phosphorus stress can severely limit the activity of pyruvate kinase (PK), an enzyme requiring P_i and ADP. PEPC and MDH can bypass PK (Figure 6) and thus maintain the flow of carbon from glycolysis to the TCA cycle by avoiding the use of ADP but generating free P_i (Theodorou et al., 1992).

Another alternative glycolytic pathway known to occur in P deficient plants is catalyzed by a PP_i-dependent phosphofructokinase (PFP) that can bypass the ATP-dependent phosphofructokinase (PFK). PFK was not represented in our EST collection, whereas the alternative PFP was found redundantly. Moreover, macroarrays revealed strong induction of the PFP gene in −P proteoid roots, compared to +P normal roots, indicating the significance of this bypass in P deficient proteoid roots of white lupin (Table 3, Figure 6).

Though these results emphasize the role of P-conserving alternative pathways in the carbon metabolism of −P proteoid roots from white lupin, we did not find evidence for the induction of a third alternative glycolytic pathway known in plants that is catalyzed by a non-phosphorylating NADP-dependent glyceraldehyde-3P dehydrogenase. None of the 1250 ESTs displayed homology to the alternative NADP-dependent form of this enzyme whereas the NAD-dependent form was found redundantly and, moreover, showed induction in −P proteoid roots (Table 3).

RNA gel blot analyses and macroarray data give an insight into the expression patterns of genes involved in organic acid synthesis. To address the question whether PEPC in roots of P deficient white lupin is also regulated at the post-translational level by phosphorylation or effectors of enzyme activity, PEPC was partially purified from P-deficient proteoid roots. Immunoblot analysis indicated a total protein of 440 kd with a subunit mass of 110 kd, indicative of a homotetramer structure for the enzyme. These findings are consistent with native and subunit masses of PEPC from soybean nodules (Schuller et al., 1990).

Known isoforms of PEPC, (C_4-photosynthetic, CAM-photosynthetic, C_3-photosynthetic and non-autotrophic forms) are distinguishable by K_m and V_{max} values, ion exchange chromatography, and sensitivity to malate inhibition (Ting and Osmond, 1973). The K_m PEP and K_m Mg^{2+} of *Zea mays* are lower in root isoforms than in the C_3 and C_4 leaf forms (Dong et al., 1998; Ting and Osmond, 1973). The low K_m PEP of 0.05 m*M* for the PEPC isoform from proteoid roots of white lupin corresponds well to the kinetic data determined for several other root and nodule forms of PEPC, e.g. a root form from *Z. mays* (K_m PEP 0.04 m*M*; Dong et al., 1998), a nodule form from soybean (K_m PEP 0.06 m*M*; Schuller et al., 1990) and a nodule form from yellow lupin (PEPC I, K_m PEP 0.09 m*M*, Marczewski, 1989).

In yellow lupin, a total of three different isoforms have been isolated from nodules and roots. We found little evidence for an additional PEPC isoform in roots of white lupin. When screening a proteoid root cDNA library, we identified a partial cDNA clone with high homology to LaPEPC1 in the coding region but distinct differences in the 3′ non-coding region. Hybridization with the 3′ non-coding region displayed extremely weak expression in 14 DAE −P normal and proteoid roots but not in nodules (data not shown). Activity stainings at various steps of the PEPC purification, however, displayed only single bands. Though protein gel blots did show two bands, they most likely reflect a change in phosphorylation state of the enzyme, since only the 440 kb band displayed PEPC activity. If there is an additional isoform of PEPC in roots of white lupin, extremely weak expression and high similarity to LaPEPC1 might make it difficult to detect.

It is well established that PEPC in leaves of C_4 and CAM plants is regulated by protein kinase-mediated phosphorylation of a specific serine residue near the N terminus (Schuller and Werner, 1993; Wang and

Chollet, 1993). This phosphorylation is known to decrease the sensitivity of PEPC to malate inhibition (Chollet et al., 1996; Jiao and Chollet, 1991; Schuller and Werner, 1993). PEPC- kinase and an opposing phosphatase have been isolated and characterized (Bakrim et al., 2001; Dong et al., 2001; Hartwell et al., 1999; Taybi et al., 2000; Zhang and Chollet, 1997). A time-dependent increase in PEPC activity due to phosphorylation by endogenous kinase *in vitro* has been shown when desalted nodule extracts were preincubated with ATP (Schuller and Werner, 1993). Preincubation of the partially purified PEPC from white lupin proteoid roots with ATP resulted in a comparable decrease of sensitivity to malate inhibition, suggesting *in vitro* phosphorylation of this PEPC isoform by an endogenous protein kinase. Correspondingly, a putative phosphorylation motif was present in the N-terminus of the proteoid root induced LaPEPC1 (Figure 1). Phosphorylation of PEPC had a strong influence on its response to several activating or inhibiting metabolites tested, indicating an important role of phosphorylation in the regulation of this isoform (Table 5). However, the phosphorylation state of the enzyme had little effect on the inhibition by citrate and 2-oxoglutarate. Inhibition and activation of a number of metabolites was influenced by pH. For example, Glu-6-P, Glu-1-P and Fru-6-P were stronger activators at pH 7 compared to pH 8, while malate and citrate were stronger inhibitors at pH 7 than at pH 8. Similar influence of pH has been reported for the effect of Glu-6-P and malate on PEPC from maize leaves and soybean nodules (Frank et al., 2001; Israel and Jackson, 1982; Schuller et al., 1990), a finding that led to the suggestion of a pH stat function of soybean nodule PEPC (Israel and Jackson, 1982).

Even though there are some differences in the degree of activation and inhibition, the effects of the analyzed metabolites in regard to pH and phosphorylation state of PEPC are similar to those described in PEPC nodule forms of soybean and lupin (Marczewski, 1989; McCloud et al., 2001; Schuller et al., 1990). These findings indicate that regulation of proteoid root-enhanced PEPC activity occurs in a general metabolic manner in response to metabolites, pH and phosphorylation and does not respond specifically to P-availability. This result is consistent with findings from Moraes and Plaxton (2000), who reported that an increase of PEPC activity under P deficiency in *Brassica napus* suspension cells was not correlated with a change in regulatory properties of PEPC.

The results presented here indicate that regulation of a proteoid root-enhanced PEPC isoform in response to P deficiency might occur mainly at the level of gene expression. Correspondingly, most genes encoding for enzymes involved in the glycolytic pathway, including two isoform(s) of MDH, displayed significant induction in −P proteoid roots on the level of transcription. Furthermore, *in situ* hybridization revealed a fine-tuning of the coordinated expression of PEPC and MDH in a tissue-dependent manner that might reflect changes in the composition of excreted organic acids during proteoid root development.

Acknowledgements

This research was supported in part by US Department of Agriculture National-Research Initiative Competitive Grant No.USDA-NRI-CGP 98-35100-6098. The authors wish to thank Kate Plaisance for advice on use of the FPLC, enzyme purification, and analysis of enzyme kinetic and inhibition data.

References

Aono T, Kanada N, Ijima A and Oyaizu H 2001 The response of the phosphate uptake system and the organic acid exudation system to phosphate starvation in *Sesbania rostrata*. Plant Cell Physiol. 15,1253–1264.

Avio L, Sbrana C and Giovannetti M 1990 The response of different species of *Lupinus* to VAM endophytes. Symbiosis 42, 321–323.

Bakrim N, Brulfert J, Vidal J and Chollet R 2001 Phosphoe*nol*pyruvate carboxylase kinase is controlled by a similar signaling cascade in CAM and C_4 plants. Biochem. Biophys. Res. Com. 286, 1158–1162.

Bradford M M 1976 A rapid and sensitive method for the quantification of microgram quantities of protein utilizing the principle of protein-dye binding. Anal. Biochem. 72:248–254.

Braum S and Helmke P A 1995 White lupin utilizes soil phosphorus that is unavailable to soybean. Plant Soil 176, 95–100.

Braun W and Abraham R 1989 Modified diffusion blotting for rapid and efficient protein transfer with PhastSystem. Electrophoresis 10, 249–253.

Chollet R, Vidal J and O'Leary M H 1996 Phospho*enol*pyruvate carboxylase: a ubiquitous, highly regulated enzyme in plants. Annu. Rev. Plant Physiol. Plant Mol. Biol. 47, 273–298.

Cretin C, Luchetta P, Joly C, Miginiac-Maslow M, Decottignies P, Jacquot J, Vidal J and Gadal P 1988 Identification of cDNA clone from sorghum leaf malate dehydrogenase (NADP). Eur. J. Biochem. 174, 497–501.

Devereux J, Haeberli P and Smithies O 1984 Sequence analysis programs for the VAX. Nucleic Acids Res. 12, 387–395.

Dinkelaker B, Romheld V and Marschner H 1989 Citric acid excretion and precipitation of calcium citrate in the rhizosphere of white lupin (*Lupinus albus* L.). Plant Cell Environ. 12, 285–292.

Dinkelaker B, Hengeler C and Marschner H 1995 Distribution and function of proteoid roots and other root clusters. Botanica Acta. 108, 169–276.

Dong L-Y, Masuda T, Kawaruma T, Hata S and Izui K 1998 Cloning, expression, and characterization of a root-form phosphoe*nol*pyruvate carboxylase from *Zea mays*: comparison with the C_4-form enzyme. Plant Cell Physiol. 39, 865–873.

Dong L, Ermolova N V and Chollet R 2001 Partial purification and biochemical characterization of a heteromeric protein phosphatase 2A holoenzyme from maize (*Zea mays* L.) leaves that dephosphorylates C_4 phosophoe*nol*pyruvate carboxylase. Planta 213, 379–89.

Fleming A J, Mandel T, Roth I and Kuhlemeier C 1993 The patterns of gene expression in the tomato shoot apical meristem. Plant Cell 5, 297–309.

Frank J, Claarke R J, Vater J and Holzwarth J F 2001 Influence of allosteric effectors on the kinetics and equilibrium binding of phospho*enol*pyruvate (PEP) to phospho*enol*pyruvate carboxylase (PEPC) from *Zea mays*. Biophys. Chem. 92, 53–64.

Gardner W K, Parbery D G and Barber D A 1982 The acquisition of phosphorus by *Lupinus albus* L. I. Some characteristics of the soil/root interface. Plant Soil 68, 19–32.

Gardner W K, Barber D A and Parbery D G 1983 The acquisition of phosphorus by *Lupinus albus* L. III. The probable mechanism by which phosphorus movement in the soil/root interface is enhanced. Plant Soil 70, 107–124.

Gietl C 1992 Malate dehydrogenase isoenzymes: Cellular location and role in the flow of metabolites between cytoplasm and cell organelles. Biochim. Biophys. Acta 110, 217–234.

Gilbert G A, Knight J D, Vance C P and Allan D L 1999 Acid phosphatase activity in phosphorus-deficient white lupin roots. Plant Cell Environ. 22, 801–810.

Gilbert G A, Knight J D, Vance C P and Allan D L 2000 Proteoid root development of phosphorus deficient lupin is mimicked by auxin and phosphonate. Ann. Bot. 85, 921–928.

Guex N, Henry H, Flach J, Richter H and Widmer F 1995 Glyoxysomal malate dehydrogenase and malate synthase from soybean cotyledons (*Glycine max* L.): enzyme association, antibody production and cDNA cloning. Planta 197, 369–375.

Hartwell J, Gill A, Nimmo G A, Wilkins M B, Jenkins G I and Nimmo H G 1999 Phospho*enol*pyruvate carboxylase kinase is a novel protein kinase regulated at the level of expression. Plant J. 20, 333–42.

Heukeshovewn J and Dernick R 1988 Improved silver staining procedure for fast staining in PhastSystem Development Unit. I. Staining of sodium dodecyl sulfate gels. Electrophoresis 9, 28–32.

Huber S C, Huber J L and McMichael R W Jr 1994 Control of plant enzyme activity by reversible protein phosphorylation. Int. Rev. Cytol. 149, 47–98.

Israel D W and Jackson W A 1982 Ion balance, uptake, and transport processes in N_2-fixing and nitrate- and urea-dependent soybean plants. Plant Physiol. 112, 171–178.

Jiao J, Echevarria C, Vidal J and Chollet R 1991 Protein turnover as a component in the light/dark regulation of phospho*enol*pyruvate carboxylase protein-serine kinase activity in C_4 plants. Proc. Natl. Acad. Sci. USA. 88, 2712–2715.

Johnson J F, Allan D L, Vance C P and Weiblen G 1996a Root carbon dioxide fixation by phosphorus-deficient *Lupinus albus*. Contribution to organic acid exudation by proteoid roots. Plant Physiol. 112, 19–30.

Johnson J F, Vance C P and Allan D L 1996b Phosphorus deficiency in *Lupinus albus*. Altered lateral root development and enhanced expression of phospho*enol*pyruvate carboxylase. Plant Physiol. 112, 31–41.

Kim D J and Smith S M 1994 Expression of a single gene encoding microbody NAD-malate dehydrogenase during glyoxysome and peroxisome development in cucumber. Plant Mol. Biol. 26, 1833–1841.

Lance C and Rustin P 1984 The central role of malate in plant metabolism. Physiol. Veg. 22, 625–641.

Latzko E and Kelly G J 1983 The many-faceted function of phospho*enol*pyruvate carboxylase in C_3 plants. Physiol. Veg. 21, 805–815.

Lepiniec L, Vidal J, Chollet R, Gadal P and Cretin C 1994 Phospho*enol*pyruvate carboxylase: structure, regulation and evolution. Plant Sci. 99, 111–124.

Liu J, Uhde-Stone C, Li A, Vance C and Allan D 2001 A phosphate transporter with enhanced expression in proteoid roots of white lupin (*Lupinus albus* L.), Plant Soil 237, 257–266.

Marczewski W 1989 Kinetic properties of phospho*enol*pyruvate carboxylase from lupin nodules and roots. Physiol Planta. 76, 539–543.

Marschner H V, Romheld V, Horst W J and Martin P 1986 Root-induced changes in the rhizosphere: importance for the mineral nutrition of plants. Z. Pflanzenernaehr Bodenk. 149, 441–456.

Marschner H, Romheld V and Cakmak I 1987 Root-induced changes of nutrient availability in the rhizosphere. J. Plant Nutr. 10, 1175–1184.

Marschner H 1995 Mineral Nutrition of Higher Plants, 2nd edition. Academic Press Inc., San Diego, CA.

Massonneau A, Langlade N, Leon S, Smutny J, Vogt E, Neumann G and Martinoia E 2001 Metabolic changes associated with cluster root development in white lupin (*Lupinus albus* L.): relationship between organic acid excretion, sucrose metabolism and energy status. Planta 213, 534–42.

Maxwell C A 1982 CO_2 fixation in alfalfa and birdsfoot trefoil root nodules: synthesis, transport and partitioning of compounds from $^{14}CO_2$ MS. University of Minnesota, St Paul, Minn.

McCloud S A, Smith R G and Schuller K A 2001. Partial purification and characterization of pyruvate kinase from the plant fraction of soybean root nodules. Physiol. Plant.111, 283–290.

Melzer E and O'Leary M H 1987 Anaplerotic CO_2 fixation by phospho*enol*pyruvate carboxylase in C_3 plants. Plant Physiol. 84, 58–60.

Metzler M C, Rothermel B A and Nelson T 1989 Maize NADP-malate dehydrogenase: cDNA cloning, sequence and mRNA characterization. Plant Mol. Biol. 12, 712-722.

Miller S S, Discroll B T, Gregerson R G, Gantt J S and Vance C P 1998 Alfalfa malate dehydrogenase (MDH): molecular cloning and characterization of five different forms reveals a unique nodule-enhanced MDH. Plant J. 15, 173–185.

Miller S S, Liu J, Allan D L, Menzhuber C J, Fedorova M and Vance C P 2001 Molecular control of acid phosphatase secreted into the rhizosphere of proteoid roots from phosphorus stressed white lupin. Plant Physiol. 127, 594–606.

Moraes T F and Plaxton W C 2000 Purification and characterization of phospho*enol*pyruvate carboxylase from *Brassica napus* (rapeseed) suspension cell cultures: implications for phospho*enol*pyruvate carboxylase regulation during phosphate starvation, and the integration of glycolysis with nitrogen assimilation. Eur. J. Biochem. 267, 4465–4476.

Musrati R A, Kollarova M, Mernik N and Mikulasova D 1998 Malate dehydrogenase: distribution, function and properties. Gen. Physiol. Biophys. 17, 193–210.

Rufty et al., 1990; Schjorring, 1986). This is frequently associated with a decline in the activity of nitrate reductase and with nitrate accumulation in the root tissue, which probably induces feed-back inhibition of nitrate uptake (Gniazdowska and Rychter, 2000; Pilbeam et al., 1993). Therefore, a lower demand of 2-oxo-acids (particularly of 2-oxo-glutaric acid) as precursors for the biosynthesis of amino acids may further contribute to a reduced consumption of organic acids and finally to citrate accumulation in the root tissue, when the biosynthesis of organic acids is upregulated under P-deficient conditions (Johnson et al., 1996a, Neumann et al., 1999). Accordingly, artificial inhibition of nitrate reductase by localized application of tungstate significantly increased the citrate/malate ratio in juvenile root clusters of white lupin within an incubation period 12–16 h (Figure 4). Although unspecific effects of inhibitors, particularly after long-term application cannot be ruled out, this result may give at least a first indication that impairment of nitrate assimilation under P-deficient conditions can also have an impact on the intracellular accumulation of organic acids. Further investigations would require direct measurements of nitrate uptake and assimilation as well as ^{14}C labeling studies to trace changes of the fate of carbon skeletons in different stages of cluster root development.

Carboxylate exudation and intracellular pH

The transient pulse of citrate exudation in mature root clusters suggests a controlled release mechanism for citrate anions. This assumption is further supported by the effectiveness of anion-channel inhibitors (Neumann et al., 1999). Similarly, anion-channel-mediated root exudation of malate and citrate has been recently demonstrated by patch-clamp studies with wheat and maize under aluminum stress (Kollmeier et al., 2001; Zhang et al., 2001). In white lupin, citrate release is charge-balanced by a concomitant extrusion of protons, which is responsible for rhizosphere acidification. Accordingly, up-regulation of plasmalemma H^+-ATPase was detected in plasma-membrane vesicles isolated from mature root clusters (Kania et al., 2001; Yan et al., 2001). However, the signal which triggers the transient release of citrate is still unknown. Anion channels can be modulated by carboxylate anions themselves, as demonstrated for vacuolar anion channels of CAM plants (Cerana et al., 1995) and for stomatal guard cells (Hedrich and Marten, 1993). Decreased levels of ATP, characteristic also for mature

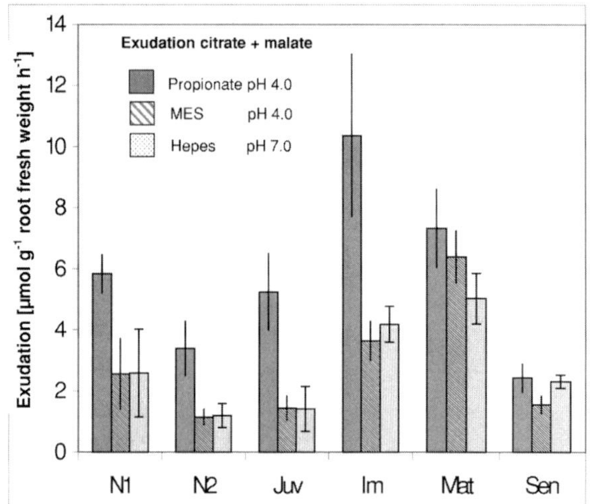

Figure 5. Citrate and malate exudation in different root zones of white lupin as affected by buffering the external pH and application of propionate. N1, 10-mm apical root zone of first-order laterals in P-sufficient plants; N2, 10-mm apical root zone of first-order laterals in P-deficient plants; juv – juvenile root clusters; Imm – immature root clusters completely developed but secretory activity not yet expressed; Mat – mature root clusters; Sen – senescent root clusters.

and senescent root clusters, activate anion channels permeable for NO_3^- (Frachisse et al., 1999; Thomine et al., 1997). Increased accumulation of lactic acid in hypoxic maize root tips induces overacidification of the cytosol (Xia and Roberts, 1994) at similar concentration levels which have been reported for citric acid accumulation in mature root clusters of white lupin (Neumann et al., 2000). To assess whether modifications of the intracellular pH may be involved in the induction of carboxylate exudation in roots of P-deficient white lupin, we used weak organic acids such as propionic, acetic or methyl trichloroacetic acid (methyl-TCA, not metabolized) for external application to the root medium at pH 4. According to the ion-trap principle, the protonated acids easily permeate the plasmalemma of the root cells and accumulate in the cytosol, due to complete dissociation at pH 7.0–7.5. This is associated with a decrease of the cytosolic pH. During an incubation period of 1 h with propionic acid, release of both, citrate and malate, was dramatically increased, particularly in apical root zones of normal lateral roots, in juvenile root clusters, and in immature cluster roots, which have reached their final length but usually exhibit only slow rates of carboxylate exudation (Figure 5A,B). Similar results were obtained with acetic acid or methyl-TCA

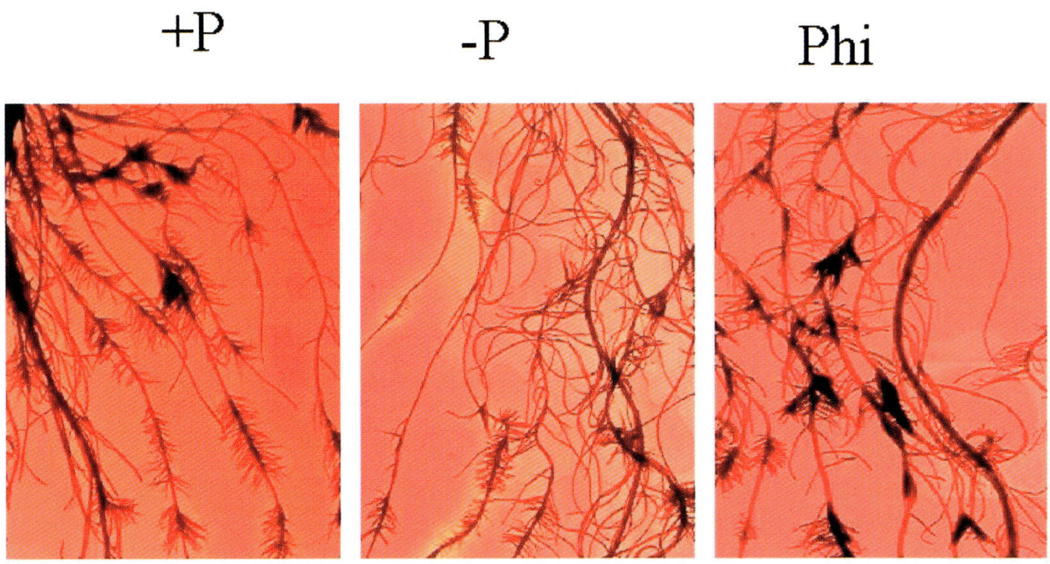

Figure 6. Effects of treatments with P_i (+P), and phosphonate (Ph_i) on root-induced pH changes in P-deficient white lupin (−P) detected by application of agar sheets with bromokresol purple as pH indicator.

(data not shown). Buffering the pH of the incubation medium to 4.0 and 7.0 with Mes or Hepes in the absence of propionate did not alter the rates of carboxylate exudation, demonstrating that increased exudation is not simply a result of a modified external pH (Figure 5A,B). These findings suggest that a drop of the cytosolic pH induced by intracellular organic acid accumulation may, in fact, be an important factor triggering the release of carboxylates in roots of P-deficient plants. In contrast, there were no consistent effects in mature and senescent clusters (Figure 5A,B). This may be explained by a low cytosolic pH in these root zones already before propionate application, induced by the high levels of intracellular citric acid accumulation. To test whether the induction of carboxylate exudation was really a pH effect, roots of P-deficient plants were preincubated for 24 h with phosphonate (H_3PO_3, Ph_i), which is taken up but not metabolized by plant roots, to achieve pH buffering of the cytosol (Carswell et al., 1996, 1997). Accordingly, neither carboxylate exudation (not shown) nor proton extrusion was induced by propionate application in roots of Ph_i-treated plants (Figure 6). Similar responses were observed when roots of P-deficient plants were preincubated for 24 h with phosphate (P_i) instead of Phi (Figure 6). In the phosphate treatments, ATP levels similar to those of P-sufficient plants (data not shown) indicate rapid P_i metabolism and recovery from P-deficient conditions, counteracting cytosolic acidification. No such effects were detected in Ph_i-treated plants. A direct confirmation for a role of the intracellular pH in triggering root exudation of carboxylates would require NMR studies for direct pH determination in the cytosol of different root zones and a membrane-physiological approach with patch-clamp experiments or isolated plasma membrane vesicles to characterize pH dependency of the transport mechanism. It is also not clear which factors are responsible for the rapid decrease of citrate exudation in senescent root clusters, despite high internal concentrations of citric acid. Possible explanations could be restabilization of the cytosolic pH after the pulse of proton and carboxylate exudation or energy limitation of the PM-H^+-ATPase due to the extremely slow rates of respiration and low ATP concentrations in these root zones.

Watt and Evans (1999a,b) reported a diurnal rhythm of citrate release from cluster roots of white lupin, suggesting the involvement of additional factors controlling the exudation process, apart of the cytosolic pH. It was demonstrated by ^{14}C labeling studies, that citric acid is synthesized in the cluster root tissue (Johnson et al., 1996b) associated with increased expression and activities of enzymes involved in carbohydrate catabolism (e.g., sucrose synthase, fructokinase, phosphoglucomutase) and biosynthesis of carboxylates such as PEP carboxylase and the malate dehydrogenase (Johnson et al., 1996a; Massonneau et al., 2001). Diurnal fluctuations in citrate relase (Watt and Evans, 1999a,b) might therefore reflect a diurnal rhythm of carbohydrate supply from the leaves as precursors for the synthesis of citric acid.

Concluding remarks

A detailed understanding of physiological mechanisms involved in the regulation of root exudation as an adaptive response for acquisition of sparingly soluble nutrients or detoxification of toxic elements in the rhizosphere is a prerequisite for attempts to manipulate resistance of crop plants to adverse soil chemical conditions by root-induced changes in rhizosphere chemistry.

Citrate is among the most potent root-borne chelators involved in P mobilization and Al detoxicication, provided it is released in sufficient amounts (Jones, 1998; Neumann and Römheld, 2000). As a basis for more detailed investigations, this study presents further evidence for the hypothesis that down-regulation of various metabolic sequences involved in citrate catabolism is an important factor for accumulation of citric acid in cluster roots of P-deficient white lupin, which precedes the burst of citrate exudation. Moreover, there is a first indication that acidification of the cytosol induced by excessive intracellular accumulation of citric acid could be part of the signal-transduction chain, which induces the transient release of citrate and protons into the rhizosphere.

Acknowledgements

The authors wish to thank the German Research Foundation, Bundesamt für Bildung und Wissenschaft, EC Nr. QLK-2000-00479, BBW Nr. 00.0413, and the Swiss National Foundation, NCCR Plant Survival for financial support, and Mrs J. Smutny for technical assistance.

References

Aono T, Kanada N, Ijima A and Oxaizu H 2001 The response of the phosphate uptake system and the organic acid exudation system to phosphate starvation in *Sesbania rostrata*. Plant Cell Physiol. 42: 11, 1253–1264.

Carswell C, Grant B R, Theodorou M E, Harris J, Niere J O and Plaxton W C, 1996 The fungicide phosphonate disrupts the phosphate-starvation response in *Brassica nigra* seedlings. Plant Physiol. 110, 105–110

Carswell M C, Grant B R and Plaxton W C (1997) Disruption of phosphate-starvation response of oilseed rape suspension cells by the fungicide phosphonate. Planta 203, 67–74

Cerana R, Giromini L Colombo R 1995 Malate-regulated channels permeable to anions in vacuoles of *Arabidopsis thaliana*. Aust J. Plant Physiol. 22, 115–121

Dinkelaker B, Römheld V and Marschner H 1989 Citric acid excretion and precipitation of calcium citrate in the rhizosphere of white lupin (*Lupinus albus* L.). Plant Cell Environ. 12, 285–292.

Elthon T E, Nickels R L and McIntosh L 1989 Monoclonal antibodies to the alternative oxidase of higher plant mitochondria. Plant Physiol. 89, 1311–1317.

Franchisse J M, Thomine S, Colcombet J, Guern J, and Barbier-Brygoo H 1999 Sulfate is both a substrate and an activator of the voltage-dependent anion channel of *Arabidopsis* hypocotyl cells. Plant Physiol 121, 253–262.

Gardner P R and Fridovich I 1992 Inactivation-reactivation of aconitase in *Escherichia coli*. J. Biol. Chem. 267:13, 8757–8763.

Gardner W K, Barber D A, and Parbery D G 1983 The acquisition of phosphorus by *Lupinus albus* L. III. The probable mechanism, by which phosphorus movement in the soil-root interface is enhanced. Plant Soil 70, 107–114.

Gniazdowska A and Rychter A M 2000 Nitrate uptake by bean (*Phaseolus vulgaris* L.) roots under phosphate deficiency. Plant Soil 226, 79–85.

Hedrich R and Marten I 1993 Malate-induced feedback regulation of plasma membrane anion channels could provide a CO_2 sensor to guard cells. EMBO J. 12, 897–901

Hoefnagel M H N, van Iren F and Libbenga K R 1993 In suspension cultures of *Catharanthus roseus* the cyanide-resistant pathway is engaged in respiration by excess sugar in combination with phosphate or nitrogen starvation. Physiol. Plant. 87, 297–304.

Johnson J F, Allan D L and Vance C P 1994 Phosphorus stress-induced proteoid roots show altered metabolism in *Lupinus albus*. Plant Physiol. 104, 657–665.

Johnson J F, Vance C P and Allan D L 1996a Phosphorus deficiency in *Lupinus albus* — Altered lateral root development and enhanced expression of phospho*enol*pyruvate carboxylase. Plant Physiol. 112, 31–41.

Johnson J F, Allan D L, Vance C P and Weiblen G 1996b Root carbon dioxide fixation by phosphorus-deficient *Lupinus albus* — Contribution to organic acid exudation by proteoid roots. Plant Physiol. 112, 19–30.

Jones D L 1998 Organic acids in the rhizosphere — a critical review. Plant Soil 205, 25–44.

Kania A, Neumann G, Cesco S, Pinton R and Römheld V 2001 Use of plasma membrane vesicles for examination of phosphorus deficiency-induced root excretion of citrate in cluster roots of white lupin (*Lupinus albus* L.). *In* Plant Nutrition: Food security and sustainability of agro-ecosystems through basic and applied research. XIV International Plant Nutrition Colloquium. pp. 546–547. Kluwer Academic Publishers, Dordrecht.

Khyse-Andersen J 1984 Electroblotting of multiple gels: a simple apparatus without buffer tank for rapid transfer of proteins from polyacrylamide to nitrocellulose. Biochem. Biophys. Methods 10, 203–209.

Kollmeier M, Dietrich P, Bauer C S, Horst W J and Hedrich R 2001 Aluminum activates a citrate-permeable anion channel in the aluminum-sensitive zone of the maize root apex. A comparison between an aluminum-sensitive and an aluminum-resistant cultivar. Plant Physiol. 126, 397–410.

Laemmli U K 1970 Cleavage of structural proteins during the assembly of the head of bacteriophage T4. Nature 227, 680–685.

Lambers H, Scheurwater I and Atkin O K 1996 Respiratory patterns in roots in relation to their functioning. *In* Plant Roots: The Hidden Half. Eds. Y Waisel, A Estel and U Kafkafi. pp. 323–362. Marcel Dekker, New York.

Lambers H, Scheurwater I and Millenaar F 1997 Variation in carbon utilization in root respiration and exudation as dependent on a species' potential growth rate and nutrient supply. *In* Radical Biology: Advances and Perspectives of the Function of Plant Roots. Eds. H E Flores, J P Lynch and D Eissenstat. pp. 116–130. American Society of Plant Physiologists, Rockville, Maryland.

Langlade N, Messerli G, Smutny J, Neumann G, Martinoia E and Massonneau A 2002 ATP citrate lyase: cloning, heterologous expression and its role in organic acid excretion. Plant Cell Environm. (In press).

Lee R B 1982 Selectivity and kinetics of ion uptake by barley plants following nutrient deficiency. Ann. Bot. 50, 429–449.

Massonneau A, Langlade N, Leon S, Smutny J, Vogt E, Neumann G and Martinoia E 2001 Metabolic changes associated with cluster root development in white lupin (*Lupinus albus* L.): relationship between organic acid excretion, sucrose metabolism and energy status. Planta 213, 534–542.

McIntosh L 1994 Molecular biology of the alternative oxidase. Plant Physiol. 105, 781–786.

Millar A H, Atkin O K, Menz R I, Henry B, Farquhar G and Day D A 1998 Analysis of respiratory chain regulation in roots of soybean seedlings. Plant Physiol. 117, 1083–1093.

Møller I M, Berczi A, van der Plas L H W and Lambers H 1988 Measurement of the activity and capacity of the alternative pathway in intact plant tissues: identification of problems and possible solutions. Physiol. Plant. 72, 642–649.

Neumann G, Massonneau A, Martinoia E and Römheld V 1999 Physiological adaptations to phosphorus deficiency during proteoid root development in white lupin. Planta 208, 373–382.

Neumann G, Massonneau A, Langlade N, Dinkelaker B, Hengeler C, Römheld V and Martinoia E 2000 Physiological aspects of cluster root function and development in phosphorus-deficient white lupin (*Lupinus albus* L.). Ann. Bot. 85, 909–919.

Neumann G and Römheld V 2001 The release of root exudates as affected by the plant's physiological status. *In* Rhizosphere: Biochemistry and Organic Substances at the Soil–Plant Interface. Eds. R Pinton, Z Varanini and P Nannipieri. pp. 41–93. Marcel Dekker, New York.

Pilbeam D J, Cakamak I, Marschner H and Kirkby E A 1993 Effect of withdrawal of phosphorus on nitrate assimilation and PEP carboxylase activity in tomato. Plant Soil 154, 111–117.

Plaxton W C 1998 Metabolic Aspects of Phosphate Starvation in Plants. *In* Phosphorus in Plant Biology: Regulatory Roles in Molecular, Cellular, Organismic, and Ecosystem Processes. Eds. J P Lynch and J Deikman. pp. 41–93. American Society of Plant Physiologists, Rockville, Maryland.

Ratledge C, Bowater M D V and Taylor P N 1997 Correlation of ATP/citrate lyase activity with lipid accumulation in developing seeds of *Brassica napus* L. Lipids. 32, 7–12.

Rufty T W, MacKown C T and Israel D W 1990 Phosphorus stress effects on assimilation of nitrate. Plant Physiol. 94, 328–333.

Rychter A M and Mikulska M 1990 The relationship between phosphate status and cyanide-resistant respiration in bean roots. Physiol. Plant. 79, 663–667.

Schjorring J K 1986 Nitrate and ammonium absorption by plants growing at a sufficient or insufficient level of phosphorus in nutrient solution. *In* Fundamental, Ecological and Agricultural Aspects of Nitrogen Metabolism in Higher Plants. Eds H Lambers, J J Neetson and I Stulen. pp. 53–58. Martinus Nijhoff, Dordrecht.

Takita E, Koyama H and Hara T 1999 Organic acid metabolism in aluminum-phosphate utilizing cells of carrot (*Daucus carota* L.). Plant Cell Physiol. 40:(5), 489–495.

Thomine S, Lelievre F, Boufflet M, Guern J and Barbier-Brygoo H 1997 Anion-channel blockers interfere with auxin responses in dark-grown *Arabidopsis* hypocotyls. Plant Physiol. 115, 533–542.

Vanlerberghe G C and McIntosh L 1996 Signals regulating the expression of the nuclear gene encoding alternative oxidase of plant mitochondria. Plant Physiol. 111, 589–595.

Verniquet F, Gaillard J, Neuburger M and Douce R 1991 Rapid inactivation of plant aconitase by hydrogen peroxide. Biochem. J. 276, 643–64

Watt M and Evans J R 1999a Linking development and determinancy with organic acid efflux from proteoid roots of white lupin grown with low phosphorus and ambient or elevated atmospheric CO_2 concentration. Plant Physiol. 120, 705–716.

Watt M and Evans J R 1999b Proteoid roots. Physiology and development. Plant Physiol. 121, 317–323.

Yan F, Zhu Y, Müller C and Schubert S 2001 Adaptation of plasma membrane H^+-ATPase of proteoid roots of white lupin (*Lupinus albus* L.) to phosphorus deficiency. *In* Plant Nutrition: Food Security and Sustainability of Agro-ecosystems Through Basic and Applied Research. XIV International Plant Nutrition Colloquium. pp. 186–187. Kluwer Academic Publishers, Dordrecht.

Xia JH and Roberts JKM 1994 Improved cytoplasmic pH regulation, increased lactate efflux, and reduced cytoplasmic lactate levels are biochemical traits expressed in root tips of whole maize seedlings acclimated to a low-oxygen environment. Plant Physiol. 105, 651–657.

Zhang W H et al. (2001) Aluminium activates malate-permeable channels in the apical cells of wheat roots. Plant Physiol. 125, 1459–1472

Plant and Soil **248:** 129–136, 2003.
© 2003 *Kluwer Academic Publishers. Printed in the Netherlands.*

Secreted acid phosphatase is expressed in cluster roots of lupin in response to phosphorus deficiency

Jun Wasaki[1], Takuya Yamamura, Takuro Shinano & Mitsuru Osaki
Graduate School of Agriculture, Hokkaido University, N9W9, Kita-ku, Sapporo, 060-8589, Japan. [1]*Corresponding author**

Received 11 January 2002. Accepted in revised form 17 April 2002

Key words: acid phosphatase, cluster root, exudation, phosphorus deficiency, white lupin

Abstract

The roots of white lupin (*Lupinus albus* L. cv. Kievskij mutant) secrete acid phosphatase, S-APase, when they grow under conditions of low available phosphorus (P). S-APases hydrolyze organic phosphate compounds in the rhizosphere and supply inorganic phosphate to the plants. Low phosphorus availability also induces vigorous growth of cluster roots. In this study, the function of cluster roots was investigated with reference to S-APase secretion. White lupins were grown in hydroponic culture in a greenhouse under P-deficient and P-sufficient conditions. S-APase in the excised roots after treatment was detected by staining with 4-methylumbelliferone phosphate (MUP). Gene expression of S-APase in cluster and normal roots was also investigated. Activity was greatest in the roots of plants grown under conditions of P -deficiency, particularly in cluster roots. S-APase gene expression was induced by a decrease in internal P concentrations, and was especially high in cluster roots formed under conditions of P -deficiency. It was suggested that decrease of internal P concentration stimulated both of the S-APase expression and cluster root formation.

Abbreviations: APase – acid phosphatase; MUP – 4-methylumbelliferone phosphate; *p*-NPP – *p*-nitrophenylphosphate; *p*-NP – *p*-nitrophenol; S-APase – secreted acid phosphatase

Introduction

Soil phosphates exist predominantly as insoluble inorganic phosphate and as organic phosphates, which are not directly available to plants. However, the growth of white lupin plants is more vigorous than that of other plants under conditions of low phosphorus (Gardner and Boundy, 1983; Luo et al., 1997). It is assumed that rapid secretion of S-APase (Tadano and Sakai, 1991) and citrate (Dinkelaker et al., 1989; Gardner et al., 1983) from the roots plays an important role.

When crude S-APase collected from lupin roots was injected into the rhizospheres of tomato and sugar beet plants, growth and P absorption of these plants increased (Tadano and Komatsu, 1994). These results suggest that S-APase can contribute to the acquisition of phosphorus from organic phosphate compounds. Ozawa et al. (1995) purified and characterized the S-APase protein, which is a homodimer with an estimated molecular weight of 140 kDa, 72 kDa for each subunit. S-APase has a very high affinity for *p*-nitrophenylphosphate (*p*-NPP) with a K_m of 0.027 mM, very wide substrate specificity, and is very stable at pH 4.0–9.0. Like many other secreted proteins, S-APase is glycosylated, which protects it against proteolytic enzymes and contributes to its stability over a wide pH range (Ozawa et al., 1995). These properties are consistent with the role suggested for S-APase in hydrolysis of organic P in the soil.

Wasaki et al. (1997, 1999a, 2000) reported the isolation and characterization of two cDNAs encoding APase isozymes, designated LASAP1 and LASAP2. The PSORT predictive program (Nakai and Kane-

* FAX No: +81-11-706-3845
E-mail: junw@chem.agr.hokudai.ac.jp

hisa, 1992) suggested that mature LASAP1 is located on the cell membrane and LASAP2 is secreted from cells. Since the primary sequence deduced from the LASAP2 cDNA is identical to partial amino acid sequences determined from purified S-APase, it was inferred that S-APase is encoded by LASAP2 cDNA. Accumulation of S-APase mRNA and the corresponding protein was induced only in roots under phosphate-deficient conditions. It is assumed that LASAP1 hydrolyzes phosphate ester compounds that reach the root surface or that are released from aging cells into the apoplastic space.

Under phosphate deficiency, several plants including white lupin form 'cluster' or 'proteoid' roots, which are highly dense roots arising from the secondary roots (Gardner et al., 1983; Shane et al., 2002; Skene, 1998; Watt and Evans, 1999). Cluster roots can use phosphorus efficiently via the increase in root surface area (Gardner et al., 1981) and have enhanced the secretion of organic acids (Gardner et al., 1993; Koyama et al., 2002; Shane et al., 2002), which is regulated by the expression of relevant genes, such as those for phosphoenolpyruvate carboxylase (PEPC) and citrate synthase (CS) (Johnson et al., 1994, 1996a,b). However, a relationship between cluster root formation and the secretion of S-APase in the utilization of organic phosphorus compounds in soils has not been reported. The objective of this study was to investigate the functions of S-APase secretions from cluster roots of white lupin grown under phosphorus-deficient conditions.

Materials and methods

Plant materials

White lupin (*Lupinus albus* L. cv. Kievskij mutant) plants were grown in hydroponic culture in a greenhouse, as described by Wasaki et al. (1999). The composition of the nutrient solution was 1.4 m*M* NH_4NO_3, 2.0 m*M* $NaNO_3$, 2.0 m*M* $MgSO_4$, 2.0 m*M* $CaCl_2$, 1.0 m*M* K_2SO_4, 36 μM $FeSO_4$, 17 μM $MnSO_4$, 3.1 μM $ZnSO_4$, 0.16 μM $CuSO_4$, 47 μM H_3BO_4, and 7.5 n*M* $(NH_4)_6Mo_7O_{24}$. The solution contained 65 μM phosphate (+P treatment) or 0 μM phosphate (−P treatment). Roots were collected, the moisture was removed from them using paper towels, and then cluster and normal roots were divided. For protein and RNA analysis, the roots were frozen immediately in liquid nitrogen.

For the analysis of germinating seeds, white lupin seeds were washed with 70% ethanol and then sown on solid MS medium (Murashige and Skoog, 1962) with phosphorus (1.1 m*M*). After germination for 3 days in the dark at 25 °C, the seeds were collected and frozen immediately in liquid nitrogen. All frozen samples were stored at −80 °C until analysis.

Measurement of Pi and acid phosphatase activity

The soluble fraction of lupin roots was extracted by the mothod of Wasaki et al. (1997). The extracts were used for measurement of Pi and APase activity. Inorganic phosphate concentrations were measured by the spectrophotometric method of Saheki et al. (1985). APase activity was assayed by the release of nitrophenyl (*p*-NP) from *p*-NPP, by the method of Ueki and Sato (1977). One unit of APase activity is defined as the amount of enzyme that hydrolyzes 1 μmol *p*-NPP per minute.

Staining for S-APase activity

To detect the parts of white lupin roots that secrete S-APase, plants were cultured hydroponically for 25 days without phosphorus, and the whole root systems were placed onto solid gel medium containing no phosphorus and 0.2% gellan gum.

To examine the function of cluster roots, excised roots were plated onto solid medium and incubated for 24 h. After hydroponic culture of white lupin under +P or −P condition for 25 days, the secondary roots with or without cluster roots were excised and placed onto solid medium.

S-APase activity was detected after incubation of the roots on the gel medium for 24 h at 25 °C in light and their subsequent removal from the medium.

The S-APase activity secreted onto the medium was observed by a method that uses 4-methylumbelliferone phosphate (MUP) as a substrate for S-APase, as previously reported (Wasaki et al., 1999). The hydrolyzed products of MUP were visualized by UV illumination at 260 nm. This experiment was done in triplicate.

Northern blot analysis

Total RNA was isolated from root tissues using a sodium dodecyl sulfate (SDS)– phenol method (Palmiter, 1974). Total RNA (5 μg) was denatured in a mixture of 70% (v/v) formamide and 8% (v/v) formaldehyde, and separated on a 1% agarose gel

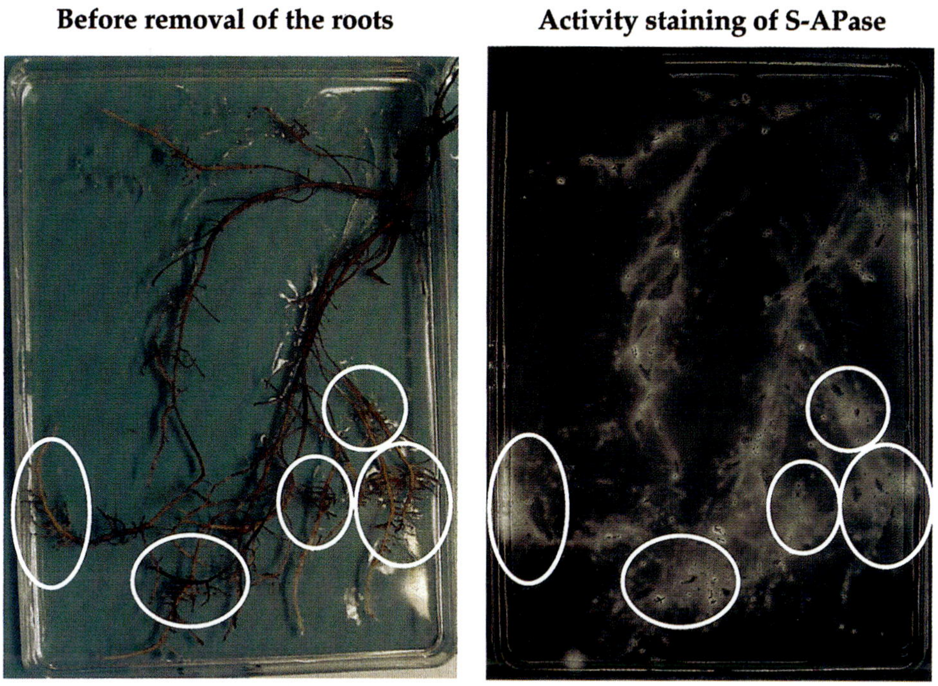

Figure 1. S-APase secretion from roots of white lupin grown under −P conditions. Lupin roots grown under −P conditions were transferred to solid medium and incubated for 24 h (left panel). Secreted APase activity was visualized (right panel). Cluster roots are indicated by white circles.

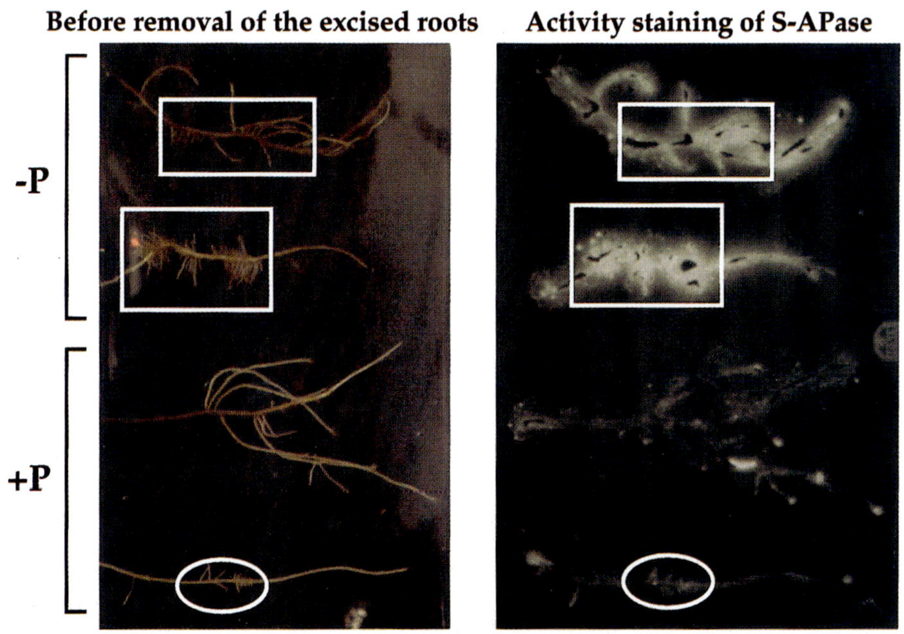

Figure 2. Pattern of S-APase secretion in white lupin roots. After the cultivation of lupins under +P or −P condition for 25 days, the secondary roots with or without cluster roots were excised and placed onto solid medium for 24 h (left panel). After removal of the roots, secreted APase was visualized (right panel). Each treatment is shown as −P or +P. The parts of the cluster roots are indicated by white frames.

containing 1.8% (v/v) formaldehyde. After electrophoresis, RNAs were transferred to a Hybond-N^+ membrane (Amersham-Pharmacia Biotech, Buckinghamshire, UK). An LASAP2RT fragment was amplified by PCR using pLASAP2RT as template (Wasaki et al., 2000), and was used as a LASAP2-specific probe (Wasaki et al., 2001). Hybridization and detection were performed using a Gene Images^{TM} labeling and detection system (Amersham-Pharmacia Biotech). The blot was hybridized and washed at 65 °C.

Results

Root regions secreting S-APase

This experiment was done in triplicate, whereas the only one replicate was shown in Figure 1 because of the negligible difference among the replications. S-APase was secreted from whole roots of white lupin (Figure 1), which was tremendously high from the cluster roots (Figure 1, circled parts). S-APase was secreted from roots markedly higher in −P plants than in +P plants (Figure 2). +P roots formed small numbers of cluster roots, however S-APase activity around +P cluster roots was similar to that around +P normal roots. On the other hand, APase activity around −P cluster roots was relatively higher than that around −P normal roots.

Expression of the S-APase gene

To clarify whether the increase in S-APase activity in cluster roots formed under −P conditions depends on the stimulation of gene expression, the accumulation of LASAP2 mRNA encoding S-APase (Wasaki et al., 2000) was estimated. Northern blot analysis of LASAP2 was performed with 5 μg total RNA (Figure 3). LASAP2 mRNA encoding S-APase accumulated only in −P roots, and was especially high in the cluster roots induced under −P conditions. LASAP2 was not detected in germinating seeds (Figure 3).

Northern analysis was used to investigate the time course of LASAP2 mRNA accumulation (Figure 4). For 7 days after treatment, the number of transcripts was only slightly higher in −P roots than in +P roots. However, at 11 days and later, LASAP2 transcripts increased notably above the 7-days levels in −P roots. Because S-APase expression seems to be related to internal phosphorus concentrations, we plotted total

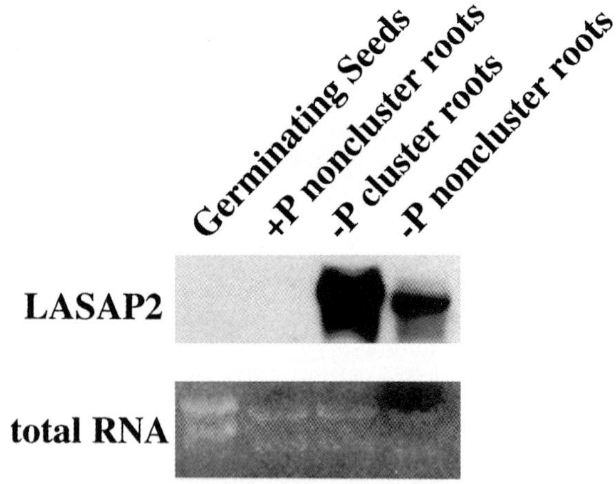

Figure 3. LASAP2 mRNA accumulation in germinating seeds and roots of white lupin. Total RNA (5 μg) isolated from germinating seeds, +P normal roots, −P cluster roots, and −P normal roots was used for northern blot analysis.

APase activity against inorganic phosphate concentrations in the lateral roots (Figure 5), and showed that Pi concentrations in the roots are inversely proportional to total APase activity.

Discussion

Wasaki et al. (1999b) have reported that white lupin seedlings secrete S-APase from the whole root system when cluster roots were not yet formed. In this paper (Figure 1), it was confirmed that S-APase was secreted from whole roots not only cluster roots. However, cluster roots secreted tremendously S-APase under phosphorus-deficient conditions compared to phosphorus-sufficient conditions.

White lupin plants form cluster roots under phosphorus-deficient conditions. Enzymes associated with organic acid synthesis, such as PEPC and CS, are activated in the cluster roots under conditions of low phosphorus supply (Johnson et al., 1994, 1996a,b). In this study, it was shown that S-APase activity and mRNA also increased in cluster roots (Figures 1–3). These results support that cluster root formation plays an important role in acquiring phosphorus from soils. However, as S-APase exudation did not increased in the +P cluster roots, the S-APase exudation was not induced by only formation of cluster roots, but phosphorus deficiency in cluster roots. Thus, even in cluster roots, phosphorus deficiency is important signal to exudates S-APase.

APase hydrolyzes the phosphate ester compounds in storage organs to supply inorganic phosphate for

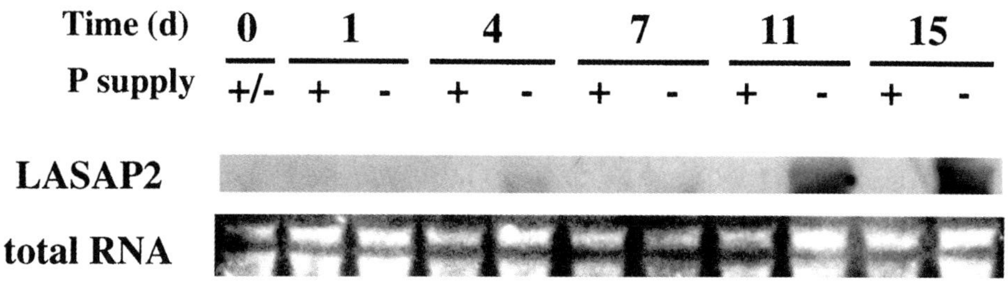

Figure 4. Time course of LASAP2 mRNA accumulation in white lupin roots grown hydroponically with or without phosphorus. Total RNA (5 μg) was used for northern blot analysis. Days after transfer and the P conditions are indicated above the lanes.

the growth of seedlings (Bhagara and Sachar, 1987; Biswass and Cundiff, 1991; Bosse and Köck, 1998; Gibson and Ullah, 1988; Haraguchi et al., 1990; Okuda et al., 1987; Tamura et al., 1982). Furthermore, a purple acid phosphatase, the primary structure of which is similar to that of LASAP1 and LASAP2, has been purified from germinating seeds of the kidney bean (Beck et al., 1986; Klabunde et al., 1994). Several phosphatases have been classified by molecular phylogenetic analysis using recently published data and our unpublished data. Figure 6 shows a phylogenetic tree based on sequence data of homologs from *Arabidopsis* of the purple acid phosphatase and phosphatases isolated from several other plants. Several APases belong to Group I, and some members of this group have widely ranging substrate specificities. Group II contains a soybean phytase and two *Arabidopsis* homologs. Group III contains a PEP-specific phosphatase from onion isolated by our group (Shinano et al., 2001). LASAP1 and LASAP2 both belong to Group I (Figure 6). The greatest difference between them is in their location in the tissues. Wasaki et al. (2000) first confirmed that LASAP2 is the secreted APase, because (i) its deduced amino acid sequence contains several partial amino acid sequences completely homologous to purified S-APase; and (ii) LASAP2 is theoretically predicted to be a secreted protein, whereas LASAP1 is predicted to be a membrane protein. Recently, Miller et al. (2001) reported the expression of a membrane-bound form of APase and a secreted APase. They found S-APase transcripts predominantly in P-deficient proteoid roots, whereas the membrane-bound form of APase (LASAP1) was detected at its highest level in P-deficient normal roots and at a lower but significant level in leaves, stems, and proteoid roots of P-deficient plants (Miller et al., 2001).

LASAP2 was not expressed in germinating seeds (Figure 3). This fact confirms a previous report (Wasaki et al., 2000) that LASAP2 mRNA does not accumulate in all the organs of seedlings. LASAP2 may not be important in using stored phosphorus during germination. On the other hand, we observed a large amount of LASAP2 mRNA in cluster roots (Figure 3), suggesting that the high S-APase activity in these parts of the plant originated from the induction of cluster roots by −P treatment. Uhde-Stone et al. (2002) also reported that the S-APase expression was altered markedly in −P proteoid roots in the case of macroarray analysis. As mentioned above, in the cluster roots induced by phosphorus deficiency, the expression of several enzymes for the acquisition of phosphorus enhanced. We propose that mechanisms specific to cluster roots exist to regulate the gene functions that contribute to the adaptation of the plant to phosphorus-deficient conditions.

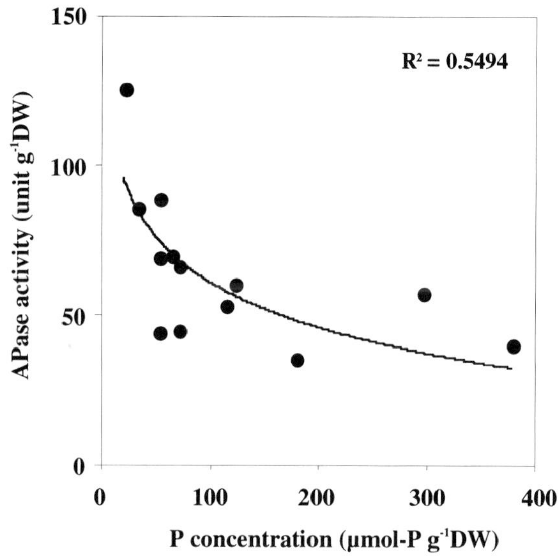

Figure 5. Relationship between P concentration and total APase activity in white lupin roots.

The time course of LASAP2 mRNA accumulation (Figure 4) was similar to the time course of protein

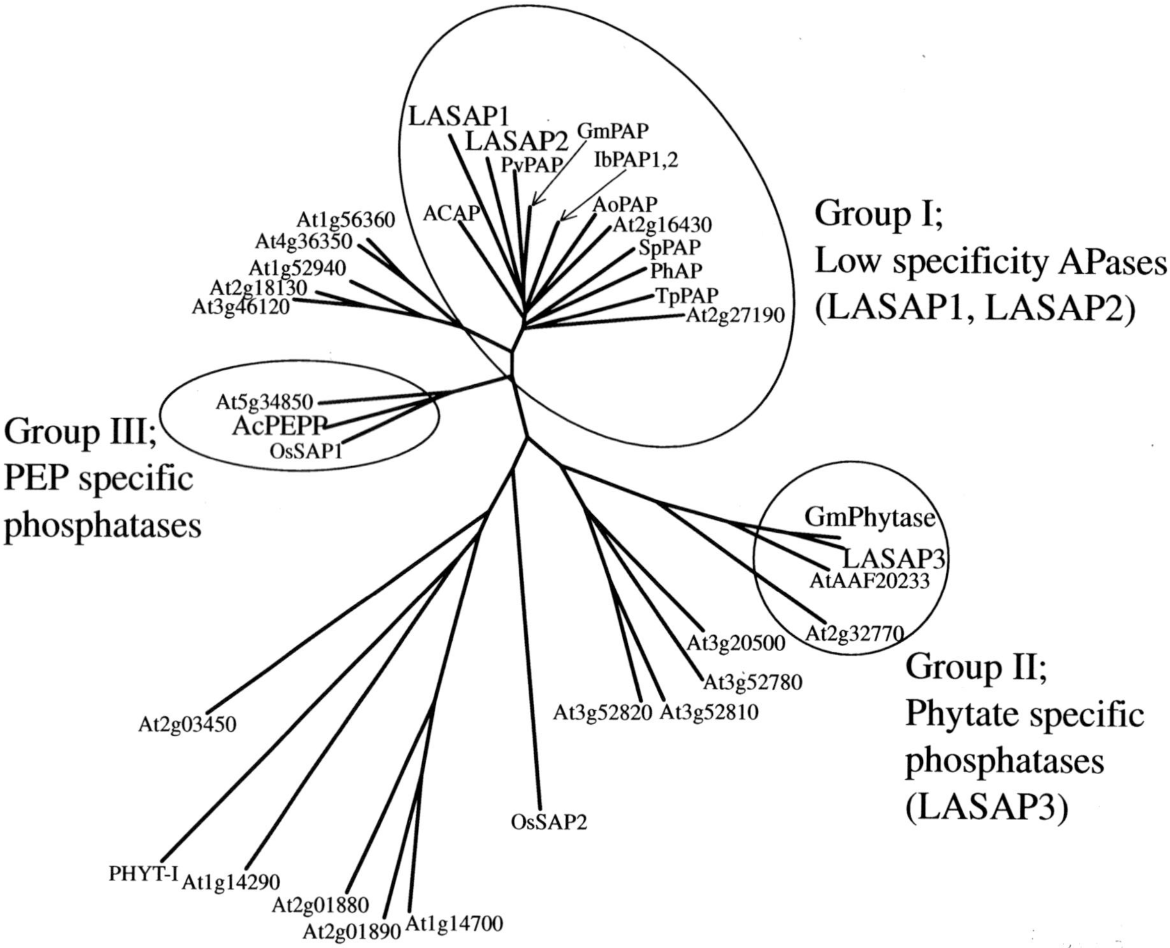

Figure 6. Molecular phylogenetic tree constructed from the genomic sequences of phosphatases from several plants, and their homologs from *Arabidopsis thaliana*. The tree was drawn using the ClustalW program.

accumulation in crude extracts and the activity of S-APase in culture solutions (Ozawa et al., 1995; Wasaki et al., 1997). Ozawa et al. (1995) showed that S-APase activity in culture solutions increases dramatically with a decrease in internal phosphorus concentrations. Considering the time course of mRNA accumulation, protein synthesis, and the secretion of S-APase, the internal phosphorus concentration of plants is probably a key signal for the regulation of S-APase expression. Figure 5 shows the relationship between total APase activity and internal phosphorus concentrations in lupin roots. In other experiments, we have shown that the crude soluble protein from roots contains nine APase isozymes, and that S-APase has the highest activity of these isozymes (data not shown). Because the ratio of constituent enzymes was stable throughout the experiment, the results shown in Figure 5 support the proposition that LASAP2 expression is regulated by internal phosphorus concentrations.

A previous study found that S-APase secretion from excised roots is induced within 12 h of transfer onto −P medium, and that the P concentration in the excised roots is similar after +P and −P treatments. From these data was drawn the conclusion that one of the key signals for S-APase secretion is external P concentration (Wasaki et al., 1999). Immunohistochemical analysis has revealed that S-APase accumulates on the epidermis of +P roots (Wasaki et al., 1997). From these facts, we propose the following scheme to describe the mechanism regulating APase secretion (Figure 7). (i) Under P-sufficient conditions, S-APase is synthesized at low levels and localized on the epidermis of roots; (ii) when the plant experiences slight P-deficiency, epidermis-localized S-APase is

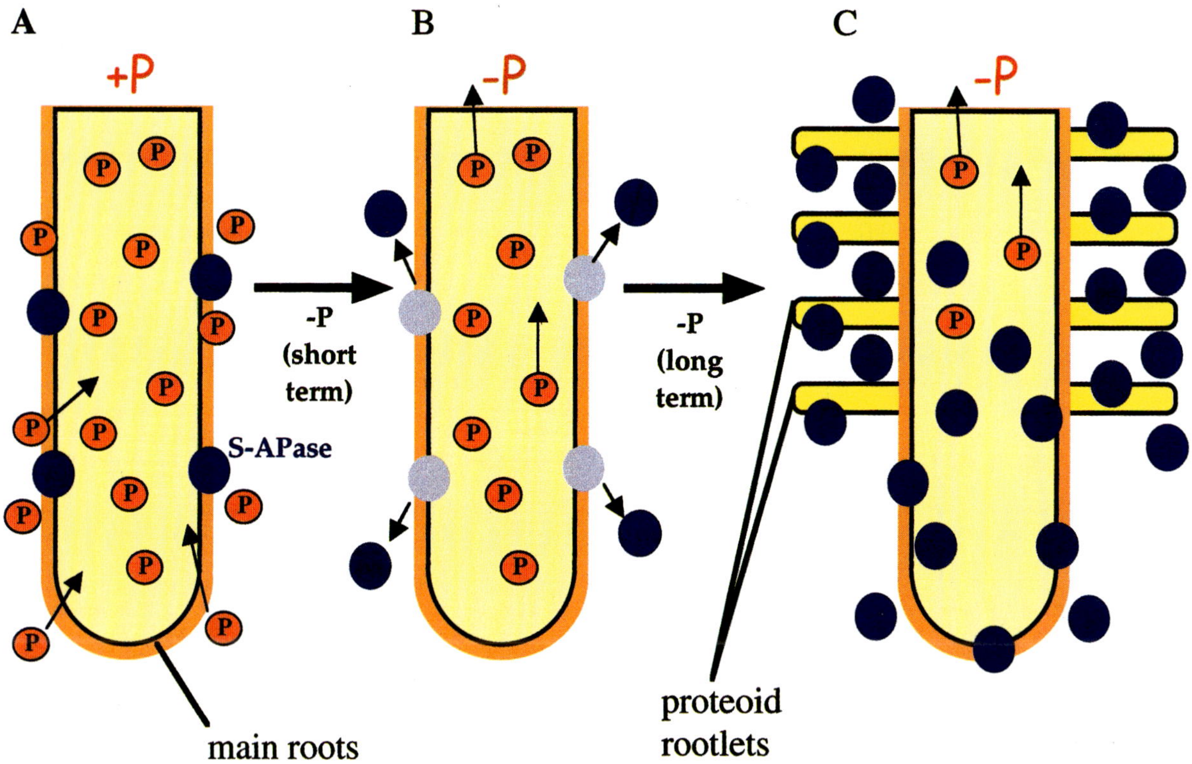

Figure 7. Schematic model of S-APase expression and cluster root formation. (A) Under +P conditions, APase collects around the epidermal cells of roots, and only a small amount of APase is secreted. (B) After −P treatment, APase is quickly released from roots within 12 h. However, APase secretion remains at a low level for a long time. (C) After Pi is sufficiently decreased in tissues, cluster roots form and APase secretion significantly increases.

secreted immediately; and (iii) if internal P concentrations decrease by a significant amount, expression of LASAP2 is induced, and S-APase production and secretion is activated to enhance the acquisition of Pi from organic P in the soil. Although there are no appreciable data available, cluster root formation was initiated around 9 days after plants were transferred to −P medium. It seems that cluster root formation plays an important role in acquiring Pi from organic P compounds, because the time at which cluster roots began to form approximated the dramatic increase in LASAP2 expression (Figure 7).

Further study of the role of cluster roots in LASAP2 expression is in progress, using histochemical localization of LASAP2 mRNA to clarify the details of S-APase function in the acquisition of phosphorus from organic phosphorus compounds in the rhizosphere.

Acknowledgements

This research has been partly supported by the Core Research for Evolutionary Science and Technology (CREST) of the Japan Science and Technology Corporation (JST) and the rice genome project MA-2111 of the Ministry of Agriculture, Forestry and Fisheries (MAFF).

References

Beck J L, McConachie L A, Summors A C, Arnold W N, Dejersey J and Zerner B 1986 Properties of a purple phosphatase from red kidney bean: a zinc–iron metalloenzyme. Biochim. Biophys. Acta 869, 61–68.

Bhargawa R and Sachar R C 1987 Induction of acid phosphatase in cotton seedlings: enzyme purification, subunit structure and kinetic properties. Phytochemistry 26, 1293–1297.

Biswass T K and Cundiff C 1991 Multiple forms of acid phosphatase in germinating seeds of *Vigna sinensis*. Phytochemistry 30, 2119–2125.

Bosse D and Köck M 1998 Influence of phosphate starvation on phosphohydrolases during development of tomato seedlings. Plant Cell Environ. 21, 325–332.

Dell B, Kuo J and Thompson G J 1980 Development of proteoid roots in *Hakea oblique* R.Br. (Proteaceae) grown in water culture. Aust. J. Bot. 28, 27–37.

Dinkelarker B, Römheld V and Marschner H 1989 Citric acid excretion and precipitation of calcium citrate in the rhizosphere of white lupin (*Lupinus albus* L.). Plant Cell Environ. 12, 285–292.

Gardner W K, Parbery D G and Barber D A 1981 Proteoid root morphology and function in *Lupinus albus*. Plant Soil 60, 143–147.

Gardner W K and Boundy K A 1983 The acquisition of phosphorus by *Lupinus albus* L. IV. The effect of interplanting wheat and white lupin on the growth and mineral composition of the two species. Plant Soil 70, 391–402.

Gardner W K, Barber D A and Parbey D G 1983 The acquisition of phosphorus by *Lupinus albus* L. III. The probable mechanism by which phosphorus movement in the soil/root interface is enhanced. Plant Soil 70, 107–124.

Gibson D M and Ullah A H J 1988 Purification and characterization of phytase from cotyledons of germinating soybean seeds. Arch. Biochem. Biophys. 260, 503–513.

Haraguchi H, Yamauchi D and Minamikawa T 1990 Multiple forms of acid phosphatase in cotyledons of *Vigna mungo* seedlings: immunological detection and quantitation. Plant Cell Physiol. 31, 917–923.

Johnson J F, Allan D L and Vance C P 1994 Phosphorus stress-induced proteoid roots show altered metabolism in *Lupinus albus*. Plant Physiol. 104, 657–665.

Johnson J F, Allan D L, Vance C P and Weiben G 1996a Root carbon dioxide fixation by phosphorus-deficient *Lupinus albus*. Plant Physiol. 112, 19–30.

Johnson J F, Vance C P and Allan D L 1996b Phosphorus deficiency in *Lupinus albus*. Altered lateral root development and enhanced expression of phospho*enol*pyruvate caboxylase. Plant Physiol. 112, 31–41.

Kihara T, Ohno T, Koyama H, Sawafuji T and Hara T 2002 Characterization of NADP-isocitrate dehydrogenase expression in a carrot mutant cell line with enhanced citrate excretion. Plant Soil 248, 145–153.

Klabunde T, Stahl B, Suerbaum H, Hahner S, Karas M, Hillenkamp F, Krebs B and Witzel H 1994 The amino acid sequence of the red kidney bean Fe(III)-Zn(II) purple acid phosphatase. Eur. J. Biochem. 226, 369–375.

Li M, Shinano T and Tadano T 1997 Distribution of exudates of lupin roots in the rhizosphere under phosphorus deficient conditions. Soil Sci. Plant Nutr. 43, 237–245.

Luo H M, Osaki M and Tadano T 1997 Mechanisms of differential tolerance of crop plants to high aluminum and low phosphorus growth conditions. *In* Plant Nutrition – for Sustainable Food Production and Environment. Eds. T Ando, K Fujita, T Mae, H Matsumoto, S Mori and J Sekiya. pp. 473–474. Kluwer Academic Publishers, Dordrecht, The Netherlands.

Miller S S, Liu J, Allan D L, Menzhuber C J, Fedorova M and Vance C P 2001 Molecular control of acid phosphatase secretion into the rhizosphere of proteoid roots from phosphorus-stressed white lupin. Plant Physiol. 127, 594–606.

Murashige T and Skoog F 1962 A revised medium for rapid growth and bio-assays with tobacco tissue cultures. Physiol. Plant. 15, 473–497.

Nakai K and Kanehisa M 1992 A knowledge base for predicting protein localization sites in eucaryotic cells. Genomics 14, 897–911.

Okuda S, Kaneko J, Ogawa T, Yamaguchi T, Izaki K and Takahashi H 1987 Increase in enzyme activities in embryonic axes of soybean seeds during germination. Agric. Biol. Chem. 51, 109–113.

Ozawa K, Osaki M, Matsui H, Honma M and Tadano T 1995 Purification and properties of acid phosphatase secreted from lupin roots under phosphorus-deficiency conditions. Soil Sci. Plant Nutr. 41, 461–469.

Palmiter R D 1974 Magnesium precipitation of ribonucleoprotein complexes. Expedient techniques for the isolation of undergraded polysomes and messenger ribonucleic acid. Biochemistry 13, 3606–3615.

Saheki S, Takeda A and Shimizu T 1985 Assay of inorganic phosphate in the mild pH range, suitable for measurement of glycogen phosphorylase activity. Anal. Biochem. 148, 277–281.

Shane M W, De Vos M, De Roock S and Lambers H 2002 Cluster root growth and citrate exudation in *Lupinus albus* L. (white lupin): separating the influences of root and shoot phosphorus. Plant Cell Environ. (In press)

Shinano T, Yonetani R, Ushihara N, Adachi H, Wasaki J, Matsui H and Osaki M 2001 Characteristics of phosphoenolpyruvate phosphatase purified from *Allium cepa*. Plant Sci. 161, 861–869.

Skene K R 1998 Cluster roots: some ecological considerations. J. Ecol. 86, 1060–1064.

Tadano T and Komatsu K 1994 Utilization of organic phosphorus in soil by plant roots. Proc. Trans. 15th World Congr. Soil Sci. 9, 521–522.

Tadano T and Sakai H 1991 Secretion of acid phosphatase by the roots of several crop species under phosphorus-deficient conditions. Soil Sci. Plant Nutr. 37, 129–140.

Tamura T, Minamikawa T and Koshiba T 1982 Multiple forms of acid phosphatase in cotyledons of *Vigna mungo* seedlings. J. Exp. Bot. 33, 1332–1339.

Uhde-Stone C, Zinn K, Ramirez M, Allan D L and Vance C P 2002 Phosphorus deficiency in white lupin: ESTs related to proteoid root development and biochemistry. Plant Soil 248, 99–116.

Ueki K and Sato S 1977 Regulation of phosphatase synthesis by orthophosphate in cultured tobacco cells. Plant Cell Physiol. 18, 1253–1263.

Wasaki J, Ando M, Ozawa K, Omura M, Osaki M, Ito H, Matsui H and Tadano T 1997 Properties of secretory acid phosphatase from lupin roots under phosphorus-deficient conditions. Soil Sci. Plant Nutr. 43, 981–986.

Wasaki J, Omura M, Osaki M, Ito H, Matsui H, Shinano T and Tadano T 1999a Structure of a cDNA for an acid phosphatase from phosphate-deficient lupin (*Lupinus albus* L.) roots. Soil Sci. Plant Nutr. 45, 439–449.

Wasaki J, Omura M, Ando M, Shinano T, Osaki M and Tadano T 1999b Secreting portion of acid phosphatase in roots of lupin (*Lupinus albus* L.) and a key signal for the secretion from the roots. Soil Sci. Plant Nutr. 45, 937–945

Wasaki J, Omura M, Ando M, Dateki H, Shinano T, Osaki M, Ito H, Matsui H and Tadano T 2000 Molecular cloning and root specific expression of secretory acid phosphatase from phosphate deficient lupin (*Lupinus albus* L.). Soil Sci. Plant Nutr. 46, 427–437.

Wasaki J, Dateki H, Shinano T, Osaki M and Tadano T 2001 Characterization of secretory acid phosphatase gene of lupin roots and the transformations of the gene into tobacco plants. *In* Plant Nutrition – Food Security and Sustainability of Agro-Ecosystems Through Basic and Applied Research. Eds. W J Horst, M K Schenk, A Bürkert, N Claassen, F Flessa, W B Frommer, H Goldbach, H-W Olfs, V Römheld, B Sattelmacher, U Schmidhalter, S Schubert, N von Wirén and L Wittenmayer. pp. 36–37. Kluwer Academic Publishers, Dordrecht, The Netherlands.

Watt M and Evans J R 1999 Proteoid roots. Physiology and development. Plant Physiol. 121, 317–323.

Plant and Soil **248:** 137–144, 2003.
© 2003 *Kluwer Academic Publishers. Printed in the Netherlands.*

Effects of altered citrate synthase and isocitrate dehydrogenase expression on internal citrate concentrations and citrate efflux from tobacco (*Nicotiana tabacum* L.) roots

Emmanuel Delhaize[1], Peter R Ryan, Peter J Hocking & Alan E Richardson
CSIRO Plant Industry, GPO Box 1600, Canberra ACT 2601, Australia. [1]*Corresponding author**

Received 14 December 2001. Accepted in revised form 22 April 2002.

Key words: aluminium, citrate, citrate synthase, efflux, isocitrate dehydrogenase, resistance

Abstract

To assess the effectiveness of manipulating citrate metabolism with the aim of increasing citrate efflux from roots, we generated transgenic tobacco (*Nicotiana tabacum* L.) lines that either overexpressed mitochondrial citrate synthase (EC 4.1.3.7) activity or had reduced activity of cytosolic isocitrate dehydrogenase (EC 1.1.1.42). Despite increases in citrate synthase activities in transgenic lines of up to 5-fold, neither internal citrate concentrations nor citrate efflux were increased compared to controls suggesting that, in tobacco, citrate synthase activity does not directly determine citrate accumulation and efflux. Consistent with a lack of effect on citrate efflux, the increase in citrate synthase activity did not enhance the aluminium resistance of the transgenic lines. Preliminary data collected on two transgenic lines with cytosolic isocitrate dehydrogenase activities reduced to one-tenth and one third of the control for shoot and root tissues respectively, showed that while these changes in activities were associated with a 1.5-fold increase in internal citrate concentrations of both types of tissue, citrate efflux from roots was not increased. Further work is needed to establish whether the increase in internal citrate concentration is associated with enhanced aluminium resistance of these lines. We conclude that in tobacco internal citrate concentrations and citrate efflux are largely insensitive to large changes in either mitochondrial citrate synthase or cytosolic isocitrate dehydrogenase activities and suggest that other factors, such as transport out of the roots, control citrate efflux.

Abbreviations: CS – citrate synthase; ICDH – isocitrate dehydrogenase

Introduction

The efflux of organic-acid anions from roots has been implicated in the Al resistance and phosphate acquisition mechanisms of plants (Ryan et al., 2001). In the case of Al resistance, efflux from roots occurs either immediately after exposure to Al or after a lag phase of several hours. The rapid response has been attributed to the activation by Al of an anion channel on the plasma membrane that is permeable to organic-acid anions. By contrast, the lag phase observed in some species before maximal efflux is achieved may represent *de novo* synthesis of proteins induced by Al exposure. These proteins could be involved in the biosynthesis of organic-acid anions or in their transport out of the cell. Less is known about the mechanisms that control efflux of organic-acid anions from roots when plants experience P deficiency where the process may be linked to root development (Ryan et al., 2001). However, it is likely that anion channels are also involved in the transport of the organic-acid anions out of P-deficient roots and the observation that anion-channel antagonists inhibit citrate efflux from proteoid roots of P-deficient white lupin (*Lupinus albus* L.) is consistent with this concept (Neumann et al., 1999).

While a mechanism is required to transport organic acid anions out of root cells, plants also require the enzymatic pathways to synthesise sufficient amounts of the organic-acid anions. Since they have essential roles in plant metabolism, virtually all plants pos-

* FAX No: 61-2-62465000. E-mail: e.delhaize@csiro.au

sess pathways for the synthesis of citrate and malate, two of the organic-acid anions commonly found in plant exudates. Citrate synthase (CS; EC 4.1.3.7) condenses acetylCoA with oxaloacetate to form citrate and the mitochondrial enzyme plays an essential role in the supply of citrate for the tricarboxylic acid cycle. There is evidence to indicate that increased CS activity is associated with an enhanced ability to accumulate citrate within plant tissues which in turn results in greater efflux of citrate from plant cells. For example, the activities of several enzymes involved in organic acid biosynthesis, including CS, are enhanced in proteoid or cluster roots of white lupin that form under low or moderate P-supply and exude large amounts of citrate (Johnson et al., 1994). Cell lines of carrot (*Daucus carota* L.) selected for their ability to use Al-phosphate as a P source have enhanced citrate efflux and this attribute is associated with greater CS and lower NADP-isocitrate dehydogenase (ICDH; EC 1.1.1.42) activities compared to the parental cell line (Kihara et al., 2002; Takita et al., 1999). The activity of CS in root tips of rye (*Secale cereale* L.) increases about 30% after 4 h exposure to Al at which time citrate efflux begins to increase (Li et al., 2000). Consistent with these observations of increased CS activity in roots exuding citrate, several studies have shown a relationship between internal concentrations of citrate and citrate efflux (Hoffland et al., 1989; Kamh et al., 1999; Mugai et al., 2000; Neumann and Römheld, 1999).

However, other studies suggest that CS activity does not directly regulate internal citrate concentrations or citrate efflux from roots. For example, an acid-containing *Citrus* variety accumulates citrate in its fruit but the level of CS activity during fruit development is similar to that observed for a variety that accumulates little citrate (Sadka et al., 2001). Other studies have shown that citrate efflux is poorly correlated with internal citrate concentrations or CS activity in P-deficient root tissues which suggests that citrate efflux is not always controlled by internal citrate concentrations (Keerthisinghe et al., 1998; Neumann et al., 2000; Watt and Evans, 1999).

Despite the conflicting conclusions regarding the role of CS in accumulation of citrate in tissues, several laboratories have sought to increase citrate efflux by manipulating citrate biosynthesis. de la Fuente et al. (1997) expressed a *Pseudomonas aeruginosa* CS gene in tobacco (*Nicotiana tabacum* L.) where the CS protein was targeted to the cytosol and reported up to a 10-fold increase in internal citrate concentration and an associated 4-fold increase in citrate efflux from roots. The increased citrate efflux was reported to enhance both the Al resistance (de la Fuente et al., 1997) and P-acquisition efficiency of the transgenic lines (López-Bucio et al., 2000). By contrast, Delhaize et al. (2000) were unable to repeat the findings of increased citrate accumulation, increased citrate efflux or enhanced Al resistance for the same transgenic plants or for ones that expressed the bacterial CS gene to a much higher level. These findings suggest that the apparent beneficial effects of the CS gene were either due to other factors or that the expression of the phenotypes was sensitive to environmental conditions. In a similar approach, Koyama et al. (1999, 2000) over-expressed plant CS genes in carrot cell lines and intact *Arabidopsis thaliana* plants and observed increased internal citrate concentrations and enhanced citrate efflux. In both of these cases the full coding region of the CS genes, including the transit peptides for targeting to the mitochondria, were used and it is likely that the CS proteins accumulated in mitochondria. Tesfaye et al. (2001) over-expressed a gene encoding a nodule-enhanced form of malate dehydrogenase in alfalfa (*Medicago sativum* L.) and showed that this increased the accumulation and efflux of a range of organic acids including citrate. An increase in Al resistance was associated with the greater organic acid efflux further illustrating the potential for enhancing organic acid efflux by manipulating metabolic processes.

Reducing the catabolism of citrate is an alternative approach to increase accumulation of citrate in plant tissues. The cytosolic form of NADP-ICDH is probably the major pathway for the catabolism of citrate after its transport out of mitochondria and conversion to isocitrate by aconitase in the cytosol. The cytosolic NADP-ICDH represents the major form of ICDH in plant tissues and is implicated in having a role in converting isocitrate to 2-oxoglutarate for subsequent biosynthesis of amino acids (Lancien et al., 2000). Kruse et al. (1998) used antisense technology to reduce cytosolic NADP-ICDH activity in potato shoots to less than 10% of control plants and found that leaves of the antisensed plants had 2.5-fold greater isocitrate concentrations and 2-fold greater citrate concentrations than the control plants. The observations of Takita et al. (1999) and Kihara et al. (2002) that carrot cell lines with high citrate efflux have reduced NADP-ICDH activity along with enhanced CS activity are consistent with a role for the cytosolic NADP-

ICDH in contributing towards the control of citrate concentrations in plant cells.

To assess the effectiveness of manipulating organic acid metabolism in tobacco (*Nicotiana tabacum* L.), with the aim of increasing internal citrate concentrations and citrate efflux, we overexpressed the mitochondrial CS gene and used post-transcriptional gene silencing to reduce the activity of the cytosolic form of NADP-ICDH. We report on the effectiveness of these modifications towards increasing citrate concentrations in tissues and citrate efflux from roots of tobacco.

Materials and methods

Plant material

All tobacco transgenic lines were generated in the Wisconsin 38 cultivar and used the cauliflower mosaic virus promoter of the gene encoding the 35S transcript (CaMV 35S) to drive expression of the introduced genes. Plants were transformed with the binary vectors described below using *Agrobacterium*-mediated transformation (Horsch et al., 1985). For the CS overexpressing lines (CS lines), primary transformants (TO) were assayed for leaf CS activity and a selection of plants with elevated CS activities were grown on soil and T1 seed collected. The lines with reduced cytosolic NADP-ICDH activities (ASIC lines) were clonally propagated as primary transformants. Plants were selected on MS agar (Murashige and Skoog, 1962) with kanamycin (50 – 100 mg/l) prior to all experiments except for the Al resistance assays.

Cloning and preparation of plasmid constructs

The coding region of the tobacco mitochondrial CS gene (GenBank accession X84226) was amplified from tobacco root RNA by RT-PCR using the forward primer CCGGGTACCTCTAGAATGGTGTTCTATCGCC and the reverse primer GCGAAGCTTTCATGCTTTCTTGCAATGG. Inclusion of *Kpn*I and *Xba*I restriction sites in the forward primer and a *Hin*dIII site in the reverse primer (shown underlined) allowed the PCR product to be inserted into the *Kpn*I/*Hin*dIII site of pBluescript SK (Stratagene, La Jolla CA, USA). After sequencing to confirm that the PCR product encoded the tobacco mitochondrial CS, the DNA was sub-cloned into the *Kpn*I/*Hin*dIII sites of pART7 (Gleave, 1992) between the CaMV 35S promoter and octopine synthase terminator. The *Not*I fragment from pART7 that contained the CS gene was introduced into the *Not*I site of the binary vector pPLEX502, originally described as pPLEX101 (Surin et al., 1998). The coding region of the tobacco cytosolic NADP-ICDH (GenBank accession X77944) was amplified from total RNA by RT-PCR using the forward primer GCGAAGCTTATGACTTTCGACAAAATCAAGGT and the reverse primer GCGGGTACCTTAGGCCTTTGCTTTCAGAAGTC. Inclusion of a *Hin*dIII site in the forward primer and a *Kpn*I site in the reverse primer allowed the PCR product to be cloned into the *Kpn*I/*Hin*dIII sites of pBluescript SK. After sequencing confirmed that the PCR product encompassed the coding region of the tobacco cytosolic NADP-ICDH, the plasmid was used as a template for amplification of an approximately 0.47 kb fragment with the forward primer ATTCTAGACTCGAGAAACCCTATCGTTGAGATGGACG and the reverse primer ATTAAGCTTGAATTCTGTTCCTTCTGGCACAAACACC. The region amplified was designed to specifically reduce the expression of the cytosolic NADP-ICDH while not affecting expression of the mitochondrial NADP-ICDH. Inclusion of the restriction sites *Xba*I and *Xho*I in the forward primer and *Hin*dIII and *Eco*RI in the reverse primer (shown underlined) allowed the fragment to be cloned in the forward and reverse orientation either side of an intron in the vector pHANNIBAL (Wesley et al., 2001) that can be used for effective gene silencing in plants. The *Not*I fragment that contained the inverse repeat of NADP-ICDH under the control of the 35S CaMV promoter was introduced into the *Not*I site of pPLEX502. Samples were extracted for RNA using a method described by Rerie et al. (1991).

Plant growth

For hydroponic culture, seedlings were first selected for kanamycin (50 mg/l) resistance on MS agar plates (Murashige and Skoog, 1962) and after 20 to to 30 day growth were transferred to hydroponic culture as described previously (Delhaize et al., 2000). The nutrient solution was as described by Delhaize et al. (1993) except that 500 μM $Ca(NO_3)_2$ was replaced with 500 μM $CaCl_2$ and the pH adjusted to 4.3. Nutrient solution was changed twice over the 11 to 15 days of growth. Plants were grown in a controlled environment cabinet at constant temperature (23 °C) and photon flux density of approximately 500 μmol m^{-2} s^{-1} with a 16-h-light/8-h-dark period.

Plant analysis

Internal citrate concentrations, citrate efflux and Al resistance were determined as described previously (Delhaize et al., 2000). Citrate synthase activity in plant tissues was assayed using the methods of Landschütze et al. (1995) except that the extraction buffer did not include α_2-macroglobin and the extraction procedure was modified as follows. Leaf (a plug of 7 mm diameter removed with a cork borer from the youngest fully expanded leaf) or root tip (four 6 mm root tips) samples were homogenised in 200 μl of extraction buffer and then immediately centrifuged for 10 s in a microfuge (16,000 *g*). A sample of the supernatant solution (10–40 μl) was then assayed using the method described by Srere (1967). To differentiate between mitochondrial and cytosolic CS activity, 0.4 M mannitol was included in the extraction buffer and after homogenisation in two volumes of buffer, samples were filtered through one layer of miracloth (Calbiochem, La Jolla, CA, USA). The filtrate was centrifuged for 5 min at 3500 *g* and the resulting supernatant solution was centrifuged at 11,200 *g* for 30 min. The supernatant solution represented the cytosolic fraction while the pellet contained a crude mitochondrial fraction. The pellet was resuspended in CS extraction buffer and homogenised with 15-strokes of a ground-glass hand-held homogeniser. Subsamples of the supernatant (100 μl) and pellet (50 μl) were then assayed for CS activity. Samples for NADP-ICDH activity were homogenised in 100 μl of the same buffer used for CS assays except that it included 5 mM dithiothreitol. The homogenates were centrifuged for 6 min at 16,000 *g* and 50 μl of the supernatant solution assayed for NADP-ICDH activity using the method described by Fieuw et al. (1995). Protein was assayed by the method of Bradford (1976). Southern blots, northern blots and sequencing were undertaken using procedures described by Sambrook et al. (1989).

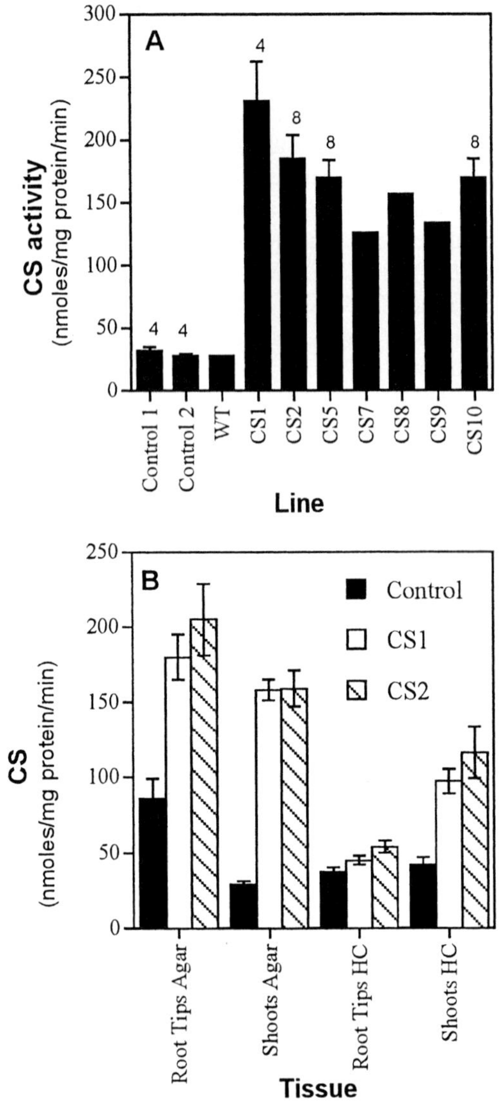

Figure 1. (A) Citrate synthase activity in leaf tissue of a range of transgenic tobacco lines overexpressing the tobacco mitochondrial CS gene (CS lines) and in control and wild-type (WT) lines. The control lines were transformed with the vector alone. Seedlings were grown on agar under tissue culture conditions with kanamycin selection for 25 days except for the wild-type lines where the kanamycin was omitted. The values above the columns indicate the number of seedlings assayed for each line where $n > 2$, and the bars indicate the standard errors of the means. (B) Citrate synthase activities in shoots and root tips of control and two selected CS lines grown on agar with kanamycin selection for 28 days or grown by hydroponic culture (HC) in a constant environment cabinet for 11 days after transfer from kanamycin selection on agar. The error bars indicate the standard errors of the means ($n = 4$).

Results and discussion

Overexpression of a mitochondrial citrate synthase gene in tobacco

Expression of the tobacco mitochondrial CS gene in tobacco under the control of the 35S CaMV promoter resulted in a range of independent transgenic lines with elevated CS activity in shoots compared with control and wild type lines (Figure 1A). When grown on agar, shoots of the CS overexpressing lines had about 5.5-fold greater CS activity than the control line while in root tips the increase was about 2-fold (Figure 1B). When plants were grown hydroponically in a controlled growth room, the increases in CS activities of the CS lines were not as marked as for agar-grown plants with approximately a 1.2- to 1.5-fold increase in root tips and a 2- to 2.5-fold increase in shoots. Estimates of the CS activities in cytosolic and mitochondrial fractions from leaves of agar-grown plants showed that not only had CS activity of the mitochondrial fraction increased 4-fold (control: 35.6±4.2; CS2: 143.4±8.1 nmol/mg protein/min; n=3) but also that CS activity in the cytosolic fraction had increased approx 7-fold (control: 9.0±0.8 nmol/mg protein/min; CS2: 67.0±3.9 nmol/mg protein/min; n=3). The increase in CS activity in the cytosolic fraction was unexpected as the coding region of the gene had retained the mitochondrial targeting sequence. It is possible that overexpression of the CS resulted in misdirection of a proportion of the protein to the cytosol or alternatively, the mitochondria were damaged during the fractionation. In the latter case, the elevated cytosolic CS could represent leakage of CS from the mitochondria. In any case, our results show that mitochondrial CS activity was greatly elevated in the CS transgenic lines compared with the control line.

Despite large increases in CS activities in shoots of agar-grown CS1 and CS2 lines over the control line, this did not translate in any detectable increase in internal citrate concentrations (Figure 2A). Similarly, neither shoots nor roots of hydroponically grown CS lines showed any increase in citrate concentrations when compared with the control line (Figure 2A). Although we did not observe an increase in internal citrate in CS over-expressing lines, it is possible that there was an increased turnover of citrate where increased biosynthesis was matched by increased degradation. Interestingly, citrate concentrations in shoots of hydroponically grown plants were about 10-fold greater than those of agar-grown shoots even though the CS activities of the agar-grown shoots were similar (control plants) or about 50% greater (CS1 and CS2 lines) than those of the hydroponically grown plants.

Previously, it was shown that tobacco possesses a mechanism for Al-activated efflux of citrate from root apices (Delhaize et al., 2000). We confirm that observation here and show that overexpression of CS did not enhance citrate efflux either in the absence or presence of Al (Figure 2B). The Al resistance of the CS lines assessed at two Al concentrations was not

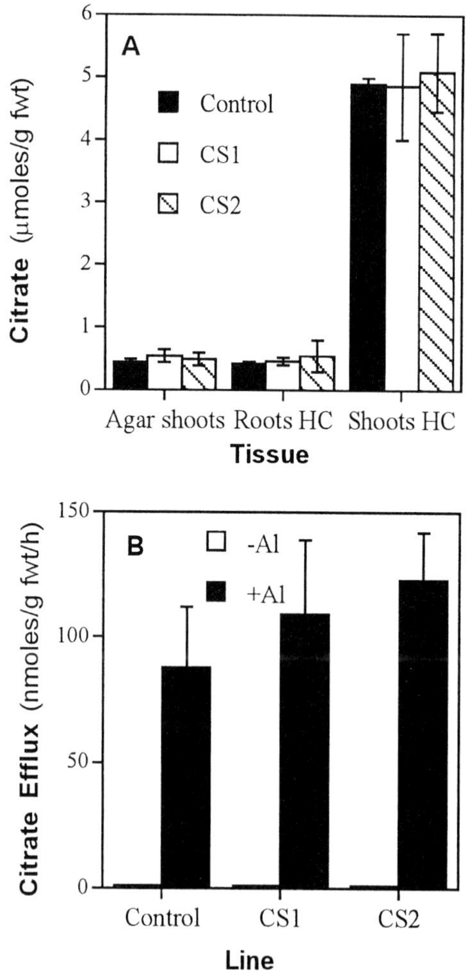

Figure 2. (A) Citrate concentrations in shoots of two CS lines and the control line grown on agar with kanamycin selection (28 days) and in both shoots and roots of hydroponically grown plants. Seedlings were grown by hydroponic culture (HC) in a controlled environment cabinet for 11 days after transfer from kanamycin selection on agar. The error bars indicate the standard errors of the means (n = 4). (B) Citrate efflux from roots of two CS lines and the control line. The efflux was measured from single plants with their roots immersed in 0.2 mM $CaCl_2$ either in the absence or in the presence of 50 μM $AlCl_3$. The samples were taken over the 3–6 h interval after the start of the experiment to avoid the lag phase that occurs over the initial 3 h (Delhaize et al., 2000). Seedlings were grown hydroponically in a controlled environment cabinet for 16 days after transfer from kanamycin selection on agar. The error bars indicate the standard errors of the means (n = 3).

increased above that of the control which is consistent with the absence of an increase in citrate efflux (Figure 3). Although it was not possible to pre-select the seedlings on kanamycin before testing for Al resistance, the data of Figure 3 do not indicate any increase of Al resistance in the two CS lines where, on average, 3 out

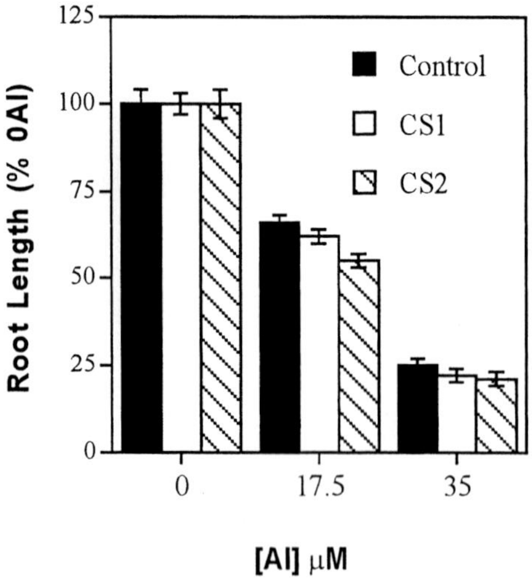

Figure 3. Aluminium resistance of two CS lines and a control line. Seedlings were grown on nylon mesh in each of three solutions as described previously (Delhaize et al., 2000) and for each line root elongation is expressed as a percentage of the control plants grown in the absence of Al (0Al). The error bars denote the standard errors of the means ($n = 29 - 37$). Note that the seedlings were of the T1 generation and were not pre-selected on kanamycin. Therefore, these data would have included null segregants at a rate of approximately 1 null to 3 transformed seedlings for a single insert.

of 4 seedlings would have been transgenic for a single insert.

Our findings contrast to those of Koyama et al. (1999, 2000) who found that overexpression of plant CS genes increased citrate efflux by up to 1.7-fold in intact Arabidopsis plants and by up to 4-fold from carrot cells grown in culture. The increase in CS activity reported for the highest expressing Arabidopsis seedlings was 3 to 4-fold greater than the controls which is similar to the increases reported here (Figure 2A). The discrepant findings might be attributed to the different species used or to different growth conditions. It is conceivable that over-expression of CS could have resulted in greater accumulation and efflux of citrate under specific growth conditions. Indeed, we found that in tobacco both internal citrate concentrations and CS activities were markedly affected by growth conditions (Figures 1 and 2). Our findings are consistent with those of Ruijter et al. (2000) who showed that in *Aspergillus niger*, overexpression of the mitochondrial CS gene increased CS activity by up to 11-fold, but did not increase either internal citrate concentrations or citrate efflux. We conclude that mitochondrial CS activity in tobacco is unlikely to limit the production of citrate for efflux, a similar conclusion to that made by Delhaize et al. (2000) when a *Pseudomonas aeruginosa* CS gene was expressed in tobacco and the CS protein was directed to the cytosol. Evidence to indicate that CS activity is usually in excess of normal cellular requirements in plant tissues, are the observations of Landschütze et al. (1995) who, using antisense techniques, were able to reduce mitochondrial CS activity in potato (*Solanum tuberosum* L.) plants to 10% of the wild type. Despite the large reduction in CS, the only phenotypic difference observed was abortion of ovules at flowering and the authors concluded that in most tissues, CS activity is in excess of requirements and does not limit production of citrate.

Silencing of the gene in tobacco encoding cytosolic NADP-isocitrate dehydrogenase

We used post-transcriptional gene silencing to reduce the expression of the cytosolic NADP-ICDH gene in tobacco. The use of a vector to generate inverted repeats of the target sequence separated by an intron effectively reduced NADP-ICDH activity in leaves of a range of transgenic plants (Figure 4A). The reduction of NADP-ICDH activity to about 10% of the control in 7 out of 8 transgenic lines compares favorably to the report of Kruse et al. (1998) who used conventional antisense technology to reduce NADP-ICDH activity in leaves of potato plants to a similar level in 4 out of over 70 T0 plants analysed. A more detailed study of two of the primary transgenic lines (ASIC1 and ASIC2) showed a similar reduction in NADP-ICDH activities in leaves of plants grown hydroponically in a controlled growth room (Figure 4B). Root tissue also had reduced NADP-ICDH activity but the residual activity was about 30% of the control suggesting that either other isoforms of NADP-ICDH, such as the mitochondrial form, may be more abundant in root tissue or that the silencing of the NADP-ICDH gene was not complete. Kruse et al. (1998) also reported a less dramatic reduction in NADP-ICDH activity (approx. 50%) in tubers of potato lines that had the greatest reduction in shoot NADP-ICDH activity.

Analysis of internal citrate concentrations in shoots and roots showed that in both cases the ASIC2 line had approximately 1.5-fold greater citrate concentrations than the control line (Figure 5). For the ASIC1 line, the increase was only statistically significant for roots. This increase in citrate concentration is similar to that reported by Kruse et al. (1998) who found approximately a 2-fold increase in citrate concentrations in leaves of potato when NADP-ICDH activity

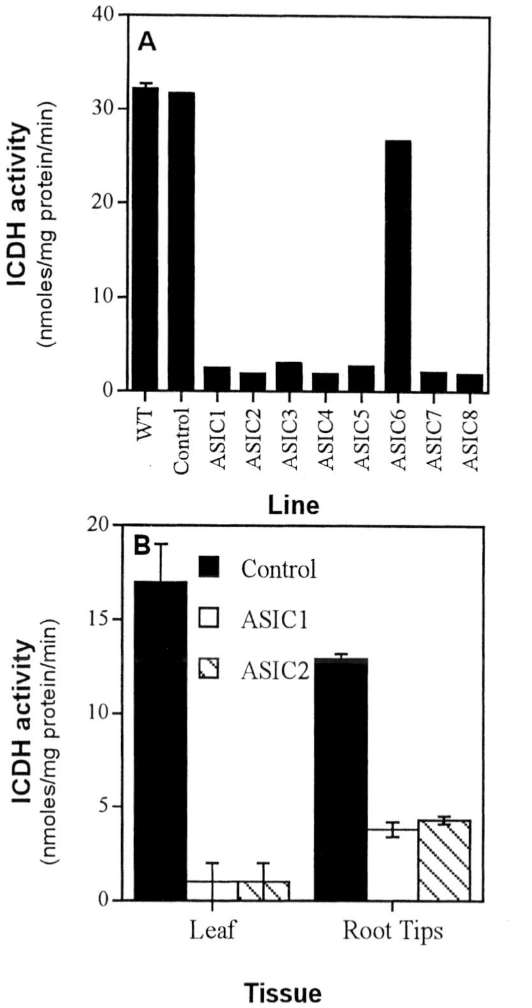

Figure 4. (A) Activity of NADP-ICDH in leaf tissue of a range of transgenic tobacco lines transformed with a construct to silence the endogenous gene encoding the cytosolic NADP-ICDH gene (ASIC lines) and in control and wild-type (WT) lines. A single plant of each primary transformant was assayed and two different wild-type (WT) plants were assayed (error bar shows the range). (B) Activity of NADP-ICDH in leaf tissue and root tips of two ASIC lines and a control line grown hydroponically in a controlled environment cabinet for 17 days after transfer from kanamycin selection on agar. The error bars indicate the standard errors of the means ($n = 3$).

was reduced to less than 10% of the control. Citrate efflux was assessed in the T1 generation of two other ASIC lines (ASIC4 and ASIC7) that showed low NADP-ICDH activity (Figure 4A). The phenotype of low leaf NADP-ICDH activity was confirmed in the kanamycin-resistant T1 generation of these lines (21.4±1.3 nmol/mg protein/min for the control, n = 6; 3.4±1.6 nmol/mg protein/min for ASIC4, n = 12; and 2.6±0.5 nmol/mg protein/mn for ASIC7, n = 10). For the ASIC4 and ASIC7 lines only seedlings that showed less than 20% of the NADP-ICDH activity of

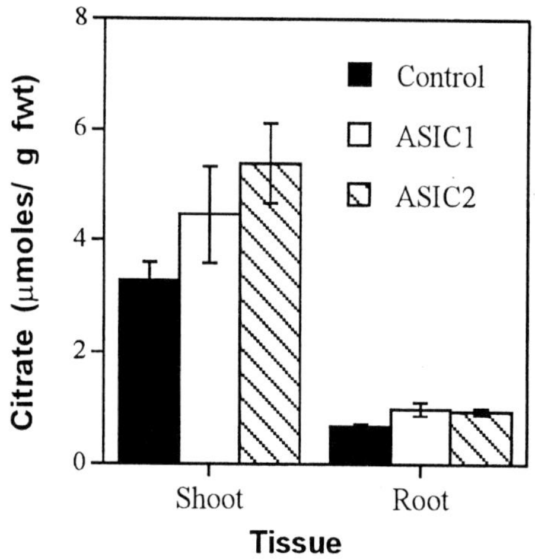

Figure 5. Citrate concentrations in shoots and roots of two ASIC lines and a control line grown hydroponically in a controlled environment cabinet for 18 days after transfer from kanamycin selection on agar. The error bars indicate the standard errors of the means (n = 3 or 4).

the control line were assessed for citrate efflux. Despite the large differences in NADH-ICDH activity, the citrate efflux of the various genotypes were similar to one another either in the presence (33±6 nmoles/g fwt/h for control; 41±6 nmoles/g fwt/h for ASIC4; and 38±7 nmoles/g fwt/h for ASIC7; n = 3) or absence (below the limits of detection for all; n = 3) of Al treatment.

Conclusions

Mitochondrial CS and cytosolic NADP-ICDH are two enzymes that have been suggested to have a key role in controlling internal citrate concentrations in plants. However, despite achieving marked increases in CS activity and large reductions in NADP-ICDH activity by genetic manipulations of tobacco, internal citrate concentrations and citrate efflux were largely resilient to these modifications. We conclude that CS activity does not limit the accumulation of internal citrate and suggest that the transport of citrate across the plasma membrane is a more likely site for regulation of citrate efflux from tobacco roots.

Acknowledgements

The authors are grateful to the Grains Research and Development Corporation of Australia for funding this work.

References

Bradford M M 1976 A rapid and sensitive method for the quantitation of microgram quantities of protein utilizing the principle of protein-dye binding. Anal. Biochem. 72, 248–254.

de la Fuente J M, Ramírez-Rodríguez V, Cabrera-Ponce J L and Herrera-Estrella L 1997 Aluminum tolerance in transgenic plants by alteration of citrate synthesis. Science 276, 1566–1568.

Delhaize E, Craig S, Beaton C D, Bennet R J, Jagadish V C and Randall P J 1993 Aluminum tolerance in wheat (*Triticum aestivum* L.) I. Uptake and distribution of aluminum in root apices. Plant Physiol. 103, 685–693.

Delhaize E, Hebb D M and Ryan P R 2000 Expression of a *Pseudomonas aeruginosa* citrate synthase gene in tobacco is not associated with either enhanced citrate accumulation or efflux. Plant Physiol. 125, 2059–2067.

Fieuw S, Müller-Röber B, Gálvez S and Willmitzer L 1995 Cloning and expression analysis of the cytosolic $NADP^+$ dependent isocitrate dehydrogenase from potato. Plant Physiol. 107, 905–913.

Gleave A P 1992 A versatile binary vector system with a T-DNA organisational structure conducive to efficient integration of cloned DNA into the plant genome. Plant Mol. Biol. 20, 1203–1207.

Hoffland E, Findenegg G R and Nelemans J A 1989 Solubilization of rock phosphate by rape. II. Local root exudation of organic acids as a response to P-starvation. Plant Soil 113, 161–165.

Horsch R B, Fry J E, Hoffman N L, Eichholtz D, Rogers S G and Fraley R T 1985 A simple and general method for transferring genes into plants. Science 227, 1229–1231.

Johnson J F, Allan D L and Vance C P 1994 Phosphorus stress-induced proteoid roots show altered metabolism in *Lupinus albus*. Plant Physiol. 104, 657–665.

Kamh M, Horst W J, Amer F, Mostafa H and Maier P 1999 Mobilization of soil and fertilizer phosphate by cover crops. Plant Soil 211, 19–27.

Keerthisinghe G, Hocking P J, Ryan P R and Delhaize E 1998 Effect of phosphorus supply on the formation and function of proteoid roots of white lupin (*Lupinus albus* L.). Plant Cell Environ. 21, 467–478.

Kihara T, Ohono T, Koyama H, Sawafuji T and Hara T 2002 Characterization of NADP-isocitrate dehydrogenase expression in a carrot mutant cell line with enhanced citrate excretion. Plant Soil (In This Issue).

Koyama H, Kawamura A, Kihara T, Hara T, Takita E and Shibata D 2000 Overexpression of mitochondrial citrate synthase in *Arabidopsis thaliana* improved growth on a phosphorus limited soil. Plant Cell Physiol. 41, 1030–1037.

Koyama H, Takita F, Kawamura A, Hara T and Shibata D 1999 Over expression of mitochondrial citrate synthase gene improves the growth of carrot cells in Al-phosphate medium. Plant Cell Physiol. 40, 482–488.

Kruse A, Fieuw S, Heineke D and Müller-Röber B 1998 Antisense inhibition of cytosolic NADP-dependent isocitrate dehydrogenase in transgenic potato plants. Planta 205, 82–91.

Lancien M, Gadal P and Hodges M 2000 Enzyme redundancy and the importance of 2-oxoglutarate in higher plant ammonium assimilation. Plant Physiol. 123, 817–824.

Landschütze V, Willmitzer L and Müller-Röber B 1995 Inhibition of flower formation by antisense repression of mitochondrial citrate synthase in transgenic potato plants leads to a specific disintegration of the ovary tissues of flowers. EMBO J. 14, 660–666.

Li X F, Ma J F and Matsumoto H 2000 Pattern of Al-induced secretion of organic acids differs between rye and wheat. Plant Physiol. 123, 1537–1543.

López-Bucio J, Martinez de la Vega O, Guevara-García A and Herrera-Estrella L 2000 Enhanced phosphorus uptake in transgenic tobacco plants that overproduce citrate. Nature Biotech. 18, 450–453.

Mugai E N, Agong S G and Matsumoto H 2000 Aluminium tolerance mechanisms in *Phaseolus vulgaris* L.: citrate sythase activity and TTC reduction are well correlated with citrate secretion. Soil Sci. Plant Nutr. 46, 939–950.

Murashige T and Skoog F 1962 A revised medium for a rapid growth and bioassay with tobacco tissue culture. Physiol. Plant. 15, 473–497.

Neumann G, Massonneau A, Langlade N, Dinkelaker B, Hengeler C, Römheld V and Martinoia E 2000 Physiological aspects of cluster root function and development in phosphorus-deficient white lupin (*Lupinus albus* L.). Ann. Bot. 85, 909–919.

Neumann G, Massonneau A, Martinoia E and Römheld V 1999 Physiological adaptations to phosphorus deficiency during proteoid root development in white lupin. Planta 208, 373–382.

Neumann G and Römheld V 1999 Root excretion of carboxylic acids and protons in phosphorus-deficient plants. Plant Soil 211, 121–130.

Rerie W G, Whitecross M and Higgins T J V 1991 Developmental and environmental regulation of pea legumin genes in transgenic tobacco. Mol. Gen. Genet. 225, 148–157.

Ruijter G, Panneman H, Xu D-B and Visser J 2000 Properties of *Aspergillus niger* citrate synthase and effects of *citA* overexpression on citric acid production. FEMS Lett. 184, 35–40.

Ryan P R, Delhaize E and Jones D L 2001 Function and mechanism of organic anion exudation from plant roots. Annu. Rev. Plant Physiol. Plant Mol. Biol. 52, 527–560.

Sadka A, Dahan E, Or E, Roose M L, Marsh K B and Cohen L 2001 Comparative analysis of mitochondrial citrate synthase gene structure, transcript level and enzymatic activity in acidless and acid-containing *Citrus* varieties. Aust. J. Plant Physiol. 28, 383–390.

Sambrook J, Fritsch E F and Maniatis T 1989. Molecular Cloning: A Laboratory Manual, 2nd edn, Cold Spring Harbor Laboratory, Cold Spring Harbor, NY.

Srere PA 1967 Citrate synthase. Methods Enzymol. 13, 3–11.

Surin B, Boevink P, Keese P, Chu P, Larkin P, Llewellyn D, Kahn R, Ellacott G and Waterhouse P 1998. A suite of promoters and terminators for plant biotechnology. *In* Agricultural Biotechnology: Laboratory, Field and Market. Proceedings of the 4th Asia-Pacific Conference on Agricultural Biotechnology Ed. P J Larkin. pp. 121–122. Canberra UTC Publishing, Darwin, Australia.

Takita E, Koyama H and Hara T 1999 Organic acid metabolism in aluminum-phosphate utilizing cells of carrot (*Daucus carota* L.). Plant Cell Physiol. 40, 489–495.

Tesfaye M, Temple S J, Allan D L, Vance C P and Samac D A (2001) Overexpression of malate dehydrogenase in transgenic alfalfa enhances organic acid synthesis and confers tolerance to aluminum. Plant Physiol. 127, 1836–1844.

Watt M and Evans J R 1999 Linking development and determinacy with organic acid efflux from proteoid roots of white lupin grown with low phosphorus and ambient or elevated atmospheric CO_2 concentration. Plant Physiol. 120, 705–716.

Wesley S V, Helliwell C A, Smith N A, Wang M-B, Rouse D T, Liu Q, Gooding P S, Singh S P, Abbott D, Stoutjesdijk P A, Robinson S P, Gleave A P, Green A G and Waterhouse P M 2001 Construct design for efficient, effective and high-throughput gene silencing in plants. Plant J. 27, 581–590.

Plant and Soil **248:** 145–153, 2003.
© 2003 *Kluwer Academic Publishers. Printed in the Netherlands.*

Characterization of NADP-isocitrate dehydrogenase expression in a carrot mutant cell line with enhanced citrate excretion

Tomonori Kihara[1], Takashi Ohno[1], Hiroyuki Koyama[1,3], Takuya Sawafuji[2] & Tetsuo Hara[1]
[1]*Laboratory of Plant Cell Technology, Faculty of Agriculture, Gifu University, 1-1, Yanagido, Gifu, 501-1193, Japan.* [2]*Laboratory of Plant Nutrition, Faculty of Agriculture, Tohoku University, 1-1 Tsutsumidohri-Amamiyamachi, Aoba, Sendai, 911-8555, Japan.* [3]*Corresponding Author**

Received 5 December 2001. Accepted in revised form 21 March 2002

Key words: Al-phosphate, citrate excretion, *Daucus carota*, isocitrate dehydrogenase

Abstract

Citrate in the cytosol is converted to isocitrate, by the action of aconitase, and then isocitrate is converted to 2-oxoglutarate by the action of NADP-specific isocitrate dehydrogenase (NADP-ICDH). This pathway is modified in a mutant carrot (*Daucus carota* L.) cell line [designated as insoluble phosphate grower (IPG): Koyama et al., 1992 Plant Cell Physiol. 33, 173–177], which releases large amounts of the citrate and grows faster than the wild-type cells in Al-phosphate medium. In the current study, the activities of aconitase in mitochondria and in crude extracts were similar in the mutant cells and the wild-type cells, suggesting that the activity of the cytosolic aconitase was similar among the cell types. By contrast, the NADP-ICDH activity in the crude extracts of the mutant cells was almost half that of the wild-type cells. Western blot analysis revealed that the cytosolic NADP-ICDH was the most abundant isoenzyme in both cell types. However the NADP-ICDH amount was lower in the mutant cells than that in the wild-type cells. Transcripts level of *DcICDH1*, which appeared to be a unique isoform encoding the cytosolic enzyme in carrot, was lower in the mutant cells than that of the wild-type cells. Thus, we inferred that the lower transcript level of the cytosolic isoform caused the lower NADP-ICDH activities in the mutant cells, compared with the wild-type cells.

Abbreviations: CS – citrate synthase; ICDH – isocitrate dehydrogenase; IPG – insoluble phosphate grower; WT – wild type

Introduction

Since phosphorus is a severely limiting and nonrenewable resource, improvement of the ability of P acquisition is an important goal in sustainable agriculture. Plants have a range of strategies to improve P acquisition from soils when deprived of P, such as high-affinity P_i-transport systems (Koyama et al., 1992), the excretion of organic anions including citrate and malate (Lipton et al., 1987), and the excretion of acid phosphatases (Goldstein et al., 1988: Wasaki et al., 2002). Genetic manipulation in terms of such traits is one approach to increase the ability of P acquisition in the crop plants (Vance et al., 2001). As an example, an enhanced citrate excretion from the roots of tobacco (de la Fuente et al., 1997 López-Bucio et al., 2000) and *Arabidopsis thaliana* (Koyama et al., 2000) was established by introducing citrate synthase (CS) genes from bacteria and plants, respectively. The transgenic plants showed superior P_i-uptake and growth when grown on P_i-limited soils, possibly due to an enhanced citrate excretion. However, several reports using the same approach indicate that the effectiveness of this strategy may be limited (Delhaize et al., 2001: Delhaize et al., 2002). For example, the transgenic *A.*

* FAX No: +81-582932911
E-mail: koyama@cc.gifu-u.ac.jp

Nucleotide sequences for *DcICDH1* and *DcICDH2* were deposited in the DDBJ, EMBL and Genbank under accession number AB019327 and AB019328, respectively.

thaliana excreted about 1.5-fold more citrate from the roots than from the wild-type, but the excretion level is much lower than that from some plant species with a greater ability to excrete citrate, such as white lupin (Koyama et al., 2000). One possible explanation why transgenic plants that over-express CS did not show improved acquisition of P_i might be a lack of other factors that play important roles in the exudation of citrate. Identification of other factors relating to the high ability to excrete citrate at molecular level would be necessary for the development of plants that show a greater level of citrate excretion than from the CS transgenic plants. Since the mechanism of citrate excretion is likely to involve several reactions, studies on mutant cell lines or plants is a useful approach.

We selected a mutant carrot cell line (designated as IPG; insoluble phosphate grower), which releases large amounts of citrate into an Al-phosphate medium, and grows faster than the wild-type (WT) cells (Koyama et al., 1990, 1992). This IPG cell line showed a typical modified organic acid metabolism that would allow production of citrate at greater level than that of wild-type cells, such as increased biosynthesis and decreased catabolism of citrate (Takita et al., 1999a). In terms of citrate biosynthesis, we found an elevated level of transcripts encoding mitochondrial CS in the IPG cells (Takita et al., 1999b). Since carrot cells are transformable, we can evaluate the effect of this factor on the increased citrate excretion in the IPG cells using transgenic approaches. The CS transgenic cell lines carrying a mitochondrial CS from *A. thaliana*, whose CS activities were comparable to that in IPG cells, released two-fold more citrate into an Al-phosphate medium than WT cells did. However, this was three times less than that from IPG cells (Koyama et al., 1999b). Thus, we speculated that other changes, such as a decreased citrate catabolism, would enable them to attain the greater citrate excretion rate in the IPG cells than that of the CS transgenic cell line.

In a previous study, we found a decreased NADP-specific isocitrate dehydrogenase (NADP-ICDH) activity in the IPG cells compared to the WT cells (Takita et al., 1999a). Since NADP-ICDH catalyzes a segment of the conversion from citrate to 2-oxoglutarate, which in turn could increase citrate accumulation, this could be another important factor that enhanced citrate excretion from IPG cells. The conversion step of citrate to 2-oxoglutarate, which is localized in the cytosol involves aconitase and NADP-ICDH, and is thought to play important roles in the supply of C skeleton for amino acid synthesis (Chen, 1998). An inhibition of this pathway has been reported to coincide with the increased citrate synthesis (accumulation) in citrus fruit (Sadka et al., 2000b) and the proteoid roots of lupin (Kania et al., 2002; Neumann et al., 1998). A similar mechanism might exist in the increased citrate excretion from the IPG cells. In higher plants, there are two types of aconitase isoenzymes and at least three isoenzymes of NADP-ICDH. To fully understand the role of cytosolic citrate catabolism in citrate excretion from IPG cells, it is important to determine the function of the cytosolic isoenzymes separately from the other isoenzymes.

In the present study, we performed a series of organelle fractionation and molecular biological studies to determine citrate catabolism in the cytosol of IPG cells and WT cells. We found no difference in aconitase activity between cell types, whereas the cytosolic NADP-ICDH level was significant lower in the IPG cells than in the WT cells. Although in vivo NADP-ICDH forms hetero- or homo-dimer in the cytosol, which varies among plant species, our results reveal that the cytosolic NADP-ICDH in carrot is a homodimer. We conclude that the decreased activity of the cytosolic NADP-ICDH in IPG cells is caused by the lower transcript level for this enzyme.

Materials and methods

Plant materials and culture conditions

The mutant cell line (IPG) of carrot (*Daucus carota* L. cv. MS Yonsun) that grows normally in Al-phosphate medium by enhanced citrate excretion and wild-type cells (WT) used in this study were identical to those used in our previous studies (Takita et al., 1999a, b). IPG cells were obtained from by subculturing wild-type cells in Al-phosphate medium as described previously (Koyama et al., 1988). Both cell lines were maintained in regular R2 medium (Ohira et al., 1973) containing 2 m*M* NaH_2PO_4 as the phosphorus source for at least 1 year prior this experiment. Each 0.3 g of fresh cell was used as inoculums, and grown for 5–9 days in R2 medium containing either NaH_2PO_4 or Al-phosphate as a sole phosphorus source. Cells were collected by vacuum filtration on a Miracloth (Calbiochem-Novabiochem, La Jolla, CA, USA) and washed thoroughly with distilled water, then frozen in liquid nitrogen and kept in a deep freezer (−80°C) until extracted.

Biochemical assays

Crude extracts containing soluble proteins were prepared as described previously (Takita et al., 1999a). The extract was desalted using a Bio-Gel P-6 (Bio-Rad) column prior to determination of enzyme activity if required. NADP-ICDH was assayed according to the method of Chen et al. (1988), and aconitase was measured using the method of Cooper and Beevers (1969). Citrate synthase was assayed spectrophotometrically according to Srere (1967) by monitoring the reduction of Acetyl-CoA in the presence of DTNB at A_{412}. Protein was quantified colorimetrically using the method of Bradford (1976). Western blot analysis for NADP-ICDH polypeptides was performed using anti-serum raised against DcICDH1 protein, and with the same protocol as described previously (Takita et al., 1999a). Quantitative image analysis was performed using the U.S. National Institutes of Health (NIH) image analysis program obtained through NIH web page (http://rsb.info.nih.gov/nih-image)

Fractionation of organelles

Mitochondria enriched fraction was obtained by the method described previously (Takita et al., 1998). Peroxisomes were separated from protoplasts by the method of Nishimura et al. (1976) with minor modifications. Protoplasts were prepared from 20 g of cells in 150 mL of enzyme solution containing 0.5 *M* mannitol, 0.5% cellulase onozuka RS (Yakult, Japan), 1% driselase (Kyowa Hakkou, Japan) and 0.01% pectolyase Y-23 (Seishin Pharmaceutical, Japan), then washed twice by centrifugation (2000 × g, 10 min) in 0.5 *M* mannitol. The washed protoplasts were suspended in 15 mL of 0.05 *M* Tricene-NaOH (pH 7.5) containing 0.5 *M* sucrose, and then ruptured by passing two layers of Miracloth (Calbiochem-Novabiochem) to break plasma membrane. The ruptured protoplasts (15 mL) were applied on top of 50 mL of the linear sucrose gradient (35–60%, w/w) containing 0.02 *M* Tricine-NaOH (pH 7.5), then centrifuged 85,000× *g* for 2.5 h. Each 1.5 mL of fraction was collected from the bottom, and then used for enzyme assay. To assess the purity of preparations the activities of fumarase (Fum), a marker enzyme for mitochondria, and alcohol dehydrogenase (ADH), a marker enzyme for cytosol were measured according to the methods of Cooper and Beevers (1969), and Hart (1970), respectively. The activities of catalase (CAT), a marker enzyme for the peroxisomes, and cytochrome-C oxidase (COX) for the mitochondria were assayed using the method of Chance and Herbert (1950) and Hodges and Leonard (1974), respectively.

Isolation and analysis of cDNAs encoding NADP-ICDH

Partial cDNA fragments encoding NADP-ICDH were obtained by RT-PCR using degenerated primers. Degenerated primers and PCR protocols were identical to those used in our previous study of isolating NADP-ICDH cDNAs from rice (Koyama et al., 1999a). Both 3′ and 5′ ends were obtained using a RACE PCR system purchased from Takara (Ohtsu, Japan). Sequence analyses were performed using a series of computer softwares from GENOMNET (www.genome.ad.jp).

Quantification of DcICDH1 transcripts by quantitative RT-PCR

The quantitative RT-PCR for DcICDH1 was carried out according to the method of Canel et al. (1996) with minor modifications as described previously for the quantification of transcripts for mitochondrial CS from IPG and WT cells (Takita et al., 1999b), where specific primers for *DcICDH1* (ICDH1qF1: 5′-CTGAAGAGTTCATTGATGCAGTAGC-3′, ICDH1qR1: 5′-CTCTTCCAGTCACATCATAGGTAAGC-3′) were used. The PCR products 230 bp long were analyzed by Southern blot and then visualized using a kit (DIG High prime DNA labeling and detection system: Boehringer Manheim). Transcript level of actin was determined by using specific primers for carrot actin (AcF1: 5′-GTGACGTTGATATCAGAAAGG-3′, AcR1: 5′-CGTCATATTCACCCTTCGAG-3′) as a control.

Results and discussion

Comparison of mitochondrial aconitase and NADP-ICDH activities between IPG and WT cells

Citrate in the cytosol is first converted into isocitrate by the action of aconitase, and then into 2-oxoglutarate by the action of NADP-specific isocitrate dehydrogenase (NADP-ICDH). Subsequently, 2-oxoglutarate is utilized as C skeleton for the glutamine synthase/glutamate synthase pathway (Chen and Gadal, 1990). Decreased activity of these enzymes was associated with citrate accumulation during the development of citrus fruits (Sadka et al., 2000a, b). Thus,

Table 1. NADP–ICDH and aconitase activities relative to fumarase activity in crude extracts and mitochondria of the IPG and the WT cells

Cell type	Fraction	Aconitase activity	NADP-ICDH activity
		(expressed as a ratio of fumarase activity)	
IPG	Crude	1.98 ± 0.15	1.15 ± 0.10
	Mitochondria	1.30 ± 0.20	0.08 ± 0.03
WT	Crude	1.85 ± 0.08	2.33 ± 0.23
	Mitochondria	1.26 ± 0.21	0.10 ± 0.04

Fresh cells (20 g) harvested at 7 d were ground gently with a glass homogenizer and centrifuged with Percoll (40%w/v) in the presence of 0.25 *M* sucrose. Activity of aconitase and NADP-ICDH are expressed as relative values to the activity of fumalase, a mitochondrial marker enzyme. Absolute fumarase activities in IPG and WT cells in the crude extracts were 0.149 ± 0.02 units (per mg protein per min) and 0.145 ± 0.006 units, respectively. Means ± SD are shown (*n*=3).

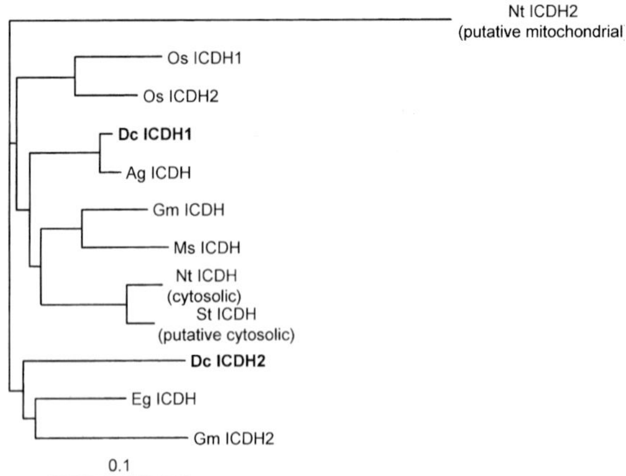

Figure 1. Phylogenetic tree of NADP specific isocitrate dehydrogenases. The carrot DcICDH1 and the DcICDH2 were compared with sequences of other plant NADP-ICDHs. Tobacco (*NtICDH*1, P50218 and *NtICDH2,* X96728), rice (*OsICDH1,* AF155333 and *OsICDH2,* AF155334), celery (*AgICDH,* Y12540), soybean (*GmICDH1,* AF055582 and *GmICDH2* AF095445), alfalfa (*MsICDH* Q40345), potato (*St ICDH,* P50217) and *Eucalyptus* (*EgICDH,* U80912). NtICDH2 has been characterized as a mitochondrial isoenzyme (Gálvez et al., 1998). On the other hand, *GmICDH2* contains a typical peroxisomal targeting signal (SKL) at the C-terminus. Proteins, other than DcICDH2, were predicted as cytosolic isoenzymes by PSORTII program. The scale bar corresponds to 0.1 amino acid substitution per residue.

the inhibition of these enzymes is thought to be involved in producing a citrate surplus, which in turn allows excretion of large amounts of citrate from the IPG cells. In the crude cell extracts, we found that NADP-ICDH activities in IPG cells were lower than those in WT cells, throughout the 9-day culture period in Al-phosphate medium (Takita et al., 1999a). However, the activities of aconitase were similar between cell types. The activities found in the crude extracts of both enzymes consisted of isoenzymes which are localized in different intracellular compartments. Higher plants contain aconitase activities in the cytosol and mitochondria (Sadka et al., 2000a), whereas NADP-ICDH activities exist in the cytosol, mitochondria and peroxisomes (Gálvez et al., 1999). To further characterize these enzymes, which may be responsible for higher citrate productivity of IPG cells, we performed organelle fractionation studies.

As described above, two types of aconitase have been found in higher plants. While one is localized in the mitochondria, the other is found in the cytosol. Thus, we could separate the activity of each aconitase by measuring the activity in the crude extracts and that in the purified mitochondria. In this experiment, the apparent specific activities from the mitochondria were affected by the intactness and the purity of mitochondria and these factors varied between each preparation. To compare IPG and WT cells, we used the relative activities, calculated as the ratio of the specific activities of aconitase and NADP-ICDH to that of fumarase (Fum), a mitochondrial matrix marker enzyme. We found no significant difference in the specific activities of Fum between cell types in crude extracts (Table 1). Both IPG and WT cells showed similar values in terms of relative aconitase activities for the crude extracts and for the mitochondria. Thus, we could infer that the cytosolic aconitase activity was similar in both cell types. We also compared NADP-ICDH activities between each cell type. The NADP-ICDH activity in the crude extract of IPG cells was about two-fold lower than that in the WT (Table 1). By contrast, the mitochondrial NADP-ICDH activity of IPG cells was similar to that of WT. Thus, the decrease in NADP-ICDH activity in IPG cells would have been due to other isoenzymes localized in the cytosol and the peroxisomes.

Isolation of a cDNA encoding cytosolic NADP-ICDH from carrot cells

To further examine the nature of the lower NADP-ICDH activity in IPG cells, we isolated cDNAs encoding NADP-ICDHs of carrot by RT-PCR using degenerate primers. These degenerate primers were obtained from reported ICDHs, and were used for isolating cDNAs encoding NADP-ICDH from rice in previous work (Koyama et al., 1998). We cloned and sequenced 50 independent partial clones encoding carrot NADP-ICDH. The nucleotide sequences of almost all clones were highly homologous to a putative cytosolic NADP-ICDH of celery (Accession number Y12540), whereas only one clone showed a different sequence compared to others. Full-sized cDNA consisted of 1512 bp with an ORF encoding 412 amino acids for the common clone and 1488 bp with an ORF encoding 416 amino acids for the unique clone were isolated by 3 prime and 5 prime RACE procedures. These cDNAs were designated as*DcICDH1* and *DcICDH2*, respectively. The amino acid sequence that was deduced from *DcICDH2* contained a typical peroxisomal targeting signal (SKL) at the C-terminus, but that from *DcICDH1* contained no consensus sequence for an organelle targeting signals. Analysis of the amino acid sequences using PSORTII, which predicts intracellular localization of proteins, gave the highest score for the cytosolic localization for *DcICDH1* (0.609) and a peroxisomal localization for *DcICDH2* (0.667). Phylogenetic analysis for plant NADP-ICDH proteins showed three clusters consisting of mitochondrial, putative cytosolic and putative peroxisomal proteins (Figure 1). From this analysis, *DcICDH1* and *DcICDH2* were categorized as the cytosolic and the peroxisomal proteins. We therefore inferred that *DcICDH1* encodes the cytosolic isoform of NADP-ICDH.

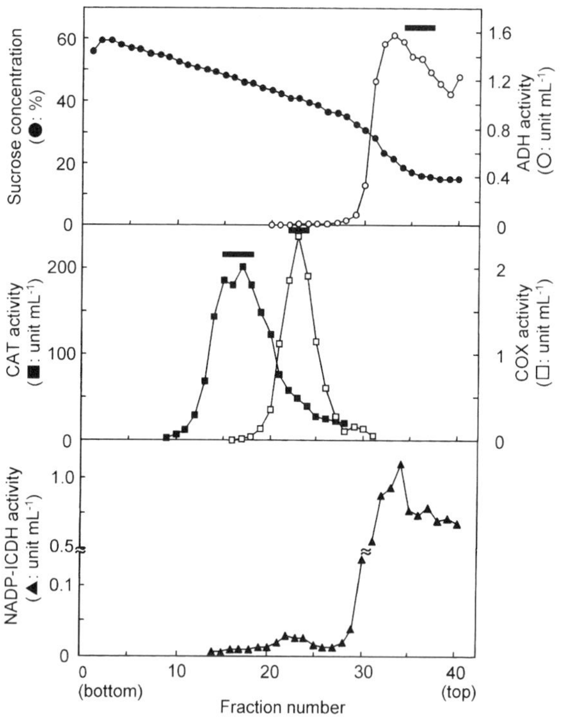

Figure 2. Organelle fractionation from carrot cultured cells. Cell organelles were purified from 9-d-cultured carrot cells by a sucrose density gradient centrifugation. Fractions were collected using a peristaltic pump, and the sucrose concentration was determined (●). The activities of NADP-specific ICDH (▲) and marker enzymes of the cytosol [alcohol dehydrogenase: ADH, ○], mitochondria [Cyt-c oxidase: COX, □] and peroxisomes [catalase: CAT, ■] in each fraction are shown. Horizontal bars indicate the fractions used for western blot analyses in Figure 3A.

Western blot analysis for NADP-ICDH isoforms in WT and IPG cells

To determine the nature of the lower activity of NADP-ICDH in IPG cells, we performed a series of western blot analyses using an anti-serum raised against the DcICDH1 protein. Since NADP-ICDH isoforms generally showed very high homology at the amino acid level, we expected cross-reactivity between the DcICDH1 anti-serum used in this study and other isoforms in carrot. As shown in Figure 2, marker enzymes of each fraction [ADH (alcohol dehydrogenase), cytosol; COX (Cyt-C oxidase), mitochondria; CAT (catalase), peroxisomes] were separated by a sucrose gradient centrifugation. Approximately 95% of the NADP-ICDH activities were found in fractions containing ADH activities, whereas only 4% and 1% of the activity was found in the fractions containing COX and CAT, respectively. Activities of other marker enzymes were absent in the cytosolic fraction. However, the mitochondrial and the peroxisomal fractions used in the western blot analysis contained marker enzyme activity from other intracellar compartments. As expected, each fraction gave signals having different pI values (Figure 3A). In the cytosol, we found only one strong signal at pI 6.5, but both the mitochondrial and the peroxisomal fractions gave two signals each. The mitochondrial fractions provided a strong signal at pI 6.9, and a weak signal at pI 6.5 was probably caused by cytosolic contamination. On the other hand, the peroxisomal fractions gave two signals with same pI value for the proteins that showed strong signal in the mitochondrial fraction, and a unique signal having a lower pI value than those of others. From these results, we could infer that the strong signal [cytosol and mitochondria] and the unique signal [peroxisomes] were due to the specific isoforms for each intracellular compartment.

Using the same experimental system, we obtained similar patterns with the crude extracts from WT and

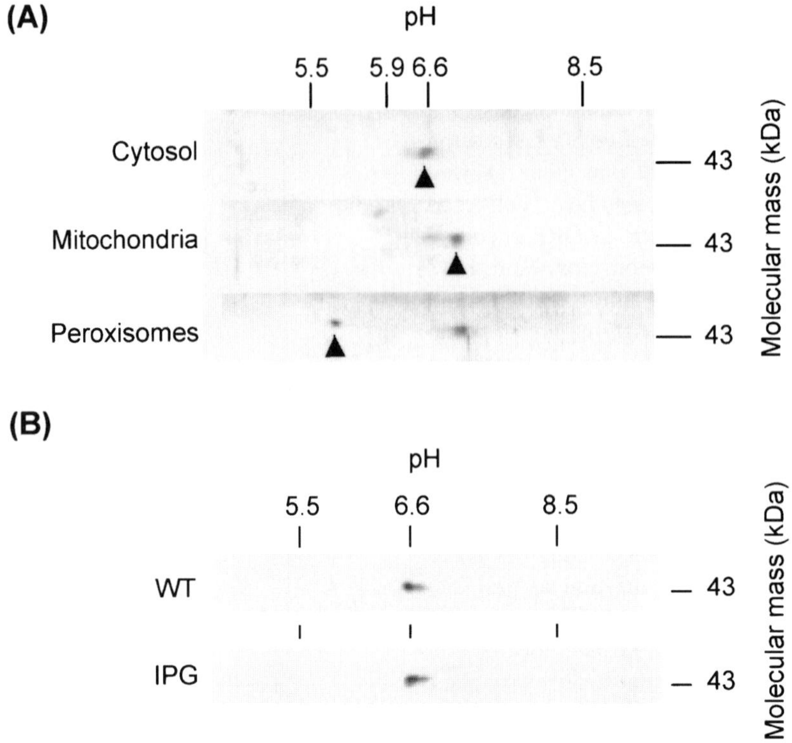

Figure 3. Western blot analyses for NADP-ICDH proteins in carrot. Proteins in the organelle fractions (A: see Figure 2) and the crude extracts (B) from the IPG and the WT cells were separated by IEF-SDS-PAGE, and then incubated with anti-DcICDH1 antiserum. The reactions were visualized using alkaline-phosphatase conjugated anti-rabbit goat antibodies. Arrow heads shown in panel A indicate the unique or the strongest spots for each organelle fraction.

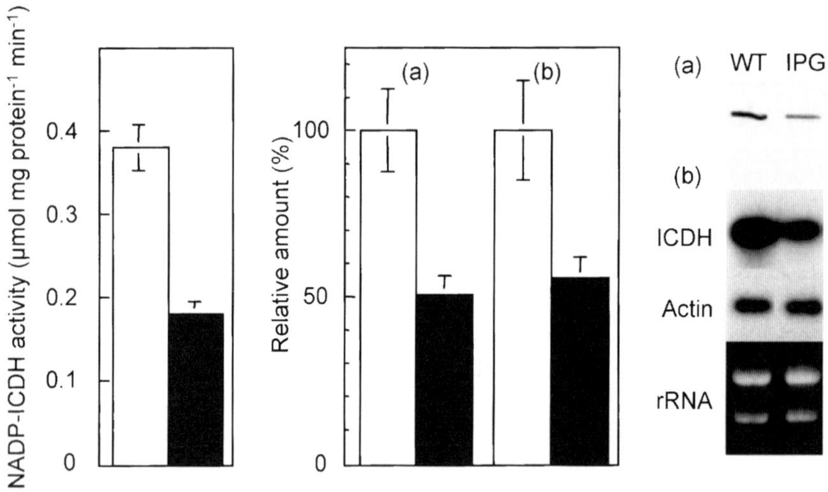

Figure 4. Analysis of protein and transcript levels of the cytosolic NADP-ICDH in WT and IPG cells, which were grown for 7 days in the Al-phosphate medium. The NADP-ICDH activity of the WT (open bar) and the IPG (closed bar) are shown in the left panel. The protein levels (a) and the transcripts levels of *DcICDH1* (b) were determined by the western blot and the quantitative RT-PCR (QRT-PCR), respectively. Transcripts level of actin was measured using the same RNA samples, as a control experiment of the QRT-PCR. Ribosomal RNAs in the total RNA (1 μg each) used for QRT-PCR was visualized to estimate the quality of RNA. Relative values for both the proteins and the DcICDH1 transcripts were determined using NIH image (the right panel).

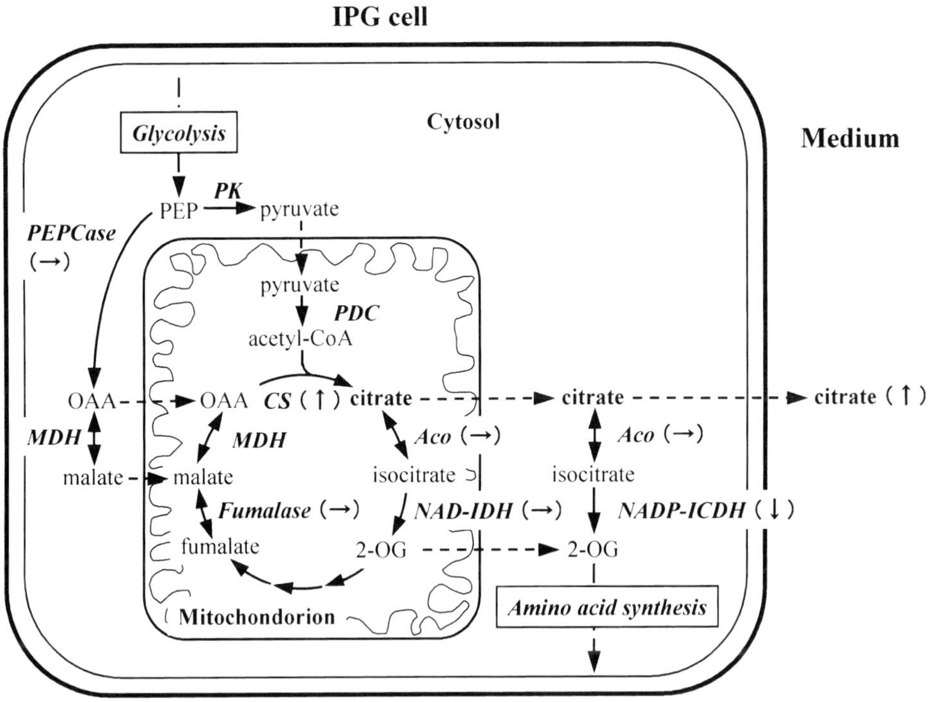

Figure 5. Diagrammatic representation of alterations of organic acid metabolism in IPG cells. The allows, (↑), (↓) and (→), indicate the activity of enzymes in IPG cells with an increase, a decrease or an equal compare to WT cells. Aco, aconitase; CS, citrate synthase; MDH, malate dehydrogenase; NAD-ICDH, NAD-isocitrate dehydrogenase; NADP-ICDH, NADP-isocitrate dehydrogenase; OAA, oxaloacetate; PEP, phospho*enol*pyruvate; PEPCase, phospho*enol*pyruvate carboxylase; PDC, pyruvate dehydrogenase complex; PK, pyruvate kinase; 2-OG, 2-oxoglutarate.

IPG cells. As described above, the lower NADP-ICDH activity of IPG cells was due to a difference in the cytosolic or the peroxisomal enzymes, since no difference of the mitochondrial NADP-ICDH activities between cell types was found (Table 1). As shown in Figure 3B, both IPG and WT cells gave a strong signal and a weak signal whose pI values corresponded to those of the cytosolic isoform and the mitochondrial isoform respectively; no signal was detected for the peroxisomal isoform (Figure 3B). Thus, we concluded that the lower NADP-ICDH activity of ICDH in the IPG cells would have been caused by lower activity of the cytosolic isoform.

Comparison of the protein and transcript levels of the cytosolic NADP-ICDH in WT and IPG cells

The cytosolic NADP-ICDH is believed to function as a dimer and in different plant species this dimer consists of either mono- or hetero-isoforms (Chen, 1998). However, western blot analysis clearly showed that the cytosolic NADP-ICDH in carrot is a homodimer (Figure 3A). As described above, one cDNA clone, *DcICDH1*, was predicted to encode a cytosolic isoform. The theoretical pI value of DcICDH1 protein deduced from the cDNA is 6.53, and this is very similar to the experimental pI value (6.5) of the cytosolic isoform (Figure 3A). These results strongly suggest that *DcICDH1* encodes a unique cytosolic NADP-ICDH isoform of carrot. To fully understand the mechanism of the lower NADP-ICDH activity in IPG cells than that in WT cells, we compared the protein level and the transcript level of *DcICDH1* between both cell types.

Since western blot analyses using a two-dimensional PAGE for the crude extracts revealed that the cytosolic NADP-ICDH was more abundant than the other isoforms (Figure 3A, B), we quantified the amount of cytosolic NADP-ICDH protein by a western blot analysis followed by SDS-PAGE. In addition, transcript levels of *DcICDH1* were accurately determined by quantitative competitive RT-PCR using specific primers encoding the 3 prime untranslated region. Using these procedures, we measured protein and transcript levels in both WT and IPG cells after 7-

day culture in Al-phosphate medium. The activity of NADP-ICDH in the IPG cells was 0.17 unit mg^{-1} protein, whereas that in the WT cells was 0.38 unit mg^{-1} protein, which corresponded to 2.2-fold of that in IPG cells. Under the given conditions, both the protein level of NADP-ICDH and the transcripts levels of *DcICDH1* were about two-fold greater in WT cells than those in IPG cells (Figure 4). Thus, these results strongly suggest that decreased transcripts level of the cytosolic isoform is responsible for the lower NADP-ICDH activity in IPG cells, compared to WT cells.

In the present study, we performed a series of organelle fractionation and molecular biological studies for characterizing the citrate catabolism in the IPG cells. Our results suggest that a lower cytosolic NADP-ICDH activity would contribute towards increase citrate productivity in the IPG cells, and this may in turn increase citrate excretion. This hypothesis is supported by a transgenic study of potato antisense for the cytosolic NADP-ICDH (Kruse et al., 1998). In this study, the citrate concentration in the leaves was higher in the transgenic plants than that of the control plants, although citrate excretion from the roots was not investigated. A similar approach would be important to evaluate the role of NADP-ICDH on the citrate excretion from the IPG cells. Unfortunately our attempt in antisense reduction was ineffective in inhibiting NADP-ICDH activity in WT cells (data not shown), and thus we need to use a more efficient system such as RNAi (Wesley et al., 2001) to create transgenic cell line carrying similar NADP-ICDH activity to that in the IPG cells.

In conclusion, the present study, as well as our previous report on the mitochondrial CS (Takita et al., 1999b), indicates that the IPG cells show a modified citrate metabolism in which at least two enzymes are regulated at the transcript level (Figure 5). Also, this cell line has shown the superior phosphate uptake (Koyama et al., 1992) due to increased transcripts levels for a high affinity Pi-transporter (Koyama et al., unpublished). It would be interesting to determine whether these changes at the transcript levels are caused by independent or coupled mutations. Further investigations are clearly needed to clarify this issue.

Acknowledgements

This study was financially supported by a Grant-in-Aid from the Ministry of Education and Sciences of Japan for author Koyama (12760042).

References

Bradford M M 1976 A rapid and sensitive method for the quantitation of microgram quantities of protein utilizing the principle of protein-dye binding. Anal. Biochem. 72, 248–254.

Chance B and Herbert D 1950 The enzyme-substrate compounds of bacterial catalases and peroxidases. Biochem. J. 46, 402–414.

Canel C, Bailey-Serres J N and Roose M L 1996 Molecular characterization of the mitochondrial citrate synthase gene of an acidless pummelo (*Citrus maxima*). Plant Mol. Biol. 31, 143–147

Chen R 1998 Plant NADP-dependent isocitrate dehydrogenases are predominantly localized in the cytosol. Planta 207, 280–285.

Chen R D and Gadal P 1990 Do the mitochondria provide the 2-oxoglutarate needed for glutamate synthesis in higher plant chloroplasts? Plant Physiol. Biochem. 28, 141–145.

Chen R, le Maréchal P, Vidal J, Jacquot J P and Gadal P 1988 Purification and comparative properties of the cytosolic isocitrate dehydrogenases (NADP) from pea (*Pisum sativum*) roots and green leaves. Eur. J. Biochem. 175, 565–572.

Cooper, T G and Beevers H 1969 Mitochondria and glyoxysomes from castor bean endosperm. Enzyme constituents and catalytic capacity. J. Biol. Chem. 244, 3507–3513.

de la Fuente J M, Ramirez-Rodrigez V, Cabrera-Ponce J S and Herrera-Esterella L 1997 Aluminum tolerance in transgenic plants by alteration in citrate synthesis. Science 276, 1566–1568.

Delhaize E, Hebb D M and Ryan P R 2001 Expression of a *Pseudomonas aeruginosa* citrate synthase gene in tobacco is not associated with either enhanced citrate accumulation or efflux. Plant Physiol. 125, 2059–2067.

Delhaize E, Ryan P R, Hocking P J and Richardson A E 2002 Effects of altered citrate synthase and isocitrate dehydrogenase expression on internal citrate concentrations of tobacco (*Nicotiana tabacum* L.) Plant Soil 248, 137-144.

Gálvez S, Roche O, Bismuth E, Brown S, Gadal P and Hodges M 1998 Mitochondrial localization of a NADP-dependent isocitrate dehydrogenase isoenzyme by using the green fluorescent protein as a marker. Proc. Natl. Acad. Sci. 95, 7813–7818.

Goldstein A H, Baertlein D A and McDaniel R G 1988 Phosphate starvation inducible metabolism in *Lycoperisicon esculentum*. I. Excretion of acid phosphatase by tomato plants and suspension cultured cells. Plant Physiol. 87, 711–715.

Hart G E 1970 Evidence for triplicate genes for alcohol dehydrogenase in hexaploid wheat. Proc. Natl. Acad. Sci. USA 66, 1136–1141.

Hodges T K and Leonard R T 1974 Purification of a plasma membrane-bound adenosine triphosphate from plant roots. Methods. Enzymol. 32, 392–406.

Kania A, Langlade N, Massonneau A, Martinoia E and Neumann G 2002 Phosphorus deficiency-induced modifications in citrate catabolism and in cytosolic pH as related to citrate exudation in cluster roots of white lupin. Plant Soil 248, 117–127.

Koyama H, Kawamura A, Kihara T, Hara T, Takita E and Shibata D 2000 Overexpression of mitochondrial citrate synthase in *Arabidopsis thaliana* improved growth on a phosphorus-limited soil. Plant Cell Physiol. 41, 1030–1037.

Koyama H, Kihara T, Takita E and Hara T 1999a Isolation of two cDNA clones encoding NADP-Specific isocitrate dehydrogenase from rice (Accession Nos. AF155333 and AF155334). (PGR99-120) Plant Physiol. 120, 1207

Koyama H, Takita E, Kawamura A, Hara T and Shibata D 1999b Over expression of mitochondrial citrate synthase gene improves the growth of carrot cells in Al-phosphate medium. Plant Cell Physiol. 40, 482–488.

Koyama H, Ojima K and Yamaya T 1990 Utilization of anhydrous aluminum phosphate as a sole source of phosphate by a selected carrot cell line. Plant Cell Physiol. 31, 173–177.

Koyama H, Ojima K and Yamaya T 1992 Characteristics of aluminum-phosphate-adapted carrot cells: uptake and utilization of phosphate. Plant Cell Physiol. 33, 171–176.

Kruse A, Fieuw S, Heineke D and Müller-Röber B1998 Antisense inhibition of cytosolic NADP-dependent isocitrate dehydrogenase in transgenic potato plants. Planta 205, 82–91.

Lipton D S, Blanchar R W and Blevins D G 1987 Citrate, malate and succinate concentrations in exudate from P-sufficient and P-stressed *Medicago sativa* L. seedlings. Plant Physiol. 85, 315–317.

López-Bucio J, de la Vega O M, Guevara-García A and Herrera-Estrella L 2000 Enhanced phosphorus uptake in transgenic tobacco plants that overproduce citrate. Nature Biotechnology 18, 450–453.

Nishimura M, Graham D and Akazawa T 1976 Isolation of intact chloroplasts and other cell organelles from spinach leaf protoplasts. Plant Physiol. 58, 309–314.

Neumann G, Massonneau A, Martinoia E and Roheld A 1999 Physiological adaptations to phosphorus deficiency during proteoid root development in white lupin. Planta 208, 373–382.

Ohira K, Ojima K and Fujiwara A 1973 Studies on the nutrition of rice cell culture. Plant Cell Physiol. 14, 1113–1121.

Sadka A, Dahan E, Cohen L and Marsh K B 2000a Aconitase activity and expression during the development of lemon fruit. Physiol. Plant. 108, 255–262.

Sadka A, Dahan E, Or E and Cohen L 2000b NADP(+)-isocitrate dehydrogenase gene expression and isozyme activity during citrus fruit development. Plant Sci. 158, 173–181.

Srere P A 1967 Citrate synthase. Methods Enzymol. 13, 3–11

Takita E, Koyama H and Hara T 1999a Organic acid metabolism in aluminum-phosphate utilizing cells of carrot (*Daucus carota* L.). Plant Cell Physiol. 40, 489–495.

Takita E, Koyama H, Shirano Y, Shibata D and Hara T 1999b Structure and expression of the mitochondrial citrate synthase gene in carrot cells utilizing Al-phosphate. Soil Sci. Plant Nutr. 45, 197–205.

Vance C P 2001 Symbiotic nitrogen fixation and phosphorus acquisition. Plant nutrition in a world of declining renewable resources. Plant Physiol. 127, 390–397.

Wasaki J, Yamamura T, Shinano T and Osaki M 2002 Secreted acid phosphatase is expressed in cluster roots of lupin in response to phosphorus deficiency. Plant Soil 248, 129–136.

Wesley S V, Helliwell C A, Smith N A, Wang M, Rouse D T, Liu Q, Gooding P S, Singh S P, Abbott D, Stoutjesdijk P A, Robinson S P, Gleave A P, Green A G and Waterhouse P M 2001 Construct design for efficient, effective and high-throuput gene silencing in plants. Plant J. 27, 581–590.

The influence of organic acid exudation and cluster roots on P and N acquisition

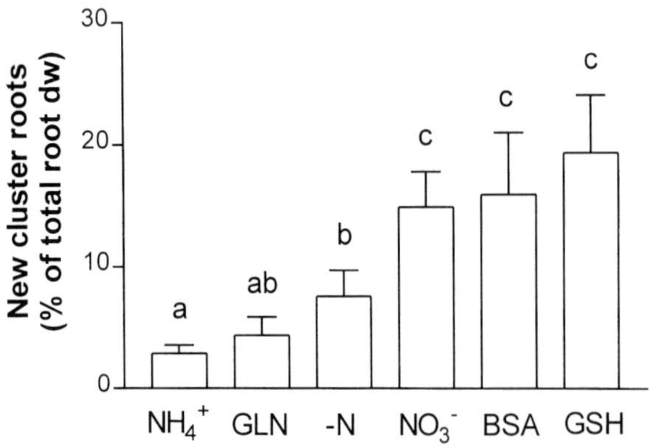

Figure 1. Cluster root formation of 1-year-old *Hakea* seedlings grown in non-sterile sand culture receiving different N sources at 0.1 m*M* N every 3 days for 10 weeks. Values are averages of five replicates, standard deviations are shown. Different letters denote significant differences between treatments at the $P < 0.05$ level (ANOVA using ln-transformed data, Tukey's post-hoc test; data shown here represent untransformed data). NH_4^+, ammonium; GLN, glutamine; –N, no N added; NO_3^-, nitrate; BSA, protein; GSH, glutathione.

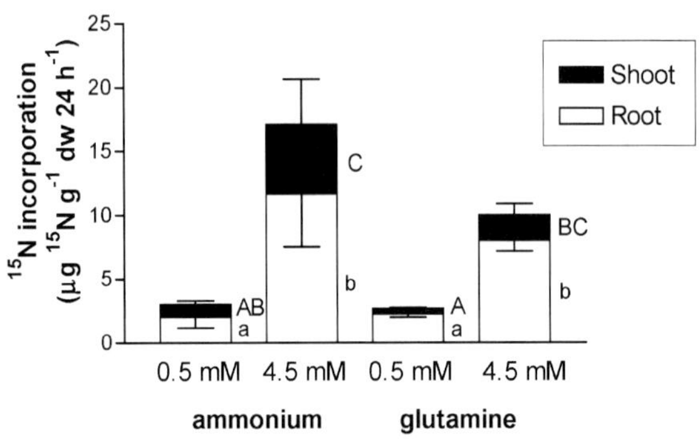

Figure 2. ^{15}N incorporation into roots and shoots of Hakea seedlings grown in non-sterile sand culture during 24 h after addition of ^{15}N-labelled ammonium or glutamine. Averages of three replicates and standard deviations are shown. Different letters denote significant differences between roots and shoots in 0.5 and 4.5 m*M* ammonium and glutamine treatments ($P < 0.05$, ANOVA on ln-transformed data, Tukey's post-hoc test; data shown here represent untransformed data).

was not assessed in this experiment whether pH effects contributed to the observed differences in cluster root formation of plants supplied with different N sources.

Seedlings receiving 4.5 m*M* ^{15}N labelled ammonium or glutamine incorporated 5.7 and 3.8 times more ^{15}N label per g dw than seedlings receiving 0.5 m*M* ^{15}N labelled ammonium or glutamine, respectively (Figure 2). Significantly ($P < 0.05$) more ^{15}N was incorporated into roots supplied with 4.5 m*M* NH_4^+ or glutamine compared to plants supplied with 0.5 m*M* NH_4^+ or glutamine, whereas shoots in the 4.5

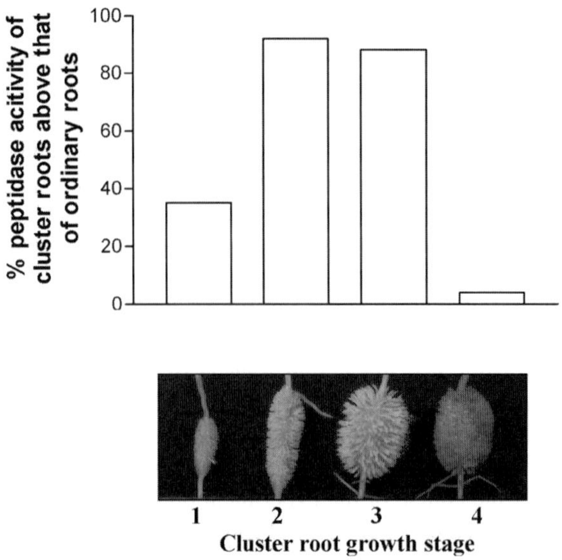

Figure 3. Peptidase activity of cluster roots of different growth stages compared to normal roots. Entire cluster roots and roots were ground up in Na-acetate buffer. Peptidase activity was assessed using azocasein. Peptidase activity of cluster roots is expressed as the percentage of activity above that of young ordinary roots. Cluster root growth stages: stage 1, recently formed (0–2 days old), stage 2, young and developing (~4 days old), stage 3, mature (~7 days old) and stage 4, senescing (>10 days old).

m*M* glutamine treatment incorporated similar amounts of ^{15}N label than shoots in the 0.5 m*M* NH_4^+ treatment.

Cluster roots accounted for approximately 75% of the total root dry weight in this experiment, which was the highest proportion of root dry weight compared to all other experiments where cluster roots accounted mostly for between 20 and 30% of total root dry weight. This indicates that different experimental conditions result in considerable differences in cluster root formation. A possible explanation could be that plants received very low levels of all nutrients while other experimental conditions here subjected plants to N limitations but supplied adequate amounts of other nutrients.

Axenically grown seedlings were able to acquire N from all offered N sources (Table 1). However, the minus-N seedlings acquired 2.8 mg of N from an unknown source. Our previous experiments with myrtaceous species did not indicate that germination on nutrient agar or growth on vermiculite-sand introduces N contamination. Gaseous forms of N appear to be the most likely source of N contamination since the growth chamber received ambient air and was situated in an urban area.

Table 1. Dry weight of axenic *Hakea* seedlings grown with different N sources and two concentrations after 6 months. Averages (and standard deviations) of 2–3 seedlings from separate containers are shown. Prior to the experiment, one cotyledon was removed from each seed to reduce stored N from an average of 0.42 mg N per seed to 0.21 mg N per seed. Nitrogen uptake indicates that seedlings obtained N from sources other than seed N[a]

N source	Concentration (m*M*)	Dry weight (mg)				N uptake (mg N)
		Shoot	Roots	Cluster roots	Cluster root (% of total root dw)	
–N	0	216	37	7	18.9	2.8
NH_4^+	0.5	291	32	0	–	7.22 (3.97)
	4.5	342	37	0	–	7.28 (1.68)
NO_3^-	0.5	271	36	6	16.7	8.36 (0.73)
	4.5	256	23	10	43.5	8.26
GLN	0.5	280	31	6	19.4	6.03
	4.5	314	39	0	–	11.18
GSH	0.5	276	33	12	36.4	5.50 (1.09)
	4.5	241	29	10	34.5	9.10
BSA	0.5	279	48	6	12.5	3.46 (1.12)
	4.5	289	42	5	11.9	9.55

[a] Abbreviations: –N, no N added; NH_4^+, ammonium; NO_3^-, nitrate; GLN, glutamine; GSH, glutathione; BSA, protein

Seedlings in the 0.5 m*M* N treatments (receiving a total of 7 mg of N) acquired between 3.46 (BSA) and 8.36 (NO_3^-) mg N. Seedlings in the 4.5 m*M* N treatments (total of 63 mg N) took up between 5.50 (NH_4^+) and 11.18 (GLN) mg N. Due to the loss of replicates through contamination, it was not possible to assess the data statistically.

Hydroponically-grown non-axenic seedlings were used to determine peptidase activity associated with cluster roots and roots. Peptidase activity associated with ground-up cluster roots varied in cluster roots of different ages (Figure 3) using degradation of the dry-protein azocasein as an indicator for peptidase activity. Developing cluster roots had over 30% higher peptidase activity than ordinary roots, which consisted of 5 cm of young white root material including the root tip. Peptidase activity increased in young and mature cluster roots to 90% higher activity compared to ordinary roots. In contrast, senescing cluster roots had similar peptidase activities as ordinary roots. However, grinding of root tissues determines total peptidase activity of the tissue, but does not identify whether peptidases are of intracellular origin, are exuded by cluster roots, or are produced by microbes associated with the surface of cluster roots.

Therefore, a technique was devised in which cluster roots were gently centrifuged to obtain compounds associated with the surface of the cluster root. The resulting solution was separated on a PAGE gel containing gelatine to separate proteins and determine proteolytic activity in the same gel (Figure 4). The size of a number of putative peptidases ranged from approximately 80–55 kD. Activity of peptidases A, C and D decreased during cluster root development, while peptidase B activity increased throughout cluster root development. When microbial sludge from the hydroponic growth containers was analysed, no bands were apparent where peptidase B, C, and D occurred (data not shown), suggesting that these peptidases are associated with cluster roots.

To further investigate whether organic forms of N are taken up by roots and cluster roots, partial cDNA clones of putative amino acid and peptidase transporters were obtained from *Hakea* total RNA. One partial sequence encoding a putative amino acid transporter (HaAAT4-1) and one sequence encoding a putative peptidase transporter (HaPepT1) were cloned. The genes were confirmed to originate from *Hakea actities* by Southern analysis. Expression of *HaAAT4-1* and *HaPepT1* in *Hakea* tissue was determined by north-

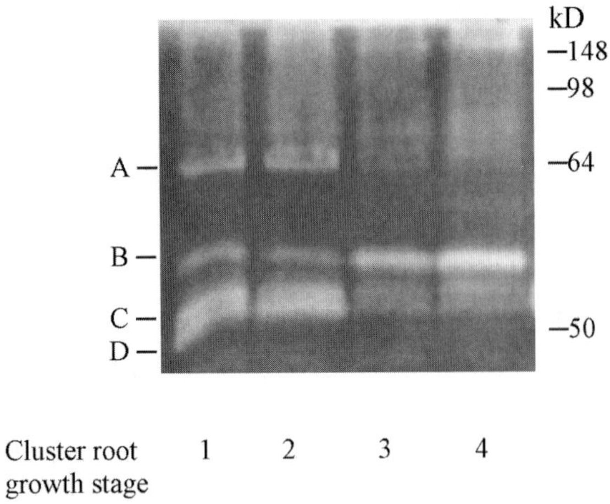

Figure 4. In-gel peptidase activity of cluster root 'exudates' at different growth stages (see Figure 2). Letters denote conspicuous peptidase bands that changed during cluster root development.

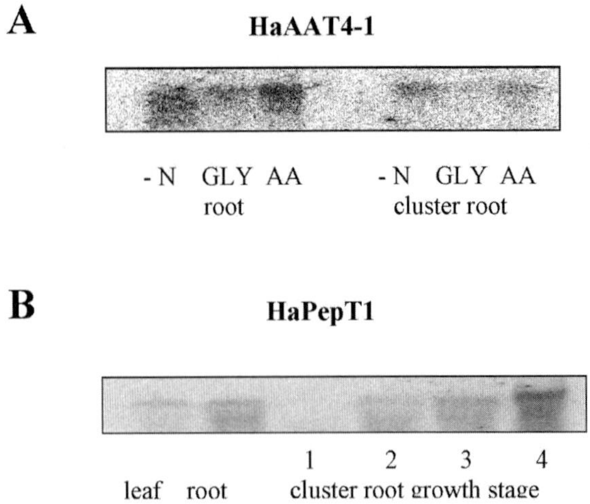

Figure 5. (A) Northern blot analysis of *Hakea actities* amino acid transporter *HaAAT4-1*. *Hakea* plants were starved of nitrogen for 2 weeks (–N) before receiving 200 μM glycine (GLY) or 200 μM of amino acid mixture (arginine, glutamate and leucine, AA). RNA was extracted from roots and cluster roots 4 h after N addition. Different quantities of total RNA were loaded for roots and cluster roots, so that RNR expression of roots and cluster roots cannot be compared. However, RNA expression is comparable between treatments of the same tissue. (B) Northern blot analysis of *Hakea actities* peptide transporter *HaPepT1* expressed in leaf, roots, and cluster roots throughout cluster root development. Comparable total RNA levels were loaded in each lane (data not shown).

ern blot analysis. Presence of *HaAAT4-1* mRNA in both roots and cluster roots confirms expression of the amino acid transporter gene in these tissues when supplied with different forms of N (Figure 5A). Expression of *HaPepT1* was detected in leaves, roots and cluster roots (Figure 5B). Data suggests that expression of *HaPepT1* in cluster roots increases throughout cluster root development but more experimentation is required to confirm this.

Discussion

Nitrogen source affects cluster root formation

Hakea seedlings produce greater quantities of cluster roots when grown with NO_3^-, and glutathione than when grown with NH_4^+ or GLN. The different cluster root formation by plants grown with different N sources that was observed here may reflect different availability of N sources, be caused by a pH effect in the rhizosphere, or be a result of different microbial associations (Neumann and Martinoia, 2002). There is evidence that cluster root induction is regulated by plant hormones. Auxin application induced cluster roots in P-sufficient lupin and *Grevillea* (Gilbert et al., 1997, 2000; Skene and James, 2000), while kinetin application had inhibitory effect on cluster root formation (Neumann et al., 2000). Cytokinin levels were lower in NH_4^+-grown than in NO_3^--grown tobacco plants (Walch-Liu et al., 2000), and interactions between auxins and cytokinins are likely to be involved in cluster root formation in lupin (Neumann et al., 2000). Cluster root formation in *Casuarina* was dependent on pH and N source; an alkaline pH resulted in more cluster roots in plants grown with NO_3^- than a neutral pH while NH_4^+-grown plants only produced cluster roots under alkaline conditions (Diem et al., 2000). However, Diem et al. (2000) suggested that Fe and P deficiencies induced under alkaline conditions are a primary factor for cluster root formation in *Casuarina*, a suggestion that was not investigated here.

The N sources offered in our study may not have been equally available to the plants. Cluster root formation in *Hakea* species increased at low N supply, but was suppressed by higher N supply (Lamont, 1973), while cluster root formation was enhanced in *Myrica gale* and *Grevillea robusta* at higher N levels compared to no added N (Crocker and Schwinzer, 1993; Moore and Keraitis, 1966). *Hakea actities* took up and assimilated NO_3^-, although at a low rate (Schmidt and Stewart, 1999), which was consistent with observations that root and shoot tissues of *H. actities* have low nitrate reductase activities (Schmidt and Stewart, 1997). Ammonium, glycine and glutamine are readily accessible to *H. actities*, as these N sources were taken up and, as shown for NH_4^+ and glycine, rapidly assimilated by *H. actities* (Schmidt and Stewart, 1999). Further experiments need to distinguish the possible effects of N deficiency and N form on cluster root formation in *Hakea*.

Can axenic Hakea actities seedlings access complex organic N sources?

In axenic culture *H. actities* seedlings took up somewhat more N when grown with 4.5 m*M* than with 0.5 m*M* glutathione or protein as N sources, although lack of replicates does not allow statistical proof. Plants also had access to a N source other than the N supplied in the nutrient solution. It is likely that seedlings used reactive air-derived N sources as growth cabinets exchanged ambient air from a metropolitan area, so that the data of the axenic experiment are not unequivocal. Axenic seedlings took up similar amounts of N when supplied with 0.5 or 4.5 m*M* NH_4^+ or NO_3^- suggesting that plant N acquisition was affected by factors other than N sources, such as restricted growth in the growth containers or pH effects. The axenic culture conditions developed here will be useful for further studies of *Hakea* N utilisation, as sterile conditions were successfully maintained over the growth period of 6 months and allowed seedlings to grow into sizeable and healthy plants.

Peptidase activity associated with cluster roots changes with cluster root age

Peptidases catalyse the hydrolysis of polypeptide bonds and control protein synthesis and turnover in tissues. Plant peptidases have been studied in relation to a range of cellular processes, such as cytoplasm breakdown (Ye and Varner, 1996), breakdown of storage compounds (Bethke et al., 1996), and in roots (Wallace and Shannon, 1981). To the best of our knowledge, it has not been studied whether plant roots exude peptidases. Peptidase activity associated with developing and mature cluster roots was higher than that of young and senescing cluster roots and roots. Different peptidases appeared to be active at different times of cluster root development, which could indicate that peptidases have a role in the N acquisition of cluster roots. However, it is important to distinguish between catabolic peptidase activity in the cluster root and exuded peptidases with a function in protein breakdown and N acquisition in the rhizosphere. Whether the peptidases are plant or microbially derived will need to be investigated. Our current experiments are characterising the possible occurrence and function of cluster root-derived peptidases using a variety of techniques.

A range of ecto- and ericoid-mycorrhizal fungi produce peptidases allowing fungi and associated plant partners to acquire high-molecular weight forms of N (Smith and Read, 1997). Acid peptidases of ericoid mycorrhizal fungi were most active at media pH of 4–5 (Leake and Read, 1990), which is similar to the pH of wallum soil (Schmidt and Stewart, 1997). While many woody plants in low nutrient environments have ecto- and ericoid mycorrhizal associations, the lack of such fungal associations in the Proteaceae family rules out that fungal associations are involved in the acquisition of complex N acquisition. However, it is possible that cluster roots represent an alternative to ecto/ericoid mycorrhizal associations in environments with a high availability of organic forms of N forms.

Amino acid and peptide transporters are expressed in Hakea *cluster roots and roots*

Significant progress has been made in recent years in the study of cell membrane transporters, which play a fundamental role in the transfer of nutrients from soil into root cells. Nitrogen is quantitatively the most important nutrient that plants acquire from the soil, and ammonium, nitrate and amino acid transporters that are expressed in roots have received considerable attention. It has been established that high- and low-affinity N transporters are induced as a response to corresponding nutrient conditions, that transporters are expressed tissue specifically, and that transporters can be diurnally regulated (Fischer et al., 1995; Frommer et al., 1994; Kwart et al., 1993; Ninnemann et al., 1994). In order to address the possible function of cluster roots for organic N acquisition, we partially cloned an amino acid and a peptide transporter. The cloned amino acid transporter *HaAAT4* is 69% identical and 79% homologous to the *Arabidopsis* root specific amino acid transporter *AAP3*. A search of the Genbank database showed that over 40 amino acid transporters have been cloned from various plant species. The recent completion of first draft sequence of the *Arabidopsis thaliana* genome has revealed 20 genes that show high homology to amino acid transporters, and eight of them are known transporters. We expect to clone further amino acid transporter genes from *Hakea actities*, and will assess how amino acid transporter expression in *Hakea* tissues is related to N source and supply.

Several peptide transporters have been identified in plants. The *Arabidopsis thaliana* peptide transporter *AtPTR2-B* is constitutively expressed in all plant organs suggesting an important physiological role in plant growth and development (Song et al., 1997). In contrast, the barley peptide transporter *HvPTR1* is localised only on the plasma membrane of scutellar epithelial cells of the germinating barley grain and is believed to play a central role in the transport of peptides during germination (Waterworth et al., 2000; West et al., 1998). Whether peptide transporters have a function for peptide acquisition in cluster root or root membranes remains to be determined.

Here, we presented a number of techniques that can be applied to study plant acquisition of organic N sources. We cannot conclusively demonstrate that *Hakea actities* is able to access complex organic N sources without assistance from associated microbes. However, with increasing awareness that organic forms of N are present in most, if not all, environments, researchers are looking beyond NH_4^+ and NO_3^- as possible plant N sources. Cluster roots may be one root specialisation allowing plants to access organic N sources.

References

Ahmad I and Hellebust J A 1991 Enzymology of N assimilation in mycorrhiza. Methods Microbiol. 23, 181–201.

Bethke P C, Hillmer S and Jones R L 1996 Isolation of intact protein storage vacuoles from barley aleurone. Plant Physiol. 110, 521–529.

Crocker L J and Schwintzer C R 1993 Factors affecting formation of cluster roots in *Myrica gale* seedlings in water culture. Plant Soil 152, 287–298.

Diem H G, Duhoux E, Zaids H and Arahou M 2000 Cluster roots in *Casuarina*: role and relationship to soil nutrient factors. Ann. Bot. 85, 929–936.

Dinkelaker B, Hengeler C and Marschner H 1995 Distribution and function of proteoid roots and other root clusters. Botanica Acta 108, 183–200.

Chapin F S, Moilanen L and Kielland K 1993 Preferential use of organic nitrogen for growth by a non-mycorrhizal arctic sedge. Nature 361, 150–153.

Fischer W-N, Kwart M, Hummel S and Frommer W B 1995 Substrate specificity and expression profile of amino acid transporters (AAPs) in *Arabidopsis*. J. Biol. Chem. 270, 16315–16320.

Frommer W B, Kwart M, Hirner B, Fischer W N, Hummel S and Ninnemann O 1994 Transporters for nitrogenous compounds in plants. Plant Molecular Biol. 26, 1651–1670.

Gilbert G A, Knight J D, Vance C and Allan D L 2000 Proteoid root development of phosphorus deficient lupin is mimicked by auxin and phosphonate. Ann. Bot. 85, 921–928.

Gilbert G A, Knight J D and Vance C 1997 Does auxin play a role in the adaptations of while lupin root to phosphate deficiency? Plant Physiol. 114 (Suppl), 67.

Grierson P F and Attiwill P M 1989 Chemical characterisation of the proteoid root mat of *Banksia integrifolia* L.f. Aust. J. Bot. 37, 137–143.

Kielland K 1994 Amino acids absorption by arctic plants: implications for plant nutrition and nitrogen cycling. Ecology 75, 2373–2383.

Kwart M, Hirner B, Hummel S and Frommer W B 1993 Differential expression of two related amino acid transporters with differing substrate specificity in *Arabidopsis thaliana*. Plant J. 4, 993–1002.

Lamont B 1993 Why are hairy root clusters so abundant in the most nutrient-impoverished soils of Australia? *In* Plant Nutrition – From Genetic Engineering to Field Practice. Ed N J Barrow. pp. 309–312. Kluwer Academic Publishers, Dordrecht, The Netherlands.

Lamont B 1982 Mechanisms for enhancing nutrient uptake in plants with particular reference to mediterranean South Africa and Western Australia. Bot. Rev. 48, 597–689.

Lamont B 1973 Factors affecting the distribution of proteoid roots within the root system of two *Hakea* species. Aust. J. Bot. 21, 165–187.

Leake J R and Read D J 1990 Proteinase activity in mycorrhizal fungi. New Phytol. 115, 243–250.

Mason M and Schmidt S Rapid isolation of high quality RNA and genomic DNA from *Hakea actities*. Func. Plant Biol., in press.

Moore C W F and Keraitis K 1966 Nutrition of *Grevillea robusta*. Aust. J. Bot. 14, 151–163.

Näsholm T, Ekblad A, Nordin A, Giesler R, Högberg M and Högberg P 1998 Boreal forest plants take up organic nitrogen. Nature 392, 914–916.

Neumann G and Martinoia E 2002 Cluster roots – an underground adaptation for survival in extreme environments. Trends Plant Sci. 7, 162–167.

Neumann G, Massonneau A, Langlade N, Dinkelaker B, Hengeler C, Römheld V and Martinoia E 2000 Physiological aspects of cluster root function and development in phosphorus-deficient white lupin (*Lupinus albus* L.) Ann. Bot. 85, 909–919.

Ninnemann O W, Jauniauex J C and Frommer W B 1994 Identification of a high affinity ammonium transporter from plants. EMBO J. 13, 3464–3471.

Pate J S and Jeschke W D 1993 Mineral uptake and transport in xylem and phloem in the proteaceous tree *Banksia prionotes*. *In* Plant Nutrition – From Genetic Engineering to Field Practice. Ed. N J Barrow. pp. 313–316. Kluwer Academic Publishers, Dordrecht, The Netherlands.

Roelofs R F R, Rengel Z, Cawthray G R, Dixon K W and Lambers H 2001 Exudation of carboxylates in Australian Proteaceae: chemical composition. Plant Cell Environ. 24, 891–903.

Schmidt S and Stewart G R 1999 Glycine metabolism by plant roots and its occurrence in Australian plant communities, Aust. J. Plant Physiol. 26, 253–264.

Schmidt S and Stewart G R 1997 Waterlogging and fire impacts on nitrogen availability and utilization in a subtropical wet heathland (wallum). Plant Cell Environ. 20, 1231–1241.

Skene K R and James W M 2000 A comparison of the effects of auxin on cluster root initiation and development in *Grevillea robusta* Cunn. Ex. R.Br. (Proteaceae) and in the genus *Lupinus* (Leguminoseae). Plant Soil 219, 221–229.

Smith S E and Read D J 1997 Mycorrhizal Symbiosis, 2nd edn. Academic Press, London.

Song W, Koh S, Czako M, Marton L, Drenkard E, Becker J M and Stacey G 1997 Antisense expression of the peptide transport gene AtPTR2-B delays flowering and arrests seed development in transgenic *Arabidopsis* plants. Plant Physiol. 114, 927–935.

Turnbull M H, Goodall R and Stewart G R 1995 The impact of mycorrhizal colonization upon nitrogen source utilization and metabolism in seedlings of *Eucalyptus grandis* Hill ex Maiden and *Eucalyptus maculata* Hook. Plant Cell Environ. 18, 1386–1394.

Walch-Lui P, Neumann G, Bangerth F and Engels C 2000 Rapid effects of nitrogen form on leaf morphogenesis in tobacco. J. Exp. Bot. 343, 227–237.

Wallace W and Shannon J D 1981 Proteolytic activity and nitrate reductase inactivation in maize roots. Aust. J. Plant Physiol. 8, 211–219.

Waterworth W M, West, C E and Bray C M 2000 The barley scutellar peptide transporter: biochemical characterization and localization to the plasma membrane. J. Exp. Bot. 51, 1201–1209.

Watt M and Evans J R 1999 Linking development and determinacy with organic acid efflux from proteoid roots of white lupin grown with low phosphorus and ambient or elevated atmospheric CO_2 concentrations. Plant Physiol. 120, 705–716.

West C E, Waterworth W M, Stephens S M, Smith C P and Bray C M 1998 Cloning and functional characterisation of a peptide transporter expressed in the scutellum of barley grain during the early stages of germination. Plant J. 15, 221–229.

Ye Z-H and Varner J E 1996 Induction of cysteine and serine proteinases during xylogenesis in *Zinnia elegans*. Plant Mol. Biol. 30, 1233–1246.

Plant and Soil **248:** 167–185, 2003.
© 2003 Kluwer Academic Publishers. Printed in the Netherlands.

Lupinus luteus cv. Wodjil takes up more phosphorus and cadmium than *Lupinus angustifolius* cv. Kalya

Ross F. Brennan[1,3] & Mike D.A. Bolland[2]
[1]*Department of Agriculture, 444 Albany Highway, Albany, Western Australia 6330, Australia.* [2]*Department of Agriculture, PO Box 1231, Bunbury, Western Australia 6231, Australia and Plant Biology, The Univesity of Western Australia, 35 Stirling Highway, Crawley, Western Australia 6009, Australia.* [3]*Corresponding author**

Received 3 November 2001. Accepted in revised form 14 May 2002.

Abstract

A field experiment on an acidic lateritic ironstone gravel sand in south-western Australia compared how *Lupinus luteus* L. cv. Wodjil and *L. angustifolius* L. cv. Kalya used different sources of fertilizer phosphorus (P) to produce shoots and seed (grain). The sources of P were triple superphosphate, highly reactive North Carolina apatite phosphate rock and low reactive Queensland (Duchess) apatite phosphate rock, all applied 14 years previously, and triple superphosphate applied in the current year. The fertilizers contained different concentrations of cadmium (Cd) as an impurity. Concentrations of Cd were measured in lupin shoots and grain to compare how the two species took up cadmium applied as the different fertilizers. *L. luteus* used all sources of P more effectively than *L. angustifolius* to produce dried shoots and grain. Per unit of applied P as each source, the concentration of P in grain of *L. luteus* was consistently about double that in *L. angustifolius*. However, P concentrations in shoots harvested 2 months earlier were about similar, suggesting *L. luteus* transferred more P to grain. For each amount of each source of fertilizer P applied, the concentration of Cd in grain was always larger for *L. luteus*. Soil test Cd provided a good indication for when grain Cd concentration was likely to be above the maximum permissible concentration. *L. luteus* developed abundant third-order lateral roots (cluster roots?), which may have enabled *L. luteus* to take up more P and Cd from the soil than *L. angustifolius*.

Introduction

About 75% (13.5 × 10^6 ha) of the soils used for agriculture in south-western Australia are acidic to neutral, sandy-surfaced to deep sandy soils (McArthur, 1991). *Lupinus angustifolius* L. is well adapted to these soils and so was successfully domesticated, largely by JS Gladstones, as the grain legume to grow in rotation with cereal crops on these soils (Buirchell and Cowling, 1998; Nelson and Delane, 1990). Recently, *L. luteus* L. has also been studied for these soils because of better tolerance to diseases increasingly affecting *L. angustifolius*, and because *L. luteus* is better adapted to some of the more naturally acidic soils in the region (Bolland et al., 2000; French et al., 2001; Sweetingham et al., 1996).

Phosphorus (P) is the major nutrient element deficiency of *L. angustifolius* in south-western Australia (Longnecker et al., 1998). As a result of previous research, much is known about the response of *L. angustifolius* to soil and applied P (Longnecker et al., 1998). By contrast, the P requirements of *L. luteus* in the soils of the region have yet to be clearly established. It therefore makes sense to compare the P requirements of *L. luteus* with that of *L. angustifolius* in the same experiment to assess if *L. luteus* requires more or less soil and fertilizer P to produce the same shoot and grain yields. Limited such studies in south-western Australia show that *L. luteus* requires less soil and fertilizer P to produce the same yield as *L. angustifolius* (Bolland, 1997a; Bolland et al., 2000; Rahman and Gladstones, 1974; Tang et al., 1998). The experiment reported here extends these studies to assess if this was also so for fertilizer P applied 14 years previously as triple superphosphate and as two

* E-mail: rbrennan@agric.wa.gov.au

Table 1. Concentration of P and Cd in fertilizers applied in 1984 (previous) and 1998 (current)

Fertilizer[a]	Cd (mg/kg)	P (g/kg)	Cd:P (mg Cd per kg P)
TSP current	5.1	19.6	26.0
TSP previous	31.5	19.6	160.7
NCPR previous	52.0	11.9	437.0
QPR previous	3.5	9.0	38.9

[a]TSP – triple superphosphate; NCPR – North Carolina phosphate rock; QPR – Queensland (Duchess) phosphate rock.

different types of apatite phosphate rocks, or as triple superphosphate applied in the current year.

A field experiment in south-western Australia comparing the fertilizer zinc requirements of *L. luteus* and *L. angustifolius* (Brennan et al., 2001) showed that *L. luteus* cvv. Motiv and Teo had about 6 times the concentration of cadmium (Cd) in grain than *L. angustifolius* cv Gungurru. Evidently, the two *L. luteus* cultivars were better at accessing Cd from the soil and/or depositing the Cd in the grain. The concentration of Cd in the various P fertilizers used in the experiment described here was different and large amounts of fertilizer, and therefore Cd, were applied in the experiment. To confirm if *L. luteus* consistently took up more Cd than *L. angustifolius*, the concentration of Cd in shoots and grain was measured in the two lupin species studied here for all sources of P.

Materials and methods

P fertilizers

Some properties of the P fertilizers used are listed in Table 1. Further properties of the fertilizers applied 14 years previously are provided by Bolland et al. (1986).

Location, soil and climate

The experiment was on the Esperance Downs Research Station, near Gibson, Western Australia. The experiment was on a lateritic texture contrast soil, comprising a yellow sand with ironstone gravel at about 45 cm depth over a yellow mottled clay. The lateritic ironstone gravel consists of loose, discrete particles, between about 2 and 30 mm diameter, in a matrix of sand, silt and clay. The soil is known locally as a Fleming gravelly sand, and is classified as a yellow duplex soil (Stace et al., 1968), Dy5.82 (Northcote, 1987), Fluventic Xerochrept (Soil Survey Staff, 1975), a Ferralic Cambisol (FAO-UNESCO, 1988), and a Ferric Orthic Tenosol (Isbell, 1996). As measured on samples of the <2 mm fraction of the top 10 cm of soil collected before the experiment began in May 1984, some properties of the soil were as follows: soil pH (1:5 soil: 0.01 M $CaCl_2$, w/v, measured as described by Rayment and Higginson (1992)), 4.9; bicarbonate-extractable soil P (Colwell, 1963) 2 mg P mg^{-1} soil; the capacity of the soil to retain P, measured as described by Ozanne and Shaw (1967), and which is the amount of P retained by the soil as the concentration of P in the final solution is raised from 0.25 to 0.35 μg P ml^{-1}, 3.0 mg kg^{-1}; cation exchange capacity (extracted in 1 *M* NH_4Cl at pH 7.0, as described by Rayment and Higginson (1992)), 4.0 cmol (+) kg^{-1}; organic carbon, measured by the wet oxidation method described by Walkley and Black (1934), 1.0%; and total nitrogen (Kjeldahl method, as described by McKenzie and Wallace (1954)), 0.032%. Mechanical analysis of the <2 mm fraction of soil, measured by the pipette method of Day (1965), indicated percentages were as follows: coarse sand (2000–2000 μm), 33.0%; fine sand (20–200 μm), 59.5%; silt (2–20 μm), 2.0%; and clay (<2 μm); 5.5%. Percentage gravel (>2 mm) in the whole top 10 cm of soil was 69%.

The climate at Gibson is Mediterranean-type, with a cool wet typical May–October growing season, and warm mostly dry November–April. Average annual rainfall is 484 mm, with 380 mm (79%) falling during the growing season. Monthly rainfall during 1998, the year the experiment described here was done, was, from January to December: 0, 1.6, 8.2, 43.8, 39.4, 67.0, 89.8, 54.6, 57.4, 47.6, 24.6, and 15.6, giving a total of 449.6 mm, about 34 mm below average, and a growing season rainfall of 355.8 mm (79%), about 24 mm below average.

Original experiment

The original experiment was a long-term study to compare the initial and residual effectiveness of highly reactive and low reactive apatite phosphate rocks relative to superphosphate and the results have been published (Bolland et al., 1986, 1988, 1989; Bolland and Gilkes 1995). The original experiment comprised a completely randomised block of six levels of P for three different P fertilizers, together with many extra nil-P treatment plots, replicated three times. The original P treatments were applied once only on 13 June 1984 when the experiment started. The three P fertilizers, with amounts of P applied in 1984, as (g P m^{-2} [$\times 10^4$]) (kg P ha^{-1}) in brackets, were: triple superphosphate (0, 60, 120, 240, 480, 960), North Carolina phosphate rock (0, 315, 608, 1215, 2430, 4860), and Duchesss phosphate rock (0, 230, 460, 1150, 2300, 4600). The P treatments were applied by hand to the soil surface in plots 2 m wide and 30 m long. There was a 0.5-m buffer between each plot. After application, all the fertilizer treatments (including the nil-P treatments) used for the study reported here were thoroughly mixed into the top 10 cm of soil using a rotary hoe. All the plots (including the nil-P treatment plots) were sown in different years to the following crop species: barley (*Hordeum vulgare* L.) in 1984 and 1985; oat (*Avena sativa* L.) in 1986 and 1988; different crop species were grown on different sections of each plot in 1987 (see Bolland et al. (1989) for further details); subterranean clover (*Trifolium subterraneum* L.) in 1989, and naturally regenerated subterranean clover grew on the plots from 1990 to 1997. For the study reported here, no extra P fertilizer was applied to the plots since 1984. Adequate basal fertilizers were applied to all the plots (including the nil-P treatment plots) each year to ensure P was the only nutrient element to limit plant yield. Further details are provided by Bolland et al. (1986, 1988, 1989), and Bolland and Gilkes (1995).

Present study

In 1998, the opportunity was taken to use the experiment to compare how *L. luteus* L. and *L. angustifolius* L. used different sources of fertilizer P to produce shoots and grain. The P sources were four of the original six levels of triple superphosphate applied 14 years previously (0, 240, 480 and 960 g P m^{-2} ($\times 10^4$) (g P ha^{-1}), and five of the original six amounts of the two phosphate rock fertilizers applied 14 years previously. The levels of NCPR applied in 1984 used for the 1998 experiment were (g P m^{-2} [$\times 10^4$]) (kg P ha^{-1}) 0, 608, 1215, 2430 and 4860; corresponding levels for QPR were 0, 460, 1150, 2300 and 4600. In addition, seven levels of triple superphosphate (0, 5, 10, 15, 20, 40, 80 kg P m^{-2} [$\times 10^4$] [g P ha^{-1}]) were placed (drilled) with the seed of the two lupin species sown in 1998 (current TSP). The seven levels of current TSP were applied to plots to which fertilizer P had not been applied in a previous year (some of the many nil-P plots in the experiment up to that time). The P applied in 1998 was applied to plots that had been nil-P treatment plots since the start of the experiment in 1984, and so had never been treated with fertilizer P in previous years.

The different sources of P used had different amounts of Cd as an impurity (Table 1). So the 1998 study also provided the opportunity to assess, for the amounts of each source of P used, the concentration of Cd in shoots and grain for *L. luteus* and *L. angustifolius*.

Experimental details for 1998

Half of each plot (15 m length of the original 2 m wide by 30 m long plots) was sown to *L. luteus* cv. Wodjil and the other half was sown to *L. angustifolius* cv. Kalya. Eight rows of seed, 18 cm apart, were sown on 15 May down the middle 1.8 m of all plots. Both species were sown using 110 kg seed ha^{-1} at about 4 cm deep, the recommended amounts of seed and sowing depth for the species in Western Australia (Nelson and Delane, 1990). The seed was incolucated with *Bradyrhizobium lupini* strain WU 425 (as commercial 'group G' inoculum) the day before sowing. Average weight of 1 seed (mg) was 136 for *L. Luteus*, and 145 for *L. angustifolius*, which are about typical values for these lupins in Western Australia (The Crop Variety Sowing Guide, 1998).

Basal fertilizers were applied to all plots while sowing in 1998 to ensure that P was the only nutrient limiting yield. The following basal fertilizers were placed 8 cm below the seed while sowing 4 cm deep (g m^{-2} [$\times 10^4$]) (kg ha^{-1}): 25 manganese sulphate (25% Mn), 5 copper sulphate (25% Cu), 3 sodium borate (11% B), 2 zinc oxide (70% Zn) and 0.2 molybdenum trioxide (67% Mo). The following fertilizers were broadcast onto the soil surface (top-dressed) of all plots about 4 weeks after sowing (g m^{-2} [$\times 10^4$]) (kg ha^{-1}): 100 potassium chloride (muriate of potash) (50% K), 100 gypsum (18% S, 22% Ca), the usual practice of applying these fertilizers in Western Australia (Edwards, 1998; Mason, 1998).

Weeds and pests were successfully controlled during 1998 by the use of recommended pre-seeding and post-emergence herbicides and insecticides (Gilby, 1993; Nelson and Delane, 1990).

Measurements during 1998

On 26 August, samples of the top 10 cm of soil were collected from each plot, using 2.5-cm diameter metal tubes that were pushed into the soil, to measure (1)

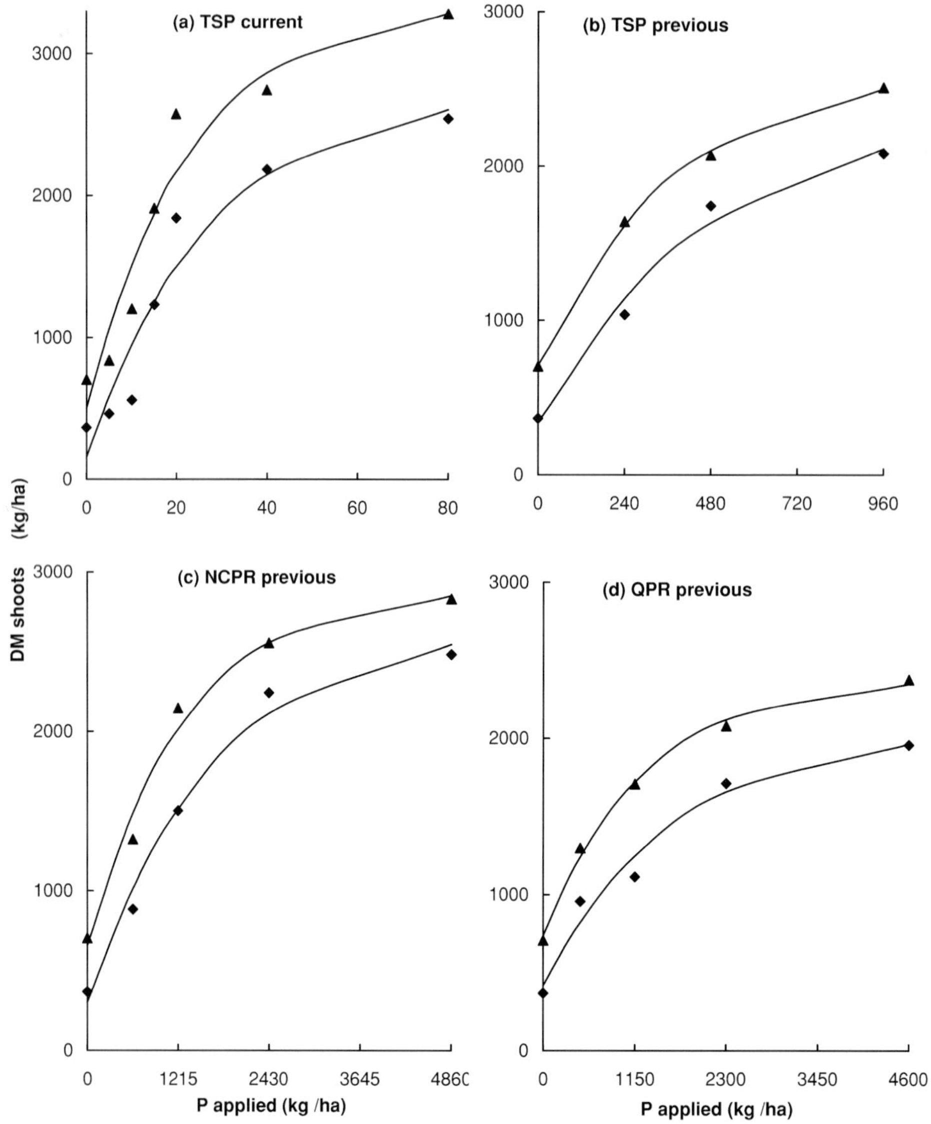

Figure 1. Relationship between yield of dried lupin shoots and the amount of fertilizer phosphorus applied as (a) triple superphosphate (TSP) applied in the current year, and for P applied 14 years previously as (b) TSP, (c) North Carolina phosphate rock (NCPR) and (d) Queensland phosphate rock (QPR). In each case *L. luteus* (▲), *L. angustifolius* (♦). SI units for kg ha^{-1} are (g m^{-2} [$\times 10^4$]).

sodium bicarbonate soil test P by the Colwell (1963) procedure, and (2) soil test Cd, using the method of Mann and Ritchie (1993), except 0.005 M $CaNO_3$ was used to extract soil Cd instead of 0.005 M KCl.

The number of lupin plants per m^2 were measured in mid June, about 28 days after sowing, by counting seedling in two adjacent rows of 0.5 m length, at 10 random positions within each half of each plot sown to each species.

Yields of lupin whole shoots were measured at 103 days after sowing by cutting plants at ground level within five random quadrats (0.5 m row by two rows) for each species within each half plot. The plant shoots were harvested at bud formation on the main stem. The plant shoots were dried at 70°C for 3 days in a force-draught oven and weighed.

Lupin grain yields were measured early December by machine-harvesting each species separately present on each plot.

Subsamples of dried shoots and grain were ground to measure concentrations of P and Cd in plant tissue. To measure P, the ground samples were digested in sulphuric acid and hydrogen peroxide (Yuen and Pollard, 1954). The P concentrations in the digest were determined colorimetrically by the molybdovanadate method (Anon., 1977). To measure concentrations of Cd, samples were digested in nitric and perchloric acids (McQuaker et al., 1979), and the concentrations of Cd in the digest were measured using inductively coupled plasma-atomic emission spectrometry.

Analysis of data

Data for the relationship between yield and the amount of P applied were fitted to a Mitscherlich equation:

$$y = a - b\exp - (cx),$$

where y is the yield (g m^{-2} [$\times 10^4$]) (kg ha^{-1}), x is the amount of P applied (g P m^{-2} [$\times 10^4$]) (kg P ha^{-1}), and a, b and c are coefficients. Coefficient a estimates the asymptote or maximum yield plateau (g m^{-2} [$\times 10^4$]) (kg ha^{-1}). Coefficient b estimates the difference between the asymptote and the intercept on the yield (y) axis at $x = 0$ (g m^{-2} [$\times 10^4$]) (kg ha^{-1}) and so estimating the maximum yield increases (responses) to applied P. Coefficient c describes the shape of the relationship, and estimates the rate at which y approaches the maximum yield plateau as the value of x increases ([m^{-2} {$\times 10^4$}][P^{-1}]) (ha kg P^{-1}). As the value of c increases, the yield response curve approaches the maximum yield plateau more rapidly (the curve is steeper) so less fertilizer P needs to be applied to produce the same yield, and vice versa. Mean data were fitted to the equation by non-linear regression using a computer program written in compiler BASIC (Barrow and Mendoza, 1990). The simplex method (Nelder and Mead, 1965) was used to locate the least square estimate of the non-linear coefficients.

Comparing how the different lupin species used soil and fertilizer P

The yield for the nil P treatment was used to compare how the different species used indigenous (native) P present in the soil, usually negligible for most south-western Australian soils (Wild, 1958). In addition, how the two lupin species used P from fertilizer P added 14 years previously or in the current year was also assessed. This was done as follows.

The relationship between yield (y axis) and the amount of P applied (x axis) was used to compare the yield response of *L. luteus* and *L. angustifolius* to each source of fertilizer P. The two lupin species mostly produced different yields for the nil-P treatment and different maximum yield plateaus. Under these circumstances it is not valid to use the c coefficient of the Mitscherlich equation to compare yield responses to previously and freshly applied P (Barrow, 1975, 1985; Barrow and Campbell, 1972). Instead, the initial slope of the relationship between yield and the amount of P applied was used to compare the yield responses of different species to the added P. For the Mitscherlich equation, as x tends to zero, dy/dx tends to bc, so that bc was used as an estimate of the initial slope (Barrow, 1975; Barrow and Campbell, 1972). For each source of P (triple superphosphate applied in 1998, and triple superphosphate and two different apatite phosphate rocks applied 14 years previously), for both shoots and grain, the response of *L. luteus* to applied P was compared to the response of *L. angustifolius* to applied P ($RR_{species}$) by dividing bc of each species by bc for *L. angustifolius*. Therefore, by definition, the $RR_{species}$ of *L. angustifolius* is always 1.00. If *L. luteus* has an $RR_{species}$ value >1.00 for a given source of P, then the species has a larger initial slope than *L. angustifolius*. This means that *L. luteus* used the added P more effectively than *L. angustifolius* and so required less P as that source of P to produce the same yield in the initial part of the response curve. However, if the $RR_{species}$ is <1.00, then *L. luteus* used P less effectively than *L. angustifolius* to produce yield and required more added P to produce the same yield in the initial part of the response curve.

Relationship between yield, and P concentration and soil test P

The relationship between yield (y axis) and the P concentration in either shoots or grain (x axis) was fitted to a Mitscherlich equation, as described for the relationship between yield and the amount of P applied above. Likewise, the Mitscherlich equation was also used to describe the relationship between grain yield (y axis) and soil test P (x axis).

Comparing differences in concentrations of P and Cd in shoots and grain of the two lupin species

For each source of P, the relationship between P concentration in shoots or grain and the amount of P applied was fitted to a linear equation:

$$y = A + Bx,$$

where y is the concentration (g P g^{-1} dry matter) in shoots or grain, x is the amount of P applied (g P m^{-2}

Table 2. Value of coefficients of the Mitscherlich equation fitted to the relationship between yield [g m^{-2} ($\times 10^4$)] (kg ha^{-1}) and the amount of P applied (g P m^{-2} [$\times 10^4$]) (kg P ha^{-1}), and initial slope (bc) and $RR_{species}$ values

Source of P[a]	Plant species[b]	Coefficients				bc	$RR_{species}$
		a	b	c (× 10 000)	R^2		
Dried shoots, 103 days after sowing							
Current TSP	*L. angustifolius*	2745	2592	366.82	0.929	95.08	1.00
	L. luteus	3376	2875	432.27	0.939	124.27	1.31
Previous TSP	*L. angustifolius*	2359	2031	22.83	0.971	4.64	1.00
	L. luteus	2660	1950	26.03	0.999	5.07	1.10
Previous NCPR	*L. angustifolius*	2685	2384	5.82	0.987	1.39	1.00
	L. luteus	2908	2248	7.57	0.985	1.70	1.22
Previous QPR	*L. angustifolius*	2064	1646	6.1	0.973	1.00	1.00
	L. luteus	2394	1658	7.85	0.997	1.30	1.30
Seed (grain)							
Current TSP	*L. angustifolius*	2392	2554	336.58	0.964	85.94	1.00
	L. luteus	1887	1878	734.87	0.991	137.83	1.63
Previous TSP	*L. angustifolius*	2358	2362	42.19	0.991	9.97	1.00
	L. luteus	1875	1821	99.18	0.999	18.10	1.82
Previous NCPR	*L. angustifolius*	2086	2119	8.36	0.994	1.77	1.00
	L. luteus	1732	1654	22.74	0.993	3.76	2.12
Previous QPR	*L. angustifolius*	2202	2227	13.29	0.973	2.96	1.00
	L. luteus	1825	1766	30.05	0.997	5.40	1.82

[a]TSP – triple superphosphate; NCPR – North Carolina phosphate rock; QPR – Queensland (Duchess) phosphate rock.
[b]*L angustifolius* cv. Kalya; *L. luteus* cv. Wodjil.

[$\times 10^4$]) (kg P ha^{-1}), and A and B are coefficients. Coefficient A is the intercept and so provides an estimate of the concentration of P taken up from native soil P where no P fertilizer was applied. Coefficient B estimates the slope of the line and so estimates the increase in P concentration in tissue per unit of applied P. Similarly, the linear equation was fitted to the relationship between Cd concentration in shoots and grain (y axis) and the amount of Cd applied as each source of fertilizer P (x axis). Cadmium was present as an impurity in all the P sources used and so different amounts of Cd were applied as different amounts of each P fertilizer were applied.

The linear equation was also use to describe the relationship between soil test P (y axis) and the amount of P applied (x axis). Similarly, the relationship between Cd concentration in grain (y axis) and soil test Cd (x axis) was fitted to the linear equation.

Results

Plant density

As measured about 1 month after sowing, plant densities of each lupin species were unaffected by the amount and source of fertilizer P applied in the current year or 14 years previously. Plant densities for each lupin species were (plants m^2, with SE in brackets (n = 23)): *L. angustifolius* 67 (9), and *L. luteus* 65 (12).

Yield of dried shoots and grain

Analysis of variance indicated highly significant effects on both shoot and grain yield due to source and amount of P applied, and lupin species ($P<0.01$). The interaction between the amount of P added for each lupin species and the source of P application was significant ($P<0.05$).

Comparison of yield responses to added fertilizer P by the two lupin species, for each source of P (triple superphosphate applied in the current year, triple superphosphate and two different phosphate rocks applied 14 years previously), were both determined using the initial slope (bc) of the relationship between plant yield and the amount of P applied. Therefore, it is stressed that $RR_{species}$ values are for the initial part of the response curve.

When no P was applied, *L. luteus* produced greater yields of dried shoots (about 60 g m^{-2}[$\times 10^4$] (600 kg ha^{-1})) than *L. angustifolius* (about 30 g m^{-2} [$\times 10^4$]) (300 kg ha^{-1}) (Figure 1). *L. angustifolius* did not produce any grain on the nil-P treatment plots. By contrast, *L. luteus* did produce about 70 kg ha^{-1} grain when no P was applied (Figure 2).

The initial slope (bc) for each P source was always higher (steeper) for *L. luteus* than for *L. angustifolius* (Figure 1, and bc and $RR_{species}$ values in Table 2). *L. luteus* used the added P about 10–30% more effectively for producing shoots than *L. angustifolius* (for which $RR_{species}$ is, by definition, 1.00).

For grain production, relative to *L. angustifolius*, *L. luteus* was about 1.5–2 times more effective at using the added P (Figure 2, see $RR_{species}$ values in Table 2). Note that the maximum yield plateau for grain production of *L. luteus* (about 180 g m^{-2} [$\times 10^4$]) (1800 kg ha^{-1}) was below the maximum yield plateau reached by *L. angustifolius* (2200 kg ha^{-1}) (Figure 2, and see the value of the a coefficient [maximum yield plateau] in Table 2).

Third-order lateral roots

When lupin roots were examined in late September, many third-order lateral roots were found on all plants of *L. luteus* examined, regardless of the source and amount of P applied. By contrast, no such roots were found on any of the *L. angustifolius* plants examined.

Concentration of phosphorus in dried shoots and grain

For both lupin species, the concentrations of P in dried shoots (Figure 3) and grain (Figure 4) increased with increasing amounts of P applied for all P sources. The coefficients of the linear equation fitted to the relationship between P concentration in tissue (y axis) and the amount of P applied (x axis) are listed in Table 3. The P concentration in dried shoots for the nil-P treatment was about 1.5 g kg^{-1} for *L. luteus* and about 1.3 g kg^{-1} P for *L. angustifolius* (Figure 3, A values in Table 3). For all sources of P, the slope (B) coefficient was always higher (steeper) for *L. luteus* than for *L. angustifolius* (Table 3). This suggests that roots of *L. luteus* were able to access more soil P and used the added P about 40% more effectively for increasing P concentration in dried shoots than *L. angustifolius.*

Table 3. Value of coefficients of the linear equation fitted to the relationship between P concentration (g kg^{-1}) in tissue and the amount of P applied (kg P m^{-2} [$\times 10^4$]) (g P ha^{-1}), and $RR_{species}$ values

Source of P[a]	Plant species[b]	Coefficients			$RR_{species}$
		A	B	R^2	
Dried shoots, 103 days after sowing					
Current TSP	*L. angustifolius*	0.130	41.0	0.98	1.00
	L. luteus	0.152	57.0	0.97	1.39
Previous TSP	*L. angustifolius*	0.134	2.6	0.99	1.00
	L. luteus	0.131	3.4	0.99	1.31
Previous NCPR	*L. angustifolius*	0.134	0.7	0.97	1.00
	L. luteus	0.180	0.9	0.97	1.28
Previous QPR	*L. angustifolius*	0.122	0.5	0.98	1.00
	L. luteus	0.144	0.9	0.98	1.80
Seed (grain)					
Current TSP	*L. angustifolius*	0.228	33.0	0.95	1.00
	L. luteus	0.364	59.0	0.96	1.79
Previous TSP	*L. angustifolius*	0.273	2.0	0.91	1.00
	L. luteus	0.354	3.0	0.95	1.50
Previous NCPR	*L. angustifolius*	0.256	0.5	0.97	1.00
	L. luteus	0.349	0.8	0.99	1.60
Previous QPR	*L. angustifolius*	0.240	0.3	0.86	1.00
	L. luteus	0.377	0.7	0.95	2.33

[a]TSP – triple superphosphate; NCPR – North Carolina phosphate rock; QPR – Queensland (Duchess) phosphate rock.
[b]*L angustifolius* cv. Kalya; *L. luteus* cv. Wodjil.

L. angustifolius produced no grain for the nil-P treatment so no P concentration data are avaialable for the nil-P treatment (Figure 4). *L. luteus* did produce about 7 g grain m^{-2} ($\times 10^4$) (70 kg grain ha^{-1}) when no P was applied, and the concentration of P in the grain was about 3.5 g kg^{-1} (Figure 4; A values of Table 3). For all sources of P, the value of the slope (B) coefficient for grain was always higher for *L. luteus* than for *L. angustifolius* (Table 3). *L. luteus* used the added P about 1.5–2.3 times more effectively for increasing P concentration in grain than *L. angustifolius*. So, compared with *L. angustifolius*, P concentrations in grain were about double for *L. luteus*.

Concentration of cadmium in dried shoots and grain

The concentrations of Cd in dried shoots (Figure 5) and grain (Figure 6) increased with increasing amounts of Cd applied for all P sources. The coefficients of the linear equation fitted to the relationship between the concentration of Cd in plant tissue (y axis)

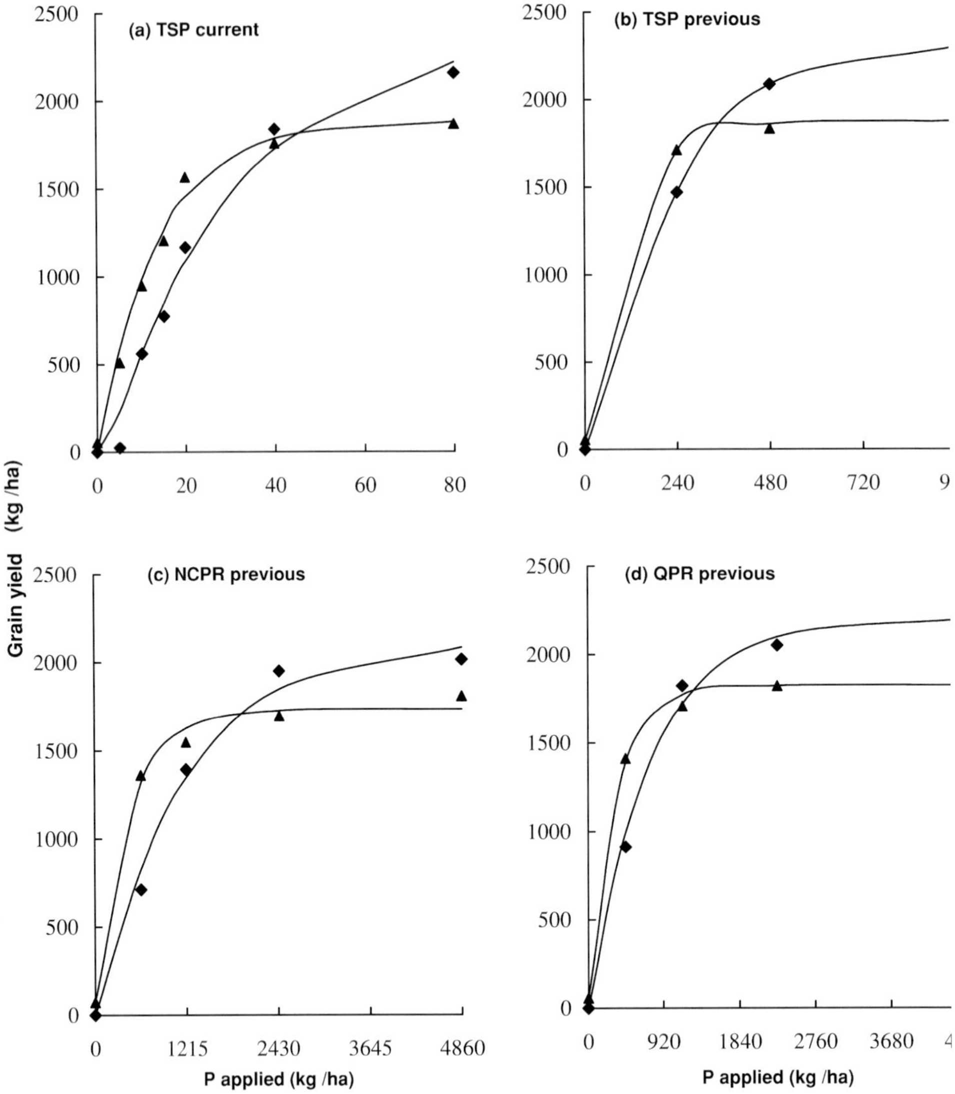

Figure 2. Relationship between grain yield of lupin and the amount of fertilizer phosphorus applied as (a) triple superphosphate (TSP) applied in the current year, and for P applied 14 years previously as (b) TSP, (c) North Carolina phosphate rock (NCPR) and (d) Queensland phosphate rock (QPR). In each case *L. luteus* (▲), *L. angustifolius* (♦). SI units for kg ha^{-1} are (g m^{-2} [$\times 10^4$]).

and the amounts of Cd applied as each P source (*x* axis) are listed in Table 4. The Cd concentrations in dried shoots for the nil-P treatment (no Cd added as fertilizer P) were about 0.050 mg kg^{-1} Cd for *L. luteus* and about 0.013 mg kg^{-1} Cd for *L. angustifolius* (Figure 5; *A* values in Table 4). For dried shoots, the value of the slope (*B*) coefficient was always higher for *L. luteus* than for *L. angustifolius* regardless of the source of P (Table 4). *L. luteus* was about 4–25 times (depending on the P source) more effective at increasing Cd concentration in dried shoots than *L. angustifolius.*

For the nil-P treatment, the concentrations of Cd in grain were about 0.05 mg kg^{-1} Cd for *L. luteus.* For *L. angustifolius,* the concentrations of Cd in grain reached about 0.05 mg kg^{-1} Cd when the highest amount of current or previous P was applied (Figure 6, Table 4). For each P source, the value of the slope (*B*) coefficient for grain was always higher for *L. luteus* than for *L. angustifolius* (Table 4). *L. luteus* was about 6 to 30 times more effective at increasing Cd concentration in grain than *L. angustifolius* (Figure 6, see RR species values in Table 4).

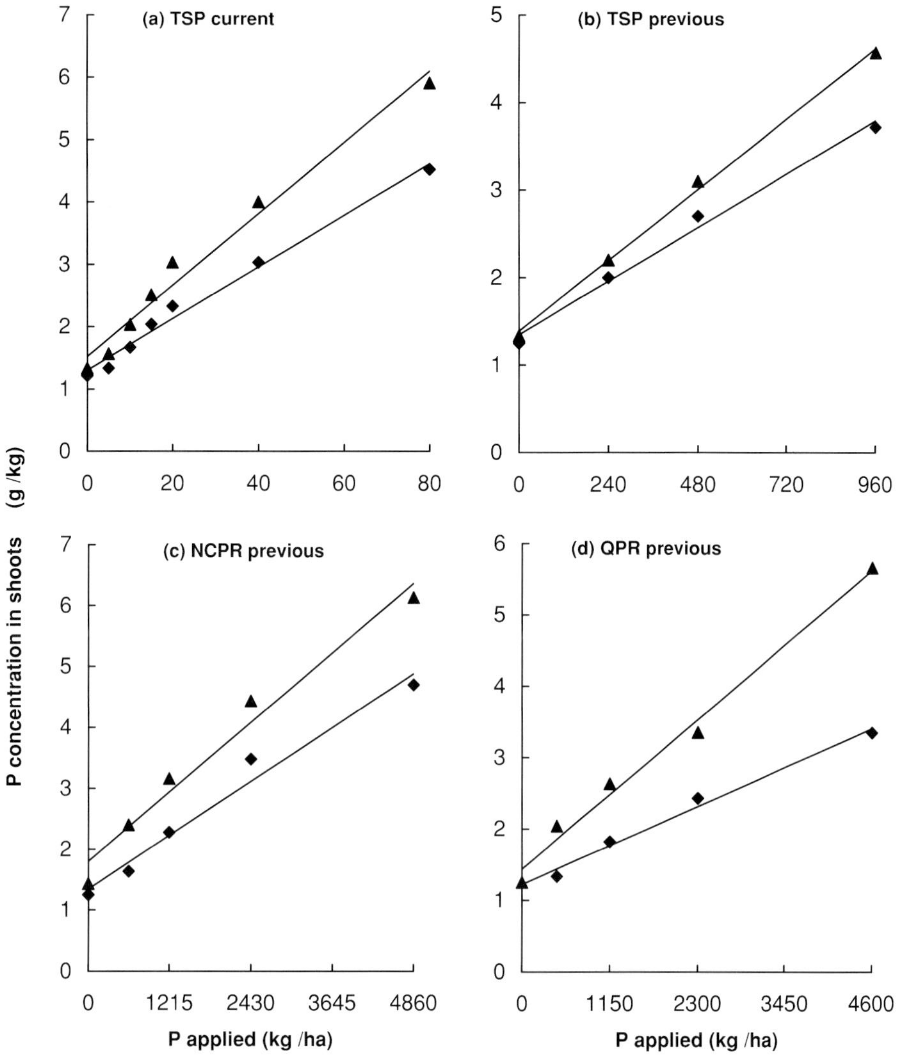

Figure 3. Relationship between phosphorus concentration of lupin shoots and the amount of fertilizer phosphorus applied as (a) triple superphosphate (TSP) applied in the current year, and for P applied 14 years previously as (b) TSP, (c) North Carolina phosphate rock (NCPR) and (d) Queensland phosphate rock (QPR). In each case *L. luteus* (▲), *L. angustifolius* (♦). SI units for g ha^{-1} are (kg m^{-2} [$\times 10^4$]).

Relationships between yield and concentration of P in tissue

Dried shoots

For each species there was a single common relationship between the yield of dried shoots (*y* axis) and the concentration of P in dried shoots (*x* axis) (internal efficiency of P use curve) (Figure 7a). To reach 200 g m^{-2} ($\times 10^4$) (200 g ha^{-1}) dried shoots both species required about 3.0 g P kg^{-1} of dried shoot. Thus, once the P was taken up from the soil, the same concentration of P in shoots produced the same yield of dried shoots regardless of the P source in the soil so that yield of dried shoots depended on the P concentration in the tissue.

Grain yield

The relationship between grain yield and P concentration in grain was different for each lupin species (Figure 7b). However, for each lupin species, the relationship was similar for all sources of P. To reach a grain yield of 150 g m^{-2} ($\times 10^4$) (1500 kg ha^{-1}), *L. luteus* required about 4.5 g P kg^{-1} grain; *L. angustifolius* required about 3.0 g P kg^{-1} grain.

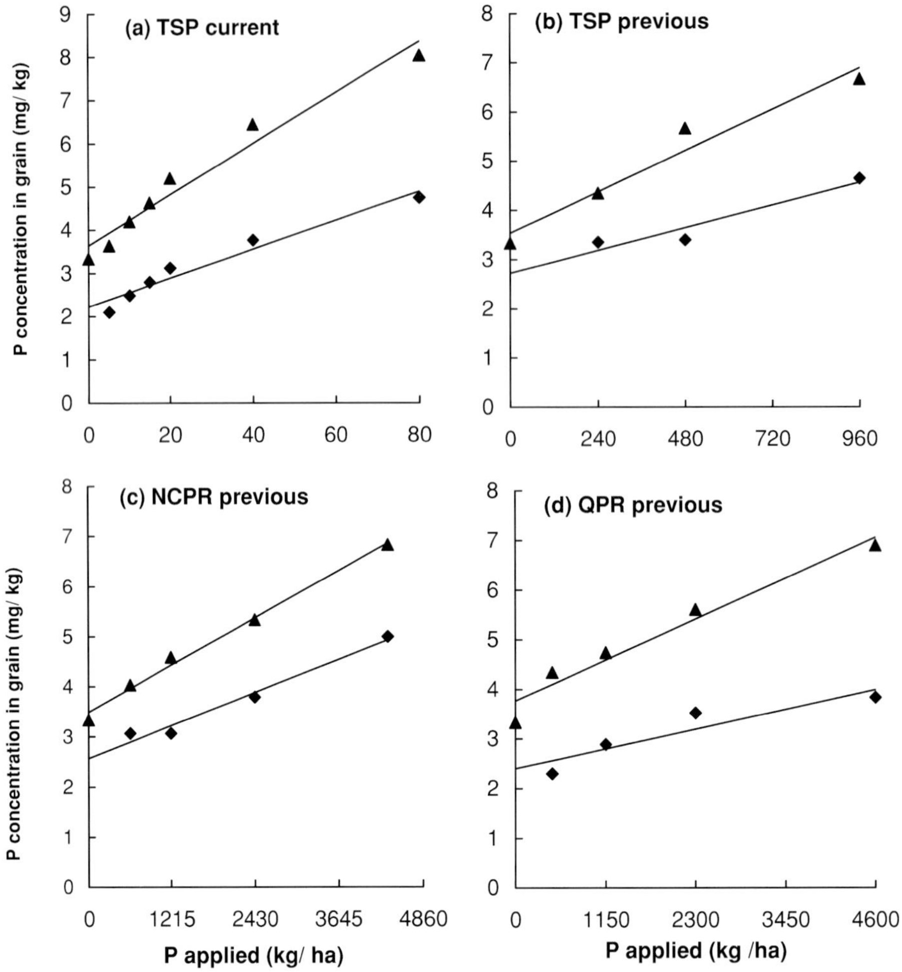

Figure 4. Relationship between phosphorus concentration in lupin grain and the amount of fertilizer phosphorus applied as (a) triple superphosphate (TSP) applied in the current year, and for P applied 14 years previously as (b) TSP, (c) North Carolina phosphate rock (NCPR) and (d) Queensland phosphate rock (QPR). In each case *L. luteus* (▲), *L. angustifolius* (◆). SI units for kg ha^{-1} are (g m^{-2} [$\times 10^4$]).

Because maximum yield plateaus for the relationship between yield of either shoots or grain and the P concentration in shoots or grain were not well defined, it was not possible to estimate the concentration of P required to produce 90% of the maximum yield (critical value).

Soil test P

The application of P increased the soil test P value for each source of P (Figure 8). For each P source, there was a different relationship between soil test P and the amount of P applied for superphosphate applied in the current year, superphosphate applied 14 years previously, and phosphate rock applied 14 years previously (Figure 8, Table 5). However, both sources of phosphate rock had a similar relationship between soil test P and the amount of P applied (Figure 8, Table 5). The value of the slope (*B*) coefficient for the relationship between soil test P and the amount of P applied was largest for triple superphosphate applied in the current year (Table 5). Relative to superphosphate in the current year, the value of the *B* coefficient decreased by about 87% for P applied 14 years previously, and by a further 89% for phosphate rock applied 14 years previously (Table 5).

The relationships between grain yield (*y* axis) and soil test for P values (*x* axis) (the soil P test calibration curve) was similar for the two species (Figure 9). To reach a grain yield of 150 g m^{-2} ($\times 10^4$) (1500 kg ha^{-1}), *L. luteus* required a soil test P value of about

Table 4. Value of coefficients of the linear equation fitted to the relationship between Cd concentration in tissue (mg kg^{-1}) and the amount of Cd applied (g Cd m^{-2} [$\times 10^4$]) (kg Cd ha^{-1}), and $RR_{species}$ values

Source of P[a]	Plant species[b]	Coefficients			$RR_{species}$
		A	*B* ($\times$1000)	R^2	
Dried shoots, 103 days after sowing					
Current TSP	*L. angustifolius*	0.011	191.1	0.967	1.00
	L. luteus	0.066	804.4	0.856	4.21
Previous TSP	*L. angustifolius*	0.019	1.3	0.873	1.00
	L. luteus	0.025	14.2	0.998	10.92
Previous NCPR	*L. angustifolius*	0.024	0.2	0.927	1.00
	L. luteus	0.123	4.6	0.923	23.00
Previous QPR	*L. angustifolius*	0.013	2.4	0.88	1.00
	L. luteus	0.035	12.3	0.966	5.13
Seed (grain)					
Current TSP	*L. angustifolius*	0.010	115.3	0.839	1.00
	L. luteus	0.025	1396.9	0.994	12.12
Previous TSP	*L. angustifolius*	0.002	4.0	0.979	1.00
	L. luteus	0.010	24.2	0.972	6.05
Previous NCPR	*L. angustifolius*	0.004	0.1	0.981	1.00
	L. luteus	0.032	3.0	0.996	30.00
Previous QPR	*L. angustifolius*	0.005	1.6	0.997	1.00
	L. luteus	0.044	35.5	0.96	22.19

[a]TSP – triple superphosphate; NCPR – North Carolina phosphate rock; QPR – Queensland (Duchess) phosphate rock.
[b]*L angustifolius* cv. Kalya; *L. luteus* cv. Wodjil.

Table 5. Value of coefficients of the linear equation fitted to the relationship between soil test P (mg P kg^{-1} soil) and the amount of P applied (g P m^{-2} [$\times 10^4$]) (kg P ha^{-1})

Source of P[a]	*A*	*B* ($\times 10^2$)	R^2
TSP current	3.95	56.11	0.996
TSP previous	4.80	7.49	0.999
NCPR	6.55	0.09	0.970
QPR	8.70	0.08	0.850

[a]TSP – Triple superphosphate; NCPR – North Carolina phosphate rock; QPR – Queensland (Duchess) phosphate rock.

Table 6. Value of coefficients of the linear equation fitted to the relationship between Cd soil test value (mg kg^{-1}) and either Cd concentration (mg kg^{-1}) or Cd content (Cd concentration multiplied by yield) (mg m^{-2} [$\times 10^4$]) (mg ha^{-1}) in grain

Species[a]	*A*	*B*	R^2
Cd concentration			
L. angustifolius	0.0078	0.0005	0.4346
L. luteus	0.0064	0.0123	0.8460
Cd content			
L. angustifolius	11.956	1.782	0.872
L. luteus	119.87	24.038	0.835

[a]*L. angustifolius* cv. Kalya; *L. luteus* cv. Wodjil.

17 mg P kg^{-1} soil while *L. angustifolius* needed about 22 mg P kg^{-1} soil (Figure 9).

Soil test Cd

The application of P increased the soil test Cd value for each source of P (data not shown). This was because, for each source of P applied, more Cd was also applied because Cd was present in all sources of P used, though the concentration of Cd varied for each P source (Table 1). The soil test Cd value for the nil-P treatment was about 0.006 mg kg^{-1} (Figure 10; *A* values in Table 6).

The relationships between Cd concentration in the grain and soil test for Cd values (i.e., soil Cd test calibration curve) was different for the two species (Figure 10). *L. luteus* had a significantly higher slope (*B* coefficient) than *L. angustifolius* (Figure 9, Table 6).

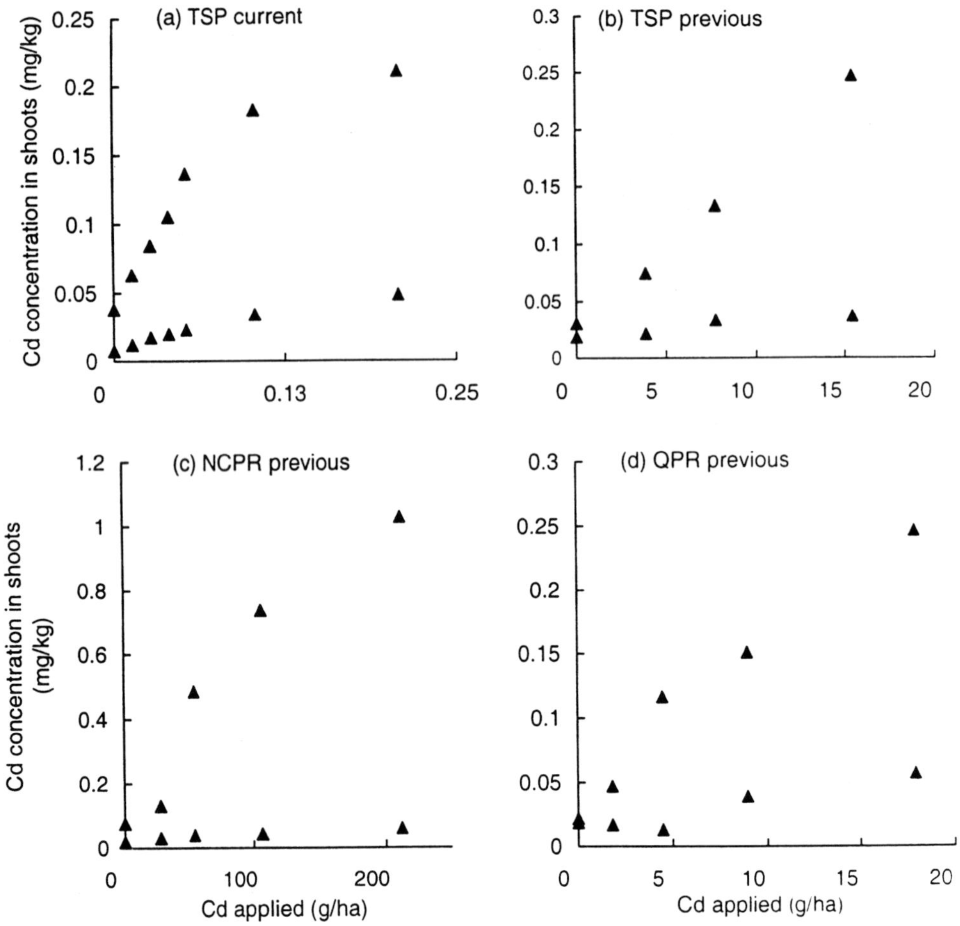

Figure 5. Relationship between cadmium concentration in lupin shoots and the amount of Cd applied as (a) triple superphosphate (TSP) applied in the current year, and for P applied 14 years previously as (b) TSP, (c) North Carolina phosphate rock (NCPR) and (d) Queensland phosphate rock (QPR). In each case *L. luteus* (▲), *L. angustifolius* (♦). SI units for g ha^{-1} are (mg m^{-2} [$\times 10^4$]).

That is, the concentration of Cd in *L. luteus* grain greatly increased with increasing soil test Cd value. As determined from the ratio of the *B* values in Table 6, the increase in Cd concentration in grain was about 25 times greater for *L. luteus* than *L. angustifolius*.

Discussion

Abundant third-order lateral roots, found only on roots of *L. luteus*, may have enabled *L. Luteus* roots to better access P and Cd from the soil than *L. angustifolius* roots. Therefore, per unit of applied P as each source, compared with *L. angustifolius*, *L. luteus* more effectively used applied P to produce dried shoots and grain, and the concentration of Cd in dried shoots and grain was much greater in *L. luteus* than *L. angustifolius*. However, P concentration in shoots was similar for both lupin species per unit of applied P as each source, but P concentrations were larger in *L. luteus* grain.

Similar to our findings in the study reported here, Bolland et al. (2000) also found abundant third-order lateral roots on secondary roots of *L. luteus*. There is debate about whether *L. luteus* develop proteoid (cluster) roots. No cluster roots were found on *L. luteus* in the studies of Bolland (1997a), Clements et al. (1993), or Skene and James (2000). However, Gerke et al. (1999) and Romer et al. (1998) found *L. luteus* did form cluster roots. Gerke et al. (1999) reports that cluster roots of *L. luteus* are less dense and possess longer rootlets. It could be that the 'third-order' lateral roots found on *L. luteus* in this study and by Bolland et al. (2000) were in fact cluster roots, but they look different to the cluster roots typically developed on two other lupin species grown in Western

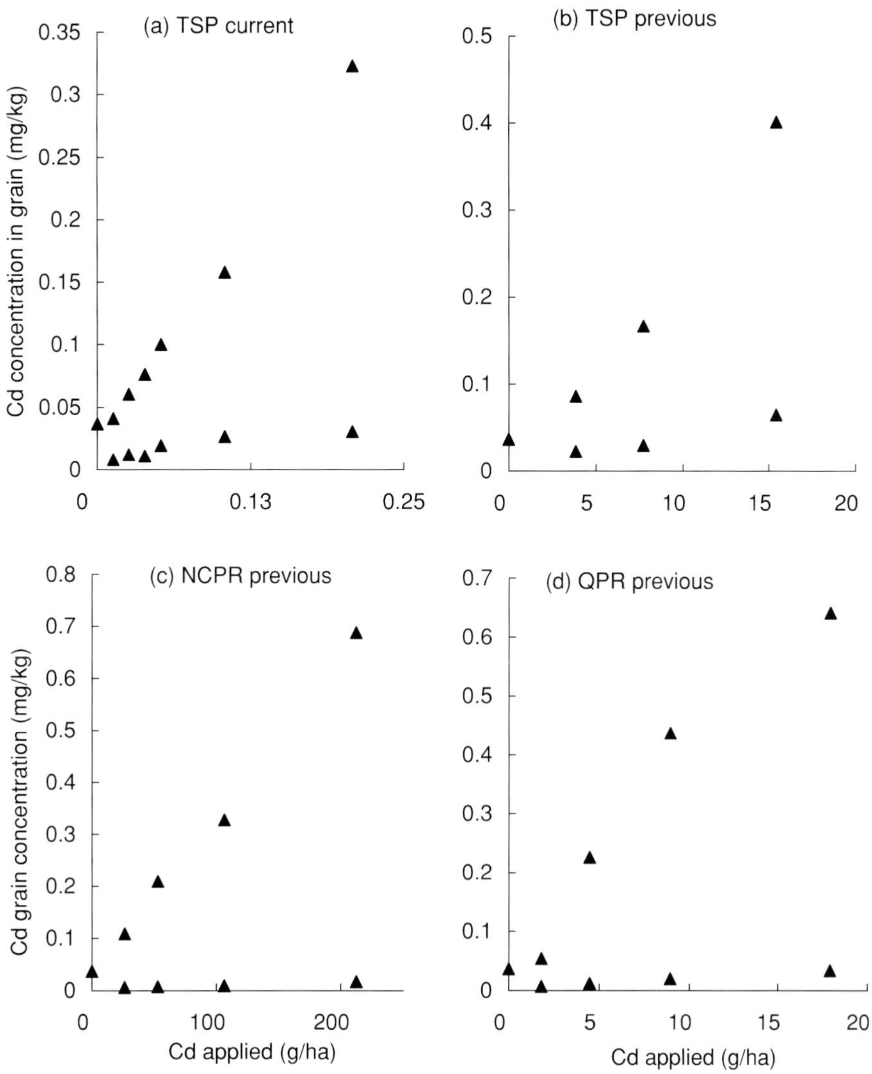

Figure 6. Relationship between cadmium concentration in lupin grain and the amount of Cd applied as (a) triple superphosphate (TSP) applied in the current year, and for P applied 14 years previously as (b) TSP, (c) North Carolina phosphate rock (NCPR) and Queensland phosphate rock (QPR). In each case *L. luteus* (▲), *L. angustifolius* (♦). SI units for g ha^{-1} are (mg m^{-2} [$\times 10^4$]).

Australia, *L. albus* L. and *L. cosentinii* L. Cluster roots of *L. albus* secrete organic acid anions as a result of P deficiency, which increases uptake of P from insoluble sources in the soil (Dinkelaker et al., 1989; Gardiner et al., 1981, 1982a, b, 1983; Gerke, 1995; Gerke et al., 1994; Hinsinger and Gilkes, 1995; White and Robson, 1989). It could be that *L. luteus* also secretes organic acid anions from third-order lateral (= cluster) roots that dissolve insoluble sources of P and Cd in the soil and so enabled the species to access more P and Cd from the soil than *L. angustifolius* in this study. Bolland et al. (2000) found that *L. luteus* cv Teo had higher concentration of potassium, sulphur, magnesium, copper and zinc in dried shoots and grain than *L. angustifolius* cv Merrit. This was attributed to the presence of the abundant third-order lateral roots (= cluster roots) found on secondary roots of *L. luteus*.

Grain of both *L. luteus* and *L. angustifolius* is used as stock feed. The higher concentrations of Cd in grain of *L. luteus* than in grain of *L. angustifolius* has practical implications for the value of *L. luteus* grain as stock feed. Reasons for the higher Cd concentration in *L. luteus* grain are not known and cannot be deduced from our study, so further research is required to determine the causes so possible remedies can be sought. We have proposed cluster roots may increase

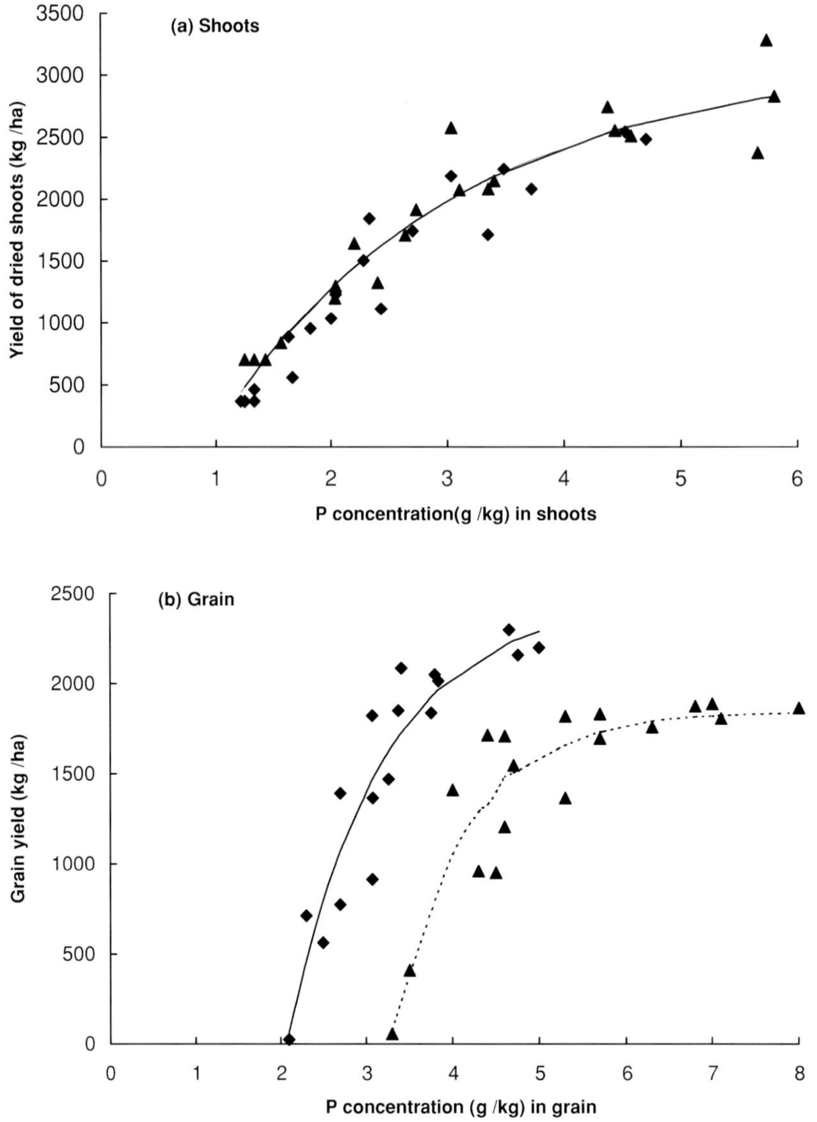

Figure 7. Relationship between lupin yield and phosphorus concentration in tissue for (a) dried shoots and (b) grain. In each case *L. luteus* (▲), *L. angustifolius* (♦).

dissolution of insoluble soil Cd and most of the dissolved Cd may be taken up by cluster roots before it is retained by the soil. However, there are several possible alternative explanations. The capacity of roots of different plant species to take up Cd from the soil may vary greatly between different species, and may have a greater impact on how much Cd is taken up by plant roots than the extent of dissolution of soil Cd by root exudates. In a recent study, Romer et al. (2000) found that both *L. albus* and *L. angustifolius* took up much less soil Cd than *Lolium multiflorum* Lam. Romer et al. (2000) found that the roots of both *L. albus* and *L. angustifolius* secreted citrate and proposed that the citrate may have formed a complex with soil Cd reducing (excluding) root uptake, thereby reducing Cd uptake by both lupin species. Secretion of citrate from plant roots may be an adaptive mechanism to alleviate aluminium toxicity in soil (Bartlett and Reigo, 1972; Gerke, 1997), and citrate may also reduce Cd uptake by plant roots (Romer et al., 2000). In our study, due to unknown reasons, maybe exclusion of Cd by *L. luteus* roots was much smaller than Cd exclusion by

L. angustifolius roots. Once Cd is taken up from the soil by the plants, translocation of Cd to grain may be greater in *L. luteus* than in *L. angustifolius*. This is supported by our P data showing similar P concentrations in shoots of both *L. luteus* and *L. angustifolius*, but evidently greater translocation of P to grain resulted in *L. luteus* grain having higher P concentrations. In our study, even though concentrations of Cd were higher in both shoots and grain of *L. luteus* than in shoots and grain of *L. angustifoius*, translocation of Cd from shoots to grain may have been higher in *L. luteus* than in *L. angustifolius*.

We studied one cultivar of two lupin species that, under the conditions of the experiment, may have differed in the way they acidified the rhizosphere affecting P and Cd acquisition, particularly from phosphate rock. For unknown reasons, *L. luteus* appears to be more tolerant of acid soils than *L. angustifolius* (French et al., 2001). Our study was undertaken on an acidic soil, which may have resulted in better root growth of *L. luteus* than of *L. angustifolius* contributing to better acquisition of P and Cd by *L. luteus* roots.

The maximum yield plateau for grain production was larger for *L. angustifolius* than *L. luteus*. There has been an intensive breeding program for 45 years in Western Australia to improve grain production of *L. angustifolius* for acidic soils in the region, and cv. Kalya is a recent cultivar from this effort. By contrast there has been little selection or breeding for *L. luteus* for acidic soils in Western Australia, so the species is unlikely to be as well adapted as *L. angustifolius* for producing grain in acidic soils in Western Australia. This may change if a more intensive selection and breeding programs is undertaken for *L. luteus* in Western Australia.

The uptake of Cd by plants is of importance for both animal and human health (Jackson and Alloway, 1992). Plants species that accumulate Cd often have the capacity to transfer Cd and other micronutrients from soil to shoots and seed (grain) (Baker and Brooks, 1989; Kuboi et al., 1986). Potato (*Solanum tuberosum* L.) also takes up Cd and accumulates Cd in the tuber (McLaughlin et al., 1995). *L. luteus* can now be added to the list of plant species that takes up and accumulates much Cd in shoots and grain.

Cadmium occurs as a natural constituent in soils and as a contaminant in some phosphorus fertilizers (Andersson, 1977; McLaughlin et al., 1996; Merry and Tiller, 1991; Singh, 1994). Phosphorus fertilizers containing Cd have resulted in increased soil

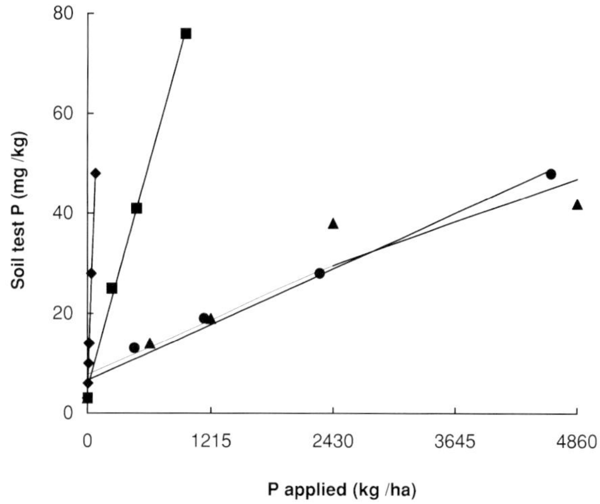

Figure 8. Relationship between the Colwell soil test phosphorus values and the amount of fertilizer phosphorus applied as (a) triple superphosphate (TSP) applied in the current year, and for P applied 14 years previously as (b) TSP, (c) North Carolina phosphate rock (NCPR) and Queensland phosphate rock (QPR). Symbols TSP fresh (♦); TSP previous (■), NCPR (●) and QPR (▲). SI units for kg ha^{-1} are (g m^{-2} [$\times 10^4$]).

Cd test values in Australia (Walkley, 1940; Williams and David, 1973, 1976). Deposition of Cd from the atmosphere is mostly negligible in Australia (Pearse, 1996; Tiller and de Vries, 1977). In our study, the high amounts of applied fertilizer also added Cd. Continued applications of smaller amounts of fertilizer P contaminated with much Cd over many years on farms in Australia also increased the Cd status of soils (McLaughlin et al., 1996).

The soil P test calibration, relating grain yield to soil test P measured by different procedures, including the Colwell (1963) procedure used in our study, frequently differ for different plant species (Bolland et al., 1989, 1994). However, the calibration was similar for the two lupin species in this study. This may not always be the case for different experiments done at different sites in the same or different years, or as measured in different years at the same site (Bolland et al., 1989).

The soil Cd test procedure used in our study gave a good prediction of when the Cd status of the soil was high enough to cause Cd concentrations in lupin grain to exceed acceptable limits. This needs to be confirmed for different soil types, seasons and plant species before soil test Cd can be used with reliability.

Recently many new crop species are being grown in rotation with wheat on farms in south-western Australia. The fertilizer P requirements of most of the

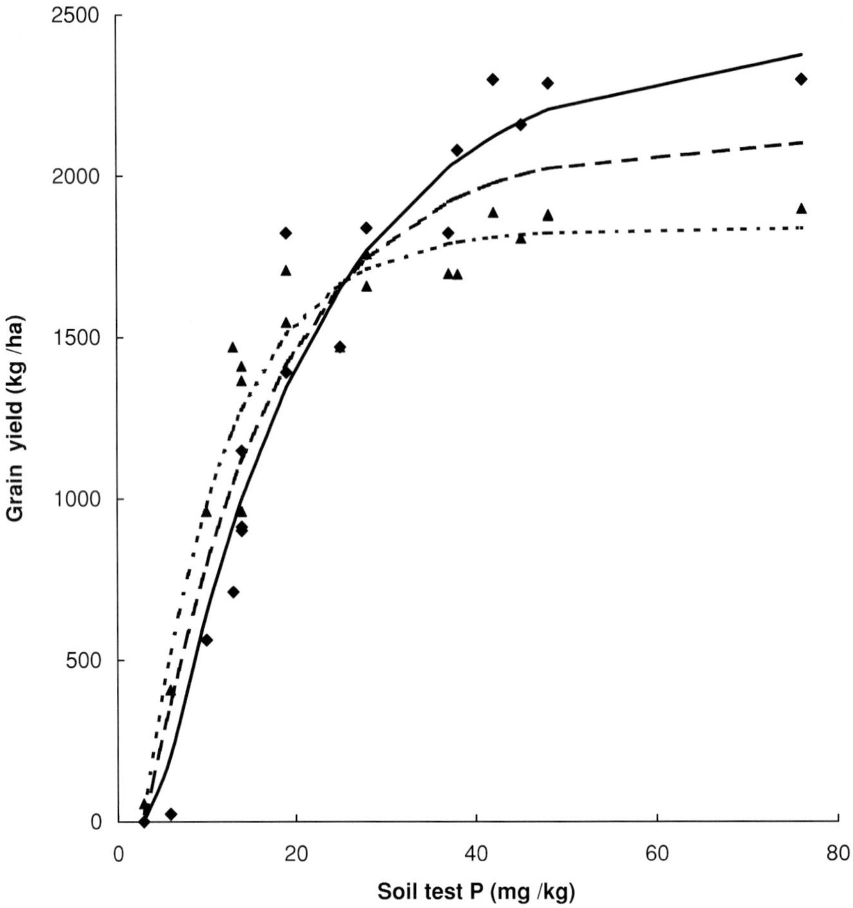

Figure 9. Relationship between lupin grain yield and Colwell soil test phosphorus values. In each case *L. luteus* (▲, short dashed line), *L. angustifolius* (♦, solid line) while the long dashed curve is the common fitted line fitted to Mitscherlich equation 1. The fitted equations are: *L. angustifolius* (♦, solid line) $y = 2419 - 3097 \exp(-0.056x)$; *L. luteus* (s, short dashed line), $y = 1838 - 2496 \exp(-0.106x)$; common fitted line (long dashes), $y = 2114 - 2672 \exp(-0.070x)$. SI units for kg ha^{-1} are (g m^{-2} [$\times 10^4$]).

new crops are not known for the soils, environments and management systems used in the region. Consequently, field experiments have been conducted to compare the yield and P uptake responses of the new species relative to either wheat or *L. angustifolius*, the two major crops grown in the region for which there is much previous local data (Bolland, 1997a, b; Bolland et al., 1999, 2000). However, reasons for the different yield and P uptake responses to applied P of the new species relative to wheat or *L. angustifolius* found in these field studies, and for *L. luteus* and *L. angustifolius* in our study reported here, cannot be deduced from the studies. Instead, possible explanations for our results have been derived from results of previous studies (Bolland, 1997a, b; Bolland et al., 1999, 2000; this paper). In addition, why *L. luteus* took up more Cd from the soil than *L. angustifolius* cannot be decuced from our field study reported here; so we have had to use results and conclusions of previous studies to provide possible explantions for our results. Parallel laboratory and glasshouse studies, and detailed studies of roots in the field studies, are required to provide explanations of the results obtained in the field. Such detailed studies have now commenced in south-western Australia, and the results of the first such study are reported by Veneklaas et al. in this issue.

Acknowledgements

Staff of Esperance Downs Research Station provided technical assistance. The Chemistry Centre (WA)

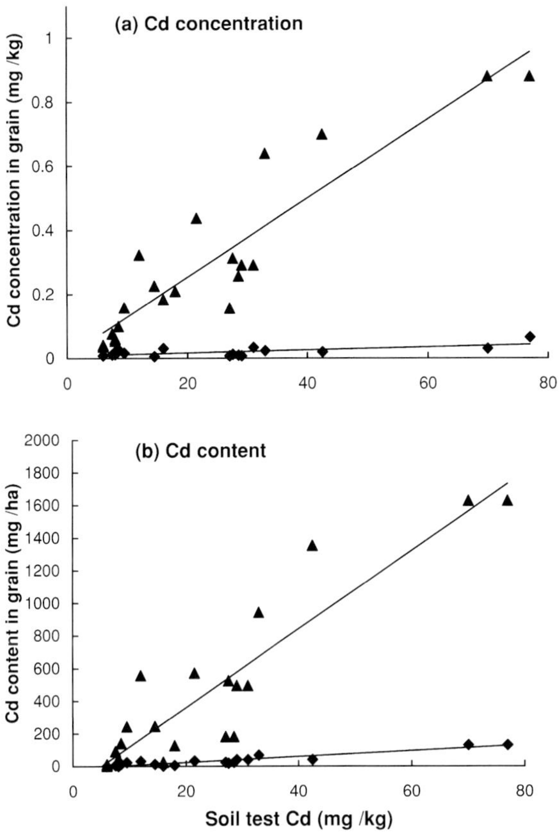

Figure 10. Relationships between (a) the cadmium concentration in lupin grain and soil test cadmium (b) the cadmium content (concentration multiplied by yield) in lupin grain and soil test cadmium. In each case *L. luteus* (▲), *L. angustifolius* (♦).

measured soil and fertilizer properties, and concentration of elements in dried shoots and grain. The Western Australian Department of Agriculture and the Grain Research and Development Corporation (GRDC) provided funds.

References

Andersson A 1977 Heavy metals in Swedish soils: On their retention, distribution and amounts. Swedish J. Agric. Res. 6, 27–36.

Anon 1977 Technicon Industrial Systems. Tarry Town, New York.

Baker AJM. and Brooks RR 1989 Terrestrial higher plants which hyperaccumulate metallic elements – a review of their distribution, ecology and phytochemistry. Biorecovery 1, 81–126.

Barrow NJ 1975 The response to phosphate of two annual pasture species. 1. Effect of the soil's ability to adsorb phosphate on comparative phosphate requirement. Aust. J. Agric. Res. 26, 137–143.

Barrow NJ 1985 Comparing the effectiveness of fertilizers. Fertil. Res. 8, 85–90.

Barrow NJ and Campbell NA 1972. Methods of measuring the residual value of fertilizers. Aust. J. Agric. Res. 12, 502–510.

Barrow NJ and Mendoza RE 1990 Equations for describing sigmoid yield responses and their application to some phosphate responses by lupins and subterranean clover. Fertil. Res. 22, 181–188.

Bartlett R and Riego R 1972 Effect of chelation on the toxicity of aluminium. Plant Soil 37, 419–423.

Bolland MDA 1997a Comparative phosphorus requirement of four lupin species. J. Plant Nutr. 20, 1239–1253.

Bolland MDA 1997b Comparative phosphorus requirement of canola and wheat. J. Plant Nutr. 20, 813–829.

Bolland MDA and Gilkes RJ 1995 Long-term residual value of North Carolina and Queensland rock phosphates compared with triple superphosphate. Fertil. Res. 41, 151–158.

Bolland MDA, Weatherley AJ, Gilkes RJ and Bowden JW 1986 Granular reactive apatite rock phosphate is not an effective phosphorus fertilizer in the short term on lateritic soils in south-western Australia. Aust. J. Exp. Agric. 26, 217–225.

Bolland MDA, Weatherley AJ and Gilkes RJ 1988. Residual effectiveness of superphosphate is greater than that of rock phosphate fertilizers for lateritic soils in south-western Australia. Aust. J. Exp. Agric. 28, 83–90.

Bolland MDA, Weatherley AJ and Gilkes RJ 1989. The long-term residual value of rock phosphate and superphosphate fertilizers for various plant species under field conditions. Fertil. Res. 20, 89–100.

Bolland MDA, Kumar V and Gilkes RJ 1994 A comparison of five soil phosphorus tests for five crop species for soil previously fertilized with superphosphate and rock phosphate. Fertil. Res. 37, 125–132.

Bolland MDA, Siddique KHM, Loss SP and Baker MJ 1999 Comparing responses of grain legumes, wheat and canola to applications of superphosphate. Nutr. Cycl. Agroecosyst. 53, 157–175.

Bolland MDA, Sweetingham MW and Jarvis RJ 2000 Effect of applied phosphorus on the growth of *Lupinus luteus*, *L. angustifolius* and *L. albus* in acidic soils in the south-west of Western Australia. Aust. J. Exp. Agric. 40, 79–92.

Brennan RF, Bolland MDA and Shea G 2001 Comparing how *Lupinus angustifolius* and *Lupinus luteus* use zinc fertilizer for seed production. Nutr. Cycl. Agroecosyst. 59, 209–217.

Buirchell BJ and Cowling WA 1998 Genetic resources in lupins. *In* Lupins as Crop Plants: Biology, Production, and Utilization. Eds. JS Gladstones, CA Atkins and J Hamblin. pp. 41–66. CAB International Press, Wallingford.

Clements JC, White PF and Buirchell BJ 1993 The root morphology of *Lupinus angustifolius* in relation to other *Lupinus* species. Aust. J. Agric. Res. 44, 1367–1375.

Colwell JD 1963 The estimation of phosphorus fertliser requirements of wheat in southern New South Wales by soil analysis. Aust. J. Exp. Agric. Ani. Husb. 3, 190–197.

Day PR 1965 Particle fractionation and particle size analysis. *In* Methods of Soil Analysis. Part 1. Agronomy Monograph No 9. Ed. CA Black. pp. 545–567. American Soil Science Society, Madison, WI.

Dinkelaker B, Romheld V and Marschner H 1989 Citric acid excretion and precipitation of calcium in the rhizosphere of white lupin (*Lupinus albus* L.). Plant Cell Environ. 12, 285–292.

Edwards NK 1998. Potassium. *In* Soilguide: A Handbook for Understanding and Managing Agricultural Soils. Ed. G Moore. pp. 176–180. Agriculture Western Australia, Bulletin 4343, South Perth.

FAO-UNESO 1988 Soil map of the world. Revised legend. World Soil Resources, Report 60, FAO, Rome.

French RJ, Sweetingham MW and Shea GG 2001 A comparison of the adaption of yellow lupin (*Lupinus luteus* L.) and narrow-leafed lupin (*L. angustifolius* L.) to acid sandplain soils in low rainfall agricultural areas of Western Australia. Aust. J. Agric. Res. 52, 945–954.

Gardner WK, Parbery DG and Barber DA 1981 Proteoid root morphology and function in *Lupinus albus*. Plant Soil 60, 143–147.

Gardner WK, Parbery DG and Barber DA 1982a The acquisition of phosphorus by *Lupinus albus* L. I. Some characteristics of the soil/root interface. Plant Soil 68, 19–32.

Gardner WK, Parbery DG and Barber DA 1982b The acquisition of phosphorus by *Lupinus albus* L. II. Effect of varying phosphorus supply and soil type on some characteristics of the soil/root interface. Plant Soil 68, 33–41.

Gardner WK, Barber DA and Parbery DG 1983 The acquisition of phosphorus by *Lupinus albus* L. III. The probable mechanism by which phosphorus movement in the soil/root interface is enhanced. Plant Soil 70, 107–124.

Gerke J 1995 Chemische Prozesse der Nahrstoffmobilisierung in der Rhizosphäre und ihre Bedeutung fur den Ubergang vom Boden in die Pflanze. Habilitationsschrift Universität Göttingen. Cuvillier Verlag, Göttingen.

Gerke J 1997 Aluminium and iron (III) species in the soil solution including organic complexes with citrate and humic substances. Z. Pflanzenernahr. Bodenkd. 160, 427–432.

Gerke J, Romer A and Jungk A 1994 The excretion of citric and malic acid by proteoid roots of *Lupins albus* L.: Effects on soil solution concentrations of phosphate, iron and aluminium in the proteoid rhizosphere in samples of an oxisol and a luvisol. Z. Pflanzenernahr. Bodenk. 157, 289–294.

Gerke J, Eckhard W, Egle K and Romer W 1999 Heavy metal acquisition by white lupin and yellow lupin. Ninth International Lupin Conference, 20–24 June 1999, Klink/Muritz, Germany.

Gilbey DJ 1993 Weed control in lupins. *In* Management of Agricultural Weeds in Western Australia. Eds. J Dodd, RJ Martin and KM Howes. pp. 106–110. Western Australian Department of Agriculture, South Perth, Bulletin 4243.

Hinsinger P and Gilkes RJ 1995 Root-induced dissolution of phosphate rock in the rhizosphere of lupins grown in alkaline soils. Aust. J. Soil Res 33, 477–489.

Isbell RF 1996 A classification system for Australian soils. CSIRO, Division of Soils, CSIRO Melbourne.

Jackson AP and Alloway BJ 1992 The transfer of cadmium from agricultural soils to the human food chain. *In* Biogeochemistry of Trace Elements. Ed D C Adriano. pp. 109–158. Lewis Publishers, Boca Raton, FL.

Kuboi Y, Noguchi A and Yazaki J 1986 Family-dependent cadmium accumulation in higher plants. Plant Soil 92, 405–415.

Longnecker N, Brennan RF and Robson AD 1998 Lupin nutrition. *In* Lupins as Crop Plants: Biology, Production and Utilization. Eds. JS Gladstones, C Atkins and J Hamblin. pp. 121–148. CAB International, Wallingford.

Mann SS and Richie GSP 1993 The influence of pH on the forms of cadmium in four West Australian soils. Aust. J. Soil Res. 31, 255–270.

Mason MG 1998. Nitrogen *In* Soilguide: A Handbook for Understanding and Managing Agricultural Soils. Ed. G Moore. pp. 164–167. Agriculture Western Australia, Bulletin 4343, South Perth.

McArthur WM 1991 Reference soils of south-western Australia. Australian Society of Soil Science Incorporated, Department of Agriculture, Western Australia, Perth, Australia.

McKenzie HA and Wallace HS 1954 Kjeldahl determination of nitrogen. Aust. J. Chem. 7, 55–70.

McLaughlin MJ, Maier NA, Freeman K, Tiller KG, Williams CMJ and Smart MK 1995. Effects of potassic and phosphatic fertilizer type, phosphatic fertilizer cadmium content and additions of zinc on cadmium uptake by commercial potato crops. Fertil. Res. 40, 63–70.

McLaughlin MJ, Tiller KG, Naidu R and Stevens DP 1996 Review: The behaviour and environmental impact of contaminants in fertilizers. Aust. J. Soil Res. 34, 1–54.

McQuaker NR, Brown DF, and Kluckner PD 1979 Digestion of environmental materials for analysis by inductively coupled plasma-atomic emission spectrometry. Anal. Chem. 51, 1082–1084.

Merry RH and Tiller KG 1991 Distribution of cadmium and lead in an agricultural region near Adelaide, South Australia. Water Air Soil Pollut. 57–58, 171–180.

Nelder JA and Mead R 1965 A simplex method for function minimisation. Compr. J. 7, 308–313.

Nelson P and Delane RJ 1990 Producing lupins in Western Australia. Western Australian Department of Agriculture, South Perth, Bulletin 4179.

Northcote KH 1979 A Factual Key for the Recognition of Australian Soils. 4th Edition. Rellim Technical Publications, Glenside, South Australia.

Ozanne PG and Shaw TC 1967 Phosphate sorption by soils as a measure of the phosphate requirement for pasture growth. Aust. J. Agric. Res. 18, 601–612.

Pearse J 1996 Cadmium: Some aspects of risk reduction. Netherlands Ministry of Housing, Spatial Planning and the Environment, nr 1996/32. Zoetermeer, Veiligheid, Straling, The Netherlands.

Rahman MS and Gladstones JS 1974 Differences among *Lupinus* species in field response to superphosphate. Aust. J. Exp. Agric. Anim. Husb. 14, 214–223.

Rayment GE and Higginson FR 1992 Australian Laboratory Handbook of Soil and Water Chemical Methods. Inkata Press, Melbourne.

Romer W, Castaneda-Ortiz N and Gerke J 1998 Zum Phosphataneignungsvermogem von Gelblupine (*Lupinus luteus* L.) ind Kichererbse (*Cicer arietinum* L.) auf zwei sauren, P-armen Boden Portugals. *In* Pflanzenernährung, Wurzelleistung und Exsudation. 8. Borkheider Seminar zur Okophysiologie des Wurzelraumes. Ed. W Merbach. pp. 143–149. B.G. Teubner Verlagsgesellschaft Stuttgart, Leipzig.

Romer W, Kang DK, Egle K, Gerke J and Keller H 2000 The acquisition of cadmium by *Lupinus albus* L. and *Lupinus angustifolius* L. and *Lolium multiflorum* Lam. J. Plant Nutr. Soil 163, 623–628.

Singh BR 1994 Trace element availability to plants in agricultural soils, with special emphasis on fertilizer inputs. Environ. Rev. 2, 133–146.

Skene KR and James WM 2000 A comparison of the effects of auxin on cluster root initiation and development in *Grevillea robusta* Cunn. ex R. Br. (Proteacea) and in the genus *Lupinus* (Leguminosae). Plant Soil 219, 221–229.

Soil Survey Staff 1987 Keys to Soil Taxonomy (third printing). SMSS Technical Monograph No. 6. Ithaca, New York.

Stace HCT, Hubble GD, Brewer R, Northcote KH, Sleemen JR, Mulcahy MJ, and Hallsworth EG 1968 A Handbook of Australian Soils. Rellim Technical Publications, Glenside, South Australia.

Sweetingham MW, French RJ, Dracup M, MacLeod WJ, Carr SJ, Davies C and Shea G 1996 Yellow lupins: a new grain crop for acid soils. Aust. Grain 6 (3), 6–8.

Tang C, Fang RY and Raphael C 1998 Factors affecting soil acidification under legumes. II. Effect of phosphorus supply. Aust. J. Agric. Res. 49, 657–664.

The Crop Variety Sowing Guide 1998 Compiled and edited by J Garlinge and D Robartson. Agriculture Western Australia, South Perth, Bulletin 4341.

Tiller KJG and de Vries MPC 1977 Contamination of soils and vegetables near the lead-zinc smelter, Port Pirie, by cadmium, lead and zinc. Search 8, 78–79.

Veneklaas EJ, Stevens J, Cawthray GR, Turner S, Grigg AM and Lambers H (2002) Chickpea and white lupin rhizosphere carboxylates vary with soil properties and enhance phosphorus uptake. Plant Soil 248, 187–197.

Walkley A 1940 The zinc content of some Australian fertilizers. J. Coun. Sci. Indust. Res. Aust. 13, 225–232.

Walkley A and Black IA 1934 An examination of the Degtjarreff method for determining soil organic matter and a proposed modification of the chromic acid titration method. Soil Sci. 37, 29–38.

Williams CH and David DJ 1973 The effect of superphosphate on the cadmium content of soils and plants. Aust. J. Soil Res. 11, 43–56.

Williams CH and David DJ 1976 The accumulation in soil of cadmium residues from phosphate fertilizers and the effect on the cadmium content of plants. Soil Sci. 121, 86–93.

White PF and Robson AD 1989 Rhizosphere acidification and Fe^{3+} reduction in lupins and peas; iron deficiency in lupins is not due to its ability to reduce Fe^{3+}. Plant Soil 119, 163–175.

Wild A 1958 The phosphate content of Australian soils. Aust. J. Agric. Res. 9, 193–204.

Yuen SH and Pollard AG 1954 Determination of nitrogen in agricultural materials by the Nessler reagent. II. Micro-determinations in plant tissue and soil extracts. J. Sci. Food Agric. 5, 364–369.

Plant and Soil **248:** 187–197, 2003.
© 2003 *Kluwer Academic Publishers. Printed in the Netherlands.*

Chickpea and white lupin rhizosphere carboxylates vary with soil properties and enhance phosphorus uptake

Erik J. Veneklaas[1,3], Jason Stevens[1,2], Gregory R. Cawthray[1], Stephen Turner[1], Alasdair M. Grigg[1] & Hans Lambers[1]

[1]*School of Plant Biology, The University of Western Australia, 35 Stirling Highway, Crawley WA 6009, Australia.* [2]*Present address: School of Earth and Geographical Sciences, The University of Western Australia, 35 Stirling Highway, Crawley WA 6009, Australia.* [3]*Corresponding author**

Received 26 February 2002. Accepted in revised form 18 July 2002

Key words: carboxylates, chickpea, phosphorus, rhizosphere, root exudates, root morphology, wheat, white lupin

Abstract

Chickpea and white lupin roots are able to exude large amounts of carboxylates, but the resulting concentrations in the rhizosphere vary widely. We grew chickpea in pots in eleven different Western Australian soils, all with low phosphorus concentrations. While final plant mass varied more than two-fold and phosphorus content almost five-fold, there were only minor changes in root morphological traits that potentially enhance phosphorus uptake (e.g., the proportion of plant mass allocated to roots, or the length of roots per unit root mass). In contrast, the concentration of carboxylates (mainly malonate, citrate and malate, extracted using a 0.2 m*M* $CaCl_2$ solution) varied ten-fold (averaging 2.3 μmol g^{-1} dry rhizosphere soil, approximately equivalent to a soil solution concentration of 23 m*M*). Plant phosphorus uptake was positively correlated with the concentration of carboxylates in the rhizosphere, and it was consistently higher in soils with a smaller capacity to sorb phosphorus. Phosphorus content was not correlated with bicarbonate-extractable phosphorus or any other single soil trait. These results suggest that exuded carboxylates increased the availability of phosphorus to the plant, however, the factors that affected root exudation rates are not known. When grown in the same six soils, three commonly used Western Australian chickpea cultivars had very similar rhizosphere carboxylate concentrations (extracted using a 0.2 m*M* $CaCl_2$ solution), suggesting that there is little genetic variation for this trait in chickpea. Variation in the concentration of carboxylates in the rhizosphere of white lupin did not parallel that of chickpea across the six soils. However, in both species the proportion of citrate decreased and that of malate increased at lower soil pH. We conclude that patterns of variation in root exudates need to be understood to optimise the use of this trait in enhancing crop phosphorus uptake.

Introduction

The roots of certain crop species, including white lupin (*Lupinus albus* L.) and chickpea (*Cicer arietinum* L.) exude large amounts of low-molecular-weight organic anions (carboxylates) which enhance the availability of soil phosphorus to the plant (Ae et al., 1990; Dinkelaker et al., 1989; Gardner et al., 1983; Gerke et al., 1994; Hocking et al., 1997; Ohwaki and Hirata, 1992). Organic anions can complex metal cations, which bind phosphates (Al^{3+}, Fe^{3+}, Ca^{2+}), or displace phosphate from the soil matrix (Jones, 1998). Where carboxylate exudation is associated with proton extrusion, the lower pH may itself contribute to greater P availability, if the soil pH is relatively high. However, cotransport of protons is certainly not associated with organic anion release in all species in all circumstances (Roelofs et al., 2001, Ryan et al., 2001). The effectiveness of carboxylates depends on their number of carboxyl groups and molecular structure. Tricarboxylates (citrate) are generally more effective

* FAX No: +61-8-93801108.
E-mail: evenekla@cyllene.uwa.edu.au

than dicarboxylates (e.g., malate, malonate) due to stronger ligand binding. Soil properties also have large effects on the effectivity of carboxylates. For example, the stability of organic anion – metal complexes depends strongly on pH (Jones, 1998), and there is large variation in the mechanisms and strength with which phosphate is held in the soil.

Root exudate patterns may need to be different on different soils for optimal effects on P availability. It is known that species differ in the carboxylates they exude, e.g., predominantly citrate in white lupin, malonate in chickpea, malate in wheat, a wide range of carboxylates in Proteaceae species (Ryan et al., 2001; Roelofs et al., 2001), but it is not clear if that affects the suitability of these species for certain soils or, more specifically, their ability to mobilise soil P. Knowledge about variation in exudation in response to soil P status is virtually limited to observations that lupin root exudation rates are highest at low P availability (Keerthisinghe et al., 1998). However, Lambers et al. (2002) found that *Banksia grandis*, an Australian Proteaceae species, modified its exudation pattern when grown on sand containing either poorly soluble Al-phosphate or Fe-phosphate as the only P source, and grew equally well on either P source. Studies of Al^{3+} resistance have shown clearly that plants respond directly to the presence of Al^{3+} by releasing organic anions, and that genotypes of wheat, maize and soybean vary enormously in this trait (Ryan et al., 2001). If such genetic variation were associated with the ability to respond to P deficiency by producing the appropriate root exudates to enhance P availability, this could be exploited to improve the efficiency of P fertiliser use.

The present paper aims at assessing soil-induced variation in rhizosphere carboxylates of chickpea and lupin, and correlating this with P uptake. Chickpea is a promising and rapidly expanding new crop in Western Australia (Siddique and Sykes, 1997), where it is grown in rotation with wheat. Western Australian soils generally have very low levels of P, and much P fertiliser is ineffective due to high P sorption capacities (Bolland et al., 1994; Brennan et al., 1994). Although iron oxides are often present in high concentrations, the sorption of P is most closely associated with aluminium (oxides and organic complexes; Gilkes and Hughes, 1994). We report the results of two pot experiments, in which we determined plant growth and P uptake as well as rhizosphere chemistry. In the first experiment, focusing on soil-induced variation in root exudates, we grew one chickpea cultivar on 11 soils, covering a range of Western Australian soil types considered suitable for chickpea. In the second experiment, we used six of these soils but grew three chickpea cultivars as well as white lupin and wheat, to establish inherent and soil-induced variation in root exudates amongst cultivars and species.

Material and methods

Experiment 1

Chickpea (*Cicer arietinum* L.) cultivar 'Sona' was grown in a pot experiment in a greenhouse in 11 different soils. Sowing date was the 12th of August 2000.

Experiment 2

Three cultivars of chickpea, and one cultivar of white lupin (*Lupinus albus* L.) and wheat (*Triticum aestivum* L.) were grown in pots in the greenhouse in six different soils. The chickpea cultivars were 'Sona', 'Tyson' (both desi type, small-seeded, chickpea) and 'Kaniva' (kabuli type, large-seeded, chickpea). The white lupin cultivar was 'Kiev Mutant', and the wheat cultivar was 'Westonia'. Sowing date was the 31st of August 2000.

Soils

The soils used in the experiments were sampled at 11 locations in the medium-rainfall area (mean rainfall about 400 mm) of southwestern Australia. Topsoil was collected from unfertilised land, and passed through a 4 mm sieve. The soils were loamy to clayey, and all were low in bicarbonate-extractable P (<15 μg g^{-1}).

Growing of plants

Measured amounts of soil (3.5–4.0 kg, depending on bulk density) were placed in deep ($150\times150\times250$ mm) black plastic bags. The bags were perforated for drainage and aeration. Moisture content at field capacity was determined using a pressure plate, and pots were held at around 50% field capacity during the experiment by watering three times per week and weighing regularly.

Nutrients were added as a nutrient solution at a rate increasing from 50 ml (start of experiment) to 200 ml per week (end of experiment). The nutrient solution contained (μM): $Ca(NO_3)_2$, 200; K_2SO_4, 100;

$MgSO_4$, 54; H_3BO_3, 2.4; $MnSO_4$, 0.24; $ZnSO_4$, 0.1; Na_2MoO_4, 0.03; and $CuSO_4$, 0.02.

Four seeds were sown at 3 cm depth, and thinned to two per pot after establishment. Seeds were not inoculated with rhizobia as varying degrees of inoculation might have introduced an undesirable variation in plant growth. We chose to supply the plants with sufficient nitrate-N. No root nodules were observed on the chickpea plants, however, some were present on the lupin roots.

For each genotype and soil type there were four replicate pots. For each soil type, there were four unplanted control pots. Replicate pots were located on different glasshouse benches, and distributed randomly within the bench. The experiment was carried out in a glasshouse in the spring of 2000 (August–November) with temperatures between 10 and 25 °C (min-max).

Plant measurements

We harvested plants at two developmental stages, but show results for the second harvest only. At that time, chickpea had its first flowers and growing pods, while white lupin inflorescences and wheat ears were developing. Results for plant P concentrations and rhizosphere chemistry were very similar for the first harvest. Harvest days were, for Experiment 1 (chickpea), 51 days after sowing (DAS); for Experiment 2, for chickpea 63–67 DAS; for white lupin 68–69 DAS; for wheat 55–61 DAS.

Shoots were cut at the soil surface. Roots with adhering rhizosphere soil were processed for rhizosphere extracts (see below), whereafter they were washed. No attempt was made to separate the roots of the two plants that were growing in each pot. Representative subsamples of the root system were subjected to image analysis to determine root length and root diameter distributions (WinRhizo system, Regent Instruments, Quebec, Canada). For white lupin, the root system was separated in non-cluster and cluster roots, which were analysed separately. Checks confirmed the ability of the scanner to measure these thin roots. Finally, dry mass of all plant fractions was determined after drying at 70 °C. Whole shoots and roots of each replicate were ground in a stainless steel ball mill for P analysis. Two-hundred and fifty mg of each sample was digested in a nitric acid / perchloric acid mixture. The P content of the digest was determined using the molybdo-vanado-phosphate method (Kitson and Mellon, 1944).

Rhizosphere chemistry

At the time of harvest, the plastic bags containing soil plus root system were cut open and placed on the bench. Root systems were held by the stem base and gently lifted out of the soil. Roots were carefully shaken to remove excess soil, and clumps of soil trapped between roots were taken out. The soil that remained adhering to the roots was defined as rhizosphere soil. Despite the care taken, some root breakage was unavoidable, yet these were included with the rest of the root system for further processing. The root system was transferred to a beaker of suitable size, in which the rhizosphere soil was washed off the roots by gently shaking the beaker after adding a measured amount of 0.2 m*M* $CaCl_2$ solution. $CaCl_2$ was used to ensure membrane integrity. The roots were then taken out of the soil extract and processed further for diameter, length and mass determination (see above). The mass of rhizosphere soil was determined by weighing, taking into account the amount of solution added. A subsample of the extract was filtered through 0.22 μm Pal Gelman Acrodisc® syringe filters into 1 mL HPLC vials and frozen. The rest of the rhizosphere soil extract was frozen. The amount of rhizosphere soil sampled by our method was 39 ± 4 g soil g^{-1} root (both on a dry mass basis) for chickpea in Experiment 1. The amount of solution used to extract the rhizosphere was 1.3 ± 0.1 ml g^{-1} soil. Similar amounts of soil were sampled from the bulk soil of the pots in which the plants had grown, and from control pots in which no plants had grown. The pH of the soil extracts was measured at a later date at room temperature.

Soil analyses

General soil analyses were performed by CSBP Futurefarm analytical laboratories (Bibra Lake, Australia). Soil samples were air-dried and passed through a 2-mm sieve. For total P, soils were digested by a Kjeldahl procedure (sulfuric acid-potassium sulfate-copper sulfate). Bicarbonate-extractable P was determined after extraction with 0.5 *M* sodium bicarbonate at pH 8.5 (Colwell, 1963). Phosphorus Retention Index, PRI, was estimated as the ratio between sorbed P and solution P after equilibration of 2.5 g soil with 50 mL of 0.02 *M* KCl containing 10 mg P L^{-1} (Allen and Jeffery, 1990; Bolland et al., 1994). P in the extracts/digests was determined colourimetrically using the phosphomolybdenum blue method (Murphy and

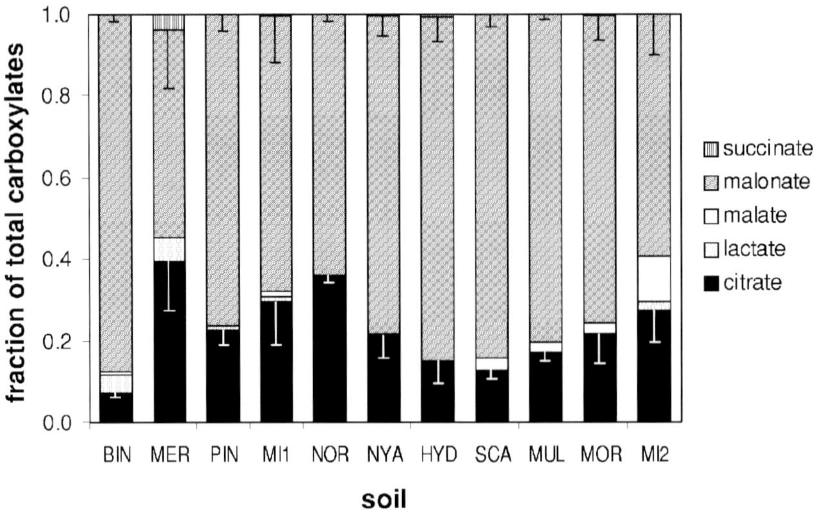

Figure 3. Composition of carboxylates (fraction of total, extracted using a 0.2 m*M* $CaCl_2$ solution) in the rhizosphere of chickpea cultivar 'Sona' grown in different Western Australian soils. Standard errors are shown for the two major carboxylates, malonate and citrate.

In the absence of correlations between P uptake and common indices of P availability, it is pertinent to look for plant traits that may have affected the ability to acquire P from these soils. The variation in P uptake was not associated with variation in the size, length or diameter of roots. Root mass represented, on average, 48% of total plant mass, and was not correlated with growth or plant P content. Specific root length (m root per g root dry mass, Figure 1) was very similar in all soils, and mean root diameter only varied between 0.71 and 0.85 mm.

In contrast to the limited morphological plasticity, the variation in the chemistry of the rhizosphere was very large. The soil around the roots had a pH that was usually lower, but sometimes higher, than that of the bulk soil (Table 1), and contained high concentrations of $CaCl_2$-extractable organic anions (up to 4.7 μmol g^{-1} dry soil). Comparing soils, these concentrations varied by an order of magnitude (Figure 2). This was evident, irrespective of the way the concentrations were expressed: soil-based and root-based concentrations were well-correlated ($r^2 = 0.78$) because variability in the sampled amounts of rhizosphere soil per root mass was small. Carboxylate composition varied as well, with malonate representing 56–86% of the total, citrate 8–37%, and malate 0–5%, plus small amounts of succinate and lactate (Figure 3). No measurable amounts of carboxylates were found in the bulk soil surrounding the rhizosphere soil, or in control pots without plants.

Given the great variability in rhizosphere chemistry, it is important to explore the possible correlation of plant P uptake with the plant-modified soil properties, rather than with the bulk soil properties. Plants did indeed acquire more P when the concentration of carboxylates in the rhizosphere was higher (Figure 4). This relationship further depended on the capacity of soils to sorb P. At a given rhizosphere concentration of carboxylates, plants took up more P from low-sorbing soils (Phosphorus Retention Index <10) than from higher-sorbing soils (PRI >10) (Figure 4). Multiple regression analysis showed an r^2 of 0.32 for plant P content and rhizosphere carboxylate concentration (μmol g^{-1} root dry mass) alone, while inclusion of PRI raised the r^2 to 0.85 ($P = 0.002$ and 0.01 for carboxylate concentration and PRI, respectively). Expression of rhizosphere carboxylate concentration in terms of μmol g^{-1} dry soil, gave very similar results (data not shown). Bulk soil pH or rhizosphere pH did not improve any of the regressions. For these regressions one soil was omitted from the analysis. This soil (Mingenew 2, Table 1) had a PRI >10 and plants took up larger amounts of P than expected; it was also rather exceptional in having the highest pH and extractable Ca, suggesting that P may be held by a different mechanism.

The positive effect of carboxylates on P availability is supported by results of soil extractions. The amounts of P extracted by 0.5 m*M* solutions of malate, malonate and citrate were (averaged for 11 soils, and

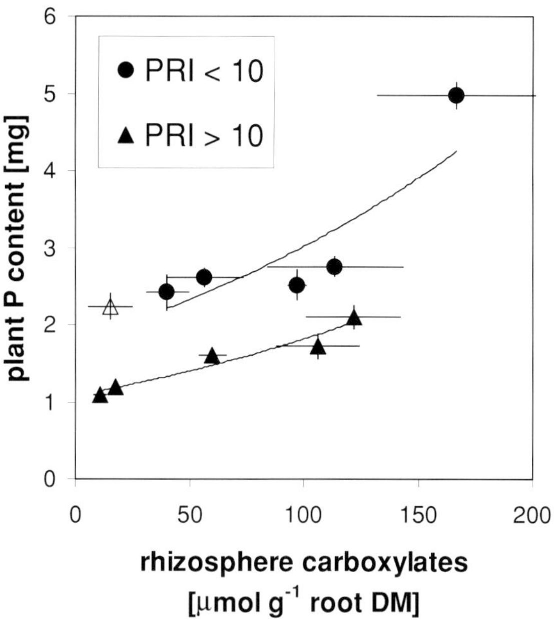

Figure 4. Relationship between the total amount of phosphorus (mg P per two plants in pot) and the concentration of carboxylates (μmol total carboxylates per g root dry mass, extracted using a 0.2 m*M* $CaCl_2$ solution) in the rhizosphere of chickpea cultivar 'Sona' grown in different Western Australian soils. Soils represented by circles had a Phosphorus Retention Index (PRI) smaller than 10; soils represented by triangles had a PRI greater than 10. The exceptional case of the open triangle is discussed in the text. Curves fitted by eye.

compared with water extractions) 111, 134 and 216%, respectively.

Comparison of chickpea cultivars, white lupin and wheat

Growth and P uptake of chickpea cultivars 'Sona', 'Tyson' and 'Kaniva', as well as that of white lupin and wheat, were greatly affected by soil type. The three chickpea cultivars showed very similar responses to the soils, but the responses of both white lupin and wheat were different from those of the chickpea cultivars. Plant P concentrations were generally low. They varied between 0.7 and 1.6 mg g^{-1} in chickpea, 0.7 and 1.3 mg g^{-1} in lupin, and 0.7 and 2.5 mg g^{-1} in wheat. Low P concentrations were invariably associated with poor growth. Plant mass at the end of the experiment varied between 0.8 and 3.3 g in chickpea, 2.0 and 5.2 g in lupin, and 0.6 and 1.7 g in wheat.

Similar to the previous experiment, morphological plasticity of the root system was small. Mean root diameter varied in most cases by less than 5% across soils, with mean values of 0.66 mm for the chickpeas, 0.66 mm for lupin non-cluster roots, 0.39 mm for lupin cluster roots, and 0.22 mm for wheat roots. Cluster roots represented 21–43% of the total root mass in white lupin, but this variation was not correlated with plant P status or any soil property.

Amounts and composition of carboxylates in the rhizosphere showed large differences amongst species and soils (Figure 5). Wheat had low rhizosphere concentrations of $CaCl_2$-extractable carboxylates (mainly lactate and malate), whereas white lupin and chickpea had high concentrations. Malonate was the dominant carboxylate in chickpea root exudates, followed by citrate and malate. Citrate plus malate made up 95–98% of the total amount of carboxylates in lupin, but their relative contributions varied between 10 and 90% in different soils (Figure 5). Interestingly, the proportion of malate was higher in soils with low pH (pH<5.5), whereas the proportion of citrate was higher in soils with higher pH values (Figure 6a). Although chickpea exuded much smaller amounts of these compounds, a similar shift in the relative contributions of malate and citrate was apparent (Figure 6b). No carboxylates were detected in bulk soil samples, with the exception of small amounts of citrate in lupin pots.

Discussion

Our results show that rhizosphere chemistry was strongly and differentially modified by plant roots of chickpea and lupin, and that this affected P availability to the plant. In the following discussion we will address the complexity of the interactions between plant roots and soils in the context of our findings.

Rhizosphere chemistry, rather than root morphology, showed the greatest response to variation in soil properties. Despite the fact that P availability was low and variable amongst the soils (Table 1), and despite the demonstrated value of thinner and longer roots in low-P soils (Lambers et al., 1998), none of the species studied responded to different soils by changing mean root diameter (Figure 1). Lupin was the only species that demonstrated some plasticity in root morphology, by changing the proportion of cluster roots. As cluster roots have rootlets that are finer than the laterals of non-cluster roots, mean root diameter and length were affected, but this did not lead to a greater exploration of soil volume, as these rootlets are very closely spaced. Comparing species in Experiment 2, wheat had the smallest mean root diameter, and therefore

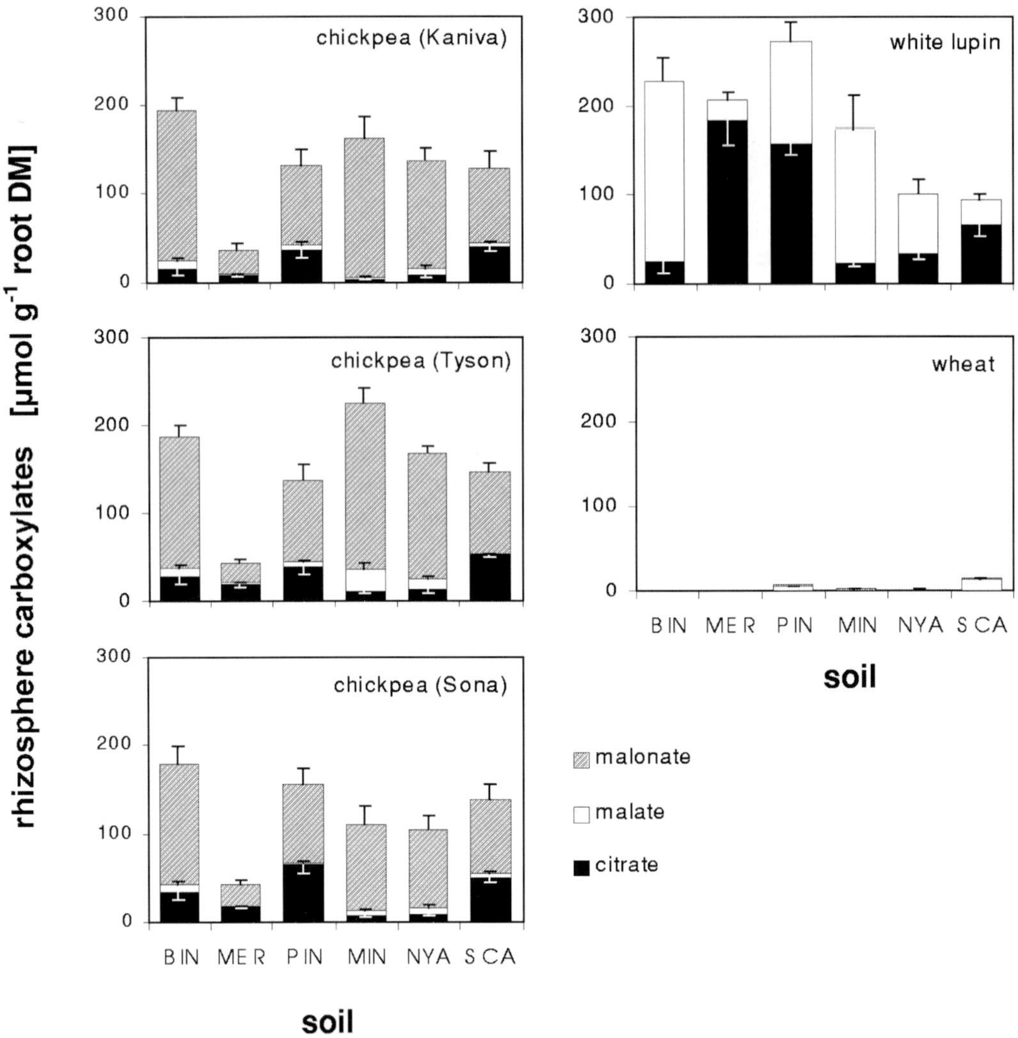

Figure 5. Carboxylates (μmol per g root dry mass, extracted using a 0.2 m*M* $CaCl_2$ solution) in the rhizosphere of three chickpea cultivars ('Kaniva', 'Tyson' and 'Sona'), white lupin and wheat, grown in six different Western Australian soils. Bars represent standard errors.

many long roots, which explored a large soil volume. This probably enabled it to access reasonable amounts of readily available P, lacking the root exudates of chickpea and lupin (cf. Watt and Evans, 2002). Indeed, shoot (but not root) concentrations of P were highest in wheat for five out of six soils, and relative growth rates and P uptake rates (per unit mass or length) were generally higher for wheat than for chickpea and lupin (data not shown).

Amounts and composition of carboxylates in the rhizosphere varied widely in both chickpea and lupin (Figures 2, 5). This is an important observation as no other study has shown direct comparisons of variation in root exudates of a species across a range of field-sampled soils. It must be acknowledged that this approach to sampling the rhizosphere provides an instantaneous picture of the carboxylates in the rhizosphere, which is the net result of the complex interactions between their exudation by roots, their sorption to soil particles, and their degradation through microbial activity. Although differences in these processes may have contributed to some variability in measured carboxylate concentrations, this is very unlikely to be responsible for the very large differences between soils. In our soil extractions using a low concentration (equivalent to 5 μmol g^{-1} soil) of malate, malonate or citrate, sorption of these carboxylates was, on average, 14, 23 and 21%, varying little across soils (SE 2.0,

2.1, 2.3; $n = 11$; data not shown). The degree of sorption was not correlated with soil pH. We cannot estimate the effect of microbial activity on extracted amounts of carboxylates. If microbial activities were high and inherently different between soils, quantitative differences in carboxylates across soils would be similar for chickpea and lupin which was not the case (Figure 5). However, microbial activities may be differentially influenced by the interaction between roots and soil properties, e.g., as the outcome of differences in pH, carboxylates, sugars, amino acids, phenolic compounds, enzymes, mucilage, etc.

The possibility exists that a small amount of the measured carboxylates originated from damaged root tissues, due to handling of the roots during extraction. For a number of reasons, this is unlikely to have had a large effect on our data. Firstly, we found very little carboxylates associated with wheat roots, which were treated in the same way. Secondly, we never detected succinate in the chickpea rhizosphere, even though it is present in the root tissue in reasonable concentrations (Ohwaki and Hirata, 1992). Thirdly, the presence of some citrate in the bulk soil of lupin, which had never been in contact with roots, must be due to high rhizosphere concentrations.

The proportion of carboxylates recovered by our extraction method probably differed somewhat between soils, because there was some variation in the amount of soil adhering to roots. The average amount of 39±4 g dry soil per g root dry mass (chickpea in Experiment 1) represented an estimated thickness of the rhizosphere soil sheath of 0.25±0.02 mm around the roots (assuming a cylindrical rhizosphere around average-diameter roots). Considerable variation in the thickness of the rhizosphere may be expected for roots of different diameter and age. Variation in the amount of rhizosphere soil was not correlated with carboxylate concentrations, and no measurable amounts of carboxylates were detected in non-rhizosphere soil. The possibly incomplete sampling of the rhizosphere can therefore not explain the large differences in rhizosphere carboxylate concentrations in different soils.

Across the six soils in Experiment 2, mean total $CaCl_2$-extractable carboxylate concentration in chickpea rhizosphere was 174 μmol g^{-1} root, or 8.4 μmol g^{-1} soil, while for lupin it was 185 μmol g^{-1} root, or 5.8 μmol g^{-1} soil. At a typical soil water content of 10% (0.1 ml g^{-1} dry soil), the estimated total carboxylate concentration in the soil solution around chickpea and lupin roots was therefore 84 and 58 m*M*, respectively. For lupin in calcareous soil, Dinkelaker et al. (1989) found citrate concentrations as high as 48 μmol g^{-1} soil in the rhizocylinder of cluster roots. In that case, however, citrate accumulated as Ca-citrate crystals. It should also be stressed that we sampled whole root systems, rather than the most actively exuding parts, leading to lower average concentrations. For chickpea, no estimates of rhizosphere carboxylates are known to us, but from hydroponic studies it is clear that the species exudes similar amounts as lupin, and greater amounts than many other leguminous species (Neumann and Römheld, 1999; Ohwaki and Hirata, 1992). The observed rhizosphere concentrations are sufficiently high to have a significant effect on P solubility (Gerke et al., 2000), particularly if one considers that the estimates include older, less active roots.

The differences between root exudates of chickpea, white lupin and wheat are in accordance with other studies of these species, showing a predominance of malonate in chickpea and of citrate and malate in lupin, with wheat exuding very little of any of these compounds (Neumann and Römheld, 1999; Ohwaki and Hirata, 1992). The similarity in the responses of exudation patterns to different soil types by different chickpea cultivars (Figure 5) has not been reported before. This observation confirms our unpublished data for eight chickpea cultivars (four desi type and four kabuli type) grown in river sand. We conclude that there is little genetic variation in root exudation amongst commonly used chickpea cultivars.

The strikingly large differences in amounts of root exudates of chickpea (Figures 2, 5) and lupin (Figure 5) on different soils are not correlated with any single soil property tested by us. Furthermore, key plant traits like tissue P concentrations or growth rate do not provide clues as to what factors regulate root exudation rates. Potentially important plant and soil factors may need to be more precisely defined, e.g., soil solution P concentration near root tips, various pools of P in tissues, cation-anion balance. Similarly, regulation of changes in the composition of root exudates is poorly understood. Our results suggest that soil pH may be a factor in determining the relative proportions of citrate and malate (Figure 6). Since lupin and chickpea showed a similar pattern, a single biochemical mechanism may be responsible. An apparently direct effect of a soil factor on root exudate pattern was also demonstrated in *Banksia grandis*, a Western Australian Proteaceae (Lambers et al., 2002). *B. grandis* that were supplied with Al-phosphate ex-

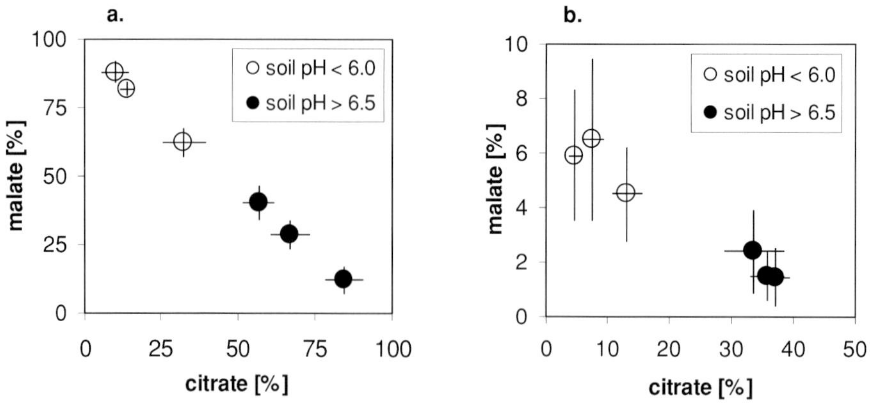

Figure 6. The relative amounts of citrate and malate (% of total carboxylates, extracted using a 0.2 m*M* $CaCl_2$ solution) in the rhizosphere of: (a) white lupin; (b) chickpea, grown in six Western Australian soils. The pH was measured in the bulk soil at the end of the experiment (0.2 m*M* $CaCl_2$ extracts). The chickpea data are means for three cultivars. Bars represent standard errors.

uded di- and tricarboxylates, whereas those supplied with Fe-phosphate exuded these same compounds plus significant amounts of monocarboxylates, while the P status of the plant was unaffected. To elucidate the regulation of exudation rates and composition in crops like lupin and chickpea, experimental variation of soil factors like P pools, Al, Fe or pH is needed.

The fact that chickpea plants with greater rhizosphere carboxylate concentrations took up more P (Figure 4) does not necessarily imply causality. However, given the demonstrated ability of solutions of citrate, malonate and malate to mobilise P in these soils, it is very likely that the exudates enhanced P uptake rather than the reverse. Alternative explanations, like that linking anion exudation with the maintenance of plant cation-anion balance (Tang et al., 1998), need to be examined. Also, the question remains why, if root exudates are functional in enhancing P uptake, exudation seems to be less abundant in plants with a low P status (cf. Figure 4 and the observation that growth and plant P concentration were correlated). It is possible that root exudation rates are low under severe P stress, maximum under moderate stress and low again at high P availability. Soil acidification under chickpea was found to have such a maximum at intermediate P availability (20 μg P g^{-1} soil) in a study by Tang et al. (1998). The extrusion of protons and organic anions is, however, biochemically separated (Jones, 1998; Tang et al., 1998), and trends in carboxylate exudation are therefore not necessarily the same as trends in acidification. Experimental determination of the response of exudation rates to different levels of P in the same soil is required to address this issue.

The amounts of carboxylates exuded by chickpea and white lupin, and the demonstrated ability of these compounds to solubilise P in soils, suggest that these species are of great potential value as crops in soils with relatively large amounts of sorbed P. Our experiments showed a large variation in rhizosphere carboxylates, which could not be explained by any of the measured soil properties. Utilisation of root exudate-producing crops to enhance soil P use will require the identification of those physiological or soil chemical factors that stimulate or inhibit the exudation of organic anions.

Acknowledgements

This research was supported by a grant from the Grains Research Committee of Western Australia. CSBP Futurefarm provided soil analyses. Mike Bolland, Mike Baker and K.H.M. Siddique (Department of Agriculture, Western Australia) were very helpful with soil, seeds and advice. Koen Bröker, Elsa Martínez-Ferri, Chris Szota, Claude Midgley and Mohammad Nuruzzaman assisted in the laboratory. We thank three anonymous reviewers for valuable feedback, as well as the participants of the Perth workshop on cluster roots and phosphorus acquisition.

References

Ae N, Arihara J, Okada K, Yoshihara T and Johansen C 1990 Phosphorus uptake by pigeon pea and its role in cropping systems of the Indian subcontinent. Science 248, 477–480.

Allen D G and Jeffery R C 1990 Methods for Analysis of Phosphorus in Western Australian Soils. Report on Investigation no. 37. Chemistry Centre of Western Australia, Perth.

Bolland M D A, Wilson I R and Allen D G 1994 Effect of P buffer capacity and P retention index of soils on soil test P, soil test P calibrations and yield response curvature. Aust. J. Soil Res. 32, 503–517.

Brennan R F, Bolland M D A, Feffery R C and Allen D G 1994 Phosphorus adsorption by a range of Western Australian soils related to soil properties. Comm. Soil Sci. Plant Anal. 25, 2785–2795.

Colwell J D 1963. The estimation of phosphorus fertilizer requirements of wheat in southern New South Wales by soil analysis. Aust. J. Exp. Agric. Anim. Husb. 3, 190–198.

Dinkelaker B, Römheld V and Marschner H 1989 Citric acid excretion and precipitation of calcium citrate in the rhizosphere of white lupin (*Lupinus albus* L.). Plant Cell Environ. 12, 285–292.

Gardner W K, Barber D A and Parberry D G 1983 The acquisition of phosphorus by *Lupinus albus* L. III. The probable mechanism by which phosphorus movement in the soil / root interface is enhanced. Plant Soil 70, 107–124.

Gerke J, Römer W and Jungk A 1994 The excretion of citric and malic acid by proteoid roots of *Lupinus albus* L.; effect on soil solution concentrations of phosphate, iron, and aluminum in the proteoid rhizosphere in samples of an oxisol and a luvisol. Z. Pflanzenernähr. Bodenk. 157, 289–294.

Gerke J, Beissner L and Römer W 2000 The quantitative effect of chemical phosphate mobilization by carboxylate anions on P uptake by a single root. I. The basic concept and determination of soil parameters. J. Plant Nutr. Soil Sc. 163, 207–212.

Gilkes R J and Hughes J C 1994 Sodium fluoride pH of south-western Australian soils as an indicator of P-sorption. Aust. J. Soil Res. 32, 755–766.

Hocking P J, Keerthisinghe G, Smith F W and Randall P J 1997 Comparison of the ability of different crop species to access poorly-available soil phosphorus. *In* Plant Nutrition for Sustainable Food Production and Environment. Proc. 13th Int. Plant Nutrition Colloq. Eds. T Ando, K Fujita, T Mae, H Matsumoto, S Mori and J Sekiya. pp. 305–308. Kluwer Academic Publishers, Dordrecht.

Jones D L 1998 Organic acids in the rhizosphere – a critical review. PlantSoil 205, 25–44.

Keerthisinghe G, Hocking P J, Ryan P R and Delhaize E 1998 Effect of phosphorus supply on the formation of proteoid roots of white lupin (*Lupinus albus* L.). Plant Cell Environ. 21, 467–478.

Kitson R E and Mellon M G 1944 Colorimetric determination of phosphorus as molybdovanadophosphoric acid. Ind. Eng. Chem. Anal. Ed. 16, 379.

Lambers H, Chapin F S III and Pons T L 1998 Plant Physiological Ecology. Springer-Verlag, New York.

Lambers H, Juniper D, Cawthray G R, Veneklaas E J and Martínez-Ferri E (2002) The pattern of carboxylate exudation in *Banksia grandis* (Proteaceae) is affected by the form of phosphate added to the soil. Plant Soil 238, 111–122.

Motomizu S, Wakimoto T and Toei K 1983 Spectrophotometric determination of phosphate in river waters with molybdate blue and malachite green. Analyst 108, 361–367.

Murphy J and Riley J P 1962 A modified single solution method for the determination of phosphorus in natural waters. Anal. Chem. Acta. 27, 3106.

Neumann G and Römheld V 1999 Root excretion of carboxylic acids and protons in phosphorus-deficient plants. Plant Soil. 211, 121–130.

Ohwaki Y and Hirata H 1992 Differences in carboxylic acid exudation among P-starved leguminous crops in relation to carboxylic acid contents in plant tissues and phospholipid level in roots. Soil Sci. Plant Nutr. 38, 235–243.

Parfitt R L and Childs C W 1988 Estimation of forms of Fe and Al: A review, and analysis of contrasting soils by dissolution and Moessbauer methods. Aust. J. Soil Res. 26, 121–144.

Roelofs R F R, Rengel Z, Cawthray G R, Dixon K W and Lambers H 2001 Exudation of carboxylates in Australian Proteaceae: chemical composition. Plant Cell Environ. 24, 891–904.

Ryan P R, Delhaize E and Jones D L 2001 Function and mechanism of organic anion exudation from plant roots. Annu. Rev. Plant Physiol. Plant Mol. Biol. 52, 527–560.

Siddique K H M and Sykes J 1997 Pulse production in Australia past, present and future. Aust. J. Exp. Agric. 37, 103–111.

Tang C, Fang R Y and Raphael C 1998 Factors affecting soil acidification under legumes. II. Effect of phosphorus supply. Aust. J. Agric. Res. 49, 657–664.

Watt M and Evans J R (2002) Phosphorus acquisition from soil by white lupin and soybean – species with contrasting root development. Plant Soil 248, 271–283.

Plant and Soil **248:** 199–206, 2003.
© 2003 *Kluwer Academic Publishers. Printed in the Netherlands.*

Role of phosphorus nutrition in development of cluster roots and release of carboxylates in soil-grown *Lupinus albus*

J. Shen[1,2], Z. Rengel[2,3], C. Tang[2] & F. Zhang[1]
[1]*Laboratory of Plant Nutrition, China Agricultural University, Beijing 100094, P. R. China.* [2]*Soil Science and Plant Nutrition, The University of Western Australia, 35 Stirling Hwy, Crawley WA 6009, Australia.* [3]*Corresponding author**

Received 24 April 2002. Accepted 26 April 2002.

Key words: citrate exudation, cluster root formation, *Lupinus albus*, P limitation, root growth, soil culture

Abstract

The present study examined the effect of phosphorus (P) limitation on cluster root formation and exudation of carboxylates by N_2-fixing white lupin (*Lupinus albus* L. cv. Kiev) grown in a P-deficient sandy soil. Plants received 10 (limited P) or 200 μg P g^{-1} soil as $FePO_4$ (adequate P) and were grown in a phytotron at 20/12 °C (12/12 h) for 76 days in soil columns. Cluster root formation was assessed and root exudates were collected at 9-day intervals. Shoot and root dry weights were higher in plants grown in the adequate-P compared to the limited-P treatment for 67 days. No clear difference in the total root length was observed between two P treatments before day 58. However, the specific root length increased rapidly from 17 m g^{-1} DW at day 40 to 28 m g^{-1} at day 49 in the P-limited plants, but decreased in the P-adequate plants. The effect of P limitation on enhancement of cluster root formation was observed from day 40 and reached the maximum at day 58. The number of cluster roots was negatively correlated with the P concentration in both roots and shoots. Phosphorus limitation increased exudation of citrate from day 40. The exudation of citrate displayed a cyclic pattern throughout the experiment, and appeared related to internal P concentration in plants, particularly P concentration in shoots. The sorption of exogenously added citrate in the soil was also examined. The amount of extractable citrate remained unchanged for 2 h, but decreased thereafter, suggesting that the soil had a low capacity to sorb citrate, and the rate of its decomposition by microorganisms was slow. Collecting solution leached through a soil column is a simple and reliable method to acquire root exudates from white lupin grown in soil. The results suggest that formation of cluster roots and exudation of citrate in white lupin are regulated by P concentration in shoots.

Introduction

White lupin is considered a model plant for studying cluster root formation and root exudation. Under P deficiency, white lupin increases P acquisition through formation of cluster roots (may represent more than 60% of the total root biomass, Keerthisinghe et al., 1998) and release of a large amount of citrate from them (Gardner et al., 1981, 1983; Johnson et al., 1994, 1996a; Marschner, 1995; Neumann et al., 1999), representing as much as 11 – 23% of the total plant dry weight (Dinkelaker et al., 1989). Exudation of citrate may mobilize sparingly soluble Fe, Al and Ca phosphates (Dinkelaker et al., 1989; Marschner, 1995) through anion exchange, chelation of bridging metal cations from organic matter in the soil, and by suppressing re-adsorption and precipitation of inorganic P (Gerke et al., 1994).

Only limited information is available on the regulation of cluster root formation and root exudation. Foliar P application suppressed cluster root formation (Marschner et al., 1987). Moreover, high internal P concentration decreased the formation of cluster roots and exudation of citrate (De Vos et al., 2001; Gilbert et al., 1998a; Keerthisinghe et al., 1998).

* FAX No.: +61-8-9380-1050.
E-mail: zrengel@agric.uwa.edu.au.

Table 1. Some properties of the Lancelin soil used in the study

Properties	Values	Properties	Values
Colour	Brown–grey	Nitrate nitrogen ($\mu g\ g^{-1}$)	2
Texture	sand	Ammonium nitrogen ($\mu g\ g^{-1}$)	1
Organic carbon (mg g^{-1})	8.4	Phosphorus (Colwell) ($\mu g\ g^{-1}$)	5
Conductivity (1:5 H_2O) (S m^{-1})	0.0022	Potassium (Colwell) ($\mu g\ g^{-1}$)	21
pH ($CaCl_2$)	5.00	Sulphur ($\mu g\ g^{-1}$)	1
pH (H_2O)	5.70	Reactive iron ($\mu g\ g^{-1}$)	173

The effect of P deficiency on cluster root formation and citrate exudation has been studied frequently in white lupin (e.g., Dinkelaker et al., 1989; Gardner et al. 1983; Gerke et al., 1994), but the great majority of studies were conducted in solution culture (Keerthisinghe et al., 1998; Neumann et al., 1999; Neumann and Römheld, 1999; Sas et al., 2001; Watt and Evans, 1999b) and silica sand culture (Johnson et al., 1996a, b). Under soil conditions, some progress has been made in collecting root exudates by (i) resin-agar (Kamh et al., 1999), (ii) extracting root exudates from the rhizosphere soil (i.e., soil adhering to the roots after removal from pots and gentle shaking) (Lynch, 1990), (iii) using a rhizobox system that separated the soil into 0.5-mm thick layers starting from the seed (Li et al., 1997), or (iv) collecting the leachate (Gardner et al., 1983; Johnson et al., 1994). In contrast, little information is available on the effect of P limitation on cluster root formation and citrate exudation over time by the whole root system of white lupin grown in soil with minimum disturbance to root growth. Hence, the objective of the present study was to characterise the effect of P limitation on cluster root formation and exudation of carboxylates in soil-grown N_2-fixing white lupin in different growth stages.

Materials and methods

Soil preparation

Virgin brown sand (Uc4.22, Northcote 1971) was collected from a bushland site 15 km south-east of Lancelin, WA (31.56 S, 115.20 E). Any discernible root pieces were removed and the samples were air-dried for 48 h at 25 °C, sieved (2 mm) and thoroughly mixed. Major soil properties (cf. Colwell 1963) are listed in Table 1.

Plant cultivation

The experiment was conducted in a phytotron with natural light, controlled temperature 20/12 °C (12/12 h) and relative humidity 75–85%. Seeds of white lupin (*Lupinus albus* L. cv. Kiev) were germinated on a stainless-steel mesh in a 5-L plastic container containing aerated solution of 1 mM $CaCl_2$ and 5 μM H_3BO_3 at 20 °C in dark for 24 h. After the radicle emerged, seven seeds of uniform size were sown in each cylindrical plastic pot (diameter$\times$height = 86$\times$380 mm) filled with 3 kg air-dried soil. After germination, plants were thinned to four per pot and were inoculated with 1 mL water suspension of *Bradyrhizobium* sp. (*Lupinus*) WU 425 (containing about 10^5 cells) by adding the solution along the seedling stem. To avoid complications in the relationship between the root exudation of acidifying agents and the excess cation uptake under different N supply, the plants were reliant only on N_2 fixation for N. The soil moisture was maintained at 90% of field capacity by watering to weight with deionised water once or twice daily as required. The average evaporative loss at 52 day after sowing was 13–14 g water h^{-1} for the P-adequate treatment and 9–10 g water h^{-1} for the P-limited treatment. The pots were completely randomised and re-positioned weekly to minimise any effect of uneven environmental factors.

P treatments

Phosphorus was added to the soil as $FePO_4$ at a rate of 10 (limiting P) or 200 (adequate P) μg P g^{-1} and mixed in the soil. To ensure the supply of other nutrients, soil was also fertilized with basal nutrients at the following rates (mg pot^{-1}): K_2SO_4, 400; $MnSO_4{\cdot}H_2O$, 20; $ZnSO_4{\cdot}7H_2O$, 30; $CuSO_4{\cdot}5H_2O$, 6; H_3BO_3, 2; $Na_2MoO_4{\cdot}5H_2O$, 0.5; $CaCl_2{\cdot}2H_2O$ 500; $MgSO_4{\cdot}7H_2O$, 130; $Co_2SO_4{\cdot}7H_2O$, 1.

Collection of root exudates

Prior to collection of root exudates, soil columns were leached and the eluate collected to determine the water volume retained in soil matrix and to wash out the rhizosphere products from soil. The gravimetric analysis showed that 300 mL of solution was retained within the soil matrix.

Root exudates from the whole root system were collected on days 40, 49, 58, 67 and 76 after sowing. Root exudates were collected through a hole in the center of conically shaped bottom of the pot by percolating the soil with deionised water for 2 h (from 10:00 to 12:00 in the morning). An aliquot of 600 mL deionised water was percolated through the soil column three times. A slight suction (0.02–0.05 MPa) was applied to enhance eluate collection. The eluted solution contained water-soluble root exudates and was referred hereafter as the root exudate. The volume of eluted solution was recorded. Given that 300 mL of solution was retained in the soil matrix, the total volume for calculating carboxylate concentrations was 900 mL. Micropur (Sicheres Trinkwasser, Germany) at 0.01 g L^{-1} and three drops of concentrated H_3PO_4 were added to the collected root exudates to inhibit the activity of microorganisms. A sub-sample of 10 mL from the collected solution was kept at –20 °C for later analysis of carboxylates.

The recovery experiments were conducted by adding a range of citrate amounts to soil columns without plants. After 2-h incubation, sol columns were leached and the amounts of citrate determined in leachate as for experiments with soil columns containing plants. The amounts of citrate recovered were 13–30% lower than the amounts added. Hence, it is possible that citrate exudation by soil-grown white lupin plants reported here could have been underestimated by up to 13–30%.

Plant harvests

After each collection of root exudates, three pots of each treatment were harvested. Roots and shoots were separated. The number of cluster roots and the total root length were determined after the soil was washed from roots. Cluster roots were defined as those portions of lateral roots bearing bottle-brush-like clusters with a density of 10 or more rootlets per cm (Johnson et al., 1996a, b). Total root length per plant was determined based on the line intersect method using a Comair root scanner (Commonwealth Aircraft Corporation Ltd., Melbourne, Australia). The whole root system was cut into small segments (considering the length of cluster roots, the length that was destroyed by cutting was less than 2.0 mm) and spread to avoid overlap. The total root length per plant was calculated through scanning a reference plate with fine lines. Roots and shoots were dried in an oven at 70 °C for 3 days and weighed. After grinding, the plant material was used for determination of nutrient composition.

Carboxylate sorption to soil

The concentration-dependent sorption of citrate was tested in the same soil according to the method of Jones and Brassington (1998). To take into account microbial degradation of carboxylates, the sterilisation and non-sterilisation treatments were compared. The sterilisation of soil was done at 100 KPa and 120 °C for 20 min in an autoclave. Two milliliters of a 5-m*M* KCl solution containing increasing amounts (0 – 210.8 μg mL^{-1}) of citric acid was added to 0.4 g of soil. The soil samples were then thoroughly mixed, shaken for 15 min or 2 h on an end-over-end shaker (diameter 570 mm) at 320 rpm, and centrifuged at 16 000 *g* for 5 min. The amount of citrate remaining in solution was determined by HPLC.

Measurements

Carboxylates

The 10-mL samples of leachates containing root exudates were filtered through sterile Millex GS Millipore 0.22-μm filters and directly analysed for carboxylates using Waters HPLC equipped with a delivery pump (Waters 600E), an UV–Vis multiple wavelength detector (Waters 490E), and an auto-sampler (Waters 712 WISP) with a 20-μL loop. The chromatographic separation was conducted on a reversed-phase HPLC column (Alltima C18 5 Micron, length 250 mm, i.d. 4.6 mm). Detection of carboxylates was carried out at 215 nm. The buffer solution of 50 m*M* KH_2PO_4, adjusted to pH 2.5 with H_3PO_4, was used for isocratic elution with a flow rate of 1 mL min^{-1} at 28 °C (Sas et al., 2001). Standards were prepared for the following organic acids: oxalic, tartaric, formic, malic, malonic, lactic, acetic, maleic, citric, succinic, fumaric and aconitic. Calibration curves were constructed for the quantification of the identified compounds present in the samples. Identification of carboxylates was performed by comparing retention time and absorption spectra with those of known standards. Data acquisition and integration of chromatograms were performed using the Millennium software (1997, Waters, USA).

P concentration in plants

Concentration of total P was determined in roots and shoots after digesting plant material in a mixture of concentrated nitric and perchloric acids (Johnson and Ulrich, 1959). Phosphorus was assayed using the vanado-molybdate method. The absorption measurement at 700 nm was recorded and P concentration was calculated (Shi, 1986; Westerman, 1990).

Statistics

Fifteen replicate pots of each treatment were set up. Three replicate pots of each treatment were used for root exudate collection and plant growth and P accumulation measurements on each of five harvests. Analysis of variance and multiple comparisons with the least significant difference were conducted using the SPSS statistical software (SPSS, 1998).

Results

Plant growth and cluster root formation

There was no significant difference in plant growth between the treatments with 10 and 200 μg P g^{-1} till day 55 after sowing. At day 58, white lupin in the treatment with 10 μg P g^{-1} started to show visual symptoms of P deficiency. The effect of a limiting P supply on shoot and root dry weight was evident from day 67 and became more pronounced with time (Table 2). Shoot dry weight at day 76 in the treatment with 10 μg P g^{-1} was 58% of that at 200 μg P g^{-1}. The effect of P supply on root dry weight followed the same pattern as for shoots, albeit less pronounced. Root dry weight at day 76 in the treatment with 10 μg P g^{-1} was 65% of that at 200 μg P g^{-1}. The number of nodules and leaves and flowering time were not significantly affected by the P treatment (data not shown).

There was no significant difference in the total root length between the treatments with 10 and 200 μg P g^{-1} until day 58 after sowing. The total root length at the P-limited supply decreased by 35% at day 67 and 26% at day 76 in comparison to the adequate-P treatment (Table 3). In contrast, the specific root length increased rapidly from 17 m g^{-1} DW at day 40 to 28 m g^{-1} DW at day 49 in the P-limited plants, but decreased in the P-adequate plants (Figure 1). At

Table 3. The total root length (m plant^{-1}) of white lupin grown with 10 (limited P) and 200 (adequate P) μg P g^{-1} soil (means±SD, n=3 replicate pots)

Days after sowing	Total root length	
	P-limited	Adequate P
40	7.4±0.4d	10.0±0.9d
49	11.1±1.1d	11.0±1.2d
58	21.7±1.5c	23.7±3.3bc
67	19.6±5.3c	30.1±3.7a
76	20.4±0.3c	27.2±4.2ab

Means followed by different letters are significantly different at $P\leq$0.05.

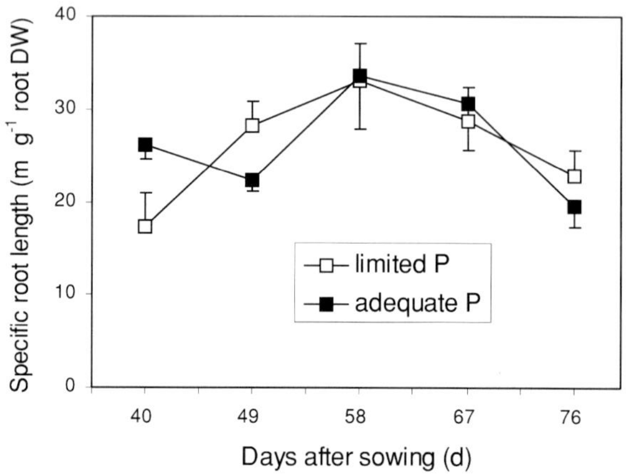

Figure 1. Specific root length (m g^{-1} root DW) of white lupin grown with 10 (limited P) or 200 (adequate P) μg P g^{-1} soil at different days after sowing. Vertical bars represent SD (n=3 replicate pots).

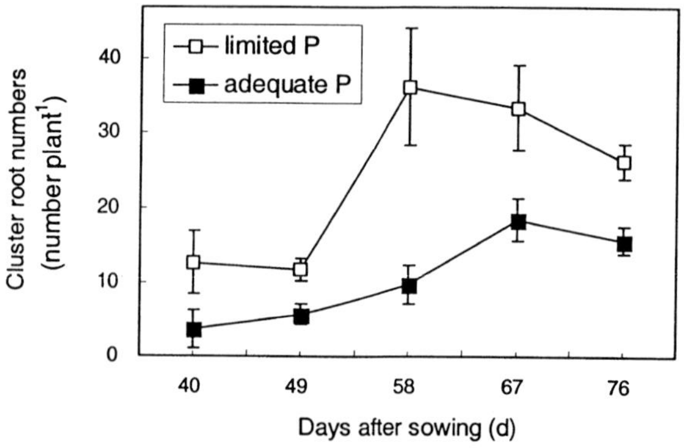

Figure 2. Number of cluster roots of white lupin grown at limited P or adequate P supply at different days after sowing. Vertical bars represent ±SD (n=3 replicate pots).

Table 2. Dry weights of shoots and roots (g plant^{-1}) of white lupin grown with 10 (limited P) and 200 (adequate P) μg P g^{-1} soil (means±SD, n=3 replicate pots)

Days after sowing	Shoot		Roots	
	P-limited	Adequate P	P-limited	Adequate P
40	0.64±0.04g	0.61±0.10g	0.44±0.07d	0.38±0.01d
49	0.94±0.09f	1.07±0.10f	0.39±0.05d	0.49±0.03d
58	1.52±0.03e	1.68±0.06de	0.66±0.05c	0.71±0.05c
67	1.87±0.01d	2.64±0.12b	0.67±0.10c	0.99±0.05b
76	2.39±0.37c	4.15±0.18a	0.90±0.11b	1.38±0.06a

Means followed by different letters for shoot or root weights are significantly different at $P \leq 0.05$.

day 49, the specific root length in the P-limited plants surpassed that in the P-adequate plants. For both P treatments, the specific root length increased steadily from day 49, reached the highest level at day 58 and decreased thereafter till day 76. There was no clear difference in the specific root length between P treatments from days 58 to 76.

From days 40 to 49, the number of cluster roots at the P-limited treatment was clearly greater than that in the adequate-P treatment. From day 49, a rapid increase in the number of cluster roots in the P-limited treatment was observed. The highest number of cluster roots was observed on day 58 for the P-limited plants and day 67 for the P-adequate plants (Figure 2). The number of cluster roots had negative relationships with the P concentrations in roots (r = –0.68) and shoots (r = –0.83) (Figure 3).

Exudation of carboxylates

In the present study, only citrate was detected in root exudate solution of white lupin. Trace amounts of succinate and maleate were detected on two occasions. No carboxylates were detected in the control pots without plants. The amount of citrate exuded by plants in the P-limiting treatment was significantly higher than that exuded by plants receiving adequate P supply. The amount of citrate in root exudates from P-limited plants showed a cyclic pattern over time (Figure 4), being higher on days 49 and 67 than on days 40, 58 and 76.

The amounts of citrate exuded showed a negative relationship to P concentration in shoots (r= –0.76) (Figure 5), but there was no clear relationship between the amounts of citrate exuded and P concentration in roots (data not shown).

Carboxylate sorption

There was no clear difference in the concentrations between the exogenously added and recovered (extracted) citrate after a 15 or 120-min sorption, with recovery decreasing substantially after 4 days. No significant difference in the concentrations of citrate between the treatments with and without soil sterilisation was observed (Figure 6).

Discussion

Impact of the limited P supply on plant growth

The decrease in white lupin growth was not evident in the P-limited treatment until day 58 after sowing. However, the response of white lupin to P deficiency generally occurs after 2 weeks in solution culture (Neumann et al., 1999; Neumann and Römheld, 1999). In the study described here, phosphorus was applied to the soil as sparingly soluble $FePO_4$ at a rate of 10 μg P g^{-1} soil in the P-limited treatment. The P-deficiency symptoms did not appear in the present soil study as early as in previously reported studies with solution cultures because the sparingly soluble P could be mobilized from soil in contrast to the solution culture containing no added P at all, or containing a small amount of soluble P that could be depleted quickly. Therefore, a major difference exists in growth conditions between soil culture described here and solution cultures reported previously, with the former being more realistic in terms of forms, amounts and dynamics of P supply.

Effect of the limited P supply on root development

The present study showed that a limiting P supply increased the number of cluster roots of white lupin in

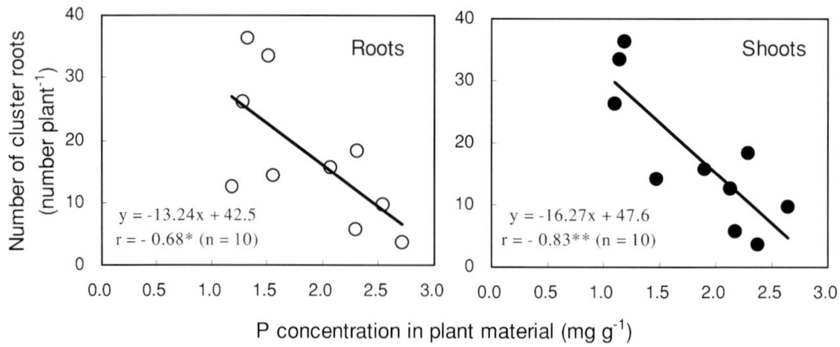

Figure 3. Relationships between the number of cluster roots and P concentrations in roots and shoots of white lupin grown at limited P or adequate P supply during the growth period monitored. Each point represents a mean of three replicate pots (10 means=2 P treatments × 5 harvests).

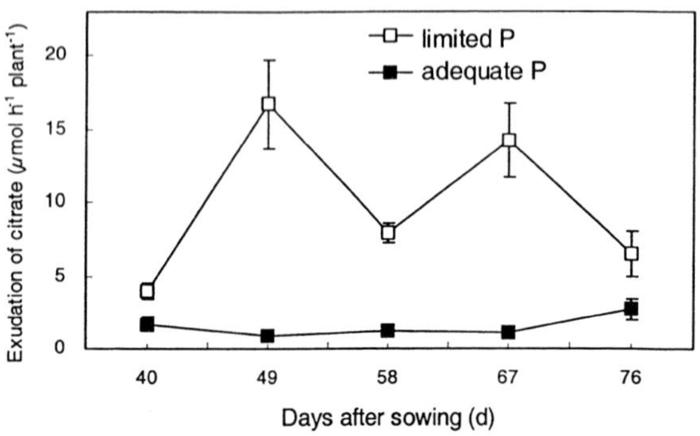

Figure 4. Citrate exudation rates (μmol h^{-1} plant^{-1}) for white lupin grown at limited P or adequate P supply at different days after sowing. Vertical bars represent ±SD (n=3 replicate pots). There were no carboxylates detected in leachate from the control pots without plants.

soil culture. The formation of cluster roots under a limiting P supply occurred earlier than the effect on shoot and root dry weight. The results support the conclusion that formation of cluster roots does not necessarily occur at the expense of dry matter production (cf. Keerthisinghe et al., 1998). White lupin is capable of changing its root system from one dominated by cluster roots to one that has only a few clusters, with no change in dry matter allocation between the shoot and the roots. However, the formation of cluster roots showed a dynamic pattern throughout the experiment irrespective of P supply. The greatest number of cluster roots occurred prior to flowering, suggesting that dry matter allocation between roots and shoots at flowering can profoundly influence the formation of cluster roots, even though the formation of cluster roots has little effect on dry matter allocation.

A considerable number of cluster roots was found in the adequate-P treatment at the late growth stage, even though the number was lower than that in the P-limited treatment. The results were consistent with previous findings that P-adequate plants also form cluster roots (Johnson et al., 1994; Marschner et al., 1987; Sas et al., 2001). One explanation is that the partial depletion of available soil P over time, or a mismatch between dynamics of supply and demand, might have induced mild P limitation and enhanced formation of cluster roots. Cluster root formation as a response to mild P deficiency has been reported previously (Gerke et al., 1994; Keerthisinghe et al., 1998; Marschner et al., 1986).

The increased number of cluster roots induced by P limitation did not influence the total root length per plant. One reason is that the increased number of cluster roots and the inhibition of non-cluster root growth are possibly induced by P limitation simultaneously, causing a balance in the total root length. However, the increased number of cluster roots induced by P limitation enhanced the root length per unit of root dry weight (specific root length), indicating that the formation of cluster roots increased the proportion of fine roots in the whole root system and thus expanded the root surface area for nutrient uptake (Marschner, 1995).

The concentrations of total P in shoots or roots of white lupin showed a clear negative correlation with the number of cluster roots. The results are consistent with suggestions that internal P concentrations regulate cluster root formation in white lupin (Marschner, 1995). Our results also indicated that the initiation of cluster roots could be more related to P concentration in shoots than in roots, which is consistent with the results of the split-root experiment (De Vos et al., 2001). The signals mediating cluster root development need further elucidation, including work on possible

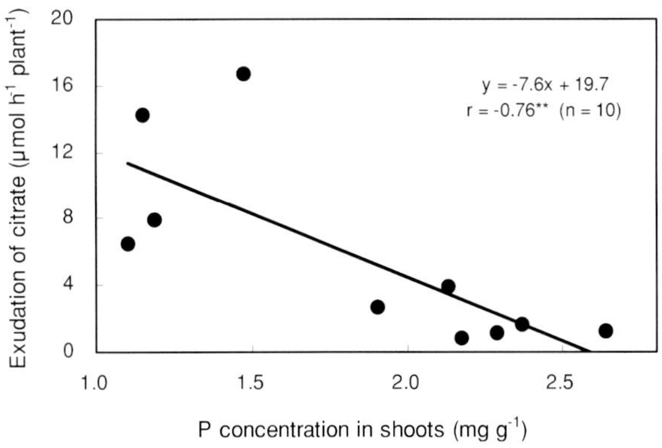

Figure 5. Relationships between citrate exudation rates and P concentration in shoots of white lupin at limited P or adequate P supply during the growth period monitored. Each point represents a mean of three replicate pots (10 points=2 P treatments × 5 harvests).

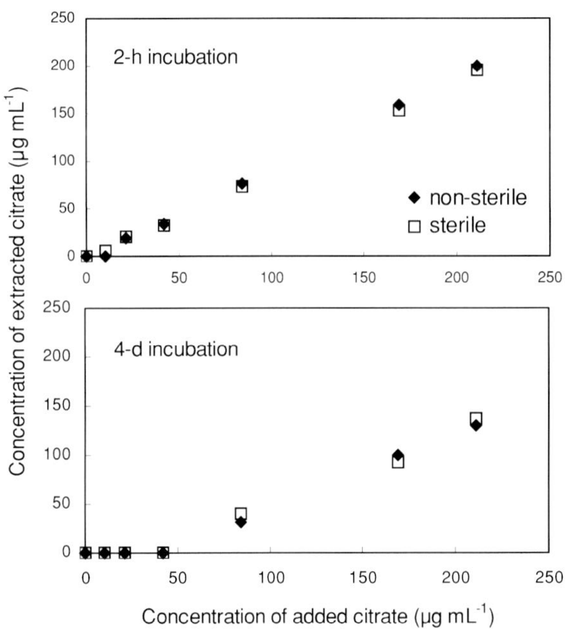

Figure 6. Relationships between concentrations of exogenously added and recovered (extracted) citrate from the sterilised or non-sterilised soil.

auxin induction (Gilbert et al., 1998a, b; Watt and Evans, 1999a).

Effect of P limitation on carboxylate exudation

Carboxylate exudation is another important mechanism for acclimation of white lupin to P limitation. The pattern of carboxylate exudation with time was clearly different from that of proton release (data not shown), providing a good support for the point that different mechanisms are involved in proton release and carboxylate exudation (Jones, 1998; Kochian, 1995; Ryan et al., 1995; Sas et al., 2001). Under the P-limited supply, the release of citrate showed a cyclic pattern, leading to two peaks of citrate exudation on days 49 and 67. This pattern of citrate exudation with two clear pulses was related to the low P concentration in plants, although a clear cyclic pattern in P concentrations in shoots was not observed. The possible explanation for the results is that P limitation increased exudation of citrate, which resulted in the slightly increased P uptake by mobilizing sparingly soluble P in soil, which, in turn, decreased carboxylate exudation. The results suggest a negative feedback in regulating exudation of citrate by white lupin via P concentration, particularly that in shoots (Figure 5), which is consistent with previous reports (De Vos et al., 2001).

Contrary to the reports from Jones and Brassington (1998), the results from carboxylate sorption experiment showed that the soil used in the present study had a low capacity to sorb citrate, and the rate of its decomposition by microorganisms was slow, suggesting that the sorption and the degradation of carboxylates are strongly dependent on the specific soil properties.

In conclusion, leaching soil column is a simple and reliable method for collecting root exudates from soil-grown white lupin plants reacting to P deficiency. Cluster root formation and citrate exudation are affected by P concentration in plants, particularly P concentration in shoots. The underpinning regulatory mechanisms need further investigation.

Acknowledgments

This study was supported by grants from the Major State Basic Research Development Program (Project No. G1999011709), by the National Natural Science Foundation of China (No. 30000102) and by AUSAid.

References

Colwell JD 1963 The estimation of the phosphorus fertilizer requirements of wheat in southern New South Wales by soil analysis. Austr. J. Exp. Agric. Anim. Husb. 3, 190–197.

De Vos M, Shane M W, Cawthray G R, Veneklaas E J and Lambers H 2001 Internal phosphorus concentration modifies the initiation, growth and functioning of cluster roots in *Hakea prostrata* R. Br. *In* Plant Nutrition — Food Security and Sustainability of Agro-Ecosystems. Eds. W J Horst et al. pp. 578–579. Kluwer Academic Publishers, Dordrecht.

Dinkelaker B, Römheld V and Marschner H 1989 Citric acid excretion and precipitation of calcium in the rhizosphere of white lupin (*Lupinus albus* L.). Plant Cell Environ. 12, 285–292.

Gardner W K, Parbery K G and Barber D A 1981 Proteoid root morphology and function in *Lupinus albus*. Plant Soil 60, 143–147.

Gardner W K, Barber D A and Parbery K G 1983 The acquisition of phosphorus by *Lupinus albus* L. III. The probable mechanism by which phosphorus movement in the soil/root interface is enhanced. Plant Soil 70, 107–124.

Gerke J, Römer W and Jungk A 1994 The excretion of citric and malic acid by proteoid roots of *Lupinus albus* L. Effect on soil solution concentrations of phosphate, iron and aluminium in the proteoid rhizosphere in samples of an oxisol and luvisol. Z. Pflanzenernähr. Bodenkd. 157, 289–294.

Gilbert G A, Allan D L and Vance C P 1998a Phosphorus deficiency in white lupin alters root development and metabolism. *In* Radical Biology: Advances and Perspectives on the Function of Plant Roots. Eds. H E Flores, J P Lynch and D Eissenstat. Vol. 18, pp. 92–103. American Society of Plant Physiologists, Rockville, MD.

Gilbert G A, Vance C P and Allan D L 1998b Regulation of white lupin root metabolism by phosphorus availability. *In* Phosphorus in Plant Biology: Regulatory Roles in Molecular, Cellular, Organismic, and Ecosystem Processes. Eds. J P Lynch and J Deikman. pp. 157–167. American Society of Plant Physiologists, Rockville, MD.

Johnson C M and Ulrich A 1959 Analytical methods for use in plant analysis. Calif. Agric. Exp. Stat. Bull. No.766.

Johnson J F, Allan D L and Vance C P 1994 Phosphorus stress-induced proteoid roots show altered metabolism in *Lupinus albus*. Plant Physiol. 104, 657–665.

Johnson J F, Allan D L, Vance C P and Weiblen G 1996a Root carbon dioxide fixation by phosphorus-deficient *Lupinus albus*. Contribution to organic acid exudation by proteoid roots. Plant Physiol. 112, 19–30.

Johnson J F, Vance C P and Allan D L 1996b Phosphorus deficiency in *Lupinus albus*. Altered lateral root development and enhanced expression of phospho*enol*pyruvate carboxylase. Plant Physiol. 112, 31–41.

Jones D L 1998 Organic acids in the rhizosphere — A critical review. Plant Soil 205, 25–44.

Jones D L and Brassington D S 1998 Sorption of organic acids in acid soils and its implications in the rhizosphere. Eur. J. Soil Sci. 49, 447–455.

Kamh M, Horst W J, Amer F, Mostafa H and Maier P 1999 Mobilization of soil and fertilizer phosphate by cover crops. Plant Soil 211, 19–27.

Keerthisinghe G, Hocking P J, Ryan P R and Delhaize E 1998 Effect of phosphorus supply on the formation and function of proteoid roots of white lupin (*Lupinus albus* L.). Plant Cell Environ. 21, 467–478.

Kochian L V 1995 Cellular mechanisms of aluminum toxicity and resistance in plants. Annu. Rev. Plant Physiol. Plant Mol. Biol. 46, 237–260.

Li M, Shinano T and Tadano T 1997 Distribution of exudates of lupin roots in the rhizosphere under phosphorus deficient conditions. Soil Sci. Plant Nutr. 43, 237–245.

Lynch J M 1990 Soil Rhizosphere. Wiley, New York.

Northcote K H 1971 A Factual Key for the Recognition of Australian Soils. Rellim, Glenside, South Australia.

Marschner H 1995 Mineral Nutrition of Higher Plants. 2nd Edition. Academic Press, London.

Marschner H, Römheld V, Horst W J and Martin P 1986 Root induced changes in the rhizosphere: importance for the mineral nutrition of plants. Z. Pflanzenernähr. Bodenkd. 149, 441–456.

Marschner H, Römheld V and Cakmak I 1987 Root induced changes of nutrient availability in the rhizosphere: J. Plant Nutr. 10, 1175–1184.

Neumann G and Römheld V 1999 Root excretion of carboxylic acids and protons in phosphorus-deficient plants. Plant Soil 211, 121–130.

Neumann G, Massonneau A, Martinoia E and Römheld V 1999 Physiological adaptations to phosphorus deficiency during proteoid root development in white lupin. Planta 208, 373–382.

Ryan P R, Delhaize E and Randall P J 1995 Malate efflux from root apices and tolerance to aluminium are highly correlated in wheat. Aust. J. Plant Physiol. 122, 531–536.

Sas L, Tang C and Rengel Z 2001 Excess cation uptake, and extrusion of proton and organic acid anions by *Lupinus albus* under phosphorus deficiency. Plant Sci. 160, 1191–1198.

Shi R 1986 Soil and Agricultural Chemistry Analysis. China Agricultural Press. Beijing.

SPSS 1998 Standard version. SPSS Inc.

Watt M and Evans J R 1999a Proteoid roots: Physiology and development. Plant Physiol. 121, 317–323.

Watt M and Evans J R 1999b Linking development and determinacy with organic acid efflux from proteoid roots of white lupin growth with low phosphorus and ambient or elevated atmospheric CO_2 concentration. Plant Physiol. 120, 705–716.

Westerman R L 1990 Soil Testing and Plant Analysis. 3rd edn. American Society of Agronomy and Soil Science Society of America, Madison, WI.

The influence of plant nutrition on plant growth and development of cluster roots

Plant and Soil **248:** 209–219, 2003.
© 2003 Kluwer Academic Publishers. Printed in the Netherlands.

Effects of external phosphorus supply on internal phosphorus concentration and the initiation, growth and exudation of cluster roots in *Hakea prostrata* R.Br.

Michael W. Shane[1,3], Martin de Vos[1,2], Sytze de Roock[1,2], Gregory R. Cawthray[1] & Hans Lambers[1,2]

[1]*School of Plant Biology, Faculty of Natural and Agricultural Sciences, The University of Western Australia, Crawley, WA 6009, Australia.* [2]*Plant Ecophysiology, Sorbonnelaan 16, 3584 CA, Utrecht University, The Netherlands*

[3]*Corresponding author**

Received 6 February 2002. Accepted in revised form 20 June 2002.

Key words: carboxylate exudation, cluster roots, *Hakea prostrata*, proteoid roots, phosphorus, Proteaceae, split-root design

Abstract

The response of internal phosphorus concentration, cluster-root initiation, and growth and carboxylate exudation to different external P supplies was investigated in *Hakea prostrata* R.Br. using a split-root design. After removal of most of the taproot, equal amounts of laterals were allowed to grow in two separate pots fastened together at the top, so that the separate root halves could be exposed to different conditions. Plants were grown for 10 weeks in this system; one root half was supplied with 1 μM P while the other halves were supplied with 0, 1, 25 or 75 μM P. Higher concentrations of P supplied to one root half significantly increased the P concentration of those roots and in the shoots. The P concentrations in root halves supplied with 1 μM P were invariably low, regardless of the P concentration supplied to the other root half. Cluster root initiation was completely suppressed on root halves supplied with 25 or 75 μM P, whereas it continued on the other halves supplied with 1 μM P indicating that cluster-root initiation was regulated by local root P concentration. Cluster-root growth (dry mass increment) on root halves supplied with 1 μM P was significantly reduced when the other half was either deprived of P or supplied with 25 or 75 μM P. Cluster-root growth was favoured by a low shoot P status at a root P supply that was adequate for increased growth of roots and shoots without increased tissue P concentrations. The differences in cluster-root growth on root halves with the same P supply suggest that decreased cluster-root growth was systemically regulated. Carboxylate-exudation rates from cluster roots on root halves supplied with 1 μM P were the same, whether the other root half was supplied with 1, 25 or 75 μM P, but were approximately 30 times faster when the other half was deprived of P. Estimates of root P-uptake rates suggest a rather limited capacity for down-regulating P uptake when phosphate was readily available.

Introduction

For plants that naturally occur on ancient, weathered soils, the availability of P is often limiting for plant growth (Handreck, 1997). Adaptations of species occurring on such soils include efficient P-acquisition mechanisms, the ability to function with very low concentrations of leaf P and the capacity to retranslocate a large fraction of P from senescing leaves before leaf shedding (Lambers et al., 1998). Common plant responses to a shortage of P include an increase in root mass ratio (Brouwer, 1963), root hair length (Bates and Lynch, 1996; Jungk, 1996), root hair density (Ma et al., 2001) and infection by mycorrhizal fungi (if the species can form a mycorrhizal symbiosis) (Reddell et

* FAX No: +61-8-93801108
E-mail: mshane@agric.uwa.edu.au

al., 1997). More specialised responses include exudation of carboxylates by roots (Dinkelaker et al., 1989; Gardner et al., 1982; Grierson, 1992) and formation of 'proteoid' or 'cluster' roots (Dinkelaker et al., 1995; Lamont, 1982; Purnell, 1960).

Many plants respond to a low availability of P with an increased exudation of carboxylates. Carboxylates play an important role in nutrient mobilisation from inorganic sources in the soil (Gardner et al., 1983; Gerke, 1994; Hinsinger, 1998; Hoffland et al., 1989; Jones, 1998). The addition of carboxylates to soil, such as citrate, has been shown to mobilise nutrients, e.g., P and Fe (Jones and Darrah, 1994; Lambers et al., 2002). Plant growth and P uptake of *Lotus pedunculatus* increased due to P mobilisation from the action of exuded carboxylates on calcium phosphate (Bolan et al., 1997). Similarly, P uptake by *Oryza sativa* was enhanced due to its mobilisation by root-exuded citrate (Kirk et al., 1999). Very fast rates of carboxylate exudation are exhibited by plants that produce proteoid or cluster roots (Keerthisinghe et al., 1998; Roelofs et al., 2001). The carboxylate composition of cluster root exudates from *Banksia grandis* depends on soil type and the form of P in the soil (Lambers et al., 2002). When P was supplied as aluminium phosphate, the major carboxylates were di- and tricarboxylates. When P was supplied as iron phosphate, a mixture of mono-, di- and tricarboxylates were exuded (Lambers et al., 2002).

Cluster roots were first described for Proteaceae, and hence called 'proteoid' roots (Purnell, 1960). Such roots are typical for most members of the Proteaceae (Jeffrey, 1967; Lamont, 1972; Purnell, 1960). Since Purnell's discovery, 'proteoid' roots have been found in a wide range of species and genera (Dinkelaker et al., 1995; Neumann and Martinoia, 2002), and hence the term 'cluster' root is commonly used. Lamont (1993) and Dinkelaker et al. (1995) argued that cluster roots are an important adaptation for nutrient uptake, especially in environments with a low availability of nutrients. With the exception of actinorhizal species, most species that form cluster roots are non-mycorrhizal (Neumann and Martinoia, 2002). While mycorrhizal hyphae increase the soil volume that is exploited by roots, cluster roots explore a comparatively small soil volume. However, the dense rootlet morphology and abundant root hairs associated with cluster roots, combined with exudation of citrate and other carboxylates, lead to an intense chemical extraction of their rhizosphere (Marschner, 1995), solubilising complexed nutrients making them available for plant uptake (Gardner et al., 1983; Lambers et al., 2002). In members of the Proteaceae (e.g., Grose, 1989; Handreck, 1991; Lamont, 2002) and other families (Keerthisinghe et al., 1998; Reddell et al., 1997) the formation of cluster roots is suppressed by increased P supply.

The cluster roots of *Hakea undulata* grown in P-deficient soil exuded the carboxylates malate and citrate (Dinkelaker et al., 1995). Carboxylate exudation began about 7 days after rootlet initiation and continued for only 2–3 days. An exudative burst of citrate also occurs from cluster roots of hydroponically grown white lupin (Watt and Evans, 1999). Cluster roots are thought to be short-lived structures and estimates of their lifespan range from 1 week to 3 months amongst the different species. The 'bottlebrush-like' cluster root morphology of *Hakea prostrata* R.Br is characteristic of most Proteaceae (Lamont, 1972) and exudation of carboxylates may play an important role in the acquisition of P and other mineral nutrients from P-binding, nutrient-impoverished soils.

The aims of the present investigation were to determine whether the initiation, growth and exudation of cluster roots are driven by external or internal P concentrations, and if internal whether by root or shoot.

Materials and methods

Plant growth and culture

Six-month-old soil-grown *Hakea prostrata* R.Br. (harsh hakea) seedlings were obtained from APACE nursery (North Fremantle, Australia). Soil was removed from roots by washing with water and the root systems were trimmed to leave approximately 3 cm of taproot with associated laterals. Each plant was transferred to a split-root container constructed from pairs of 4-L black plastic pots fastened together. Shoots were gently supported in the center of a rectangular gray-foam lid that made a light-tight seal over each container. Each root half was independently aerated through black tubing inserted through the lid.

Each pot contained 3.5 L of continuously aerated nutrient solution (pH 5.8) containing (in μM): 1.0 PO_4^{3-}; 400 NO_3^-, 200 Ca^{2+}, 200 K^+, 154 SO_4^{2-}, 54 Mg^{2+}, 20 Cl^-, 2.0 Fe-EDTA, 0.24 Mn^{2+}, 0.10 Zn^{2+}, 0.02 Cu^{2+}, 2.4 H_3BO_3, 0.3 Mo^{4+} made up in deionised water. Plants were grown for 8 weeks (under a shade cloth for the first 4 weeks) in a glasshouse at min/max 20/32 °C, and root temperature was maintained at 18–20 °C with the pots in a temperature-controlled tank. The complete nutrient solution was replenished each week for the first 4 weeks, and thereafter every 3 days for an additional 4 weeks. Equal numbers of lateral roots (three to four) on each side of the taproot were maintained by regular pruning of new roots.

Thirteen plants of uniform size and with equal numbers of lateral roots on each root half were selected, and transferred to a controlled-environment chamber, maintained at 25/20 °C with a 12/12-h day/night cycle, 700 μmol quanta m^{-2} s^{-1} (Multi-Hi-Ace Metal Halide lamps, Iwasaki Electric Co., Ltd, Tokyo) at shoot height, and 60% relative humidity. Plants were grown for 4 weeks and then were randomly assigned to one of four (a–d) P treatments (supplied as KH_2PO_4) in the basal nutrient solution. All plants had one root half supplied with 1 μM P while the other half was either: (a) deprived of P (n = 4), or supplied with (b) 1 μM (n = 3), (c) 25 μM (n = 3), or (d) 75 μM P (n = 3). The complete nutrient solution in each pot was changed daily for the duration of the experiment (10 weeks). Roots were observed daily during nutrient solution changes, and the numbers of newly initiated cluster roots were recorded each week for each root half for the first 8 weeks.

Collection of cluster root exudates

Exudates were collected from 6–9 day-old cluster roots on root halves (in their split-root treatment containers) supplied with 1 μM P after 8 and 9 weeks of split-root cultivation. Exudate collection cells were constructed to fit mature cluster roots with a minimal disturbance to root systems. The cells were elliptical in shape (similar to that of the clusters), sealed at the bottom and open at the top with a narrow slit at each end. A cluster was gently lifted into the cell and the unbranched root axis (proximal and distal from each cluster) passed through the notch at each end of the cell and made watertight with Terostat®. Depending upon the size of the cluster root, exudates were collected in 3.0 – 6.0 mL of the complete nutrient solution for 2 or 3 h between 11:00–15:00 h. Each collection cell was attached to a firm but flexible support that was secured to the edge of the pot. The cell containing the cluster root could be orientated within the pot and was partly immersed into the nutrient solution such that the level of collection solution inside the cell was below that of the surrounding solution in the pot. Leaks in the cells were therefore easily detected and would result in an increased collection solution volume inside the cells. Each cell was continuously aerated during root exudate collection through two short lengths of small diameter tubing sealed into each side of the cell. After collection, exudates were immediately filtered through a Supor® membrane (0.2 μM pore size, Pall Gelman Laboratory) and stored at −20 °C until analysis.

Gas exchange measurements and plant harvests

Photosynthesis and transpiration were measured on four youngest fully expanded leaves for each plant using a LI-6400 Portable Photosynthesis System (LICOR, Lincolin, NE) at 1500 μmol quanta m^{-2} s^{-1}. Stomatal conductance and internal [CO_2] were calculated. Plants were harvested after 10 weeks of P treatments in split-root culture. All the roots were washed for 1 min in a 1 μM solution of K_2SO_4 to remove P in the nutrient solution from the roots. Each plant was separated into shoots, leaves, cluster and non-cluster roots. Samples were weighed fresh and after drying for 6 days at 80 °C.

Total nitrogen and phosphorus concentrations

Dried material was ground to a fine powder using a stainless steel ball mill. Total P concentration in tissue samples was determined using wavelength-dispersive X-ray fluorescence spectrometry (PW-1400, Philips Corporation, The Netherlands) and verified colourimetrically using a UV–vis spectrophotometer (Shimadzu Corporation, Japan) by the vanado-molybdate method (A.O.A.C., 1975) after digestion of ground material in hot concentrated HNO_3:$HClO_4$ (3:1). Total N content in leaf dry material was determined by thermal conductivity after combustion using a CHN-1000 elemental analyser (LECO® Corporation, St. Joseph, MI, USA).

Exudate analysis

Organic acids in cluster root exudates were separated (Alltima C-18 (250 mm × 4.6 mm internal diameter with 5 μM packing) and identified using Waters® HPLC (600E pump, 717 auto-injector and 996 photodiode-array detector, Milford, MA). The mobile phase was 25 mM KH_2PO_4 (pH 2.5) at a flow rate of 1 ml min^{-1}. Detection was at 210 nm; however, data from 195 to 400 nm was collected and used for spectrum matching and peak purity analysis according to Lambers et al. (2002). The sample injection volume was generally 100 μL, but for higher carboxylate

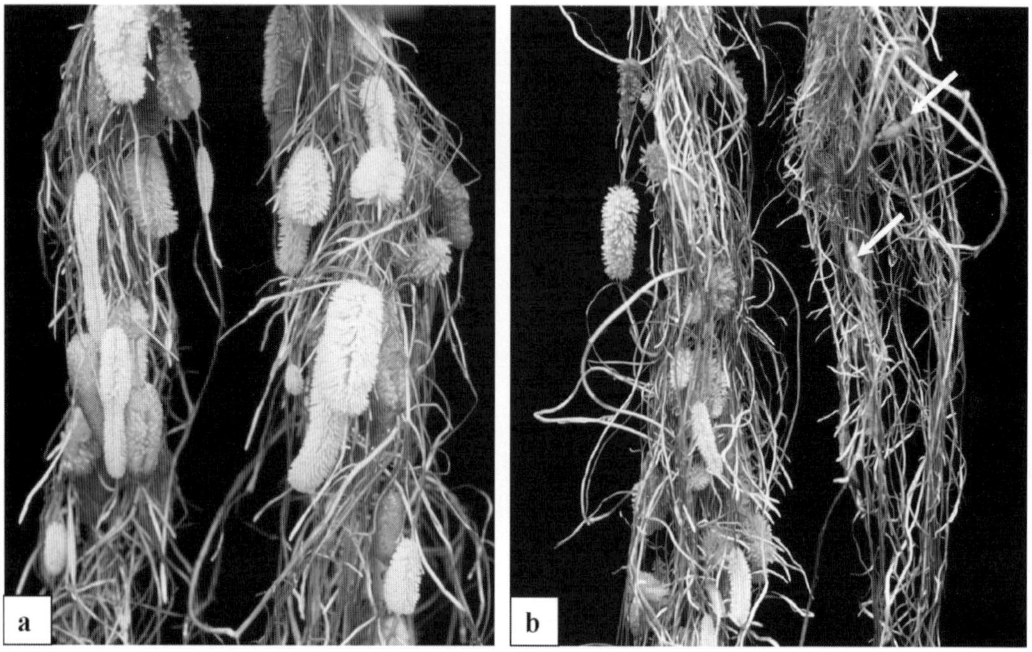

Figure 1. Root systems of hydroponically grown split-root *H. prostrata* after 12-weeks of first cutting the taproot and then a 6-weeks P treatment. (a) 1 μM P supplied to both root halves. Numerous cluster roots developed on the lateral roots of both root halves. (b) Cluster roots developed on the root half supplied with 1 μM P (left) and no new clusters were initiated on the root half supplied with 25 μM P (right). Cluster roots that were initiated before the beginning of the P treatments failed to develop (arrows) after the 25 μM P treatment was started.

concentrations it was reduced. The column was completely flushed (gradient elution using 60% methanol) prior to the subsequent sample injection to eliminate the transfer of highly polar compounds.

Data acquisition and processing used Millennium software (Waters, Milford MA, USA). Retention times of organic acid standards: tartaric-, formic-, malic-, malonic, lactic, acetic, maleic, citric, succinic, fumaric, *cis*-aconitic and *trans*-aconitic acid were used to identify organic acids in root exudates.

Several compounds (peaks) with retention times similar to organic acids were observed however, their spectra were different from those of our standards. Identification of organic acids in exudates was confirmed after spectrum purity was checked and matched with the spectral library of standards.

The rate of organic acid exudation was calculated for each cluster root using the concentration of organic acids determined using HPLC, volume of the collection solution used, the fresh mass of the cluster roots, and the collection time.

Statistics

Data were analysed with one-way analysis of variance (ANOVA) with P treatment as the main fixed factor in the model (Statgraphics Ver. 7.0, 1993). Specific differences between treatments were tested using LSD multiple range tests ($\alpha = 0.05$) following the ANOVA. To ensure normality and homogeneity of variances, data were log transformed where necessary.

Results

Effects of root trimming

The plants grew well in solution culture with a split root system. The two grouping of three to four lateral roots allowed to develop opposite each other at the taproot base became woody (diameter, approx. 5–10 mm) and development of their branch roots formed the root system of each root half (Figure 1a, b). Trimming of the taproot had no obvious effect on the morphology or development of cluster roots. We checked whether plants grown with divided root systems had altered xylem transport from root to shoot. Specifically, we investigated whether roots of one root half were connected with a particular sub-set of shoot branches. A dye solution (0.05%, w/v, basic fuchsin) was used to trace the potential water transport pathways from individual roots on each root half to the branches of

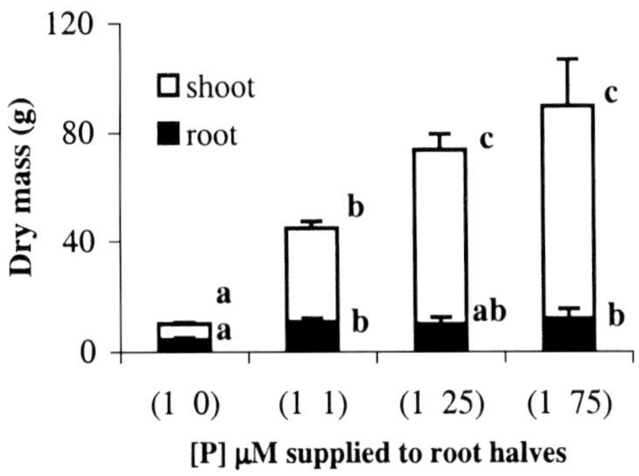

Figure 2. Influence of P supplied to root halves (in parentheses) on total shoot and root dry mass after 10 weeks of P treatments, in *H. prostrata* plants. Bars indicate standard errors; roots and shoots were tested separately, different letters denote significant differences ($P < 0.05$).

the shoots (Shane et al., 2000). Dye solution was introduced to root xylem (through their cut surface under solution) during transpiration. The pattern of dye-stained xylem walls indicated that the xylem of individual roots (on both root halves) were not exclusively connected with particular shoot branches, i.e., the dye moved through the lumen of mature xylem vessels in the roots and were continuous with the xylem of all shoot branches and their leaves (data not shown).

P supply and growth

Plants with one root half deprived of P had a significantly reduced shoot mass compared with plants having both root halves supplied with 1 μM P (Figure 2). Increasing the P supply to one root half to 25 or 75 μM P resulted in a further increase in shoot mass compared with those plants having 1 μM P supplied to both halves. Total root mass was almost doubled in plants with both root halves supplied with 1 μM P compared with those plants having one root half deprived of P. Increasing the P supply to 25 or 75 μM P to one root half had no further significant influence on total root mass (Figure 2).

When root halves were supplied with 25 or 75 μM P they immediately stopped initiating cluster roots and the growth of those already initiated ceased (Figure 1b). Cluster roots were, however, initiated in similar numbers on all root halves supplied with 1 μM P, irrespective of the concentration of P supplied to the other root half. During the first 8 weeks of split-root culture we recorded a continuous rise and fall in the numbers of cluster roots initiated on root halves. These undulations had a periodicity of approximately two weeks and were not synchronized (Figure 3) between root halves in the treatments where cluster roots were initiated on both halves (e.g., both halves supplied with 1 μM P).

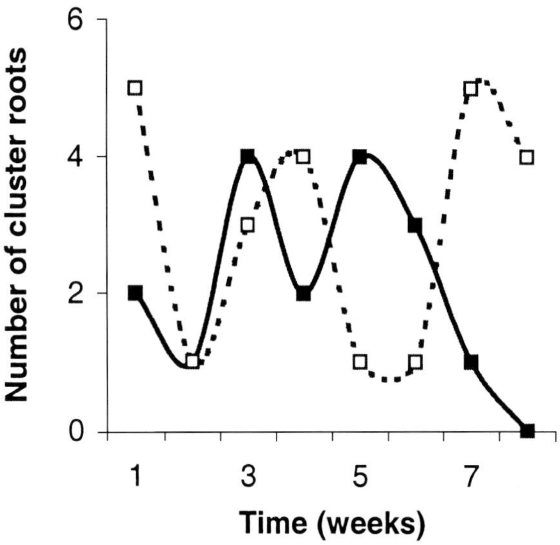

Figure 3. The rise and fall in the numbers of newly initiated cluster roots on each root half of a single *H. prostrata* plant when both root halves were supplied with 1 μM P. Each (continuous and dashed) line represents one root half.

Table 1. Influence of P supplied to root halves (in parentheses) on the percentage of total root dry mass allocated to cluster roots after 10 weeks P treatments. Different letters denote significant differences ($P < 0.05$)

	[P] μM supplied to root halves							
	(1–0)		(1–1)		(1–25)		(1–75)	
% Cluster roots	3^a	7^a	18^b	21^b	4^a	0	6^a	0

On root halves supplied with 1 μM P, the proportion of total root mass that was allocated to cluster roots was highest when the other half was supplied with 1 μM P (Table 1). It was significantly reduced when the other half was either deprived of P or supplied with 25 or 75 μM P.

Internal P concentrations

The total plant phosphorus content at final harvest increased with increasing P supply to one root half (Table 2). The P concentrations of both leaf (Figure 4a) and stem (Figure 4b) increased when one root half was supplied with 25 or 75 μM P compared with

Table 2. Influence of P supply to root halves of *H. prostrata* plants, on total plant phosphorus content and P-uptake rates after 10 weeks P treatment in split-root culture. Standard errors in parentheses

	Phosphorus concentration supplied to each root half					
	(1–1 μM)		(1–25 μM)		(1–75 μM)	
Dry mass (g)	5.2 (0.87)	5.3 (0.84)	5.1 (2.2)	4.7 (0.89)	7.3 (2.3)	4.4 (1.3)
Total P supplied to root halves (mg)[#]	7.8	7.8	7.8	195	7.8	585
Total plant P content at harvest (mg)	12.6 (1.06)		104 (2.0)		196 (42)	
% P uptake of supplied	81		51		33	
P uptake (mg g^{-1} root)[†]	2.4(0.27)		24 (3.7)		52 (12)	
P uptake / [P] supplied	2.4		1.0		0.7	

[#]Over 10 weeks.
[†]Based on the average dry mass of the 25 and 75 μM P root half; plants with 1 μM P supplied to both halves, the average mass of both halves.

plants having one root half deprived of P or with both halves supplied with 1 μM P. The P concentration of roots on halves supplied with 0, 1, 25 or 75 μM P (filled bars in Figure 4c) increased 7-fold at 25 μM P and 11-fold at 75 μM P. In contrast, the influence of P supply to these root halves on the root P concentration of the other half supplied with 1 μM P (open bars in Figure 4c) was minor. The P concentrations of the root halves supplied with 1 μM P were low, but the values were significantly higher (2-fold) when the other halves received 25 or 75 μM P compared with those when the other halves were deprived of P (Figure 4c).

P uptake by root halves

P uptake was calculated for plants with both root halves supplied with 1 μM P and those with one half supplied with 25 or 75 μM P (Table 2). Total plant P was divided by the mass of the root half giving an (under) estimate of their P uptake (mg P g^{-1} root), since the roots were growing and we divided by the final mass. The P uptake was divided by the P concentration supplied to each root half, which gives an estimate of the P uptake rates as dependent on external P concentration. As would be expected, the total P content and P uptake (Table 2) increased with increasing P supply to a root half. The rate of P uptake decreased by 58% when the P supply was increased from 1 to 25 μM, and by a further 30% when the P supply was increased from 25 to 75 μM.

Table 3. Total carboxylate exudation rates from cluster roots on root halves supplied with 1 μM P. Influence on rates by P supplied (0, 1, 25 and 75 μM) to the other half. Standard errors in parentheses; different letters denote significant differences ($P < 0.05$)

[P] μM supplied to other root half	Exudation rate (nmol g^{-1}FM s^{-1})
0	2.18 (0.35)[a]
1	0.08 (0.044)[b]
25	0.09 (0.059)[b]
75	0.02 (0.005)[b]

Carboxylate exudation

Rates of carboxylate exudation were determined for 6–9 day-old cluster roots growing on root halves supplied with 1 μM P. The term carboxylate refers to the entire spectrum of different carboxylic acid anions analysed in root exudates. The organic anion composition of cluster-root exudates was similar amongst treatments (76–97% citrate). The exudation rates are listed corresponding to the P concentration supplied to the other root half (Table 3). Rates of exudation were fast from cluster roots when the other root half was deprived of P (Table 3), and similar to those reported by Roelofs et al. (2001). When the other root half was supplied with 1, 25 or 75 μM P, the rate of exudation decreased 27-, 23- and 109-fold, respectively.

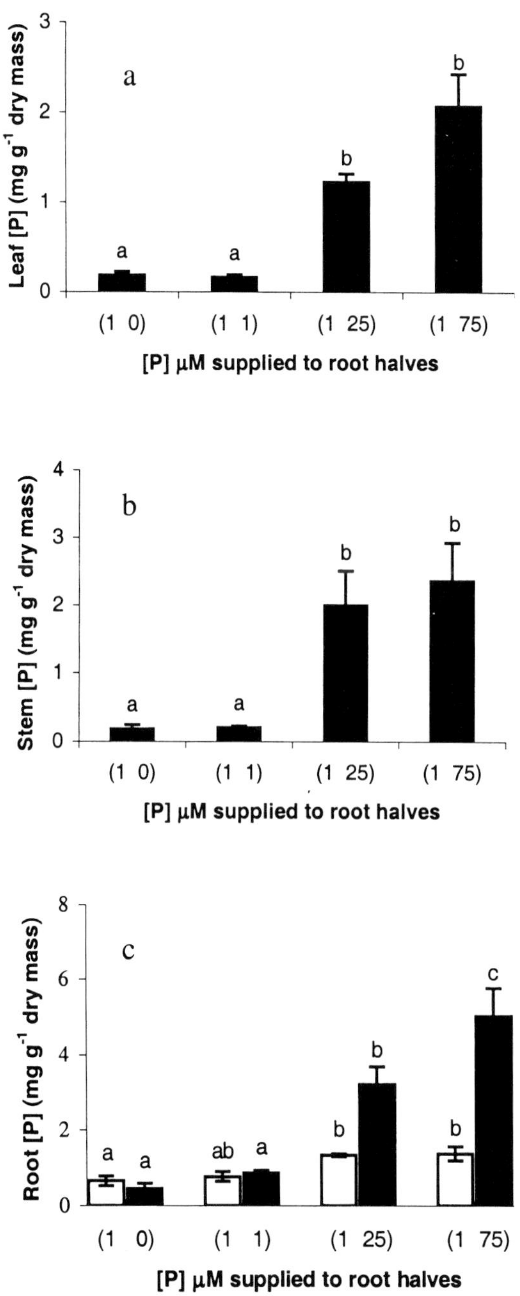

Figure 4. Influence of P supply (in parentheses) on total P concentration in (a) leaves, (b) stem and (c) root halves after 10 weeks of P treatments in split-root culture, in *H. prostrata* plants. Bars indicate standard errors; different letters denote significant differences ($P < 0.05$). The 1 μM P root halves (open bars) were tested against each other as were the 0, 1, 25 and 75 μM P root halves (filled bars).

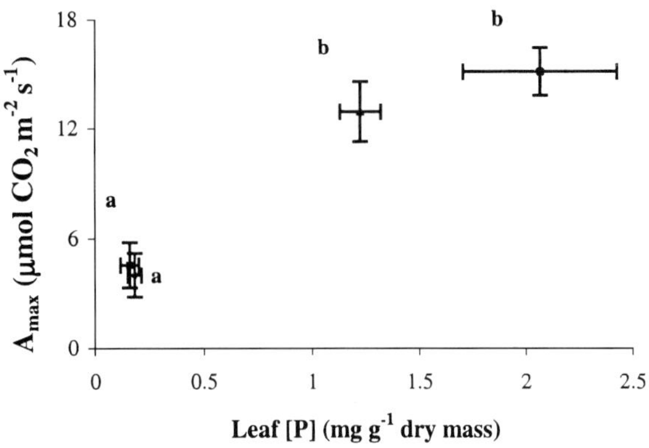

Figure 5. Influence of P supply to root halves on leaf P concentration and the light-saturated rate of photosynthesis (A_{max}) after 10 weeks of P treatments. P supply from bottom to top as from left to right in Figure 2. Bars indicate standard errors; different letters denote significant differences ($P < 0.05$) in photosynthetic rate.

Gas exchange

The rates of light-saturated photosynthesis increased with increasing leaf P concentrations (Figure 5), as has been found before (Raaimakers et al., 1995). Leaf N concentrations ranged from 11 (one half deprived) to 15 mg g^{-1} DM (one half supplied with 75 μM P) with increasing P supply, albeit not as strongly as leaf phosphorus concentrations (from 0.16 to 2.1 mg g^{-1} DM) (Figure 4a). Stomatal conductance and transpiration increased in a similar manner to photosynthesis, and the internal CO_2 concentration was unaffected (data not shown). As there was no correlation between photosynthesis and leaf N concentrations, the positive correlation between the rate of photosynthesis and leaf P concentrations was not due to the slightly increased leaf N concentrations. The increased stomatal conductance did not account for increasing rates of photosynthesis with increasing P supply.

Therefore, the slow rate of photosynthesis at low P supply was not due to increased stomatal limitation and was probably due to a slow exchange of cytosolic P_i against triose phosphate from the chloroplast (Stitt, 1997), which then decreased stomatal conductance and transpiration.

Discussion

Influence of P supply on growth

Our growth conditions produced two populations of plants, those with a relatively low-shoot P status (plants with both root halves supplied with 1 μM P

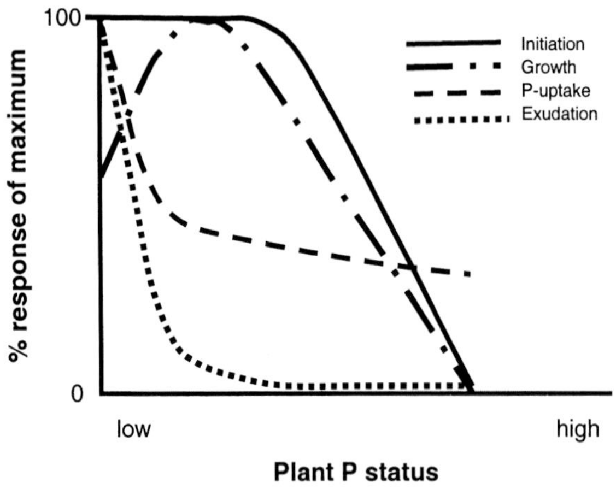

Figure 6. Hypothetical model of the effects of plant P status on cluster-root initiation, cluster-root growth, and exudation, and on P uptake in *Hakea prostrata* (Proteaceae). For further information, see text.

and plants with one-half deprived of P) and those with a high-shoot P status (plants with one half supplied with 25 or 75 μM) (Figures 4a, b and 5). The corresponding leaf P concentrations varied between 0.16 and 2.1 mg g^{-1} of the dry mass, respectively. As observed for other species (e.g., *Stylosanthes hamata*, Smith et al., 1990; *Phaseolus vulgaris*, Table 8.22 in Marschner, 1995), the ratio of shoot mass to total plant mass (the shoot mass ratio) increased with increasing P supply (Figure 2), in accordance with the 'functional equilibrium' model proposed by Brouwer (1963).

Cluster-root initiation and root P concentration

The number of cluster roots initiated on root halves supplied with 1 μM P and on those deprived of P were the same. Initiation was completely suppressed on the root halves supplied with either 25 or 75 μM P. Lamont et al. (1984) also found that in split-root *Leucadendron laureolum* (Proteaceae), root halves exposed to a soil containing a low P concentration continued to develop cluster roots while the other half exposed to a high P concentration did not.

There is ample evidence (Dinkelaker et al., 1995; Keerthisinghe et al., 1998; Lamont, 1972) demonstrating the dependence of number of cluster roots initiated on the concentration of P supplied; as the concentration is increased the number initiated decrease. It was not surprising then, that in our study similar numbers of cluster roots were initiated on root halves supplied with 1 μM P. The P concentrations of these roots were invariably low, but tended to increase with increasing P supply to the other root half (Figure 4c). What was surprising was the continued initiation of cluster roots on root halves of plants with comparatively high concentrations of P in their shoots and in the other root half. This suggests that root P concentration locally regulates cluster-root initiation, at least within the range of P supplied in the present experiments. Systemic signalling has been suggested to account for the synchronous development of cluster roots in white lupin (Watt and Evans, 1999). In *H. prostrata* local regulation was further indicated by the periodicity of initiation, which was out of phase on root halves of individual plants (Figure 3), indicating that the signal or the receptivity for the signal was generated or regulated locally.

Cluster-root growth and exudation are influenced systemically

In our study cluster-root growth was defined as the proportion of root mass developed into cluster roots. Whereas the numbers of cluster roots initiated, as described above, were the same amongst root halves supplied with 1 μM P, there were differences in cluster-root growth when either higher or lower P concentrations were supplied to the other half. We did not measure the dimensions of individual cluster roots but a decrease or increase in cluster-root mass indicates larger or smaller individual cluster roots, respectively.

Growth of cluster roots was more sensitive to changes in overall P supply than their initiation, as shown in a hypothetical model based on our present results and cited earlier work (Figure 6). Cluster-root growth was at a maximum when 1 μM P was supplied to both root halves (Table 1). It was reduced at both a lower supply (other half deprived) and at a higher supply (other half supplied with 25 or 75 μM P). This finding is in agreement with other studies demonstrating an optimal P supply for cluster-root growth (Proteaceae, *Hakea prostrata*, Lamont 1971; *Telopea speciosissima*, Grose, 1989; *Casuarinaceae, Casuarina cunninghamiana*, Reddell et al., 1997). In our plants with divided root systems, since the root P concentration of root halves supplied with 1 μM P was increased only when 25 or 75 μM P was supplied to the other root half (Figure 4c), the trigger for changes in cluster-root growth was not just a local one, but must also involve a systemic signal. Auxins are probably involved in signalling the plant P status (Gilbert et al., 2000; Torrey, 1986), and root-produced cytokinins are known to play a role in the transduction pathway between sensing the P supply and increased leaf growth (Kuiper et al., 1988, 1989). The influence of plant nitrogen on cluster root development has been examined in another split-root study with *H. prostrata* by Lamont (1973). In that study, 650 μM nitrogen added to one root half as NH_4NO_3 every 3–4 days reduced cluster-root growth on that half, compared with that of the N-deprived half, but only if the shoots were of relatively low N status. We found the leaf N concentrations were similar amongst P treatments when supplied equally to both root halves.

The decrease in cluster root growth for high-P status plants can be explained in terms of a higher shoot P concentration resulting from a higher P supply to one root half (Figure 4a). However something in addition to plant-P status is needed to explain the differences in cluster root growth amongst low-P status plants. Our low-P status plants had similar internal concentrations of P (in leaf, stem and root halves) (Figure 4), however, those plants supplied with 1 μM P to both root halves had, in addition to a greater cluster root mass (proportion), significantly larger shoot mass (Figure 2). De Groot et al. (2001) have also observed that P supply rate and tissue P concentration were not linearly related in tomato. At P supply rates that limit plant growth an increase in available P (P supply) was used to increase plant growth rate, and hence plant biomass, while total plant P increases, the concentration of P remained the same. When an inadequate supply of P is limiting the growth of both cluster and noncluster roots a slight increase in P supply would be expected to increase growth, and at least partly dilute the extra P.

The total carboxylate exudation rates from mature cluster roots on 1 μM P root halves was significantly faster when the other root half was deprived of P. Since the external P concentration (1 μM) was identical for all cluster roots on these root halves, the decrease in exudation rate must have been due to an internal signal. This internal signal was systemic and, as shown above, was not just root or shoot P concentration. A faster rate of carboxylate exudation from individual cluster roots could be considered as a functional response to obtain P when the supply of P is lowest.

Changes in the growth and exudation of cluster roots over a very narrow range in P supply without concomitantly increased tissue P concentrations may represent an important transition in the metabolism of plants under P stress. Exudation was more sensitive to plant P status than either growth or initiation, as depicted in our hypothetical model (Figure 6).

Regulation of P uptake

Table 2 gives the P-uptake rate divided by the P concentration supplied to the root half. If the same linear kinetics of P-uptake rates as described by Roelofs et al (2001) applied to all *H. prostrata* plants, irrespective of treatment, a constant value is expected. In fact, this value decreased from 2.4, when both root halves received 1 μM P, to 1.0 and 0.7 when one root half received 25 and 75 μM P, respectively (Table 2). This suggests some down-regulation of the P-uptake system when 25 μM P was supplied, with little further effects when 75 μM P was supplied (see comments in results section). In comparison with the general influence of increasing P supply on P-uptake by crop species, P uptake in *H. prostrata* appears to be relatively insensitive to plant P status (Figure 6). A relatively low capacity to control the P-uptake system might account for the frequently observed symptoms of P toxicity in species of the Proteaceae (Handreck, 1997; Lambers et al., 2002; Parks et al., 2000).

Concluding remarks

Our findings illustrate the effects of P supply on internal P concentrations that regulate cluster root growth and function. While cluster-root initiation could be explained in terms of root P concentration, the P concentrations supplied to root halves and the tissue P concentrations were inadequate to entirely explain the changes in cluster-root growth and carboxylate exudation. These differences were attributed to a change in the level of P-stress where an increase in the total amount of P supplied and its incorporation into new tissues were coordinated such that growth increased without increasing the tissue P concentration. The exudation rates of cluster roots appears to be the most sensitive to plant P status, followed, by cluster-root growth and then cluster initiation, with P uptake being the least sensitive in *H. prostrata* (Figure 6).

Acknowledgements

M.W. Shane was the recipient of an International Postgraduate Research Scholarship and a University

Postgraduate Award. M. de Vos and S. de Roock were supported, in part, by Utrecht University. Thanks to Dr. T.D. Colmer and Dr. P. Poot (University of Western Australia) for critical reading of the initial manuscript and to Dr. M.D. Cramer (University of Stellenbosch, South Africa) for helpful discussions during the writing of the manuscript. We thank M. Smirk from the School of Earth and Geographical Sciences (UWA) for his assistance with tissue P (XRFS) and N (LECO®) analysis and R. Scott and L. Cox at the UWA Combined Workshop for building the split-root containers and exudate-collection cells.

References

A. O. A. C. 1975 Official Methods of Analysis. 12th edn. Association of Official Analytical Chemists, Washington, DC.

Bates T R and Lynch J P 1996 Stimulation of root hair elongation in *Arabidopsis thaliana* by low phosphorus availability. Plant Cell Environ. 19, 529–538.

Brouwer R 1963 Some aspects of the equilibrium between overground and underground plant parts. Meded. Inst. Biol. Scheikd. Onderzoek Landbouwgewassen. 213, 31–39.

Bolan N D, Elliot J, Gregg P E H and Weil S 1997 Enhanced dissolution of phosphate rocks in the rhizosphere. Biol. Fertil. Soils 24, 169–174.

De Groot C C, Marcelis R F M, Van Den Boogaard R and Lambers H 2001 Growth and dry-mass partitioning in tomato as affected by phosphorus nutrition and light. Plant Cell Environ. 24,

Dinkelaker B, Romheld V and Marschner H 989 Citric acid excretion and precipitation of calcium citrate in the rhizosphere of white lupin (*Lupinus albus*). Plant Cell Environ. 12, 285–292.

Dinkelaker B, Hengeler C and Marschner H 1995 Distribution and function of proteoid root clusters and other root clusters. Bot. Acta 108, 183–200.

Gardner W, Parbery D and Barber D 1982 The acquisition of phosphorus by *Lupinus albus* L. I. Some characteristics of the soil/root interface. Plant Soil 68, 19–32.

Gardner W, Barber D and Parbery D 1983 The acquisition of phosphorus by *Lupinus albus* L. III. The probable mechanism by which phosphorus movement in the soil/root interface is enhanced. Plant Soil 70, 107–124.

Gerke J 1994 Kinetics of soil phosphate desorption as affected by citric acid. Z. Bodenk. Pflanzenernaehr. 157, 17–22.

Gilbert G A, Knight J D, Vance C P and Allan D L 2000 Proteoid root development of phosphorus deficient lupin is mimicked by auxin and phosphonate. Ann. Bot. 85, 921–928.

Grierson P F 1992 Organic acids in the rhizosphere of *Banksia integrifolia* L.f. Plant Soil 144, 259–265.

Grose M J 1989 Phosphorus nutrition of seedlings of the waratah *Telopea speciosissima* (Sm.) R. Br. (Proteaceae). Aust. J. Bot. 37, 313–320.

Handreck K A 1991 Interactions between iron and phosphorus in the nutrition of *Banksia ericifolia* L. f. var. ericifolia (Proteaceae) in soil-less potting media. Aust. J. Bot. 39, 373–384.

Handreck K A 1997 Phosphorus requirements of Australian native plants. Aust. J. Soil Res. 35, 241–289.

Hinsinger P 1998 How do plant roots acquire mineral nutrients? Chemical processes involved in the rhizosphere. Adv. Agron. 64, 225–265

Hoffland E, Findenegg G R and Nelemans J A 1989 Solubilization of rock phosphate by rape. II. Local root exudation of organic acids as a response to P-starvation. Plant Soil 113, 161–165.

Jeffrey D W 1967 Phosphate nutrition of Australian heath plants. I. The importance of proteoid roots in Banksia (Proteaceae). Aust. J. Bot. 15, 403–411.

Jones D L 1998 Organic acids in the rhizosphere – a critical review. Plant Soil 205, 25–44.

Jones D L and Darrah P R 1994 Role of root derived organic acids in the mobilization of nutrients from the rhizosphere. Plant Soil 166, 247–257.

Jungk A O 1996 Dynamics of nutrient movement at the soil–root interface. *In* Plant Roots: The Hidden Half, 2nd edn. Eds Y Waisel, A Eshel & U Kafkaki. pp. 529–556. Marcel Dekker, New York.

Keerthisinghe G, Hocking P, Ryan PR and Delhaize, E 1998 Proteoid roots of lupin (*Lupinus albus* L.): Effect of phosphorus supply on formation and spatial variation in citrate efflux and enzyme activity. Plant Cell Environ. 21, 467–478.

Kirk G J D, Santos E E and Santos M B 1999 Phosphate solubilization by organic anion excretion from rice roots in aerobic soil: rates of excretion and decomposition, effects on rhizosphere pH, and effects on phosphate solubility. New Phytol. 142, 185–200.

Kuiper D, Schuit J and Kuiper P J C 1988 Effect of internal and external cytokinin concentrations on root growth and root to shoot ratio of *Plantago major* ssp. *pleiosperma* at different nutrient concentrations. Plant Soil 111, 231–236.

Kuiper D, Kuiper P J C, Lambers H, Schuit J and Staal M 1989 Cytokinin concentration in relation to mineral nutrition and benzyladenine treatment in *Plantago major* ssp. *pleiosperma*. Physiol. Plant. 75, 511–517.

Lambers H, Stuart Chapin F III and Pons T L 1998 Plant Physiological Ecology. Springer, New York.

Lambers H, Juniper D, Cawthray G R, Veneklaas E J and Martinez-Ferri E 2002 The pattern of carboxylate exudation in *Banksia grandis* (Proteaceae) is affected by the form of phosphate added to the soil. Plant Soil 238, 111–122.

Lamont B 1971 The effect of soil nutrients on the production of proteoid roots by Hakea species. Aust. J. Bot. 20, 27–40.

Lamont B 1972 The morphology and anatomy of proteoid roots in the genus *Hakea*. Aust. J. Bot. 20, 155–174.

Lamont B 1973 Factors affecting the distribution of proteoid roots within the root systems of two *Hakea* species Aust. J. Bot. 21, 165–187.

Lamont B 1982 Mechanisms for enhancing nutrient uptake in plant with particular reference to Mediterranean South Africa and Western Australia. Bot. Rev. 48, 597–689.

Lamont B 1993 Why are hairy root clusters so abundant in the most nutrient-impoverished soils of Australia? Plant Soil 155/156, 269–272.

Lamont B 2002 Structure and ecology of proteoid root clusters. Plant Soil 248, 1–19.

Lamont B, Brown G and Mitchell DT 1984 Structure, environmental effects on their formation, and function of proteoid roots in *Leucadendron laureolum* (Proteaceae). New Phytol. 97, 381–390.

Ma Z, Bielenberg D G, Brown K M and Lynch J P 2001 Regulation of root hair density by phosphorus availability in *Arabidopsis thaliana*. Plant Cell Environ. 24, 459–467.

Marschner H 1995 Mineral Nutrition of Higher Plants. 2nd edn. Academic Press, London.

Neumann G and Martinoia E 2002 Cluster roots – an underground adaptation for survival in extreme environments. Trends Plant Sci. 7, 162–167.

Parks S E., Haigh A M. and Creswell G C 2000 Stem tissue phosphorus as an index of the phosphorus status of *Banksia ericifolia* L. Plant Soil 227, 59–65.

Purnell H M 1960 Studies of the family Proteaceae. 1. Anatomy and morphology of the roots of some Victorian species. Aust. J. Bot. 8, 38–50.

Raaimakers D, Boot R G A, Dijkstra P, Pot S and Pons T 1995 Photosynthetic rates in relation to leaf phosphorus content in pioneer versus climax tropical rainforest trees. Oecologia. 102, 120–125.

Reddell P, Yun Y and Shipton W A 1997 Cluster roots and mycorrhizae in *Casuarina cunninghamiana*: Their occurrence and formation in relation to phosphorus supply. Aust. J. Bot. 45, 41–51.

Roelofs R F R, Rengel Z, Cawthray G R, Dixon K W and Lambers H 2001 Exudation of carboxylates in Australian Proteaceae: chemical composition. Plant Cell Environ. 24, 891–904.

Shane M W, McCully M E and Canny M J 2000 The vascular system of maize stems revisited: Implications for water transport and xylem safety. Ann. Bot. 86, 245–258.

Smith F W, Jackson W A and Vanden Berg P J 1990 Internal phosphorus flows during development of phosphorus stress in *Stylosanthes hamata*. Aust. J. Plant Physiol. 17, 451–464.

Stitt M 1997 The flux of carbon between the chloroplast and the cytoplasm. *In* Plant Metabolism. Eds. D T Denis, D H Turpin, D D Lefebvre and D B Layzell. pp. 382–400. Longman Scientific and Technical, Singapore.

Torrey J G 1986 Endogenous and exogenous influences in the regulation of lateral root formation. *In* New Root Formation in Plants and Cuttings. Ed. AB Jackson. pp. 31–66. Marinus Nijhoff, Dordrecht.

Watt M and Evans J R 1999 Linking development and determinacy with organic acid efflux from proteoid roots of white lupin grown with low phosphorus and ambient or elevated atmospheric CO_2 concentration. Plant Physiol. 120, 705–716.

Plant and Soil **248**: 221–227, 2003.
© 2003 Kluwer Academic Publishers. Printed in the Netherlands.

Differences in cluster-root formation and carboxylate exudation in *Lupinus albus* L. under different nutrient deficiencies

Ruixia Liang & Chunjian Li[1]
Department of Plant Nutrition, China Agricultural University, Beijing 100094, PR China.
[1]*Corresponding author**

Received 24 October 2001. Accepted in revised form 12 March 2002

Key words: *Lupinus albus* L., nutrient deficiency, organic acid exudation, proteoid roots, root clusters

Abstract

In contrast with the well document role of proteoid root formation and carboxylate exudation in acclimation to P deficiency in white lupin (*Lupinus albus* L.), their role under other nutrient deficiencies and their ecological significance are still poorly understood. In the present work, differences in proteoid root formation, exudation of carboxylates by root clusters, non-proteoid and proteoid root tips by using a non-destructive method, and concentrations of organic acids in the tissues of plants grown in the absence of P, Fe or K were studied. Proton release from roots increased soon after withdrawing Fe from the medium; within three days the solution pH decreased from 6 to about 4, and this increased release in protons continued until the end of the experiment. Acidification appeared much later, on the 10th day and the 14th day after withdrawal of P and K, respectively; the extent of the acidification was also weaker than under –Fe (5.2 for –P and 5.7 for control on the 10th day; 6.0 for –K and 6.1 for control on the 14th day). Root clusters formed when plants were grown under –P and –Fe, but not under –K conditions. The root clusters developed sooner under –Fe conditions, but the number of clusters was far less than under –P. Under P deficiency, root clusters released mainly citrate, but also some malate; while the major organic acid released by root tips of both non-proteoid and proteoid roots was malate. However, under Fe deficiency, the majority of the organic acids exuded both by the root clusters and root tips was malate, whereas only a small amount of citrate was detected. The release rate of citrate by – P root clusters was greater than that by – Fe root clusters. Moreover, the release rate of malate was greater in –Fe root clusters than in –P root clusters, but the opposite was found in proteoid root tips, i.e. faster in –P than in –Fe proteoid root tips. The significances of proteoid root formation and release of organic acids in acclimation to different nutrient deficiencies for white lupin plants are discussed.

Introduction

Proteoid roots develop in white lupin (*Lupinus albus* L.) under phosphorus (P) or iron (Fe) deficiency (Gardner et al., 1982, 1983; Skene, 2001). Many species, including white lupin, show an increase in organic acid exudation by roots when grown under P or Fe starvation or Al toxicity (for a review see Jones, 1998; Neumann and Römheld, 2001). The formation and functioning of proteoid roots of white lupin in response to -P are well documented. The functions of proteoid roots can be characterized by (i) enhancing the surface area of roots to enhance the surface for nutrient uptake; (ii) secreting acid phosphatase to hydrolyze organic soil P fractions; (iii) releasing phenolics, which may have antibiotic properties; and (vi) releasing a large amount of citrate and malate to mobilize P in soils (Dinkelaker et al., 1995; Gardner et al., 1982; Gilbert et al., 1999; Marschner et al., 1987; Neumann et al., 1999). Also the Al detoxification effects by carboxylates through a complexation reaction either outside (apoplasm) or inside (symplasm) the roots, and their ecological significances have been reviewed (Ma, 2000).

Casuarina glauca is very sensitive to Fe deficiency and can form proteoid roots when grown on alkaline

* FAX No: +86 10 6289 1016. E-mail: lichj@mail.cau.edu.cn

soil (Arahou and Diem, 1997). Also lupins suffer severely from Fe deficiency when they are grown on calcareous soil (White and Robson, 1989). It is well known that the typical responses of strategy I plants, including white lupin, to Fe-deficiency are to increase net efflux of H^+ and root ferric reducing capacity, and to release chelating compounds, and/or to form transfer cells in roots (Marschner, 1995; Marschner et al., 1986). Whether formation of proteoid roots and release of organic acids also play an important role in Fe-deficient white lupin plants to acclimate to –Fe stress is not well understood.

It has been reported that plants can grow under K deficiency, without external supply of fertilizer K, showing abilities to release potassium form K-bearing minerals directly (Hinsinger and Jaillard, 1993; Hinsinger et al., 1993). Recent evidence shows that *Lolium perenne*, *Zea mays* and *Brassica campestris* can mobilize mineral K form K-bearing minerals, when they grow under P deficiency. This is contributed to the release of organic acids from the P-deficient roots (Wang et al., 2000). However, direct evidence for increased release of organic acids by plant roots under K deficiency is absent.

The objective of this study was to compare the formation of proteoid roots and amounts of organic acids exuded by white lupin roots under different nutrient deficiencies, and to discuss their ecological significances for acclimation to different nutrient-stressed environments.

Materials and methods

Plant culture and treatments

Seeds of white lupin (*Lupinus albus* L. cv. Amigo, Suedwestdeutsche Saatzucht, Rastat, Germany) were surface-sterilized with 10% hydrogen peroxide for 30 min, washed in running tap water, and then germinated in river sand under controlled conditions at 25 °C in the dark. The sand was irrigated with a saturated $CaSO_4$ solution.

When the seedling tap roots were 3 – 5 cm long (8 days after germination), they were transferred into an aerated 1/2-strength complete nutrient solution containing (m*M* for full strength): K_2SO_4 0.75, KH_2PO_4 0.25, KCl 0.1, $MgSO_4$ 0.65, $Ca(NO_3)_2$ 2, FeNaEDTA 0.1, H_3BO_3 1×10^{-2}, $MnSO_4$ 1×10^{-3}, $ZnSO_4$ 1×10^{-3}, $CuSO_4$ 4×10^{-4}, $(NH_4)_6Mo_2O_4$ 5×10^{-6}. There were four seedlings per pot containing 2 L nutrient solution. Plants were grown in a growth room at 25/18 °C and with a 14-h photoperiod at a light intensity 180–210 μmol m^{-2} s^{-1} provided by reflector sunlight dysprosium lamps (DDF 400, Nanjing, China).

After 4 days (12 days after germination), plants were treated with a full-strength solution or with one element omitted, as follows: (i) control (no omissions in nutrient); (ii) -Fe (FeNaEDTA was omitted); (iii) -P (KH_2PO_4 was replaced by 0.25 m*M* KCl); and (IV) -K (K_2SO_4, KH_2PO_4 and KCl were replaced by 0.75 m*M* Na_2SO_4, 0.25 m*M* NaH_2PO_4 and 0.05 m*M* $CaCl_2$). The nutrient solution was replaced every 3 days.

The pH of the nutrient solution was adjusted to 6.0 at each change of solution and was measured three times a day.

Number of proteoid roots

A root cluster was regarded as the lateral root axis and emergent rootlets. A proteoid root was considered as the lateral root axis plus one or several root clusters. Three pots with four plants for every treatment were used to record every three days the numbers of proteoid roots formed.

Localized collection of root exudates

Carboxylates exuded either by non-proteoid root tips, proteoid root tips or cluster rootlets were collected using a non-destructive method (Neumann and Römheld, 1999). A single plant was transferred onto filter paper and moistened with sterilized water. A piece of plastic film, which was soaked previously in methanol for 6 h and washed with sterilized water, was placed under the root tip or cluster rootlet. The root tip or mature root cluster (4–5 days after emergence of rootlets) was then positioned between two circular filter papers (0.7 cm in diameter sterilized as above). Five root tips or mature root clusters in a plant were collected as one sample. The rest of the roots were covered with moistened filter papers. During the collection, the whole root was covered with a black plastic sheet to keep them dark and moist. After 2 h, five pairs of circular filter papers for each sample were placed together into a 1.5 mL Eppendorf tube. Organic acids were extracted by addition of 400 μl bi-distilled water and subsequent centrifugation at 10 000 *g* for 10 min. The supernatant analyzed for organic acid content by High Performance Liquid Chromatography (HPLC).

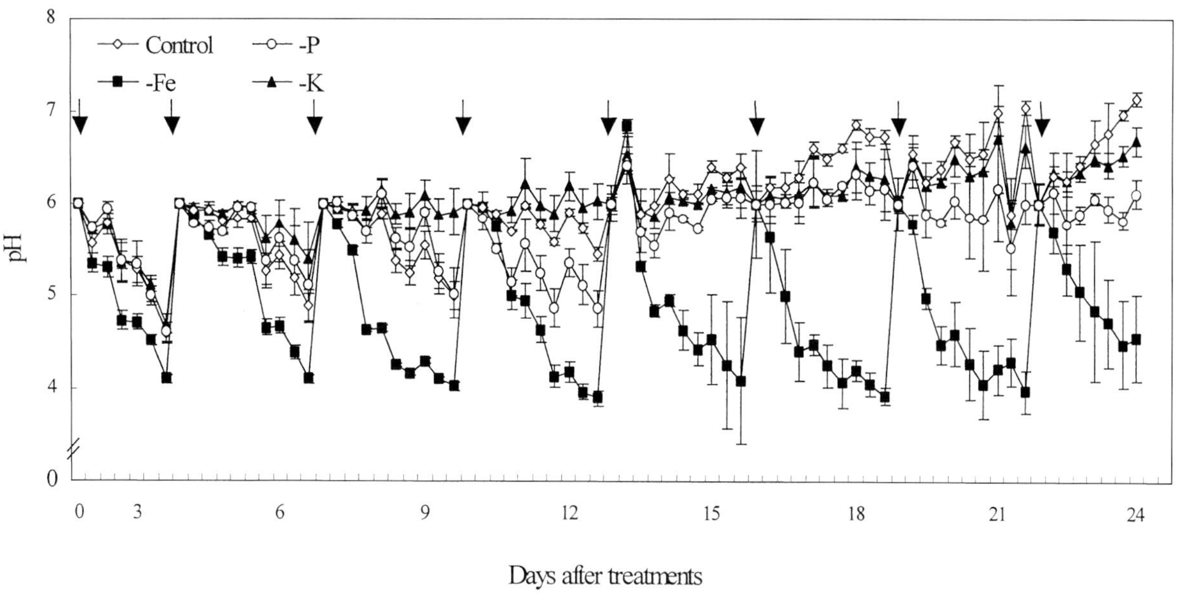

Figure 1. Changes in nutrient solution pH of white lupin plants grown under -P, -K or -Fe conditions for 24 days. The pH was measured three times per day. Arrows indicate the times when nutrient solutions were replaced. Means ± SE, n=5.

Extraction of organic acids from cluster rootlets and proteoid root tips

Cluster rootlets or proteoid root tips were excised from proteoid roots, placed into a 1.5 mL Eppendorf tube separately, weighed and then frozen below −18 °C until analyzed. For analysis, 1 mL of 440 mM H_3PO_4 solution was added to the tube, organic acids were extracted in a sonic water bath for 15 min. After centrifugation at 20 000 g (4 °C), organic acids were analyzed using HPLC.

Organic acid analyses

Organic acids were analyzed according to Neumann and Römheld (1999) using HPLC. Twenty μl of sample solution, either from the collected root exudates, or from the tissue extraction, was injected into a reversed-phase C-18 column (Silicon 250×5.6 mm, Shimadzu, Japan) in a HPLC (Shimadzu LC-10A, Japan). The column was eluted for 20 min at 26 °C with the mobile phase KH_2PO_4 (18 mM, pH 2.1) and the flow rate was 0.7 mL/min. Detection of the organic acids was at 214 nm, using a multiple wavelength spectrophotometer (190–600 UV-VIS; SPD-10A/Shimadzu, Japan). Organic acids in the samples were identified by comparison with the retention times of pure standards obtained from Sigma Chemicals Ltd, USA. Standards of the following organic acids were prepared (in the order of increasing retention time): L-malic, malonic, acetic, citric and succinic acid.

Statistics

There were three replicates for each treatment. For statistical analysis of the data, the program SAS for Windows (version 6.12) was used (SAS, 1987). Differences in rates of carboxylate exudation by different types of root tips under –P, -Fe or –K, and in organic acid concentration in root tissues were tested with ANOVA. The values for the organic acids in root tissues were squared-transformed before statistical analysis.

Results

Shoot growth

Typical Fe-deficiency symptoms appeared earlier in white lupin than P- or K-deficiency symptoms; nine days after Fe starvation, young leaves started to show mild chlorosis, and the symptoms gradually became more severe during experiment. Under K deficiency, the petiole of older leaves yellowed until about 13 days after the beginning of the treatment and was shed after 22 days. P-deficiency symptoms were present by 18 days after onset of the treatment (results not shown).

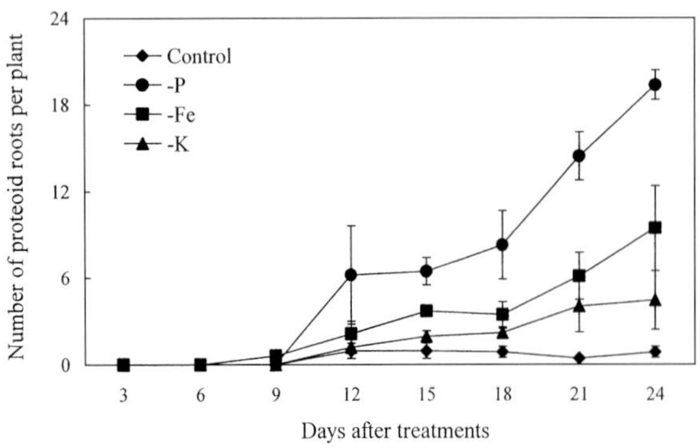

Figure 2. Effects of -P, -Fe or -K treatment on proteoid root formation in white lupin plants over a period of 24 days. Means ± SE, n=3.

Solution pH

The solution pH decreased faster on the first day of the -Fe treatment, and fell rapidly from 6 to 4.1 on the third day (Figure 1). This root response under -Fe occurred much earlier than the response of the leaves. On the 7th day after -Fe, when root clusters started to occur, the nutrient solution was acidified from 6.0 to 4.6 during the first 12 h after changing the nutrient solution, decreasing more rapidly than before. This tendency was maintained until the end of the study. Acidification did not occur until the 10th day and the 14th day after -P and –K, respectively, that is, much later than that after Fe-deficiency, and the extent of acidification was also less than under -Fe (Figure 1).

For all of the treatments the changes in pH at day and night were different. They decreased at day but increased at night (Figure 1).

Proteoid root formation

Proteoid roots were observed under both P-deficient and Fe-deficient conditions, but the time of initiation, appearance and numbers of proteoid roots per plant varied (Figure 2). Root clusters formed in the upper young lateral roots on the 7th or 8th day after the start of the -Fe treatment, but did not appear until the 12th day after the beginning of the -P treatment. Proteoid roots were much less and had a lower density of rootlets in the –Fe treatment than in the -P treatment (Figure 2). Furthermore, proteoid roots of –Fe plants were yellow, in agreement with the recent report by Hagström et al. (2001). Almost no proteoid roots were formed in K-deficient plants.

Table 1. Effects of -P, -Fe or -K on rates of carboxylate exudations from non-proteoid root tips of white lupin plants during a 24-day study period. For each carboxylate anion, values for different days and treatments with the same letter are not significantly different at $P > 0.05$. * not detected

Treatments	Rate of carboxylate exudation (10^{-9}g M^{-1} root s^{-1})		
	Malate	Malonate	Citrate
4 d			
Control	7.15 gfeh	ND*	2.43 ced
-P	5.01 gfeh	ND	2.61 ced
-Fe	1.67gh	ND	ND
-K	5.04 gfeh	ND	2.20 ced
9 d			
Control	ND	ND	2.34 ced
-P	10.87 gfed	ND	4.84 cbd
-Fe	9.33 gfed	ND	ND
-K	10.97 gfed	1.01 bc	ND
14 d			
Control	6.26 gfeh	ND	3.54 ced
-P	10.55 gfed	ND	5.86 bc
-Fe	11.01 gfed	ND	1.45 ed
-K	4.70 gfh	ND	1.23 ed
19 d			
Control	16.34 cbd	0.76 bc	4.83 cbd
-P	17.37 cb	0.81 bc	7.56 b
-Fe	14.01 ced	ND	3.97 ced
-K	12.61 cfed	ND	3.10 ced
24 d			
Control	21.22 cb	2.38 a	5.61 bc
-P	34.97 a	1.37 ab	18.85 a
-Fe	24.88 b	1.24 bc	4.42 cbd
-K	17.99 cbd	0.86 bc	5.55 bc

Organic acid exudation from non-proteoid root tips

There were no differences in the kind of carboxylates exuded by non-proteoid root tips amongst the treatments. The major carboxylate was malate (citrate was also found, especially for -P plants). The exudation rate of carboxylate increased with increasing exposure to the treatment (Table 1). In comparison with the control plants, carboxylate exudation by non-proteoid lateral root tips increased under -P, and less so under -Fe, but decreased under -K.

Table 2. Effects of –P or -Fe on rates of carboxylate exudations from cluster rootlets and proteoid root tips of white lupin plants during a 24-day study period. For each carboxylate anion, values for different days and treatments with the same letter are not significantly different at $P > 0.05$. * not detected

Treatments	Rate of carboxylate exudation (10^{-9}g M^{-1} root s^{-1})		
	Malate	Malonate	Citrate
14 d			
Cluster rootlet, -P	1.45 d	ND*	ND
Cluster rootlet, -Fe	20.72 cd	ND	ND
Proteoid root tip, -P	10.13 cd	ND	ND
Proteoid root tip, -Fe	ND	ND	ND
19 d			
Cluster rootlet, -P	6.83 cd	ND	86.38 a
Cluster rootlet, -Fe	12.31 cd	ND	11.45 d
Proteoid root tip, -P	24.60 b	1.89 b	5.83 d
Proteoid root tip, -Fe	10.63 cd	ND	3.83 d
24 d			
Cluster rootlet, -P	12.58 cd	3.06 ab	36.94 b
Cluster rootlet, -Fe	26.63 b	1.39 b	7.87 d
Proteoid root tip, -P	41.89 a	3.19 ab	21.81 c
Proteoid root tip, -Fe	17.09 cb	5.91 a	7.96 d

Organic acid exudation from cluster rootlets and proteoid root tips

Because there was almost no proteoid root formation under –K, we only compared the concentrations of organic acids in proteoid root tissues and exudates under –P and – Fe stress. The cluster rootlets of P-deficient plants released a large amount of citrate from 19 days after the treatment. Cluster rootlets of the Fe-deficient plants exuded mainly malate, and the exudative burst began earlier than that under P stress, i.e. 14 days after the –Fe treatment. The release rate of malate by cluster rootlets under –Fe was much greater than that under –P stress, but that of citrate was much greater under –P (Table 2).

In comparison with the cluster rootlets, proteoid root tips released mainly malate, both under–P and –Fe stress. The exudation rates of both malate and citrate of –P proteoid root tips were greater than those of –Fe ones (Table 2).

Tissue organic acid concentrations

The variation in organic acids in root tissues was greater than that in root exudates (Tables 2 and 3). A large amount of citrate was detected in –P cluster rootlets. There was more malate both in cluster rootlets and proteoid root tips of -P plants than in those of –Fe plants, except in the cluster rootlet tissues on day 24 after the treatment commenced (Table 3).

Discussion

White lupin showed higher tolerance to -P than to -Fe. Soon after omitting Fe from the nutrient solution, the roots started to increase the release of protons, resulting in stranger acidification of the solution. This increased release of protons was further enhanced when root clusters were initiated (Figure 1, Table 1), which may be associated with the exudation of carboxylates by the clusters (Dinkelaker et al., 1995; Neumann et al., 1999). Chlorosis in young leaves was apparent only one week after Fe starvation, while P-deficient symptoms did not appear until day 18 of the treatment.

As has been found before (Gardner et al., 1982, 1983; Neumann et al., 1999), root clusters of P-deficient white lupin plants released mainly citrate (Tables 1 and 2), while the major carboxylate released by root tips, including non-proteoid roots and proteoid roots, was malate. In comparison, the major organic acid exuded by both root clusters and root tips of Fe-deficient plants was malate. Moreover, the release rate of malate was faster in -Fe root clusters than in -P root clusters, but the opposite was found in proteoid root tips, i.e. faster in -P than in -Fe proteoid root tips. The release rate of citrate from root clusters under -P was faster than that under -Fe (Table 2). These findings disagree with the recent report that citrate release by -Fe cluster roots during the exudative burst was 10-fold higher than that in -P cluster roots (Hagström et al., 2001). The reasons for these conflicting findings are unknown.

Proteoid root formation and increase in carboxylate release were not specific responses of white lupin to P deficiency. They were also found under Fe deficiency (Figure 2, Tables 1 and 2), in agreement with the literature (for review see Skene, 2001). However, these responses might have different ecological significance for white lupin to acclimate to -P or -Fe stress. Up to 40 days after the onset of the -P treatment, root clusters were observed in almost all of the lateral roots, and there were even several clusters in one proteoid root (results not shown). These large numbers of root clusters released a major amount of organic acids, especially citrate (Gardner et al., 1982, 1983; Neumann et al., 1999). Citrate is very efficient in comparison with malate to release P from soils (Jones, 1998), and thus enable plants to acclimate to a low-P environ-

Table 3. Changes in composition and concentration of carboxylates in tissues of cluster rootlets and proteoid root tips of white lupin plants grown under -P or -Fe conditions for a-24 day period. For each carboxylate anion, values for different days and treatments with the same letter are not significantly different at $P > 0.05$. * not detected

Treatments	Carboxylate concentration (mg g^{-1} FW)				
	Malate	Malonate	Acetate	Citrate	Succinate
14 d					
Cluster rootlet,-P	0.28 ab	0.21 bc	0.66 b	0.85 b	0.90 de
Cluster rootlet,-Fe	0.12 b	0.16 bc	1.21 b	ND*	3.02 ab
Proteoid root tip,-P	1.52 a	0.37 ab	0.97 b	ND	3.98 a
Proteoid root tip,-Fe	0.59 ab	0.10 c	1.55 ab	ND	4.23 a
19 d					
Cluster rootlet,-P	0.59 ab	0.03 c	1.29 b	3.43 a	ND
Cluster rootlet,-Fe	0.13 b	0.03 c	1.61 ab	0.55 b	1.44 dc
Proteoid root tip,-P	1.34 ab	ND	0.81 b	0.43 b	0.73 de
Proteoid root tip,-Fe	0.22 ab	ND	0.93 b	0.36 b	1.26 dce
24 d					
Cluster rootlet,-P	0.29 ab	ND	0.72 b	2.70 a	0.08 e
Cluster rootlet,-Fe	1.45 ab	0.47 a	1.27 b	0.45 b	1.02 de
Proteoid root tip,-P	1.45 ab	ND	1.08 b	1.36 b	0.72 de
Proteoid root tip,-Fe	0.99 ab	0.16 bc	2.38 a	0.81 b	2.38 bc

ment. Under Fe deficiency, however, proteoid root formation and organic acid release could not play an important role for white lupin to acclimate to low-Fe environment. Firstly, only a few proteoid roots were formed before day 24. Up to 40 days after the onset of the treatment, there was no increase in the number of proteoid roots. There was commonly single cluster in each proteoid root. Secondly, the total rate of organic acid exudation under -Fe was lower than that under -P, and the majority of the organic acids exuded under -Fe stress were malate (Tables 1 and 2). Malate has a high affinity for trivalent metals such as Fe^{3+}, and thus can release Fe from soils, but it loses its capacity to complex metals when the pH is higher than 4. By contrast, citrate can complex metals until pH 8 (Jones, 1998). Therefore, proteoid root formation and carboxylate release under Fe starvation may not play an important role in white lupin to acclimate to a low-Fe environment.

White lupin roots increased to release protons soon after -Fe treatment commenced. During the first 3 days, the solution pH fell from 6 to about 4, and this increased release of protons continued until the end of the experiment (Figure 1). This is probably due to a highly up-regulated root H^+ ATPase (Alcantara et al., 1991; Rabotti and Zocchi, 1994). An increase in net efflux of $H^+$is one of the typical responses to Fe deficiency of strategy I plants (Marschner, 1995). The acidified rhizosphere is beneficial for (i) increasing the Fe^{2+} concentration in the rhizosphere (Lindsay and Schwab, 1982); and (ii) maintaining a high activity of plasma membrane ferric reductase in roots, since the activity of the ferric reductase is strongly stimulated by low pH (Marschner, 1995; Marschner et al., 1986). Although the root ferric reductase activity under -Fe stress was not measured in the present study, proton release and increases in activity of plasma membrane ferric reductase may play an important role in white lupin to acclimate to Fe deficiency.

To explain the abilities of plants to release potassium from K-bearing minerals, dynamic equilibrium reactions where the exchange equilibrium shift occurs due to plant uptake of K, and direct mobilization of mineral K by plant roots as a result of H^+ excretion and subsequent dissolution of K-bearing minerals have been suggested (Hinsinger et al., 1993, 2001). Although organic acids exuded by P-deficient roots can mobilize mineral K from K-bearing minerals (Wang et al., 2000), direct evidence for increased release of organic acids by plant roots under K deficiency is until now absent. In the present experiment, -K lupin plants did not show obvious increase in proton release (Figure 1), proteoid root formation (Figure 2) or carboxylate exudation (Tables 1 and 2), implying that releases in proton and organic acids are not a strategy, at least in white lupin to acclimate to K deficiency.

Acknowledgements

The authors are grateful to Prof. Dr. H. Lambers for critical evaluation and correction of the manuscript and to the NKBRSF (No: G1999011707) for financial support.

References

Alcantara E, de la Guardia M D and Romer F J 1991 Plasmalemma redox activity and H^+ extrusion in roots of Fe-deficient cucumber plants. Plant Physiol. 96, 1034–1037.

Dinkelaker B, Hengeler C and Marschner H 1995 Distribution and function of proteoid roots and other root clusters. Bot. Acta 108, 183–200.

Gardner W K, Parbery D G and Barber D A 1982 The acquisition of phosphorus by *Lupinus albus* L. I. Some characteristics of the soil /root interface. Plant Soil 68, 19–32.

Gardner W K, Barber D A and Parbery D G 1983 The acquisition of phosphorus by *Lupinus albus* L. III. The probable mechanism by which phosphorus movement in the soil /root interface is enhanced. Plant Soil 70, 107–124.

Gilbert G A, Knight J D, Vance C P and Allan D L 1999 Acid phosphatase activity in phosphorus-deficient white lupin roots. Plant Cell Environ. 22, 801–811.

Hagström J, James W M and Skene K R 2001 A comparison of structure, development and function in cluster roots of *Lupinus albus* L. under phosphate and iron stress. Plant Soil 232 , 81–90.

Hinsinger P, Barros O N F, Benedetti M F, Noack Y and Callot G 2001 Plant-induced weathering of a basaltic rock: Experimental evidence. Geochim. Cosmochim. Acta 65, 137–152.

Hinsinger P and Jaillard B 1993 Root-induced release of interlayer potassium and vermiculitization of phlogopite as related to potassium depletion in the rhizosphere of ryegrass. J. Soil Sci. 44, 525–534.

Hinsinger P, Elsass F, Jaillard B and Robert M 1993 Root-induced irreversible transformation of a trioctahedral mica in the rhizosphere of rape. J. Soil Sci. 44, 535–545.

Jones D L 1998 Organic acids in the rhizosphere – a critical review. Plant Soil 205, 25–44.

Lindsay W L and Schwab A P 1982 The chemistry of iron in soil and its availability to plants. J. Plant Nutr. 5, 821–840.

Ma J F 2000 Role of organic acids in detoxification of aluminium in higher plants. Plant Cell Physiol. 41, 383–390.

Marschner H 1995 Mineral Nutrition of High Plants. Academic Press, London. Marschner H, Römheld V and Kissel M 1986 Different strategies in higher plants in mobilization and uptake of iron. J. Plant Nutr. 9, 695.

Marschner H, Römheld V and Cakmak I 1987 Nutrition availability in the rhizosphere. J. Plant Nutr. 10, 1174–1184.

Neumann G and Römheld V 1999 Root excretion of carboxylic acids and protons in phosphorus-deficient plants. Plant Soil 211, 121–130.

Neumann G and Römheld V 2001 The release of root exudates as affected by the plant's physiological status. *In* The Rhizosphere, Biochemistry and Organic Substances at the Soil-Plant Interface. Eds. R Pinton, Z Varanini and P Nannipieri. pp. 41–93. Marcel Dekker, Inc. New York.

Neumann G, Massonneau A, Martinoia E and Römheld V 1999 Physiological adaptations to phosphorus deficiency during proteoid root development. Planta 208, 373–382.

Rabotti G and Zocchi G 1994 Plasma membrane-bound H^+ ATPase and reductase activities in Fe-deficient cucumber roots. Physiol. Plant. 90, 779–785.

SAS Institute Inc. 1987 SAS/STAT guide for personal computers, 6th edn. Cary, NC.

Skene K R 2001 Cluster roots: model experimental tools for key biological problems. J. Exp. Bot. 52, 479–485.

Wang J G, Zhang F S, Zhang X L and Cao Y P 2000 Release of potassium from K-bearing minerals: Effects of plant roots under P deficiency. Nutrient Cycling in Agroecosystems 56, 45–52.

White P F and Robson A D 1989 Effect of soil pH and texture on the growth and nodulation of lupins. Aust. J. Agric. Res. 40, 63–73.

Plant and Soil **248:** 229–235, 2003.
© 2003 Kluwer Academic Publishers. Printed in the Netherlands.

Is Fe deficiency rather than P deficiency the cause of cluster root formation in *Casuarina* species?

El Houssine Zaïd[1,3], Moustapha Arahou[1], Hoang G. Diem[2] & Rachida El Morabet[1]
[1]*Department of Biology, University Mohammed V, Rabat-Agdal, BP 1014 Avenue Ibn Battouta, Morocco.* [2]*CNRS, Direction des Relations Internationales, 3 Rue Michel-Ange, 7594 Paris Cedex 16, France.* [3]*Corresponding author**

Received 4 January 2002. Accepted in revised form 11 June 2002.

Key words: CAS assay, *Casuarina*, cluster roots, Fe deficiency, P deficiency, proteoid roots

Abstract

When subjected, directly (through nutritional deficiencies) or indirectly (through alkaline constraints leading to such deficiencies) to nutrient deficiencies, certain plants respond by developing special root structures called cluster roots. This phenomenon can be considered as an ecophysiological response to a specific nutrient deficiency enabling plants to enhance nutrient uptake. Experiments conducted on an alkaline and an acid soil showed that *Casuarina glauca* (Sieber ex Spreng.) produced cluster roots only in the alkaline soil and not in the acid soil. In addition, iron (Fe) and phosphorus (P) deficiencies were examined separately or together to determine their effect on cluster root formation in *C. glauca* seedlings grown hydroponically. Results from experiments carried out on three *Casuarina* species (*C. glauca*, *C. cunninghamiana* Miq. and *C. equisetifolia* L.) indicated that Fe is involved in cluster root formation. In nutrient media lacking P but containing Fe, no cluster roots formed while seedlings receiving P and lacking Fe developed cluster roots. When incubated on chrome-azurol S-agar on blue plates (CAS assay), a technique used routinely to detect the production of siderophores by micro-organisms, the root system of Fe-deficient plants exhibited orange halos around cluster roots, indicating production of a ferric-chelating agent. It is concluded that the capacity of cluster roots of *C. glauca* to chelate Fe allows the plant to grow normally on alkaline soils.

Introduction

The acquisition of water and nutrients is a major function of a plant root system. Since plant survival depends on the ability to acquire nutrients, plants have developed strategies to enhance nutrient uptake from nutrient-poor soils. Of particular interest is the occurrence of cluster roots (CR), also known as proteoid roots, typically found on plants species adapted to nutrient-poor soils (Lamont, 1993) in a wide range of species and families (Lamont, 1993; Purnell, 1960) as a means to acquire nutrients. During the last few years, evidence has accumulated which suggests that these root structures are important adaptations for nutrient acquisition (Skene, 2000). Formation of CR is favoured by P deficiency in a variety of species, including *Myrica cerifera* L. (Louis et al., 1990), *Gymnostoma papuanum* (S. More) L. Johnson (Racette et al., 1990), *M. gale* L. (Crocker and Schwintzer, 1993), *Grevillea robusta* Cunn. ex R. Br. (Skene et al., 1996) and *Lupinus albus* L. (Gilbert et al., 2000; Kerley et al., 2000; Neumann et al., 2000). However, Fe deficiency is the main cause of cluster root formation in *Casuarina glauca* (Arahou and Diem, 1997; Diem and Arahou, 1996). The recent comparative study of the structure, development and function of CR under iron and phosphorus stress made by Hagström et al. (2001) in white lupin revealed that the number of CR formed was 6-fold greater in –P than in –Fe conditions.

The aim of the present study was to determine which of the deficiencies (Fe or P deficiency) is the main cause of CR formation in three *Casuarina* spe-

* E-mail: abouyounous@caramail.com

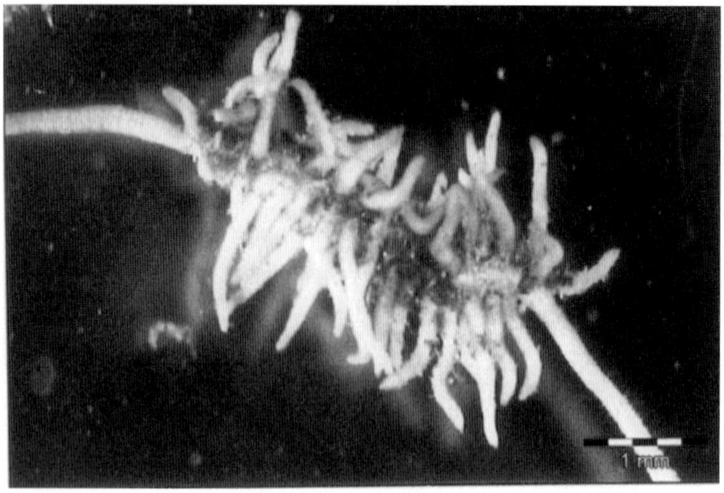

Figure 1. Details of CR from *Casuarina glauca.*

cies, and to examine briefly the means by which nutrients are acquired by theses structures.

Materials and methods

In preliminary work, *Casuarina glauca* Sieber ex Spreng. was tested for its ability to produce CR under nutritional deficiencies (Arahou and Diem, 1997; Zaïd and Arahou, 1997).

Together with this species, two others, *C. cunninghamiana* Miq. and *C. equisetifolia* L., were included in the present investigation.

Seeds of *Casuarina glauca* and *C. cunninghamiana* were collected around the city of Rabat (Morocco) whereas those of *C. equisetifolia* were kindly provided by CSIRO (Australia).

Pre-germinated seeds were cultivated on autoclaved sand and irrigated with a nutrient solution (Broughton and Dilworth, 1971) containing (μM): $CaCl_2$ (1000), KH_2PO_4 (500), $MgSO_4$ (250), K_2SO_4 (250), H_3BO_3 (2), $MnSO_4$ (1), $ZnSO_4$ (0.5), $CuSO_4$ (0.2), $CoSO_4$ (0.1), Na_2MoO_4 (0.1) and supplemented with 500 μM KNO_3. Iron was supplied as NaFe (III) EDTA, 20 μM and the pH was adjusted with NaOH (1 M) to 6.8. EDDHA (ethylenediamine dihydroxyphenylacetic acid) was supplied as an Fe-binding chelator at the concentration of 50 μM. This concentration has been reported (Arahou and Diem, 1997) as an efficient concentration causing a strong chelation of low Fe in the solution and resulting in Fe-stressed conditions.

Twelve weeks after sowing, the seedlings were removed from the sand and transferred to nutrient solution in capped plastic pots, five seedlings per pot. Each pot contained 700 ml of nutrient solution without aeration.

Experiments were carried out in a growth chamber at a relative humidity of 75% and 26/20 °C day and night temperature, 14-h day length. Light in the growth chamber was provided by mixed cool-white fluorescent tubes and incandescent bulbs. After a 2-week acclimation period in the complete nutrient solution, plants were separated into two groups according to the following experiments:

Experiment 1: effect of iron

The effect of iron ($FeCl_3$) concentrations (0.0, 0.1, 0.5, 2.0, 5.0, 10.0 and 20.0 μM) on CR formation by *Casuarina* species in the absence of P (KH_2PO_4 replaced by KCl) was studied.

Experiment 2: effect of phosphorus

This experiment was carried out to examine the effect of P concentrations (0, 0.5, 1, 2 and 4 mM as KH_2PO_4) in the absence of Fe on the CR formation.

In both experiments, the nutrient solution was renewed once a week and the plants were harvested after 8 (experiment 2) or 12 weeks (experiment 1) of treatment and the results were means of eight to 10 replicates.

Table 1. Effect of Fe concentrations on CR formation by *Casuarina* plants growing in nutrient solution containing EDDHA (50 μM) but lacking P (12 weeks of treatment)

$FeCl_3$	% of plants with CR			No. of CR per plant		
(μM)	*C. gl*	*C. cun.*	*C. eq*	*C. gl*	*C. cun.*	*C. eq*
0	90	100	60	16.7 e	30.7 d	14.5 c
0.1	90	100	40	14.0 d	25.2 cd	11.0 bc
0.5	80	100	50	9.7 cd	20.2 c	9.8 b
2	80	80	20	9.7 cd	9.0 b	7.3 b
5	70	40	0	6.8 bc	1.3 a	0.0 a
10	40	10	0	4.0 ab	0.2 a	0.0 a
20	40	0	0	4.0 ab	0.0 a	0.0 a
100	0	0	0	0.0 a	0.0 a	0.0 a

Values within a column followed by the same letter are not significantly different according to Newman–Keuls test at $P = 0.05$.

Table 2. Effect of P concentrations on CR formation by *Casuarina* plants growing in nutrient solution containing EDDHA (50 μM) but lacking Fe (8 weeks of treatment)

KH_2PO_4 (mM)	% of plants with CR		No. of CR per plant		Density of rootlets per CR	
	C. gl	*C. cun.*	*C. gl*	*C. cun.*	*C. gl*	*C. cun.*
0.0	90	100	22.6 c	23.5 b	14.6	19.6
0.5	90	100	9.3 a	16.4 a	16.5	17.5
2.0	80	70	13.0 ab	17.0 a	15.8	16.1
4.0	80	80	12.5 ab	17.2 a	26.5	21.3

Values within a column followed by the same letter are not significantly different according to Newman–Keuls test at $P = 0.05$, $n = 10$.

Experiment 3: CAS test

Plants used for this assay were those forming typical CR (Figure 1). CAS-agar was prepared as described by Schwyn and Neilands (1987). Agar (0.75%) was added to CAS solution and then autoclaved. A blue complex is formed after the chrome-azurol S reacts with iron (III) in the presence of a detergent. This blue solution was kept at 40 °C until it was used to cover the root system placed onto a plate. A positive reaction with CAS reagent is indicated by an orange colour. Indeed, when iron (III) ($FeCl_3$) is removed from the dye by a strong chelator (excreted through the root system), the blue complex turns to orange.

Results

Effect of Fe

After 3 weeks of growth under Fe deficiency, plants developed yellowish leaves (chlorosis). The magnitude of chlorosis as indicated by chlorosis indices (results not shown) indicated a strong Fe deficiency in all the species tested.

The percent of plants forming CR as well as the number of CR per plant and the density of rootlets per CR decreased as iron concentration in the nutrient solution increased (Table 1). In the absence of P, no CR formed when enough iron was supplied in the nutrient solution. In contrast, in the absence of iron, CR formed even when P was added (Table 2). These observations agree with previous results (Diem and Arahou, 1996) on *C. glauca* and show the crucial role of iron deficiency in CR formation in *Casuarina.*

The exogenous iron ($FeCl_3$) concentration required to completely inhibit formation of CR varied with plant species (Table 1): this concentration was about 100 μM for *C. glauca*, 20 μM for *C. cunninghamiana* and only 5 μM for *C. equisetifolia.*

It is concluded that under our experimental conditions, *C. equisetifolia* was the most sensitive to iron deficiency, while *C. glauca* was the least sensitive.

Effect of P

Under Fe deficiency, plants produced CR in the absence of P and continued to produce CR even when 4 mM P was added to the medium (Table 2).

Detection of iron chelation activity

In order to test whether CR manifested iron chelation activity, the CAS assay, which is a technique used routinely to detect the production of siderophores by micro-organisms, was used. When incubated on chrome-azurol S-agar on blue plates, the root system exhibited orange halos around CR (Figure 2) indicating production of a ferric-chelating agent. No such orange colour was observed in other parts of the root system.

Discussion

Alkaline soils, typically calcareous, which cover more than 30% of the Earth's surface (Chen and Barak, 1982) may cause serious problems to non-adapted plants incapable of using efficient strategies to overcome nutrient shortage and achieve the necessary metabolic activity required for their normal growth, as discussed elsewhere (Zaïd et al., 2002).

In such soils, the activity of Ca^{2+} ions is elevated (Passama et al., 1975; Salsac, 1969; Zaïd and Guerrier, 1982) while P and Fe are only slightly soluble (Lindsay, 1984) thus plants receive too little P and Fe. As a result, a decrease in P (Carter, 1987) and Fe (Tang et al., 1993) uptake is observed leading to iron chlorosis, which is a world-wide problem (Miller et al., 1984), especially for plants that are adapted to acid soils.

There are many mechanisms by which plants that are adapted to alkaline soils alleviate alkalinity stress. Indeed, in response to certain environmental signals and stresses, plants can modify their rhizosphere by exuding from their roots organic anions, which have been associated with nutrient deficiencies (Ryan et al., 2001). These plants can, through their root action, make a contribution to the mobilisation of some macro- and micronutrients. For example, morphological changes in roots are induced by low bioavailability of P and Fe (Schmidt and Schikora, 2001). Interestingly, CR formed in *L. albus* under low internal levels of plant P and Fe (Hagström et al., 2001; Liang and Li, 2002). Römheld and Marschner (1986) proposed two strategies for iron acquisition by plants while Gardner et al. (1982) and Reddell et al. (1997) suggested that the production of reducing agents, hydrogen ions, chelating agents and phosphatases as root exudates may increase phosphate mobility.

In CR-producing plants, this exudation is performed by these special root structures and the benefits that plants derive from the presence of such compounds in the rhizosphere is the mobilisation of nutrients from soil (Watt and Evans, 1999) such as mineral P, Mn, Fe, Cu and Zn (Braum and Helmke, 1995; Dinkelaker et al., 1995). Therefore, in these species, the formation of CR may be another way to overcome the constraints of alkalinity. Under low internal levels of plant P (Hagström et al., 2001; Shane et al., 2002; Shen et al., 2002) and Fe (Hagström et al., 2001), CR exude citrate. However, in *L. albus* the –Fe CR exhibited much faster (about 80-fold) rates of citrate exudation than the –P ones (Hagström et al., 2001). In the same species, Liang and Li (2002) found that the nature of the major organic acid released by CR depends on the type of deficiency: citrate was mainly excreted under –P conditions while it was malate which was mainly released under –Fe conditions.

Effect of iron

Iron plays a key role in carbon metabolism (Miller et al., 1984; Nenova and Stoyanov, 1993) as well as in nitrogen metabolism (Appleby, 1984; Rueter and Ades, 1987; Suzuki and Gadal, 1982) and contributes to the inhibition of senescence (Garg et al., 1986). One of the important roles of this micronutrient is its involvement as an essential cofactor in chlorophyll biosynthesis (Rai et al., 1984; Sharma and Sanwal, 1992). Indeed, iron is needed partially for ferredoxin (Alcaraz et al., 1986; Huang et al., 1992; Miller et al., 1982; 1984), which may be necessary in the reduction and activation of the enzyme synthetising δ amino levulinic acid (Huang et al., 1992) and could directly affect chlorophyll biosynthesis (Miller et al., 1995).

Iron occurs in soils in different forms, only some of which are available to plants. Most is in the form of Fe^{3+} (Uren, 1984) and in calcareous soils, iron mainly occurs as $Fe(OH)_4^-$ (Loeppert, 1986). Since iron must be solubilised before it can be taken up (Bienfait and Van Der Mark, 1983), roots excrete specific molecules to achieve this goal. CR produced by *C. glauca* seem to be well suited to this task.

Effect of P

In certain actinorhizal species, P deficiency rather than Fe deficiency induces the formation of CR. The

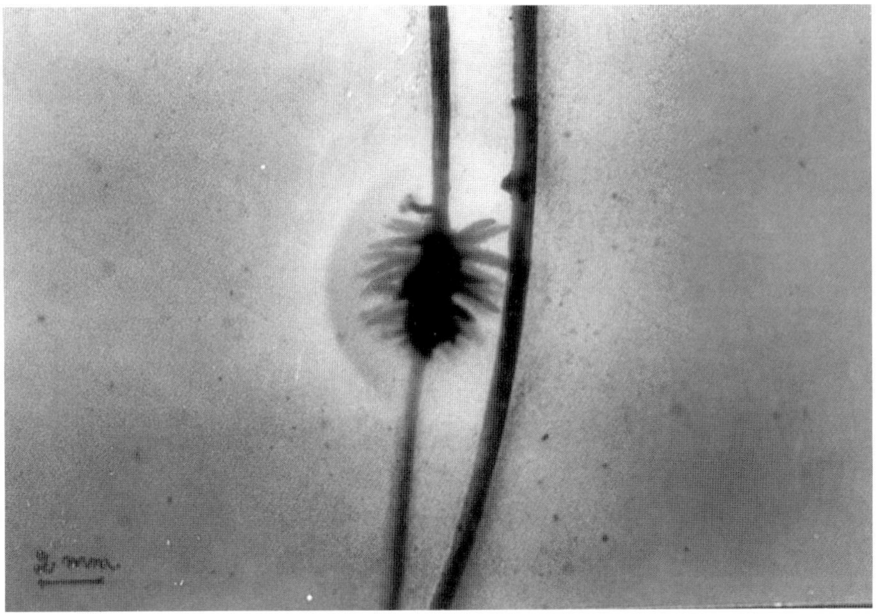

Figure 2. Development of an orange halo around the cluster roots of *Casuarina glauca* as assessed with the CAS assay. Note there is no orange colour surrounding noncluster roots.

P concentration required to completely suppress CR formation in such species depends on the plant species: for example, 1 mg l^{-1} of P is sufficient to suppress CR formation in *Myrica cerifera* (Louis et al., 1990), while much more P is needed, 8 mg l^{-1} and 16 mg l^{-1}, for CR suppression in *M. gale* (Crocker and Schwintzer, 1993) and *Gymnostoma papuanum* (Racette et al., 1990), respectively. Thus, in most species, the involvement of P nutrition in the elaboration of CR is clear (Watt and Evans, 1999). However, certain species, e.g., *Alnus incana*, Betulaceae (Hurd and Schwintzer, 1996) and *C. glauca*, Casuarinaceae (Arahou and Diem, 1997), produce CR in the presence of P.

If the formation of CR responds to the necessity to acquire exclusively P and not Fe, a production of these special root structures would be expected under P deficiency independent of iron supply, while no CR formation would occur under iron deficiency. Results presented in Tables 1 and 2 did not confirm this hypothesis. Furthermore, a selective absorption of iron and manganese by CR has been reported by Gardner et al. (1982) in *Lupinus albus*. We can thus conclude, in agreement with previous results (Diem and Arahou, 1996), that P deficiency alone is insufficient to induce the CR formation in *Casuarina* species tested since the presence of iron ($FeCl_3$) stops this formation.

The fact that CR were only produced in alkaline soil and not in acid soil (Arahou et Diem, 1997) raises the question about the reason behind this response. P is one of the major plant nutrients that is least available in soil (Raghothama, 1999) and its availability is highest between pH 5.0 and 6.0, where the monovalent $H_2PO_4^-$ anion dominates (Furihata et al., 1992). On the other hand, Fe availability decreases as the pH of the soil solution increases. Therefore, in alkaline soils, both Fe and P become unavailable. Under acid soil conditions, the availability of Fe increases sometimes resulting in nutrient disorders called iron toxicity in some plants (Audebert and Sahrawat, 2000).

The absence of CR formation on acid soils may well be due to high Fe availability in the soil solution. If, however, the low Fe availability in alkaline soils were the main cause for the occurrence of CR, no such formation should occur in the presence of Fe, despite the absence of P. This is exactly what Arahou and Diem (1997) observed in *C. glauca* and what we found in *C. cunninghamiana* and *C. equisetifolia* (Table 2). Another feature of this comparison between P and Fe acquisition should be emphasised: the occurrence of iron chlorosis in alkaline but not in acid soils and its relationship to CR formation in *C. glauca*. The findings of Diem and Arahou (1996) demonstrated a close correlation between this formation and plant chlorosis alleviation as indicated by the increase of chlorophyll content that occurs at the onset of CR formation. The aim of the CAS-test, which revealed the production of an agent specific in chelating Fe^{III} in areas adjacent to CR and making it available to plants *via* their root system resulting in their regreening, was to examine this hypothesis. However, this qualitative assay did not give any information about the magnitude of the Fe-mobilisation response. More research is needed in order to elucidate this observation.

The availability of Fe and P in the soil and their concentration in the plants can be influenced by various soil factors. Peiter et al. (2001) demonstrated that together with internal factors, environmental stimuli govern CR induction in *Lupinus albus*, while Zohlen and Tyler (1997) suggested that the leaf immobilisation of Fe in physiologically less active forms is related to the occurrence of chlorosis in *Veronica officinalis* L. Therefore, under Fe deficiency, when supplied with $NaHCO_3$ (Arahou and Diem, 1997), NO_3^- or NH_4^+ at high pH (Zaïd and Arahou, 1997), *Casuarina* plants respond by developing CR which can be considered as a means to acquire iron through their ability to excrete Fe-chelating substances (Figure 2). Due to this formation of CR, iron chlorosis is avoided; yellowing of newer plant growth and this abundant CR formation in *Casuarina* are indicator of low Fe availability. It is concluded that alkaline conditions stimulate the production of CR as stated by Peiter et al. (2001) for *Lupinus albus* and by Zaïd and Arahou (1997) and Diem *et al.* (2000) for *Casuarina*.

Acknowledgements

The authors gratefully thank Prof. Hans Lambers for inviting Prof. H. Zaid to present this work as well as for corrections he made on the last version of the manuscript. Dr. K. Pinyopusarerk (CSIRO, Forestry and Forest Products, Australia) is also acknowledged for kindly providing seeds of *Casuarina equisetifolia*.

References

Alcaraz C F, Martinez-Sanchez F, Sevilla F and Hellin E 1986 Influence of ferredoxin levels on nitrate reductase activity in iron deficient lemon leaves. J. Plant Nutr. 9, 1405–1413.

Appleby C A 1984 Leghaemoglobin and *Rhizobium* respiration. Annu. Rev. Plant Physiol. 35, 443–478.

Arahou M and Diem H G 1997 Iron deficiency induces cluster (proteoid) root formation in *Casuarina glauca*. Plant Soil 196, 71–79.

Audebert A and Sahrawat K L 2000 Mechanisms for iron toxicity tolerance in lowland rice. J. Plant Nutr. 23, 1877–1885.

Bienfait H F and Van Der Mark F 1983 Phytoferritin and its role in iron metabolism. *In* Metals and Micronutrients, Uptake and Utilization by Plants. Eds. D A Robb and W S Pierpoint. pp. 111–123. Academic Press, New York.

Braum S M and Helmke P A 1995 White lupin utilizes soil phosphorus that is unavailable to soybean. Plant Soil 176, 95–100.

Broughton W T and Dilworth M J 1971 Control of leghaemoglobin synthesis in snake beans. Biochem. J. 125, 1075–1080.

Carter M R 1987 Seedling growth and mineral nutrition of scot pine under acidic to calcareous conditions. Soil Sci. 144, 175–180.

Chen Y and Barak P 1982 Iron nutrition in calcareous soils. Adv. Agron. 35, 217–240

Crocker L J and Schwintzer C R 1993 Factors affecting formation of cluster roots in *Myrica gale* seedlings in water culture. Plant Soil 152, 287–298.

Diem H G and Arahou M 1996 A review of cluster root formation: a primary strategy of Casuarinaceae to overcome soil nutrient deficiency. *In* Recent *Casuarina* Research and Development. Eds. K. Pinyopusarerk, J W Turnbull and S J Midgley. pp. 51–58. CSIRO, Canberra.

Diem H G, Duhoux E, Zaid H and Arahou M 2000 Cluster roots in Casuarinaceae: Role and relationship to soil nutrient factors. Ann. Bot. 85, 929–936.

Dinkelaker B, Hengeler C and Marschner H 1995 Distribution and function of proteoid roots and other root clusters. Bot. Acta 108, 183–200.

Furihata T, Suzuki M, Sakurai H 1992 Kineticcharacterization of two phosphate uptake systems with different affinities in suspension-cultured *Catharanthus roseus* protoplasts. Plant Cell Physiol. 33: 1151–1157.

Gardner W K, Parbery G D and Barber D A 1982 The acquisition of phosphorus by *Lupinus albus* L. I. Some characteristics of the soil/root interface. Plant Soil 68, 19–32.

Gardner W K, Parbery G D and Barber D A 1983 The acquisition of phosphorus by *Lupinus albus* L. III. The probable mechanism by which phosphorus movement in the soil/root interface is enhanced. Plant Soil 70, 107–124.

Garg O K, Hemantaranjan A and Ramesh C 1986 Effect of iron and zinc fertilization on senescence in french bean (*Phaseolus vulgaris* L.). J. Plant Nutr. 9, 257–266.

Gilbert G A, Knight J D, Vance C P and Allan D L 2000 Proteoid root development of phosphorus deficient lupin is mimicked by auxin and phosphonate. Ann. Bot. 85, 921–928.

Hagström J, James W M and Skene K R 2001 A comparison of structure, development and function in cluster roots of *Lupinus albus* L. under phosphate and iron stress. Plant Soil 232, 81–90.

Huang I J, Welkie G and Miller G W 1992 Ferredoxin and flavodoxin analysis in tobacco response to iron stress. J. Plant Nutr. 15, 1765–1782.

Hurd T.M and Schwintzer C R 1996 Cluster roots of *Alnus incana* ssp. *rugosa* in the field and of four *Alnus* species in water culture with phosphorus and iron deficiency. Can. J. Bot. 74, 1684–1686.

Kerley S J, Leach J E, Swain J L and Huyghe C 2000 Investigation into the exploitation of heterogeneous soils by *Lupinus albus* L. and *L. pilosus* Murr. and the effect upon plant growth. Plant Soil 222, 241–253.

Lamont B B 1993 Why are hairy root clusters so abundant in the most nutrient-impoverished soils of Australia? Plant Soil 155/156, 269–272.

Liang R and Li C 2002 Differences in cluster-root formation and carboxylate exudation in *Lupinus albus* L. under different nutrient deficiencies. Plant Soil, 248, 221–227.

Lindsay W L 1984 Soil and plant relationships associated with iron deficiency with emphasis on nutrient interactions. J. Plant Nutr. 7, 489–500.

Loeppert R H 1986 Reactions of iron and carbonates in calcareous soils. J. Plant Nutr. 9, 195–214.

Louis I, Racette S and Torrey J G 1990 Occurrence of cluster roots on *Myrica cerifera* L. (Myricaceae) in water culture in relation to phosphorus nutrition. New Phytol. 115, 311–317.

Miller G W, Denney A, Pushnik J C and Ho-Yu M 1982 The formation of delta aminolevulinate a precursor of chlorophyll in barley and the role of iron. J. Plant Nutr. 5, 289–300.

Miller G W, Pushnik J C and Welkie G W 1984 Iron chlorosis, a world-wide problem, the relation of chlorophyll biosynthesis to iron. J. Plant Nutr. 7, 1–22

Miller G W, Huang I J, Welkie G W and Pushnik J C 1995 Function of iron in plants with special emphasis on chloroplasts and photosynthetic activity. *In* Iron Nutrition in Soils and Plants. Ed. J. Abadia, pp. 19–28. Kluwer Academic Publishers, Dordrecht.

Nenova V and Stoyanov I 1993 Physiological and biochemical changes in young maize plants under iron deficiency. I. Growth and photosynthesis. J. Plant Nutr. 16, 835–849.

Neumann G, Massonneau A, Langlade N, Dinkelaker B, Hengeler C, Romheld V and Martinoia E 2000 Physiological aspects of cluster root function and development in phosphorous-deficient white lupin (*Lupinus albus* L.). Ann. Bot. 85, 909–919.

Passama L, Ghorbal M H, Hamzé M, Salsac L and Wacquant J P 1975 Sur quelques facteurs écophysiologiques de différenciation entre calcicoles et calcifuges en milieu calcaire. Rev. Ecol. Biol. Sol 12, 309–327.

Purnell H M 1960 Studies of the family proteaceae. I. Anatomy and morphology of the roots of some Victorian species. Austr. J. Bot. 8, 38–50.

Racette S, Louis I and Torrey J G 1990 Cluster root formation by *Gymnostoma papuanum* (Casuarinaceae) in relation to aeration and mineral nutrient availability in water culture. Can. J. Bot. 68, 2564–2570.

Raghothama K G 1999 Phosphate acquisition. Annu. Rev. Plant Physiol. Plant Mol. Biol. 50, 665–693.

Rai R, Prasad V, Choudhury S K and Sinha N P 1984 Iron nutrition and and symbiotic N_2-fixation of lentil (*Lens culinaris*) genotypes in calcareous soils. J. Plant Nutr. 7, 399–405.

Reddell P, Yun Y and Shipton W A 1997 Cluster roots and mycorrhizae in *Casuarina cunninghamiana*: their occurence and formation in relation to phosphorus supply. Aust. J. Bot. 45, 41–51.

Römheld V and Marschner H 1986 Mobilization of iron in the rhizosphere of different plant species. Adv. Plant Nutr. 2, 155–204.

Rueter J G and Ades D R 1987 The role of iron nutrition in photosynthesis and nitrogen assimilation in *Scenedesmus quadricauda* (chlorophyllaceae). J. Phycol. 23, 452–457.

Ryan P R, Delhaize E and Jones D L 2001 Function and mechanism of organic anion exudation from plant roots. Annu. Rev. Plant Physiol. Plant Mol. Biol. 2001. 52: 527–560.

Salsac L 1969 Caractère particulier de l'absorption du calcium. Bull. Soc. Fr. Physiol. Vég. 15, 213–236.

Schmidt W and Schikora A 2001 Different pathways are involved in phosphate and iron stress-induced alterations of root epidermal cell development. Plant Physiol. 125, 2078–2084.

Schwyn B and Neilands J B 1987 Universal chemical assay for the detection and determination of siderophores. Anal. Biochem. 160, 47–56.

Shane M W, De Vos M, De Roock S, Cawthray G R and Lambers H 2002 Effect of external phosphorus supply on internal phosphorus concentration and the initiation, growth and exudation of cluster roots in *Hakea prostrata* R.Br. Plant Soil, 248, 209–219.

Sharma S and Sanwal G G 1992 Effect of Fe deficiency on the photosynthesis system of maize. J. Plant Physiol. 140, 527–530.

Shen J, Rengel Z, Tang C and Zhang F 2002 Role of phosphorus nutrition in development of cluster roots and release of carboxylates in soil-grown *Lupinus albus*. Plant Soil 248, 199–206.

Skene K R 2000 Pattern formation in cluster roots: Some developmental and evolutionary considerations. Ann. Bot. 85, 901–908.

Skene K R, Kierans M, Sprent J I and Raven J A 1996 Structural aspects of cluster root development and their possible significance for nutrient acquisition in *Grevillea robusta* (Proteaceae). Ann. Bot. 77, 443–452.

Suzuki A and Gadal P 1982 Glutamate synthase from rice leaves. Plant Physiol. 69, 848–852.

Tang C, Robson A D, Longnecker N E and Greenway H 1993 Physiological responses of lupin roots to high pH. Plant Soil 155/156, 509–512.

Uren NC 1984 Forms, reactions and availability of iron in soils. J. Plant Nutr. 7, 165–176.

Watt M and Evans J R 1999 Proteoid roots. Physiology and development. Plant Physiol. 121, 317–323

Zaïd H and Guerrier G 1982 Sur le caractère calcifuge du théier. Mise en évidence à partir des cinétiques d'absorption du calcium. Rev. Gén. Bot. 89, 161–171.

Zaïd H and Arahou M 1997 Relationship between nitrogen form and cluster roots formation in *Casuarina* plants growing in an iron-free nutrient solution. *In* 9th International Symposium on Iron Nutrition and Interactions in Plants, 20–25 July 1997, Universität Hohenheim, Stuttgart.

Zaïd H, Hemantaranjan A, Layachi R, Mehdioui R and Alem A 2002 Soil acidity and alkalinity: two main constraints for conditionning the behaviour of plants. A review of current knowledge. Adv. Plant Physiol. 5, (In press).

Zohlen A and Tyler G 1997 Differences in iron nutrition strategies of two calcifuges, *Carex pilulifera* L. and *Veronica officinalis* L. Ann. Bot. 80, 553–559.

Plant and Soil **248:** 237–246, 2003.
© 2003 *Kluwer Academic Publishers. Printed in the Netherlands.*

The formation, morphology and anatomy of cluster root of *Lupinus albus* L. as dependent on soil type and phosphorus supply

C. S. Peek[1], A. D. Robson[1] & J. Kuo[2,3]
[1]*School of Earth and Geographic Sciences,* [2]*Centre for Microscopy and Microanalysis, Faculty of Natural and Agricultural Sciences, The University of Western Australia, 35 Stirling Highway, Crawley, WA 6009, Australia*
[3]*Corresponding author**

Received 14 February 2002. Accepted in revised form 24 June 2002.

Key words: cluster roots, *Lupinus albus*, phosphorus, root anatomy, root morphology

Abstract

The effect of phosphorus supply on the formation, morphology and anatomy of cluster roots of *Lupinus albus* L. cv Ultra grown in a loam and two sandy soils was examined relative to its effect on total root length, shoot weight and the phosphorus concentration of the shoots. The loam soil was most conducive to the formation of cluster roots. Cluster roots growing in the sandy soils developed to a lesser extent on plants of an equivalent phosphorus status, suggesting that some biotic or abiotic factors independent of phosphorus supply were also operating. The presence of mature cluster rootlets on a length of lateral root increased the root surface area by 14–22 times of an equal length of lateral roots not bearing cluster rootlets. The application of phosphorus decreased cluster-root length, whereas total root length showed a steady increase. There was an inverse relationship between cluster-root production and phosphorus concentration in shoots ranging from 2 to 8.5 mg g^{-1} with the critical phosphorus level for maximum shoot growth being around 2.5 mg g^{-1}. Cluster roots formed in solution culture were not well developed in comparison with those grown in the loam soil or nutrient solution with added loam soil. The organisation of the cluster rootlet was similar to that of the lateral roots. Mature rootlets lacked an apical meristem and a vascular cambium with a reduced root cap and cortical tissue.

Introduction

Cluster roots are distinct regions of dense rootlets formed on the lateral roots of a large number of plant species. They are considered to play important roles in plant exudation and nutrient uptake. The formation of cluster roots can be initiated by a deficiency of plant nutrients, in particular phosphorous and sometimes other elements such as iron (Arahou and Diem, 1997; Dinkelaker et al., 1995; Watt and Evans, 1999a).

Amongst about 300 species in the genus *Lupinus*, only some of the species from the Old World possess cluster roots (Skene, 2000) and white lupin, *L. albus*, is the most studied agricultural plants possessing cluster roots. Gardner et al. (1982a) concluded that cluster-root formation on soil-grown white lupin was predominantly affected by the phosphorus status of the plants. The organic compounds exuded from white lupin cluster roots facilitate the absorption of nutrients (Dinkelaker et al., 1989; Gardner et al., 1983; Johnson et al., 1994; Keerthisinghe et al., 1998).

The morphology and anatomy of cluster roots and rootlets have been documented for several species of the Proteaceae (Dell et al., 1980; Lamont, 1972; Purnell, 1960; Skene et al., 1996, 1998a, b), but a detailed description of cluster roots in lupins is only available for plants grown in hydroponic cultures (Watt and Evans,1999a). In addition, comparative information on the cluster root of white lupin under solution culture and soil-grown conditions is still lacking.

* E-mail: jjskuo@cyllene.uwa.edu.au

Table 1. The response to phosphorus (P) supply of the cluster-root length, total root length and the dry weight and phosphorus concentration of the shoots of *Lupinus ablus* grown on three types of soil for 30 days (Experiment I)[1]

Variable	Swan Valley Loam						Lancelin Sand			Kojonup Sand		
P application ($g\ kg^{-1}$)	0	0.01	0.033	0.1	0.33	0.01	0.033			0.043	0.173	
Dry weight of shoots ($g\ plant^{-1}$)	0.71	0.73	0.82	0.81	0.68	**	1.49	1.76	*	1.92	1.83	NS
P concentration in shoots (% dry weight)	0.20	0.22	0.25	0.33	0.85	***	0.18	0.36	*	0.29	0.82	**
Cluster root length[1] ($cm\ plant^{-1}$)	76	72	72	64	34	***	44	24	*	23	7	**
Total root length[2] ($cm\ plant^{-1}$)	420	475	540	560	580	*	545	647	NS	712	1162	**
% cluster root/ total root length	18.0	15.2	13.3	11.4	5.9		8.1	3.7		3.2	0.6	

NS, *, **, and *** indicate treatment effect not significant, significant at the 0.05, 0.01 and 0.001 levels, respectively (after one-way AVOVA).
[1] Based on the mean of six plants from each phosphorus treatment.
[2] Cluster-root length is defined as the length of the lateral on which the cluster rootlets form.
[3] Total root length is defined as the entire root systems of the plants.

Materials and methods

Experiment I

Three types of soil from natural habitats in South Western Australia were used: (a) Swan Valley loam which was collected from a man-constructed levee on the banks of the Swan River at Guildford with a bicarbonate extractable P level of 19 $\mu g\ g^{-1}$; (b) virgin Lancelin sand with pH 5.1 ($CaCl_2$), organic C 10 $mg\ g^{-1}$, clay 3% and extractable P 5 $\mu g\ g^{-1}$; and (c) virgin Kojonup sand with pH 5.2 ($CaCl_2$), organic C 22 $mg\ g^{-1}$, clay 8% and extractable P 8 $\mu g\ g^{-1}$. The phosphorus was extracted in 0.5 M $NaHCO_3$ at 1:100 ratio (Colwell, 1963) and its levels in these three soils were classified as low. Field capacity values for the three soils were 14, 10 and 12%, respectively.

Each plastic pot of 16 cm diameter and 16 cm deep was filled with 3 kg of one of the soil types and mixed with the following nutrients: K_2SO_4, 360 mg; $ZnSO_4$, 22.5 mg; $CuSO_4{\cdot}5H_2O$, 16 mg; $Na_2MoO_4{\cdot}2H_2O$, 0.8 mg; $CoSO_4{\cdot}7H_2O$, 0.2 mg and $MnSO_4{\cdot}4H_2O$, 32.5 mg. In addition, a solution of KH_2PO_4 was applied to the soil surface, and when dry, was mixed thoroughly through the soil by shaking in a plastic shaking jar for 3 min. The rates of phosphorus supplied varied with soil types and covered a range from deficient to adequate. Swan Valley loam (0, 0.01, 0.033, 0.1, 0.33 g P kg^{-1}) had four replicates, Lancelin sand (0.01, 0.033 g P kg^{-1}) and Kojonup sand (0.043, 0.173 g P kg^{-1}) had two replicates each.

Seeds of *Lupinus albus* L. cv Ultra were germinated in an aerated beaker of deionized water for 24 h. Ten germinated seeds were put in each of the pots at a depth of 1.5 cm and inoculated with *Bradyrhizobium* sp. (*Lupinus*) WU 425. Seedlings were thinned to three per pot after 7 days. Plants were grown in a glasshouse from May to August, Perth, Western Australia and the soils were maintained at field capacity by adding deionized water every second day.

Plants were harvested 30 days after germination and all the plants were well nodulated. Shoot fresh and dry weight (70 °C for 3 days) was determined. The roots were soaked overnight after which top soil attached to the cluster roots was removed. Each individual root systems was cleared with a solution of 50% ethanol, 45% deionized water and 5% acetic acid (v/v). The total length of both the cluster roots and the entire root system of six plants from each phosphorus treatment were measured, using a sheet of graph paper sandwiched between two pieces of perspex. Cluster-root length is defined as the length of lateral on which the cluster roots form.

Experiment II

To compare the root morphology of plant grown in soil and nutrient solution, plants were grown in Swan

Valley loam and solution culture. In the soil treatment, each pot was filled with 5 kg loam soil and basic nutrients as in Experiment I without additional phosphorus supply. Five plants were grown in each of four pots. For the solution culture, each plastic container was filled with 5 l of deionized water containing the following nutrients: NH_4NO_3, 400 μM; $Ca(NO_3)_2$, 500 μM; KH_2PO_4, 10 μM; K_2SO_4, 400 μM; $MgSO_4$, 100 μM; NaCl, 0.5 μM; $FeSO_4$, 2.0 μM; H_3BO_3, 3.0 μM; $MnSO_4$, 0.5 μM; $ZnSO_4$, 0.2 μM; $CuSO_4$, 0.1 μM; $NaMoO_4$, 0.02 μM. In an additional set of the solution culture pots were added 5 g of Swan Valley loam each. Five uniform seedlings were grown to five plants per pot and each treatment had four replicates. Pots were maintained at 22 °C in a root-cooling unit. Nutrient solutions were continuously aerated and renewed weekly in the first 2 weeks and every third day thereafter. Solution pH was adjusted daily to 5.5 using 0.01 M KOH. One plant from each soil-grown pot was harvested at 14, 21, 34 and 66 days after germination and solution-grown plants were harvested at day 68 after germination. At each harvest 30 cluster roots were randomly selected and the length of the cluster-root axis, the average length of the rootlets, and the number of rootlets per cm of first-order lateral axis was measured for each individual cluster root.For both the soil and solution grown plants, shoot dry weight and the length of cluster roots was measured, and cluster roots were photographed at each harvest.

The materials used for all anatomical studies were prepared from soil-free cluster roots from solution culture with 5 g Swan Valley loam because it was extremely difficult to section on soil grown cluster roots. The freshly collected cluster roots formed on the first-order laterals and the same laterals were fixed in 2.5% (v/v) glutaraldehyde in 0.05 mM phosphate buffer, pH 7.0, for 24 h at room temperature. Fixed samples were dehydrated in solvents and embedded in glycol methacrylate (O'Brien and McCully, 1981). At least three cluster roots, cluster rootlets and laterals were sectioned transversely and longitudinally at 2 μm in thickness with glass knives on a Sorvall microtome. Sections were stained with 0.05% (v/v) toluidine blue O in benzoate buffer at pH 4.4. Stained sections were examined and photographed using a Zeiss Axioskop 2 Microscope.

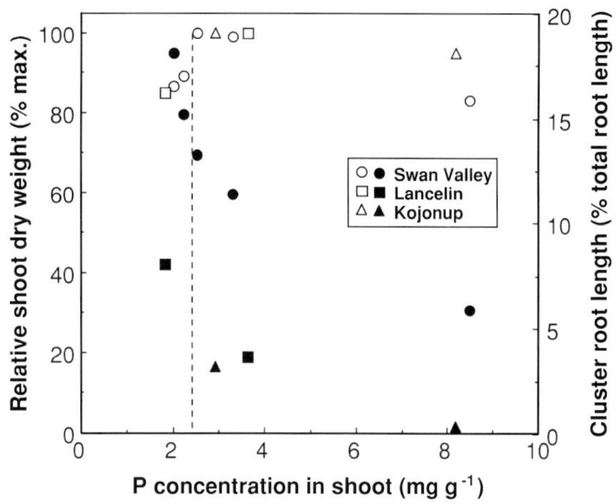

Figure 1. Relations between phosphorus concentration in shoot, relative shoot dry weight (% of maximum) and cluster-root length (% of the total root length) of *Lupinus albus* grown in three types of soil.The dashed line indicates the P level resulting in maximum shoot growth for the Swan Valley loam. Open symbols represent shoot weights and closed symbols represent cluster-root length.

Results

Shoot mass and phosphorus concentration

Phosphorus application to the Swan Valley loam increased ($P < 0.01$) shoot weight, except at the highest level (0.33 g P kg^{-1}) (Table 1). Maximum shoot weight occurred at phosphorus rate of 0.033 g P kg^{-1}. Plants grown without phosphorus application produced 86% of maximum shoot weight. Phosphorus concentration of the shoots increased steadily with increasing phosphorus rate (Table 1). A phosphorus concentration of 2.5 mg g^{-1} (dry weight basis) was associated with maximum shoot weight (Figure 1). Phosphorus application increased shoot weight ($P < 0.05$) of plants grown on the Lancelin sand, but not on the Kojonup sand (Table 1). The latter was probably caused by the higher phosphorus applications on the Kojonup sand.

Cluster root formation and phosphorus concentration

Total cluster-root length on the Swan Valley loam decreased ($P < 0.001$) at the two highest applications, whereas the length of non-cluster roots showed a steady increase with phosphorus supply (Table 1). Therefore, plants grown without phosphorus application had the greatest relative amounts of cluster roots, their length being 19% of the total root length. At the highest phosphorus supply this value had decreased to only 6% of the total root length. Increasing the phosphorus rate decreased cluster-root length less than 2-fold in the Lancelin sand, and more than 3-fold in the Kojonup sand. Total root length increased ($P < 0.01$) with an increased phosphorus application in the Kojonup sand, but not in the Lancelin sand. There was an inverse relationship between cluster-root length (as percent of total root length) and phosphorus concentration in shoots (Figure 1). However, at similar shoot phosphorus concentration, cluster-root formation was greater in the Swan Valley loam than in other two soils.

Cluster root morphology

The formation of cluster roots showed no particular distributional pattern and clusters were confined to the laterals, primarily on the first order, occasionally, on the second order, but never on the tap root. Cluster roots were formed as early as 2 weeks after germination (Table 2). The cluster roots were observed to cling tightly to both sand grains and particles of organic matter in all three soils; this was particularly noticeable in the Swan Valley loam. Furthermore, the presence of dense matt of root hairs was evident on the rootlets of the cluster roots in all soils.

The cluster roots developed sequentially along the extending lateral (Figure 2A). The rootlets of an individual cluster root appeared to be initiated more or less simultaneously. The rootlets usually developed into two, occasionally three, longitudinal rows along the parent lateral axis (Figure 2A, B). The rootlets in each row were not always placed directly adjacent to one another. The rootlets often emerged slightly to the back or the front of an adjacent rootlet; this resulted in the rootlets emerging at slightly different angles to each other (Figure 2A). The emergence of the rootlets in three longitudinal rows was adjacent to one another giving a more symmetrical appearance. The youngest cluster-root rootlets forming behind the root tip of the growing lateral were often tapered at each end (Figure 2A). The end rootlets appeared to grow more slowly than the central rootlets. The mature cluster roots were located at a greater distance from the lateral-root tip with a more uniform length (Figure 2B), although in some cases the rootlets forming at the ends, particularly the end most proximal to the root tip of the parent lateral, remained shorter. The cluster rootlets were of limited length and never gave rise to further rootlets.

Number of rootlets per cm root cluster axis was relatively constant regardless of the age of plants grown on the Swan Valley loam, but the length of cluster roots and rootlets was greater in the older plants (Table 2). The results should not, however, be interpreted as a rootlet growth curve. The sampling intervals were too widely spaced and the average rootlet length was calculated from a random sample of 30 cluster roots.

The newly emerged rootlets lacked root-hair formation but those developed from the base to tip as the rootlets increased in length (Figure 2A). The rootlets of the mature cluster roots were completely covered in a dense matt of root hairs (Figure 2B, C). The old cluster roots, from those plants harvested at approximately 66 days after germination, had desiccated, their rootlets having lost their root hairs. Therefore both young and old rootlets lacked root hairs and lost their ability to cling to particulate matter.

The rootlets from cluster roots grown on the Swan Valley loam varied in diameter from the base to tip; 460 and 300 μm, respectively with a mean of 360 μm and the rootlet was, on average, 56% smaller in diameter than the normal laterals (Table 3). The surface area per centimetre of lateral axis with the exclusion of root hairs was increased by 14–22 times when bearing mature cluster rootlets.

The morphology of cluster roots of plants grown in solution culture with added soil was visually similar to those collected from the soil experiments. Plants of similar age grown on loam soil and added loam soil solution culture had a longer cluster-root length but a similar number of rootlets per cm cluster root axis (Table 2). However, the cluster rootlets were shorter in added loam soil solution culture with a mean of 4.0 mm in comparison with 10.6 mm in the loam soil grown plants resulting approximately 3-fold decrease in the root surface area ratio in added loam soil solution culture (Table 2).

Root systems formed on plants grown in solution culture without added loam soil had longer tap roots but shorter and less frequent laterals than those grown with added loam soil. The formation of cluster roots was delayed for at least 8 days and cluster roots were not as defined as those formed on solution culture with loam soil. Furthermore, the rootlets from culture solution appeared to be more widely spaced on the parent axis and the rootlets within a root cluster were frequently less uniform in length as well as much shorter than those from loam soil grown ones. In some cases, it was difficult to distinguish the rootlets from the higher order laterals. Furthermore, root hair devel-

Table 2. The dimensions of cluster roots and rootlets of *Lupinus albus* measured at harvests of plants grown on the Swan Valley loam and solution culture with added loam soil (Expeirment II)

Rooting medium	Days after germination	Cluster root length[1] (mm)	Range of cluster root length (mm)	Length of cluster rootlets[1] (mm)	No of rootlets[2] per cm cluster root axis[1]	Ratio of surface area3 with cluster rootlets to surface area without cluster roots
Loam soil	14	11.0±1.8	1–15	2.0±0.1	46±1.0	4.0
	21	12.2±1.7	1–34	3.7±0.1	47±1.0	7.6
	34	25.2±3.1	10–60	7.1±0.3	47±0.9	14.7
	66	34.0±1.9	10–70	10.6±1.1	46±1.2	22.0
Solution culture with added loam	68	39.6±1.8	10–60	4.0±0.2	44±1.1	7.7

[1]Mean + standard error based on four plants from each harvest.
[2]Rootlet density for a cluster root with two longitudinal rows of rootlets.
[3]Root hairs not included in surface area calculation.

opment was less apparent for those grown in solution culture (Figure 2D).

Cluster root anatomy

The rootlets of the cluster roots of *L. albus* were initiated opposite each protoxylem pole of the cluster root axis (Figure 3A–C). The stele of the parent lateral was generally diarch (Figure 3A, B) resulting in the formation of two protoxylem poles. The rootlets formed along the length of these two poles on a section of lateral axis. This resulted in the formation of the two longitudinal rows of rootlets in each cluster root. The stele of the parent lateral was occasionally observed to be triarch resulting in the formation of three longitudinal rows of rootlets (Figure 3B). The rootlets in each of the two to three rows were not necessarily initiated exactly opposite the corresponding rootlets in the other row(s) (Figure 3A, B). After initiation (Figure 3A), the rootlets pushed through the cortical and epidermal tissues of the parent lateral to arise endogenously (Figure 3B).

The earliest rootlet consisted of mainly meristematic tissue (Figure 3A). Mature rootlets had few root caps at tips but lacked an apical meristem preventing the further development. Cross-sections through adjacent rootlets revealed they had similar diameter and anatomical structures (Figure 4A). The rootlets (Figure 4A–D) possessed an epidermis and four rows of cortical cells. The radial diameter of cortical cells was approximately 50% of those in the first order laterals that were bearing the cluster roots. An endodermis with Casparian bands surrounded the stele, which had large pericycle cells and two small groups of xylem and phloem tissues arrangement (Figure 4C). Each xylem and phloem group could consist of up to three files (Figure 4C). Cambium tissue in the stele was absent in rootlets, preventing the development of secondary or meta-xylem and phloem, which was present in the non-cluster lateral (Figure 4E). A comparison of various anatomical aspects of the rootlets and the first order laterals of *L. albus* is summarised in Table 3.

Discussion

The proportion of the root axis length occupied by cluster roots decreased with an increasing rate of phosphorus application to the soil, consistent with previous findings from solution culture (e.g., Johnson et al. 1994; Keerthisinghe et al., 1998; Watt and Evans, 2002) and soil studies (e.g., Gardner et al., 1983a). However, the present study also showed an inverse relationship between cluster-root production and phosphorus concentration in shoots up to 8.5 mg g^{-1}; this concentration was at least three times greater than that at which shoot growth started to decrease (Figure 1). Clearly, cluster-root production was independent of shoot growth. In solution culture, cluster-root production of the same species occurred prior to a decrease in plant growth in response to phosphorus deficiency, and was also inversely related to phosphorus concentration in shoots with a critical level of 2 mg g^{-1} (Sas et al., 2001). Marschner et al. (1987) showed that the number of clusters could be reduced by foliar application of phosphorus to *L. albus* and concluded that cluster-root

Figure 2. Cluster roots from *Lupinus albus* grown in the Swan Valley loam without phosphorus addition (A–C), and grown in solution culture without added loam (D). (A) Young (developing) cluster root showing the tapered end; note that the young rootlets lack root hairs. (B) The two longitudinal rows of rootlets can be clearly seen in a mature cluster root; note that the rootlets are covered in root hairs. (C) The development of root hairs on the rootlets of a cluster root from the base to the tip. The fully elongated rootlets are covered in a dense matter of root hairs.

Table 3. A comparison between the anatomy fo the first-order laterals and cluster rootlets of *Lupinus albus* grown on solution culture with the Swan Valley loam, + = Present; – = Absent

Variable	First-order Lateral[1]	Cluster rootlet
Root diameter (μm)	735–890	360
Epidermis	+	+
Hypodermis	+	–
Cortex		
Number of rows	6–8	4–5
Diameter (μm)	50–140 (n = 84)	36–54 (n = 46)
Vascular cambium	+	–
Endodermis	+	+
Phloem		
Number of protophloem poles	2–3	2
Metaphloem	+	–
Secondary phloem	+	–
Xylem		
Number of protoxylem poles	2–3	2
Metaxylem	+	–
Secondary xylem	+	–
Apical meristem	+	–

[1] Application to the axis of the cluster roots as well.

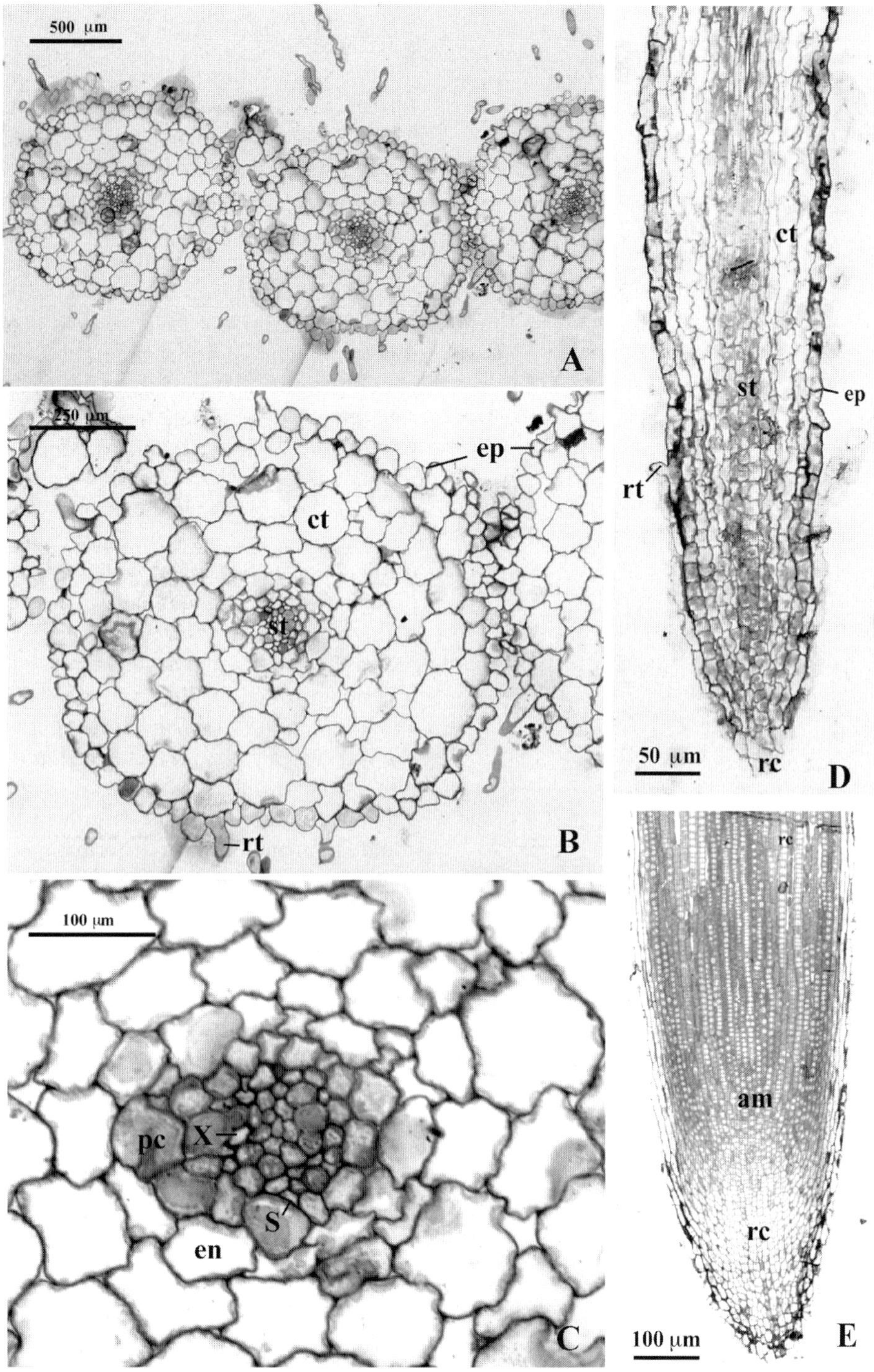

Figure 4. Transverse sections of cluster rootlets of *Lupinus albus* grown in culture solution with added Swan Valley loam. (A) Three rootlets from the same cluster-root axis showing a similar dimensions and anatomical arrangements.(B) A rootlet showing some epidermal cells (ep), bear root hairs (rh), several layers of cortical cells (ct) and a central stele (st). (C) The stele of a rootlet is surrounded by a layer of endodermis (en) and consists of pericycle (pc), two poles of protophloem (S) and protoxylem (X). (D) A longitudinal-sectional aspect of a cluster rootlet of *L. albus* showing some epidermal cells (ep) bear root hairs (rh), few layers of cortical cells (ct), a central stele (st) and a much reduced root cap (rc). Note that an apical meristem is absent. (E) A longitudinal section near root tip of a first order lateral root of *L. albus* showing an apical meristem (am) which is protected by an enlarged root cap (rc).

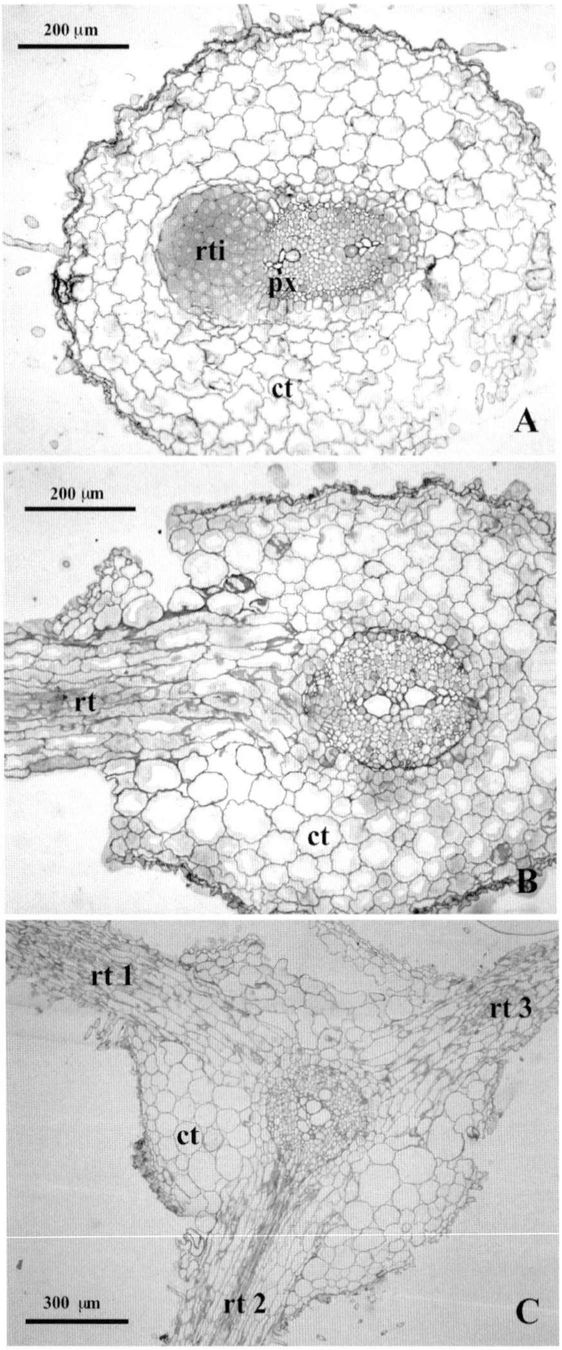

Figure 3. Transverse sections of cluster-root axes of *Lupinus albus* grown in culture solution with added Swan Valley loam. (A) A cluster-root axis with a diarch stele showing rootlet primordia (rti) initiated opposite the protoxylem (px) and extending laterally into the cortex (ct). (B) A cluster-root axis with a diarch stele showing a rootlet (rt) already passed through the cortex (ct) and extended beyond the cluster root-axis. (C) A cluster-root axis with a triarch stele showing three rootlets (rt 1, rt 2, rt 3) already extended beyond the cluster root.

formation was regulated by the internal phosphorus concentration.

The Swan Valley loam was more conducive to the formation of cluster roots than either the Lancelin or Kojonup sands for plants of equivalent phosphorus status. The greater cluster-root formation in the Swan Valley loam was not due to an increase in lateral-root production because the length of the total root system of plants grown in the Lancelin sand was, in fact, greater than that grown in the Swan Valley loam. Furthermore, the present study showed that after adding the loam soil into the pure culture solution, a root morphology similar to that of plants grown in the loam soil was achieved, suggesting that either micro-organisms or other factors in the loam may stimulate the cluster formation. Comparative organic matter contents and microbial activity of the three soils were unknown. A high earthworm activity was observed on collections of the Swan Valley loam, which might be indicative of higher organic matter contents and microbial activity. Increased microbial activity has been shown to increase cluster-root production in *L. albus*, although it is not essential for their formation (Gardner et al., 1982b). Purnell (1960) and Lamont et al. (1984) also showed that an organic patch induced local cluster root development. Alternatively, these soils might contain different amounts of growth hormones that could stimulate cluster-root formation. For example, Neumann et al. (2000) demonstrated that the application of the phytohormone IAA (indole acetic acid) to phosphorus-sufficient culture solutions increased the formation of cluster roots in *L. albus*.

The morphogenesis of the cluster roots of *Lupinus albus* such as the sequential development of cluster roots along an extending lateral, is similar to that of the cluster roots of Proteaceae (Lamont, 1972; Skene, 2000; Watt and Evans, 1999b). The number of rows of rootlets within a cluster root depends on the number of xylem poles in the stele. The length of a cluster along a cluster root axis and the final length of a rootlet also vary depending on environmental conditions and between species (Dinkelaker et al., 1989, 1995; Johnson et al., 1996a; Keerthisinghe et al., 1998). Watt and Evans (1999a) demonstrated that rootlets were shortened when *L. albus* was grown in elevated atmospheric CO_2 (700 μL L^{-1}) instead of under ambient conditions (350 μL L^{-1}). The present study showed that cluster roots of plants grown in pure culture solution were poorly formed but were prolifically produced in the loam soil or culture solution with the addition of the loam soil. Development of root hairs in *L albus* differs from that of *Grevillea robusta* (Skene et al., 1996, 1998a) and *Hakea obliqua* (Dell et al., 1980). In *G. robusta*, root hairs were produced only after the rootlet had reached its final length; they then developed back from the tip. The present study demonstrated that the young rootlets of *L. albus* developed root hairs from the base to the tip as the rootlets elongated, and the old rootlets lost their hairs, thus both the young and old cluster roots lacked the ability to cling to the soil.

Rootlet development in cluster roots is related to biochemical and physiological activities in the cluster roots (see review Dinkelaker et al., 1995; Skene, 2000; Watt and Evans, 1999b) with conflicting observations. Citrate efflux in lupins was related to the age of the rootlet but independent of the presence of root hairs (Watt and Evans, 1999a). Maximal secretion of citrate acids was found when these rootlets had reached maturity (Neumann et al., 1999) and after xylem and phloem tissues had differentiated to the tips of rootlets (Watt and Evans, 1999a). In contrast, Keerthisinghe et al (1998) reported that citrate secretion was confined to young cluster rootlets of the 1–3-cm region from the root tip and took place over 2–3 days (Roelofs et al., 2001; Watt and Evans, 1999a) and deceased in senescent cluster rootlets (Neumann et al., 1999). On the other hand, the ability of phosphorus uptake by cluster roots and non-cluster roots in *L. albus* was similar and there was no specific anatomical or structural difference in the regions of cluster rootlets along the cluster root axis (Keerthisinghe et al., 1998). Mature cluster roots were also involved in micronutrient uptake and amino acid synthesis (Gardner et al., 1982, 1983; Jeschke and Pate, 1995; Johnson et al., 1996b). However, most of the above-mentioned studies were conducted in solution culture. The present study clearly demonstrated that the formation and morphology of cluster roots in *L. albus* vary with different growth media. Whether these morphological differences also reflect biochemical and physiological performances of *L. albus* grown in different media warrants future investigations.

Acknowledgements

We thank Dr. C. Tang, Professor H. Lambers and two anonymous referees for their constructive comments on the manuscript. Appreciation is also extended Dr. A. P. Hamblin for her invaluable advice and to B. Simmons, K. Snowball and T. Marfleet for their capable technical assistance.

References

Arahou A and Diem H G 1997 Iron deficiency induces cluster (proteoid) root formation in *Casurarina glauca*. Plant Soil 196, 71–79.

Colwell J D 1963 The estimation of the phosphorus fertilizer requirements of wheat in southern New South wales by soil analysis. Aust. J. Exp. Agric. Anim. Husb. 3, 190–197.

Dell B, Kuo J and Thomson G J 1980 Development of proteiod roots in *Hakea oblique* R. Br. (Proteaceae) grown in water culture. Aust. J. Bot. 28, 27–37.

Dinkelaker B, Römheld V and Marschner H 1989 Citric acid excretion and precipitation of calcium citrate in the rhizosphere of white lupin (*Lupinus albus* L.). Plant Cell Environ. 12, 285–292.

Dinkelaker B, Hengeler C and Marschner H 1995 Distribution and function of proteoid roots and other root clusters. Bot. Acta 108, 183–200.

Gardner W K, Parbery D G and Barber D A 1982a The acquisition of phosphorus by *Lupinus albus* L. II. The effect of varying phosphorus supply and soil type on some characteristics of the soil/root surface. Plant Soil 68, 33–41.

Gardner W K, Barber D A and Parbery D G 1982b Effect of microorganisms on the formation and activity of proteoid roots of *Lupinus albus*. Aust. J. Bot. 30, 303–309.

Gardner W K, Parbery D G and Barber D A 1983 The acquisition of phosphorus by *Lupinus albus* L. III. The possible mechanism by which phosphorus in the soil/root interface is enhanced. Plant Soil 68, 33–41.

Jeschke W D and Pate JS 1995 Mineral nutrition and transport in xylem and phloem of *Banksia prionotes* (Proteaceae), a tree with dimorphic root morphology. J. Exp. Bot. 46, 895–905.

Johnson J F, Allen D L and Vance C P 1994 Phosphorus stress-induced proteoid roots show altered metabolism in *Lupinus albus*. Plant Physiol. 10, 657–665.

Johnson J F, Allen D L and Vance C P 1996a Phosphorus-deficient in *Lupinus albus*. Plant Physiol. 112, 31–40.

Johnson J F, Allen D L, Vance C P and Weiblen G 1996b Root carbon dioxide fixation by phosphorus-deficient *Lupinus albus*. Plant Physiol. 112, 19–30.

Keerthisinghe G, Hocking PJ, Ryan P R and Delhaize E 1998 Effect of phosphorus supply on the formation and function of proteoid roots of white lupin (*Lupinus albus* L.) Plant Cell Environ. 21, 467–478.

Lamont B B 1972 The morphology and anatomy of proteoid roots in the genus *Hakea*. Aust. J. Bot. 20, 155–174.

Lamont B B, Brown G and Mitchell DT 1984 Structure, environmental effects on their formation, and function of proteoid roots in *Leucadendron laureolum* (Proteaceae). New Phytol. 97, 381–390.

Marschner H, Romheld V and Cakmak I 1987 Root-induced changes in nutrient availability in the rhizosphere. J. Plant Nutr. 10, 1175–1184.

Neumann G, Massonneau A, Martinoia E and Römheld V 1999 Physiological adaptations to phosphorus deficiency during proteoid root development in white lupin. Planta 208, 373–382.

Neumann G, Massonneau A,Langlade N, Dinkelaker B, Römheld V, and Martinoia E 2000 Physiological aspects of cluster root function and development in phosphorus-deficient white lupin. Ann. Bot. 85, 909–919.

O'Brien T P and McCully M E 1981 The Study of Plant Structure — Principles and Selected Methods. Themarcarphi Pty. Ltd., Melbourne. 347 pp.

Purnell H M 1960 Studies of the family Proteaceae. I. Anatomy and morphology of the roots of some Victorian species. Aust. J. Bot. 8, 38–50.

Roelofs R F R, Rengel Z, Cawthray G R, Dixon K W and Lambers H 2001 Exudation of carboxylates in Australian Proteaceae: chemical composition. Plant Cell Environ. 24, 891–903.

Sas L, Rengel Z and Tang C 2001 Excess cation uptake, and extrusion of protons and organic acid anions by *Lupinus albus* under phosphorus deficiency. Plant Sci. 160, 1191–1198.

Skene K R 2000 Pattern formation in cluster roots: some developmental and evolutionary considerations, Ann. Bot. 85, 901–908.

Skene K R, Kiranans M, Sprent J I and Raven J A 1996 Structural aspects of cluster root development and their possible significance for nutrient acquisition in *Grevillea robusta* (Proteaceae). Ann. Bot. 77, 443–451.

Skene K R, Raven J A and Sprent J I 1998a Cluster root development in *Grevillea robusta* (Proteaceae). I. Xylem, pericycle, cortex, and epidermis development in a determinate root. New Phytol. 138, 725–732.

Skene K R, Raven J A and Sprent J I 1998b Cluster root development in *Grevillea robusta* (Proteaceae). II. The development of the endodermis in a determinate root and in an indeterminate, lateral root. New Phytol. 138, 733–742.

Watt M and Evans J R 1999a Linking development and determinacy with organic acid efflux from proteoid roots of white lupin crown with low phosphorus and ambient or elevated atmospheric CO_2 concentration. Plant Physiol. 120, 705–716.

Watt M and Evans J R 1999b Proteoid roots. Physiology and development. Plant Physiol. 121, 317–323.

Watt M and Evans J R 2002 Soil phosphorus acquisition by white lupin (*Lupinus albus* L.) and soybean (*Glycine max* L.) — species with contrasting root development. Plant Soil 248, 271–283.

Plant and Soil **248:** 247–256, 2003.
© 2003 Kluwer Academic Publishers. Printed in the Netherlands.

Localized supply of phosphorus induces root morphological and architectural changes of rice in split and stratified soil cultures

Yong He, Hong Liao & Xiaolong Yan[1]
Laboratory of Plant Nutritional Genetics and Root Biology Center, South China Agricultural University, Guangzhou 510642, China. [1]*Corresponding author**

Received 31 December 2001. Accepted in revised form 17 April 2002

Key words: localized phosphorus supply, rice, root architecture, root morphology

Abstract

A localized supply of phosphorus may affect root morphology and architecture, and thereby affect phosphorus uptake by rice plants. In the present study, we attempted to test this hypothesis using two rice cultivars representing upland and lowland ecotypes grown in specially designed split and stratified soil cultures with a low-phosphorus red soil. Our data indicate that a localized supply of phosphorus increased both total root length and root fineness, particularly in the high-phosphorus zone. In split culture, plants roots tended to preferentially grow on the high-phosphorus zone, with about 70–75% of the total root length allocated to the high-phosphorus compartment. The total root length on the high-phosphorus side in the split-phosphorus treatment was significantly longer than that in the homogenously high-phosphorus treatment, implying that a phosphorus-deficiency signal from the low-phosphorus side may stimulate the growth of the roots located in the high-phosphorus zone. In stratified soil culture, changes in root morphology and architecture were also observed as indicated by increased total root length, root fineness and relative root allocation in the high-phosphorus layers, again suggesting altered root morphology and preferential root proliferation in the high-phosphorus regions. The induced changes in root morphology and architecture by localized phosphorus supply may have both physiological significance and practical implications in that plants can meet the demand for phosphorus with parts of the roots reaching the high-phosphorus zone, hence localized fertilization methods such as side dressing or banded application of phosphorus fertilizers may both minimize phosphorus fixation by the soil and increase phosphorus uptake efficiency from the fertilizers.

Introduction

Rice (*Oryza sativa* L.) is the main food for nearly half of the world's population, and it plays an important role in human nutrition for people in developing countries, especially in China, which accounts for 22% of world's rice cultivated area and 36% world's rice grain yield (IRRI, 2000). Traditionally, rice is mostly grown in flooding paddy fields where irrigation is available. As water resources are becoming more and more limiting, upland rice provides an alternative for rice cultivation in areas with water scarcity (Jin and Ouyang, 1999). However, as compared with traditional flooding cultivation, upland rice is more strongly restricted by deficiency of nutrients, particularly phosphorus, due to changes in physical and chemical properties of the soil (Sample et al., 1980).

Phosphorus is an essential nutrient for rice growth, but phosphorus is usually deficient in tropical soils, particularly in upland soils where phosphorus is commonly bound to iron and/or aluminum oxides through chemical precipitation or physical adsorption. In order to minimize phosphorus fixation in the soil, localized application methods (such as banded or side dressing) of phosphorus fertilizers are commonly recommended in practice (Tisdale and Nelson, 1975). However, we do not know if such application methods would affect phosphorus uptake and plant growth through inducing changes in root growth and development.

* FAX No: 86-20-85281829
E-mail: xlyan@scau.edu.cn

Previous studies have shown that plant roots have great plasticity in response to phosphorus status (Snapp et al., 1995). For example, phosphorus deficiency in the soil could induce various morphological changes in plant roots, including formation of root hairs (Bates and Lynch, 1996; Föhse and Jungk, 1983; Gahoonia and Nielsen, 1998) and cluster roots (Barum, 1995; Dinkelaker and Romheld, 1989; Johnson et al., 1994), changes in root length and diameter (Föhse et al., 1991), and formation and growth of secondary lateral roots (Drew and Saker, 1978; Sun and Zhang, 2000). Adaptive changes in root architecture in response to phosphorus deficiency were also observed, including changes in basal root growth angle (Bonser et al., 1996; Liao et al., 2001) and relative distribution of roots in the topsoil (Liao et al., 2001). Nevertheless, little information is available if localized supply of phosphorus fertilizers would induce changes in root morphology and architecture and subsequently affect phosphorus uptake by rice.

In the present study, two rice cultivars representing upland and lowland ecotypes were grown in both split and stratified soil cultures that resemble those upland soils with localized supply of phosphorus fertilizers in such field practices as side dressing and banded applications (Tisdale and Nelson, 1975). The objective of this study was to determine the possible effects of localized supply of phosphorus on root morphology and architecture and subsequently phosphorus uptake of rice grown under upland conditions.

Materials and methods

Plant and soil materials

Seeds of two varieties of rice (*Oryza sativa* L.), Azucena and IR1552, were provided by Dr. Ping Wu from the College of Life Science, Zhejiang University, China. These two varieties represent different ecotypical origins: Azucena is a tropical upland rice cultivar and IR1552 is a lowland rice cultivar. The soil was a Ustisol collected from the Experimental Station at the South China Agricultural University, Guangzhou, China (Latitude: 26° 06′ N, Longitude: 113° 15′ E). The available phosphorus content of the soil was 0.806 mg/kg by Bray II extraction. The soil was mixed with cleaned quartz sand in a ratio of 2:1 to improve soil texture and aeration.

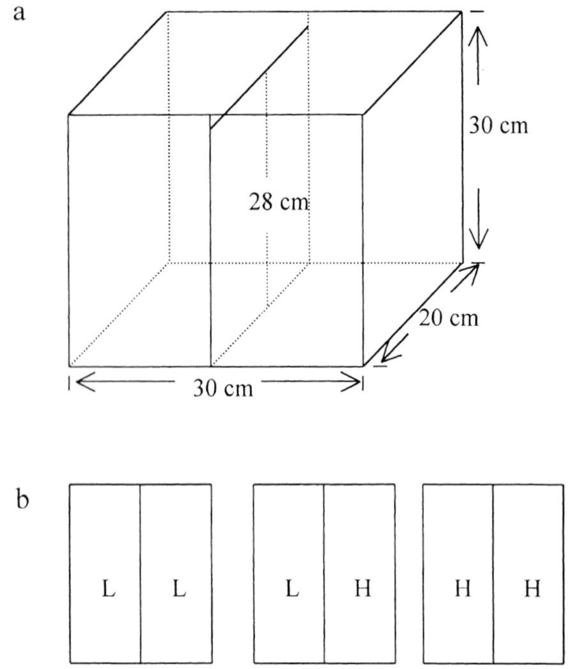

Figure 1. Schematic representation of the split-root soil culture system. (a) the growth box with two separated compartments; (b) the three phosphorus treatments in the split-root experiment: L-L, no phosphorus added in either side; L-H, one side without phosphorus addition and the other side with phosphorus added as KH_2PO_4 at 136.1 mg/kg soil; H-H, phosphorus was added as KH_2PO_4 at 136.1 mg/kg soil in both sides.

Split-root soil culture

The set up of the split soil culture system is shown in Figure 1. Growth boxes were made of transparent plexiglass at 30 cm×20 cm×30 cm (length×wide×height). The box was evenly divided into two independent compartments separated by a 3-mm-thick plexiglass partition board with a sharp edge on the top for convenient splitting of roots. The two boards on both sides were removable for final sampling of roots. During rice growth, the boxes were covered with opaque white board on the sides.

The split experiment included three treatments, which were low phosphorus-low phosphorus (no phosphorus added in either side, designated as L-L), low phosphorus-high phosphorus (one side without phosphorus addition and the other side with phosphorus added as KH_2PO_4 at 136.1 mg/kg soil, designated as L-H), and high phosphorus-high phosphorus (phosphorus was added as KH_2PO_4 at 136.1 mg/kg soil in both sides, designated as H-H). The other nutrients were added as (mg/kg soil): 236.2 $Ca(NO_3)_2.4H_2O$, 34.3 NH_4NO_3, 15.2

Fe-EDTA(Na), 0.64 $(NH_4)_6Mo_7O_{24}.4H_2O$, 410.9 $MgSO_4.7H_2O$, 1.20 H_3BO_3, 0.04 g $ZnSO_4.7H_2O$, 0.004 $CuSO_4.5H_2O$ and 1.54 $MnSO_4.H_2O$. In all low-phosphorus treatments, KH_2PO_4 was replaced by 89.4 K_2SO_4 (g/kg soil).

Rice seeds were surface sterilized with 10% NaClO for 2 min, then geminated in quartz sand at 30 °C for 5 days before transplanting. Rice seedlings were transplanted with adventitious roots equally separated into both sides of the growth boxes. Two rice seedlings were planted in each growth box. The plants were grown in a greenhouse at the South China Agricultural University, Guangzhou, China, in completely randomized block design with three replicates. Plants were irrigated with distill water to maintain the soil water content at about 70% field capacity (previously calibrated for this soil as the best water content for both rice varieties) by regular weighing of the soil box.

At 40 days after transplanting, plant shoots were harvested and dry weights were taken after oven-drying for 3 days at 60 °C. Phosphorus was analyzed colorimeterically as described by Murphy and Riley (1963). Meanwhile, plant roots were harvested from both sides of the growth box after removing the side boards. Roots were carefully recovered from the soil and rinsed with water for root measurements. All the roots (not subsamples) from each layer were spread out in a transparent plexiglass tray with a thin layer of water. Root morphological and architectural parameters, such as root length, and average root diameter were quantified with a computer image analysis software (WinRhizo Pro, Régent Instruments, Québec, Canada) after acquiring the root images from a scanner with a blue board as background. Total root length, root length per diameter class (an index to indicate fineness of the roots) and relative root allocation (the percentage of root length over the total root length) in each compartment were calculated as root morphological and architectural parameters.

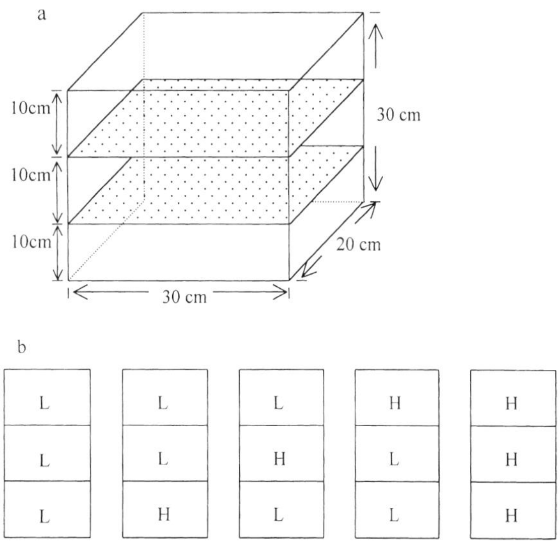

Figure 2. Schematic representation of the stratified soil culture. (a) growth box with three layers; (b) the five phosphorus treatments along soil profile: L-L-L, with homogenously low phosphorus (no phosphorus addition) along the soil profile; L-L-H, with high phosphorus in the bottom layer and low phosphorus in the top and middle layers; L-H-L, with high phosphorus in the middle layer and low phosphorus in the top and bottom layers; H-L-L, with high phosphorus (phosphorus added as KH_2PO_4 at 136.1 mg/kg soil) in the top layer (0–10 cm) and low phosphorus in the middle (10–20 cm) and bottom (20–30 cm) layers; and H-H-H, with homogenously high phosphorus along the soil profile.

Stratified soil culture

The growth boxes for the stratified soil culture were similar to those of the split culture system except that no partition board was installed in the middle (Figure 2a). The soil profile was stratified into three layers with fiberglass meshes (2 mm) installed at 10 cm intervals. Each layer was treated either with low phosphorus (no phosphorus added) or high phosphorus (phosphorus added as KH_2PO_4 at 136.1 mg/kg soil). In this experiment, there were five treatments with different combinations of phosphorus levels in the three layers (from top to bottom): (1) low phosphorus-low phosphorus-low phosphorus (designated as L-L-L, with homogenously low phosphorus along the soil profile); (2) low phosphorus-low phosphorus-high phosphorus (designated as L-L-H, with high phosphorus in the bottom layer and low phosphorus in the top and middle layers); (3) low phosphorus-high phosphorus-low phosphorus (designated as L-H-L, with high phosphorus in the middle layer and low phosphorus in the top and bottom layers); (4) high phosphorus-low phosphorus-low phosphorus (designated as H-L-L, with high phosphorus in the top layer (0–10 cm) and low phosphorus in the middle (10–20 cm) and bottom (20–30 cm) layers); and (5) high phosphorus-high phosphorus-high phosphorus (designated as H-H-H, with homogenously high phosphorus along the soil profile) (Figure 2b).

Rice seeds were germinated and plant seedlings were transplanted into the soil boxes with the above-mentioned treatments. Two rice seedlings were planted in each growth box. Nutrient and water managements were the same as those previously described for the split experiment, with the exception that water was irrigated from below with a glass tube inserted to the bottom of the box to avoid P movement among different layers causing by watering. The experiment was done in the same greenhouse with three replicates as described above. At 40 days after transplanting, plant shoots were harvested and dry weight and phosphorus content were taken as described above. Meanwhile, the roots from each stratified layer were carefully recovered from the soil and roots were scanned for morphological and architectural parameters as previously described for the split experiment.

Data analysis

The data of shoot biomass, phosphorus content and all the root parameters were analyzed by two-way ANOVA (Excel 97, Microsoft Corporation, 1985–1997). The data of root length per diameter class were analyzed by three-way ANOVA with SAS program (SAS Institute, Cary, NC). The univariate F statistics presented here were all in agreement with multivariate F statistics (Wilk's Lambda, Pillai Trace, and Hotelling-Lawly Trace).

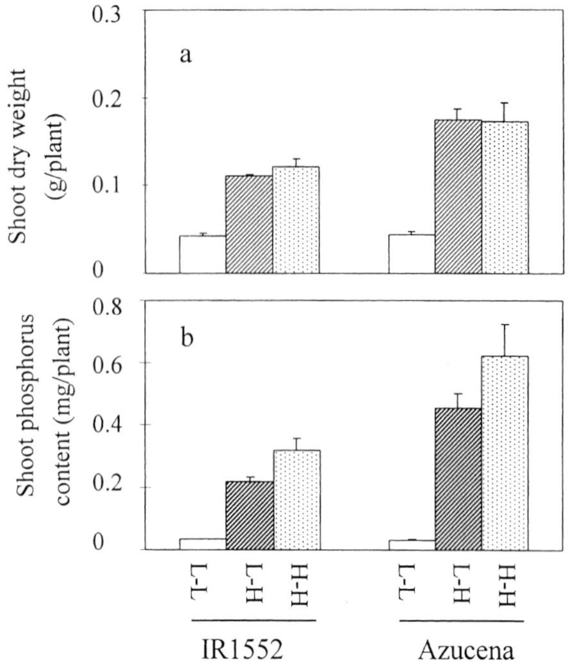

Figure 3. Shoot biomass and phosphorus content of two rice varieties in split-root soil culture. (a) shoot biomass; (b) shoot phosphorus content. L-L, no phosphorus added in either side; H-L, one side without phosphorus addition and the other side with phosphorus added as KH_2PO_4 at 136.1 mg/kg soil; H-H, phosphorus was added as KH_2PO_4 at 136.1 mg/kg soil in both sides. Each bar is the mean of three replicates with standard error.

Results

Split soil culture experiment

Plant growth and phosphorus uptake

Phosphorus treatments had great effects on plant growth and shoot phosphorus content in the split soil culture experiment. At 40 days after transplanting, the L-H and H-H treatments increased shoot biomass by about 3–4 folds and shoot phosphorus content by about 8–16 folds for both rice varieties, indicating severe phosphorus deficiency in this soil (Figure 3a, b). Interestingly, plant growth and phosphorus content of the L-H treatment did not differ significantly from that of the H-H treatment, suggesting that plants can grow normally even with supply of phosphorus to half of the roots. The two rice varieties from different ecotypic origins did not differ significantly in shoot biomass or phosphorus content in the L-L treatment, although in the H-H treatment, the upland rice variety, Azucena, seemed to have higher shoot biomass and phosphorus uptake than the lowland rice variety, IR1552 (Figure 3).

Total root length

Phosphorus significantly affected total root length of both rice varieties in the split soil culture experiment (Table 1). The total root length in both H-H and L-H treatments more than doubled that in the L-L treatment. The total root length on the low-phosphorus side of the L-H treatment was similar to that of the L-L treatment (F=1.73, nonsignificant), but the total root length on the high-phosphorus side was significantly longer than that of the H-H treatment (F=8.83, P<0.01). Both varieties showed similar response pattern, although Azucena (upland cultivar) generally had higher root length than IR1552 (lowland cultivar).

Root length per diameter class

The major portion of the root length of the two rice varieties fell into the diameter class smaller than 0.5 mm, and phosphorus treatments mainly affected root length in this class (Table 2). Root length in the <0.5 mm diameter class of the low-phosphorus treatment (L-L) was about the same as that on the low-phosphorus side of the split-phosphorus treat-

Table 1. Total root length (m/plant) of two rice varieties in different phosphorus treatments of the split root experiment

Genotype	L-L[a]	L-H [b]		H-H[c]	*F* value		
		Low P[d]	High P[e]		Genotype	P treatment	G×P
IR1552	5.07 (0.42)	5.36 (1.08)	17.36 (3.61)	12.62 (1.56)	9.903	16.053	2.841
Azucena	8.50 (0.84)	12.34 (2.79)	26.54 (4.08)	13.46 (1.97)	**	***	ns

Each value in the table is the mean of three replicates with standard error.
*0.05>P>0.01; **0.01>P> 0.001; ***P<0.001; ns, nonsignifiant at the 0.05 level.
[a]Low phosphorus-low phosphorus (no phosphorus added in either side); [b]low phosphorus-high phosphorus (one side without phosphorus addition and the other side with phosphorus added as KH_2PO_4 at 136.1 mg/kg soil); [c]high phosphorus-high phosphorus (phosphorus was added as KH_2PO_4 at 136.1 mg/kg soil in both sides); [d]root length calculated by multiplying a factor of 2 to have the same soil volume as L-L treatment; [e]root length calculated by multiplying a factor of 2 to have the same soil volume as H-H treatment.

Table 2. Root length (m/plant) per diameter class and its relative root allocation (%) of two rice varieties in different phosphorus treatments of the split root experiment

P treatment		IR1552			Azucena		
		<0.5 mm	0.5–1.0 mm	>1.0 mm	<0.5 mm	0.5–1.0 mm	>1.0 mm
L-L[a]		4.07±0.36	0.96±0.06	0.04±0.01	6.14±0.46	2.23±0.34	0.14±0.05
L-H[b]	L P[d]	3.84±0.88	1.45±0.24	0.07±0.02	7.36±1.63	3.89±0.98	1.08±0.22
	H P[e]	13.62±2.95	3.07±0.53	0.67±0.22	20.52±2.83	4.64±1.00	1.38±0.41
H-H[c]		9.59±1.20	2.66±0.37	0.37±0.04	10.13±1.36	2.77±0.57	0.55±0.11

Each value in the table is the mean of three replicates with standard error. Numbers in parenthesis are percentage of root length per diameter class relative to total root length.
[a]Low phosphorus-low phosphorus (no phosphorus added in either side); [b]low phosphorus-high phosphorus (one side without phosphorus addition and the other side with phosphorus added as KH_2PO_4 at 136.1 mg/kg soil); [c]high phosphorus-high phosphorus (phosphorus was added as KH_2PO_4 at 136.1 mg/kg soil in both sides); [d]root length calculated with the same soil volume as L-L treatment; [e]root length calculated with the same soil volume as H-H treatment.

ment (L-H) (F=0.25, nonsignificant), but significantly less than that on the high-phosphorus side of split-phosphorus treatment, indicating that more fine roots were found in the high-phosphorus regions. Interestingly, the root length with diameter finer than 0.5 mm on the high-phosphorus side of the L-H treatment were significant longer that of the H-H treatment (F=10.44, P<0.01), indicating that the enhanced growth of fine roots in the high-phosphorus zone might be induced by the low-phosphorus stress from the other half of the same root system.

Relative root allocation

Phosphorus allocation significantly affected relative root allocation in the split system. Relatively speaking, about 75% of the root length for IR1552 and 70% for Azucena, respectively, were allocated on the high-phosphorus side of the L-H treatment, while approximately same amount of roots were distributed on both sides of the L-L and H-H treatments (Figure 4).

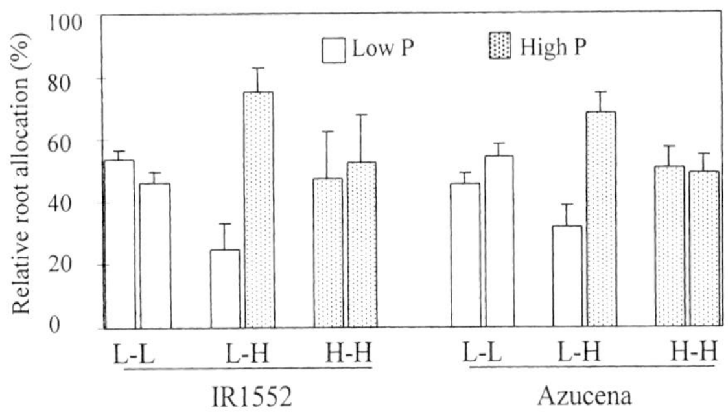

Figure 4. Relative root allocation (percentage of root length in each side over total root length) in the split-root soil culture. L-L, no phosphorus added in either sides; L-H, one side without phosphorus addition and the other side with phosphorus added as KH_2PO_4 at 136.1 mg/kg soil; H-H, phosphorus was added as KH_2PO_4 at 136.1 mg/kg soil in both sides. Each bar is the mean of three replicates with standard error.

Stratified soil culture experiment

Plant growth and phosphorus uptake

Phosphorus allocation in different layers along the soil profile significantly affected plant growth and phosphorus uptake in stratified soil culture. Although no significant difference was found for shoot bio-

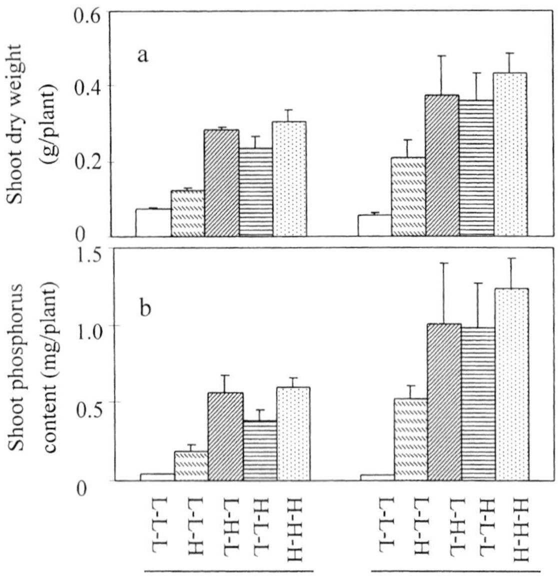

Figure 5. Shoot biomass and shoot phosphorus content of two rice varieties in the stratified soil culture. (a) shoot biomass; (b) phosphorus content. L-L-L, with homogenously low phosphorus (no phosphorus addition) along the soil profile; L-L-H, with high phosphorus in the bottom layer and low phosphorus in the top and middle layers; L-H-L, with high phosphorus in the middle layer and low phosphorus in the top and bottom layers; H-L-L, with high phosphorus (phosphorus added as KH_2PO_4 at 136.1 mg/kg soil) in the top layer (0–10 cm) and low phosphorus in the middle (10–20 cm) and bottom (20–30 cm) layers; and H-H-H, with homogenously high phosphorus along the soil profile. Each bar is the mean of three replicates with standard error.

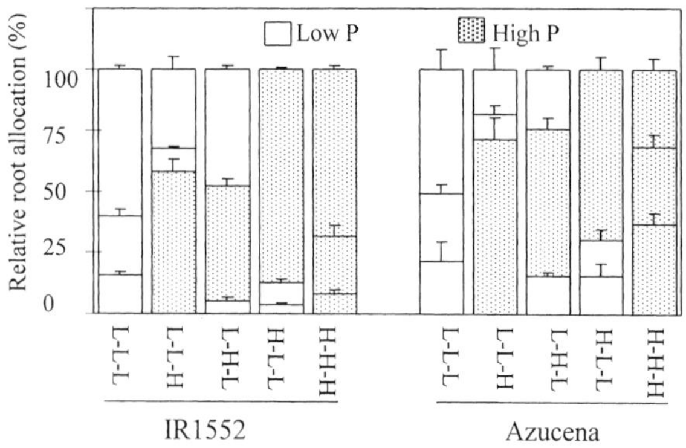

Figure 6. Relative root allocation of two rice varieties in the stratified soil culture system. Relative root allocation is the percentage of the root length per layer relative to total root length. Sections on which bars are divided mean relative root allocation of different layers. L-L-L, with homogenously low phosphorus (no phosphorus addition) along the soil profile; L-L-H, with high phosphorus in the bottom layer and low phosphorus in the top and middle layers; L-H-L, with high phosphorus in the middle layer and low phosphorus in the top and bottom layers; H-L-L, with high phosphorus (phosphorus added as KH_2PO_4 at 136.1 mg/kg soil) in the top layer (0–10 cm) and low phosphorus in the middle (10–20 cm) and bottom (20–30 cm) layers; and H-H-H, with homogenously high phosphorus along the soil profile. Each bar is the mean of three replicates with standard error.

mass among the H-L-L, L-H-L and H-H-H treatments, shoot biomass of all these treatments was much higher than that of the L-L-L and L-L-H treatments (Figure 5a). Phosphorus content represented almost the same pattern as shoot biomass, namely, phosphorus content was not significantly different among the H-L-L, L-H-L and H-H-H treatments, but much higher than that of the L-L-L and L-L-H treatments (Figure 5b). This indicates that phosphorus application in the topsoil (upper and middle layers, or 0–20 cm) had better effects on plant growth and phosphorus uptake than in the subsoil (bottom layer, or below 20 cm).

Total root length

Phosphorus allocation also significantly affected root morphological and architectural parameters in the stratified soil culture experiment. Phosphorus localized on the upper layer significantly increased total root length of both rice varieties, and no difference in total root length was found among H-L-L, L-H-L and H-H-H treatments (Table 3), suggesting that phosphorus supply to the topsoil (upper and middle layers) is more effective to induce root growth.

Root length per diameter class

We also found that phosphorus allocation may change root fineness. Table 4 shows that the total root length was always higher in the high-phosphorus zone, which also had relatively higher percentage of fine roots (diameter class less than 0.5 mm), suggesting that phosphorus allocation increases root length mainly through increasing fine root length in the layer supplied with high phosphorus (Table 4). Interestingly, high phosphorus localized in the upper layer (0–10 cm) enhanced root length in the middle layer (10–20 cm), while high phosphorus localized in the middle layer (10–20 cm) also enhanced root length in the upper layer (0–10 cm), but phosphorus localized in the bottom layer (below 20 cm) did not change the root length in the middle and upper layers. This again indicates that phosphorus located in the topsoil (0–20 cm) would better affect root growth than located in the subsoil.

Table 3. Total root length (m/plant) of two rice varieties in different phosphorus treatments of the stratified soil experiment

	L-L-L[a]	L-L-H[b]	L-H-L[c]	H-L-L[d]	H-H-H[e]	*F* value		
						Genotype	P treatment	G×P
IR1552	4.69±0.42	8.06±0.13	12.64±2.12	14.79±0.95	16.87±0.78	14.70	10.97	0.90
Azucena	5.80±0.72	16.39±1.11	20.30±1.60	23.99±7.13	21.03±1.31	***	***	ns

Each value in the table is the mean of three replicates with standard error.
*0.05>P>0.01; **0.01>P>0.001; ***P<0.001; ns, nonsignifiant at the 0.05 level.
[a]With homogenously low phosphorus (no phosphorus addition) along the soil profile; [b]with high phosphorus (phosphorus added as KH_2PO_4 at 136.1 mg/kg soil) in the top layer (0–10 cm) and low phosphorus in the middle (10–20 cm) and bottom (20–30 cm) layers; [c]with high phosphorus in the middle layer and low phosphorus in the top and bottom layers; [d]with high phosphorus in the bottom layer and low phosphorus in the top and middle layers; [e]with homogenously high phosphorus along the soil profile.

Relative root allocation

Phosphorus allocation changed rice root architecture. When phosphorous was homogenously low along the soil profile, rice roots of both varieties became shallower (i.e. greater relative root allocation in 0–10 cm and less roots below 20 cm). The upland variety, Azucena, tended to have deeper roots than lowland variety, IR1552. On the other hand, when phosphorus was supplied to a certain layer, the relative root allocation in that layer was dramatically increased. For example, as compared with that in the homogenous high-phosphorus treatment (H-H-H), the relative root allocation in the high-phosphorus layer of the L-H-L treatment increased about 2 folds for both IR1552 and Azucena (Figure 6).

Discussion

Our results indicate that phosphorus deficiency is indeed a major limiting factor for the red soil used in the present experiment. Without addition of phosphorus fertilizer, rice plants produced less than 30% of the shoot biomass with phosphorus addition in both split-root and stratified soil cultures (Figures 3 and 4). The growth of both rice varieties representing different ecotypical origins was equally restricted by phosphorus deficiency, but the upland variety, Azucena, grew better than the lowland variety, IR1552, under high phosphorus conditions, probably because the former in general had deeper root systems and therefore is more adaptive to the upland conditions provided phosphorus is no longer a limiting factor. Generally speaking, the upland variety, Azucena, was more responsive to localized phosphorus supply than the lowland variety, IR1552, probably because the former was evolved under upland conditions where phosphorus is often patchy thus developed more plasticity in root morphological and architectural adaptation to ununiform phosphorus supply.

Phosphorus availability may affect root morphology and architecture. Previous studies with common bean indicated that phosphorus stress generally tends to induce root shallowness while high phosphorus promotes root proliferation to the high-phosphorus zone (Liao et al., 2001). The effects of localized phosphorus supply on root morphology and architecture were also observed in our study with the stratified soil system. Phosphorus-rich layer generally had higher total root length (Table 3), possibly through increasing the growth of fine roots (<0.5 mm in diameter) in the layer supplied with high phosphorus (Table 4). In fact, visual observation of the root samples recovered from the soil show that localized phosphorus supply (such as the L-H-L treatment) induced formation of cluster-root-like fine roots (Figure 7). Furthermore, the relative root allocation is always higher in the soil layer with high-phosphorus supply, suggesting preferential root proliferation to the high-phosphorus zone (Figure 6). Our results from the stratified root experiment are consistent with previous findings in other species that nonuniform supply of nutrients (nitrogen and phosphorus) would induce preferential root proliferation and growth in nutrient-rich zones (Drew and Saker, 1975, 1978; Snapp et al., 1995; Zhang and Forde, 1998). In the split soil culture, phosphorus side dressing significantly affected root distribution by preferential allocation of root length on the high-phosphorus side (Table 1). The increased root length was probably attributed to the increased formation of fine roots (diameter class <0.5 mm) (Table 2). It should be noted that the total root length on the high-phosphorus side of the L-H treatment was even longer than that of the H-H treatment. Apparently, phosphorus deficiency on one side could stimulate root growth on the other side with sufficient phosphorus

Table 4. Root length (m/plant) per diameter class and its relative root allocation (%) of two rice varieties in the stratified root experiment

	IR1552			Azucena		
	<0.5 mm	0.5–1.0 mm	>1.0 mm	<0.5 mm	0.5–1.0 mm	>1.0 mm
Layer 1 (0–10 cm)						
L-L-L[a]	2.26±0.12	0.50±0.07	0.04±0.01	2.04±0.05	0.78±0.09	0.04±0.01
L-L-H[b]	1.50±0.16	1.02±0.24	0.08±0.04	1.99±0.64	0.74±0.19	0.44±0.17
L-H-L[c]	3.77±0.58	1.81±0.27	0.47±0.19	2.83±0.63	1.53±0.04	0.67±0.25
H-L-L[d]	9.19±0.72	2.74±0.28	0.97±0.27	12.41±3.90	3.04±0.85	1.32±0.38
H-H-H[e]	7.93±0.92	2.81±0.25	0.83±0.05	3.92±0.39	1.58±0.09	1.12±0.16
Layer 2 (10–20 cm)						
L-L-L	0.96±0.17	0.19±0.07	0.01±0.00	1.17±0.05	0.39±0.04	0.03±0.00
L-L-H	0.53±0.08	0.24±0.06	0.01±0.00	1.07±0.59	0.36±0.16	0.25±0.09
L-H-L	4.36±0.84	1.16±0.06	0.38±0.10	7.53±0.23	2.83±0.10	1.74±0.22
H-L-L	0.95±0.11	0.34±0.05	0.04±0.01	2.52±1.29	1.00±0.42	0.44±0.23
H-H-H	2.87±0.77	0.90±0.08	0.14±0.02	4.42±0.60	1.54±0.38	0.61±0.24
Layer 3 (20–30 cm)						
L-L-L	0.53±0.07	0.19±0.06	0.01±0.00	0.96±0.46	0.37±0.16	0.06±0.05
L-L-H	3.76±0.45	0.71±0.02	0.22±0.03	8.08±0.41	2.53±0.27	0.94±0.07
L-H-L	0.42±0.18	0.25±0.12	0.02±0.02	1.71±0.30	1.17±0.22	0.30±0.12
H-L-L	0.46±0.10	0.11±0.03	0.00±0.00	2.19±0.52	0.80±0.12	0.26±0.15
H-H-H	1.08±0.18	0.26±0.11	0.04±0.02	5.22±0.79	2.02±0.36	0.58±0.22

Each value of root length was the mean of three replicates with standard error and numbers in parenthesis are percentage of root length per diameter class relative to total root length.
[a]With homogenously low phosphorus (no phosphorus addition) along the soil profile; [b]with high phosphorus (phosphorus added as KH_2PO_4 at 136.1 mg/kg soil) in the top layer (0–10 cm) and low phosphorus in the middle (10–20 cm) and bottom (20–30 cm) layers; [c]with high phosphorus in the middle layer and low phosphorus in the top and bottom layers; [d]with high phosphorus in the bottom layer and low phosphorus in the top and middle layers; [e]with homogenously high phosphorus along the soil profile.

supply. This may imply that the phosphorus deficiency signal comes from the low-phosphorus side, while enhanced root growth is realized on the high-phosphorus side where phosphate as a growth substrate is sufficient. However, in a hydroponic study with divided roots of *Hakea prostrata* to investigate the effects of external phosphorus supplies on cluster root initiation and growth, Shane et al. (2002) found that root phosphorus concentration of root halves supplied with low phosphorus was not significantly different from that with high-phosphorus treatments, suggesting that the signal for changes in cluster-root growth was not a local one, but must have been a systemic signal. Also, by using a similar split-root approach, Burleigh and Harrison (1999) showed that the expression of a phosphorus starvation-inducible gene (*Mt4*) is repressed in the phosphorus-deficient side, implying that phosphorus fertilization on one-half of the root results in the systemic down-regulation of *Mt4* expression throughout the whole root system. Therefore, the molecular aspects of rice roots in response to localized phosphorus supply deserve further investigation.

The induced changes in root morphology and architecture by localized phosphorus supply may have physiological significance that phosphorus uptake by plants can be achieved by allocating a portion of the roots to the high-phosphorus zone (Pang et al., 2000). Indeed, almost all localized phosphorus treatments (except the L-L-H treatment where phosphorus was located in the bottom layer) had similar amounts of phosphorus uptake to that of the homogenously high-phosphorus treatment, indicating that plants can meet the demand for phosphorus with parts of the roots reaching the high-phosphorus zone (Figures 3 and 5). In other words, a partial supply with 1/2 (split-root treatment) and 1/3 (stratified soil treatment) of the phosphorus in the homogenously high-phosphorus treatment could achieve similar biomass and phosphorus uptake (Figures 3 and 5, except the L-L-H treatment where phosphorus was applied in the bottom layer). This might mean that phosphorus uptake

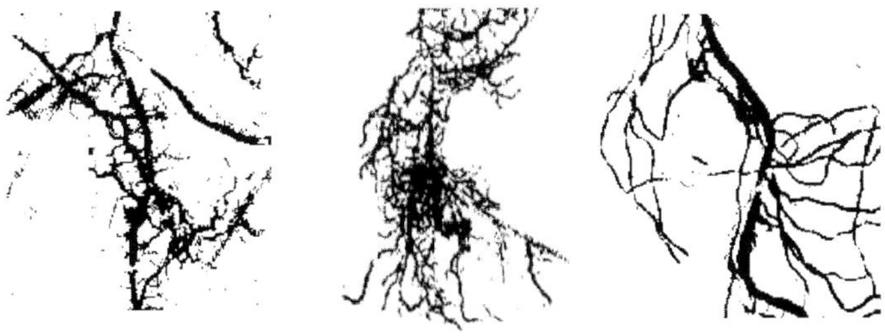

Figure 7. Representative root images selected from the stratified soil culture experiment with rice cultivar, Azucena. L-L-L, with homogenously low phosphorus (no phosphorus addition) along the soil profile; L-H-L, with high phosphorus in the middle layer and low phosphorus in the top and bottom layers (image was from high phosphorus layer); and H-H-H, with homogenously high phosphorus along the soil profile.

efficiency from the fertilizers is increased. This has practical implications that side dressing or banded application of phosphorus fertilizers may not only minimize phosphorus fixation by the soil but also increase phosphorus uptake efficiency from fertilizers as compared with other application methods such as broadcasting.

Acknowledgements

This research was supported by the National Key Basic Research Special Funds of China grant (G1999011700), the Guangdong Natural Science Foundation and the National Natural Science Foundation of China grants (39925025/30070441). The authors are grateful to Dr. Ping Wu for supply of the rice seeds and helpful discussion. We also thank Xiurong Wang, Shaoling Zheng, Ruibin Kuang, Lide Wang, Aiqin Cao, Xianglian Deng and Weifeng Zhang for their assistance.

References

Barum S M 1995 Mobilization of phosphorus and other mineral nutrient by citrate in the rhizosphere of *Lupinus albus* L. Ph.D. Thesis, University of Wisconsin, Madison.

Bates T R and Lynch J P 1996 Stimulation of root hair elongation in *Arabidopsis thaliana* by low phosphorus availability. Plant Cell Environ. 19, 529–538.

Bonser A M, Lynch J P and Snapp S 1996 Effect of phosphorus deficiency on growth angle of basal roots in *Phaseolus vulgaris*. New Phytol. 132, 281–288.

Burleigh S H and Harrison M J 1999 The down-regulation of Mt4-like genes by phosphate fertilization occurs systematically and involves phosphate translocation to the shoots. Plant Physiol. 119, 241–248.

Dinkelaker B, Romheld V and Marschner H 1989 Citric acid excretion and precipitation of calcium citrate in the rhizosphere of white lupin (*Lupinus albus* L.). Plant Cell Environ. 12, 285–292.

Drew M C and Saker R L 1975 Nutrient supply and the growth of the seminal root system in barley. II. Localized compensatory increase in lateral root growth and rates of nitrate uptake when nitrate supply is restricted to only part of the root system. J. Exp. Bot. 26, 79–90.

Drew M C and Saker R L 1978 Nutrient supply and the growth of the seminal root system in barley. III. Compensatory increase in growth of lateral roots, and in rates of phosphate uptake, in response to localized supply of phosphate. J. Exp. Bot. 29, 435–451.

Föhse D and Jungk A 1983 Influence of phosphate and nitrate supply on root formation of rape, spinach and tobacco plants. Plant Soil 74, 359–368.

Föhse D, Classssen N and Jungk A 1991 Phosphorus efficiency of plants. II. Significance of root radius, root hairs and cation-anion balance for phosphorus influx in seven plant species. Plant Soil 132, 261–272.

Gahoonia T S and Nielsen N E 1998 Direct evidence on participation of root hairs in phosphorus (^{32}P) uptake from soil. Plant Soil 198, 147–152.

IRRI 2000 World rice statistics. Los Banos. Philippines.

Jin Q Y and Ouyang Y N 1999 Discussion about several problems on development of water efficient rice cultivation in China. Chinese Rice 1, 9–12 (in Chinese).

Johnson J F, Deborah L A and Vance C P 1994 Phosphorus stress-induced proteoid roots show altered metabolism in *Lupinus albus*. Plant Physiol. 104, 657–665.

Liao H, Rubio G, Yan X L, Cao A, Brown K m and Lynch J P 2001 Effect of phosphorus availability on basal root shallowness in common bean. Plant Soil 232, 69–79.

Murphy J and Riley J 1963 A modified single solution method for the determination of phosphate in natural waters. Anal. Chem. Acta 27, 31–35.

Pang X, Li C J and Zhang F S 2000 Effects of partial-root supply of P nutrient on the growth and assimilate distribution of wheat seedlings. Acta Agro. Sin. 26, 719–724 (in Chinese with English abstract).

Sample E C, Soper R J and Racz g J 1980 Reactions of phosphate fertilizers in soils. *In* The Role of Phosphorus in Agriculture. Eds. F E Khasawneh et al. pp. 263–310. American Society of Agronomy, Madison, WI, USA.

Shane M W, de Vos M, de Roock S, Cawthray G R and Lambers H 2002 Effects of external phosphorus supply on internal phosphorus concentration and the initiation, growth and exudation of cluster roots in *Hakea prostrata* R.Br. Plant Soil 248, 209–219.

Plant N concentration decreases with P limitation (De Groot et al., 2001b; Jeschke et al., 1996; Rufty et al., 1993; Schjørring, 1986). There are several possible causes for this decrease. Firstly, the decrease of plant N concentration with P limitation might be due to a shift in dry-mass partitioning from plant organs high in N concentration to plant organs lower in N concentration. A second possibility is a decrease in the rate of N uptake, due to feedback inhibition of the uptake as a result of N accumulation in the roots (Gniazdowska and Rychter, 1999; Rufty et al., 1990, 1993; Schjørring, 1986). Furthermore, N uptake might be decreased due to decreased energy availability as suggested by a decrease in root growth and/or decreased ATP concentrations (Rufty et al., 1990, 1993). On the other hand, it has also been reported that a shortage of energy does not play a role in decreasing N uptake under P-limited conditions as suggested by the lack of growth inhibition in P-limited roots and the lack of a change in ATPase activity regardless of a decrease in ATP concentration (Buwalda and Warmenhoven, 1999; Gniazdowska and Rychter, 2000).

Cytokinin levels may also offer a possible explanation for a decrease in N uptake at P limitation. Cytokinin levels in root exudate (Dhillon, 1978; Menary and Van Staden, 1976) and shoot material (Horgan and Wareing, 1980; Kuiper et al., 1988; Salama and Wareing, 1979; Thorsteinsson and Eliasson, 1990) were shown to decrease upon P limitation. Decreased cytokinin levels might cause a decrease in nitrate reductase activity (Bueno et al., 1994; Gaudinová, 1990; Lu et al., 1992) and net protein synthesis (Klämbt, 1977). Similarly, it has been shown that nitrate reductase activity and protein synthesis decrease in P-limited plants (Andrews et al., 1999; Gniazdowska and Rychter, 2000; Jacob and Lawlor, 1991; Rufty et al., 1990, 1993). This suggests that cytokinins might be involved in regulating the effects of P limitation on plant N concentration and N uptake. Furthermore, the expression of P-starvation responsive genes could be repressed by supplying exogenous cytokinins (Martin et al., 2000). In this paper we address the question of whether the decrease in plant N concentration with P limitation is due to a shift in dry-mass partitioning or due to a decrease in N uptake. Furthermore, we investigate the role of carbohydrate metabolism and cytokinins in explaining decreased N uptake under P limiting conditions.

Several authors have reported a strict relation between plant N concentration and dry-mass partitioning to the roots, independent of applied light intensity (De Pinheiro Henriques and Marcelis, 2000; Van der Werf and Nagel, 1996). Since P limitation decreases total plant N concentration (De Groot et al., 2001b; Jeschke et al., 1996; Rufty et al., 1993; Schjørring, 1986), we investigated whether the relation between N concentration and dry-mass partitioning also holds when N concentration is varied by P supply. Furthermore, with decreasing plant N concentration it is expected that reduced-N concentration as well as nitrate-N concentration decrease. Andrews et al. (1999) showed a strong relation between dry-mass partitioning and reduced-N concentration, while Scheible et al. (1997) showed dry-mass partitioning to be strongly related with nitrate-N concentration. We tested whether the nitrate-N concentration or the reduced-N concentration, as varied by either N supply or P supply, is closer related to dry-mass partitioning to the roots.

Two detailed growth experiments, with varied N or P supply, were conducted with tomato plants in order to unravel the effects of N and P limitation on growth in this species. In this paper we will focus on the differences and similarities in the growth response to N and P limitation, and especially on their interaction.

Materials and methods

Growth of plants

Two experiments were conducted in which N supply (N experiment) or P supply (P experiment) was varied. In both experiments seeds of tomato (*Lycopersicon esculentum* Mill. cv. Capita) were germinated on moistened vermiculite at 21 °C. Eight days after sowing (DAS) plants were transferred into 2.7-dm^3 containers, one plant per container, with either N-free nutrient solution brought to 1 m*M* NO_3^- by adding KNO_3 (N experiment) or with P-free nutrient solution brought to 55 μM $H_2PO_4^-$ by adding KH_2PO_4 (P experiment). These concentrations are one-twelfth of the concentrations normally used (Steiner, 1984). Earlier (pilot) experiments have shown that tomato seedlings grow perfectly well at these concentrations. The aim of growing the seedlings on this nutrient solution for a week was to let them grow well, without getting them used to high concentrations of N or P, in order to shorten the time needed to adapt to the nutrient supply treatments. The containers were placed in a growth chamber with a photosynthetically active radiation (PAR) of 300 μmol m^{-2} s^{-1} for 16 h day^{-1}

Table 1. Concentration of the macronutrients in nutrient solution

Element	Concentration in nutrient solution (m*M*)		
	P experiment	N experiment	Free access to P and N (FA)
K	7.0	8.5	7.0
Ca	4.5	5.5	4.5
Mg	2.0	2.4	2.0
S	3.7	10.5	3.5
N (as nitrate)	12.7	0.0	12.0
P	0.005	3.0	1.0

(TL-D-HF, Philips, Eindhoven, The Netherlands) followed by 30 min of incandescent light. The relative humidity was 70% with a day/night temperature set to 23/23 °C, which resulted in a day/night temperature of 24/23 °C for the N experiment and 23/21 °C for the P experiment. The composition of the macronutrients and trace elements of the nutrient solution was as described by Steiner (1984), with the exception that in the N-free nutrient solution nitrate was replaced by phosphate and sulphate without changing the ratio of these ions and in the P-free nutrient solution phosphate was replaced by nitrate and sulphate without changing the ratio of these ions (Table 1). At the beginning of the treatments (15 DAS) the nutrient solution in the containers was replaced by either N-free or P-free solution. For the P experiment when the pH dropped below 5.5 it was readjusted to 6.0 with a 1:1 mixture of sulphuric and nitric acid. For the N experiment it was not necessary to readjust the pH during the experiment.

Nutrient treatments

At the beginning of the treatments (15 DAS) eight rates of N supply (N experiment) or seven rates of P supply (P experiment) were applied. Both experiments were performed twice. P was supplied at a growth-saturating concentration of 1 m*M* KH_2PO_4 (free access (FA) treatment), or daily at relative addition rates of 70, 120, 170, 220, 270 or 320 mg g^{-1} day^{-1}. The same treatments were applied for the N experiment, except that for the second replica of the N experiment the relative addition rate of 120 mg g^{-1} day^{-1} was replaced by 370 mg g^{-1} day^{-1}. The growth-saturating concentration in the N experiment was 12 m*M* NO_3^-.

From known N and P concentrations of tomato seedlings and from the fresh weight of the seedlings harvested at 15 DAS, the initial amount of N and P in the plants at the start of the treatments was estimated to be 1.4 and 0.24 mg, respectively. In order to shorten the time needed for the plants of the four lowest treatments to adjust to the lower supply, the plants were starved for N or P during a short period. After transferring the plants to N or P free nutrient solution, at 15 DAS, the supply of the highest treatments started immediately. The treatments 220, 170, 120 and 70 mg g^{-1} day^{-1} were starved for 1, 2, 3 and 4 days, respectively. The N or P supply of these treatments started at 16, 17, 18 and 19 DAS.

Harvests and chemical analysis

Twelve plants were harvested at the start of the treatments (15 DAS) and six plants per treatment at 27, 31 and 35 DAS. At harvest, roots were rinsed three times in tap water and blotted dry between absorbing paper. Plants were divided into leaf blades, roots and stem (including petioles). Fresh mass and dry mass (after at least 48 h at 70 °C) were determined. At 31 DAS two leaflets of the third leaf, counted from bottom to top (total number of leaves ranged from 5 to 12; leaves were counted when 5 mm long), were sampled at the end of the light period, and two at the end of the dark period. Leaf area and fresh mass of these leaflets were determined. The samples were freeze-dried, weighed and stored for determination of soluble sugars, cytokinins, nitrate and phosphate.

For cytokinins and soluble sugars determination 25 mg freeze-dried material collected at 31 DAS were extracted four times with 80% (v/v) ethanol at 85 °C. After each extraction samples were centrifuged (4 °C, 14 000×*g* for 10 min) and the supernatant collected, the residue was extracted again. The supernatant, pooled over the four extractions, was purified through Sep-Pak C18 cartridges and used for soluble sugars and cytokinin determination.

Soluble sugars (glucose, fructose and sucrose) were separated on anion exchange HPLC, using a CarboPac PA1 column (4×250 mm) preceded by a Carbopac PA1 guard column (4×50 mm) both from Dionex BV (Bavel, The Netherlands). The soluble sugars were quantified using pulsed amperometric detection with an ED40 electrochemical detector equipped with a ED40 gold electrode (Dionex, Bavel, The Netherlands).

For nitrate and phosphate determination, 10 mg freeze-dried material, collected at 31 DAS, were extracted for 2 min with 1 mL distilled water at 100 °C. After centrifugation (14 000 × *g* for 10 min) the supernatants was stored at −20 °C until analysis. Nitrate and phosphate were determined by anion chromatography with suppressed conductivity detection. Tissue concentrations of total N and total P were determined in each plant organ at 31 DAS. Total N was determined with an element analyser according to the Dumas method (ROBOPREP-CN, Biological Sample Converter, Europe Scientific, Crewe, UK). Total P was determined colourimetrically after digestion with a 1:1 (v/v) mixture of concentrated sulphuric acid and concentrated nitric acid (Humphries, 1956). Organic-P concentration was estimated as the total-P concentration minus the average phosphate-P concentration. Phosphate-P concentration decreased with decreasing P supply and was not significantly different for the end of light and at the end of the dark period. Reduced-N concentration was estimated in a similar way by subtracting nitrate-N from total-N concentration. Nitrate-N concentration decreased with decreasing N supply and showed the same pattern for the end of the light and the end of the dark period. At high N supply rates leaf nitrate concentration was significantly lower at the end of the light period than at the end of the dark period ($P<0.001$).

Cytokinin determination

Cytokinins were determined on two replicate samples per treatment. Each sample consisted of six separate samples which were pooled into one, in order to obtain samples large enough for the cytokinin determination. Free cytokinins were determined for three treatments; 170 mg g^{-1} day^{-1} (subsequently referred to as low N and low P), 320 mg g^{-1} day^{-1} (subsequently referred to as high N and high P) and the growth saturating (free access) treatment. For extraction, purification and quantification of cytokinins a slightly modified procedure as described by Vonk et al. (1986) was used. Leaf material was extracted as described above. Ethanol extracts were pooled per replicate experiment (six samples, three at the end of the light period and three at the end of the dark period, see above), resulting in two samples per treatment. The ethanol was removed by evaporation under reduced pressure and the aqueous residues were taken up in water. For the separation of free and bound cytokinins, samples were partitioned (pH 8.2) against water-saturated butan-1-ol. Butanol layers (free cytokinins) were evaporated to dryness and then purified by a combination of anionic exchange and reversed phase C18 column chromatography. Cytokinins were eluted from the C18 columns with 80% (v/v) ethanol and the eluates were evaporated to dryness. Cytokinin quantification was performed by an enzyme-linked immunoassay (ELISA). Antibodies raised against the cytokinins zeatin riboside (ZR) and isopentenyl adenosine (IPAR) were used. These antibodies bind not only the ribosides of the above-mentioned cytokinin types but also their free bases and the 9-glucosides. Combined contents of these ribosides, free bases and glucosides are calculated as cytokinin riboside equivalents. A recovery test for the whole purification and quantification procedure was carried out with $[^3H]$ZR, which showed a 85% recovery. Results were corrected for this recovery percentage. Recently, the ELISA method was confirmed with LC–MS–MS using deuterated standards (Van Rhijn et al., 2001).

Calculations and statistics

Growth was determined as relative growth rate (RGR, mg g^{-1} day^{-1}) based on dry mass. RGR on 31 DAS was calculated as $1000 * [\ln W(t_{35DAS}) - \ln W(t_{27DAS})]/[t_{35DAS} - t_{27DAS}]$; where W is dry mass (g). Both the N and the P experiment were conducted twice, each time with three replicate plants per treatment per harvest, which gives a total of six replicate plants. The three plants per treatment were arranged in a block design with three blocks. The P experiment was analysed with an ANOVA (general analysis of variance, GENSTAT 5 release 4.2). However, for the determination of leaf soluble sugar concentration, phosphate-P and nitrate-N concentration some samples were too small for analysis, which resulted in missing values. These parameters were analysed with linear mixed models (REML). When the N experiment was conducted for the second time the relative addition rate of 120 mg^{-1} day^{-1} was replaced by 370 mg g^{-1} day^{-1}, which left the block design unbalanced and could not be analysed by ANOVA. All the parameters of the N experiment have been analysed by REML. Data were analysed by ANOVA or REML using GENSTAT 5 release 4.2 at a significance level of $\alpha=0.05$ (Lawes Agricultural Trust, IACR-Rothamsted, UK).

The present experiments were designed such that the effects of plant size and applied treatments could be distinguished. We have shown before, that data analysed for a specific date did not yield different conclusions compared with data analysed for a specific plant size (De Groot et al., 2001a). This justifies, in this study, the use of plants of the same age and, therefore the data discussed here pertain to data at 31 DAS.

Results

RGR response to phosphorus and nitrogen limitation

Decreasing the P supply decreased both the organic P and the inorganic phosphate (P_i) concentration of the leaves (Figure 1a). In well nourished tomato plants P_i made up about 25% of the total P in the leaves (free access treatment), while at low P supply, P_i decreased to a minimum of 14% (Figure 1a). In contrast, only 5.5% of the N in the leaves of well nourished tomato plants was nitrate; this value decreased to 0% with decreasing N supply (Figure 1b). Total and organic leaf P concentration approached a minimum concentration at the four lowest P supply treatments (Figure 1a) while the concentrations of total and reduced leaf N continued to decrease with decreasing N supply (Figure 1b). This resulted in a steep increase of the relation between RGR and leaf P concentration (Figure 2a) and a more gradual increase in RGR with increasing leaf N concentration (Figure 2b). Both curves leveled off at non-limiting supply rates, however the plateau was much broader for the P experiment than for the N experiment. Plotting RGR against the organic leaf P concentration for the P experiment and against reduced leaf N concentration (total leaf N concentration corrected for nitrate accumulation) for the N experiment, did not change the shapes of the curves (Figure 2a, b).

Interaction between N and P uptake

N concentration of the total plant and of the different plant organs decreased with decreasing P supply rate (Figure 3a). The N concentration of the stem and roots decreased further than leaf N concentration. The decrease in N concentration with decreasing P supply could explain 10 out of the 74% decrease in RGR. In the N experiment no consistent increase or decrease was found for the P concentration of roots and leaves. However, plant P concentration and stem P concentration did increase consistently with decreasing N supply treatment ($P<0.001$, Figure 3b). The decrease in leaf N concentration with decreasing P supply was mainly due to a decrease in reduced-N concentration

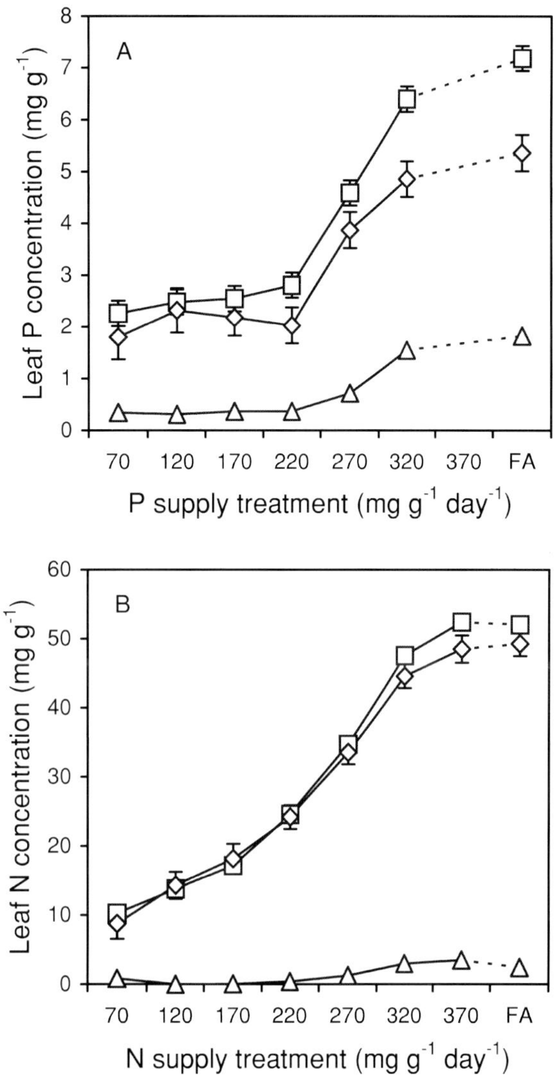

Figure 1. (A) Leaf P concentration in relation to P-supply treatment, and (B) leaf N concentration plotted against N-supply treatment. Squares=total leaf P (A) or total leaf N concentration (B); triangles = P_i concentration (A) or leaf nitrate (B); diamonds=organic P concentration (A) or leaf reduced N concentration (B). FA=Free access to N and P. Shown numbers for nitrate and phosphate concentration are averages of measurements at the end of the light and at the end of the dark period. Vertical bars indicate standard errors of means; only shown when larger than symbol (n = 6).

and to a smaller extent due to a decrease in nitrate-N concentration (average of end of light and end of dark period) (Figure 4a). This decrease in nitrate-N concentration was only due to a decrease at the end of the dark period ($P<0.01$, Figure 4b). Changes in leaf nitrate concentration at the end of the light period with decreasing P supply were not significant (Figure 4b).

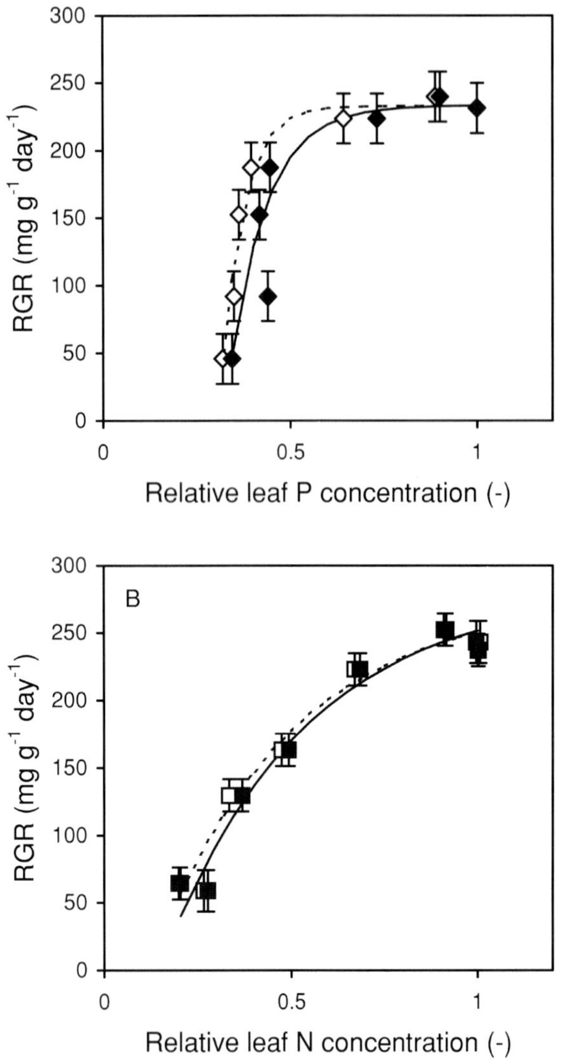

Figure 2. Relative growth rate (RGR) in relation to relative leaf P concentration for the P experiment (A) and against relative leaf N concentration for the N experiment (B). (1) Concentration of FA treatment. Open symbols=total leaf P concentration (A) or total leaf N concentration (B); closed symbols=leaf organic-P concentration (A) or reduced-N concentration (B). Vertical bars indicate standard errors of means; only shown when larger than symbol. Exponential curves were fitted, for all curves: $P < 0.001$, $r^2 > 0.86$ ($n = 6$).

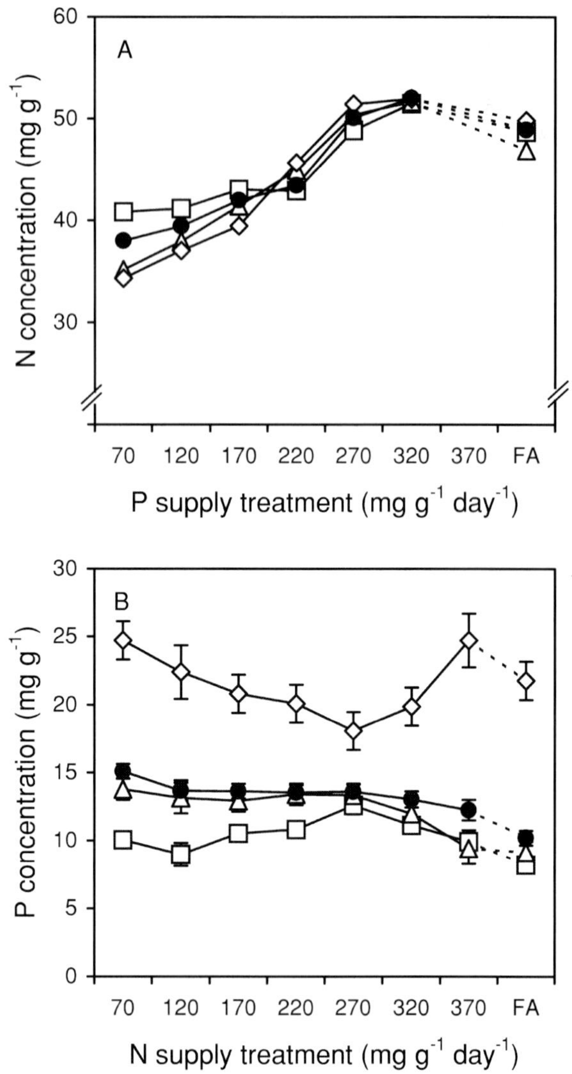

Figure 3. N concentration of the different plant organs in relation to P supply treatment (A) and P concentration of the different plant organs in relation to N supply treatment (B). Circles=total plant; squares=leaves; triangles=stem; diamonds=roots. Vertical bars indicate standard errors of means; only shown when larger than symbol ($n = 6$).

Leaf cytokinin concentration

In three selected nutrient supply treatments ZR and IPAR equivalents were determined. In general the cytokinin concentrations in the leaves were lowest at low nutrient supply (170 mg g^{-1} day^{-1}), intermediate at high nutrient supply (320 mg g^{-1} day^{-1}) and highest in the free access treatment (Table 2). The decrease in ZR equivalents with decreasing nutrient supply, was comparable for the N and P experiment (P<0.05). The concentration of IPAR equivalents decreased significantly (P<0.05) with decreasing N supply. In the P experiment no significant differences were found in IPAR equivalents between the different P treatments (Table 2).

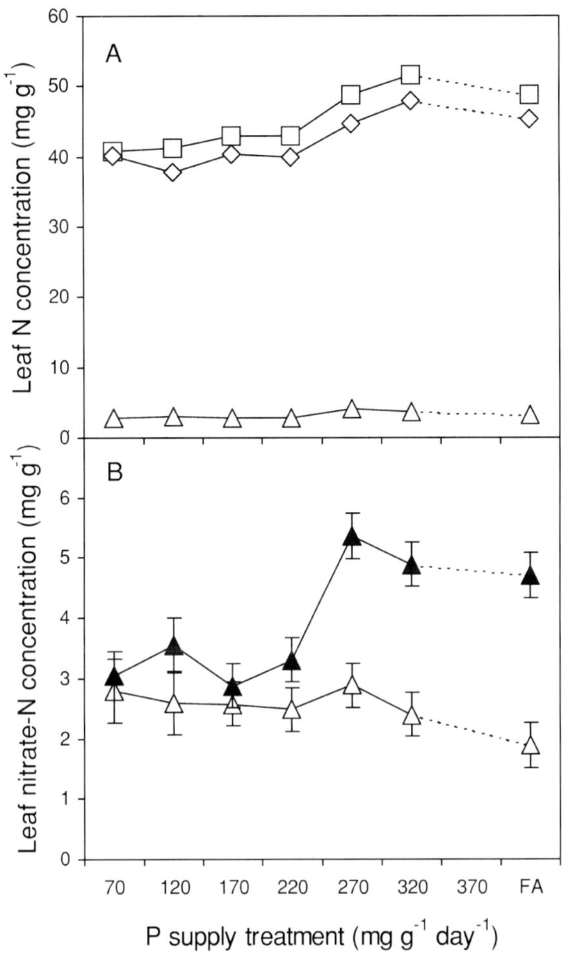

Figure 4. Leaf N concentration in relation to P-supply treatment (A). Squares=total leaf N concentration; diamonds=leaf reduced-N concentration; triangles=leaf nitrate-N concentration. Shown numbers for nitrate concentration are averages of measurements at the end of the light and at the end of the dark period. (B) shows leaf nitrate concentration at the end of the light period (open symbols) and at the end of the dark period (filled symbols) in relation to P-supply treatment (B). Vertical bars indicate standard errors of means; only shown when larger than symbol (n = 6).

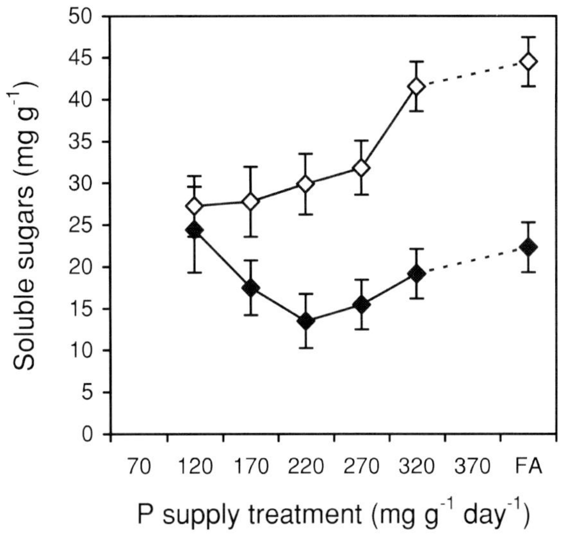

Figure 5. Leaf soluble sugar concentration at the end of the light period (open symbols) and at the end of the dark period (filled symbols) in relation to P-supply treatment. Vertical bars indicate standard errors of means; only shown when larger than symbol (n = 6).

Table 2. Concentrations (pmol g^{-1} dry mass, average of two pooled samples) of free zeatin riboside (ZR) equivalents and free isopentenyl adenine riboside (IPAR) equivalents in leaves. Within a column differences between means are indicated by different letters

Nutrient treatment	N experiment		P experiment	
	ZR	IPAR	ZR	IPAR
Low (170 mg g^{-1} day^{-1})	12.3^{a}	82.9^{a}	$<5^{a}$*	136^{a}
High (320 mg g^{-1} day^{-1})	24.4^{ab}	148^{ab}	14.0^{ab}	203^{a}
Free access	40.9^{b}	221^{b}	27.1^{b}	205^{a}
$LSD_{5\%}$	26.6	81.3	17.9	144.5

*Concentration was below the detection limit. To incorporate this information into the statistical analysis the two extreme values (0 and 5) were assigned to this treatment.

Soluble sugars

At the end of the dark period the soluble sugar concentration first decreased with P supply at mild P limitation, and increased again at more severe P limitation. However, these differences were not significant (Figure 5). The decrease in the concentration of soluble sugars at the end of the light period with decreasing P supply was significant (P<0.01, Figure 5). The difference in leaf soluble sugar concentration between the end of the dark and the end of the light period represents the nightly export or consumption of sugars. This difference decreased significantly (P<0.05) with decreasing P supply rate.

Dry-mass partitioning to the roots

Dry-mass partitioning to the roots decreased linearly with increasing leaf nitrate concentration (r^2=0.34, P<0.001) (Figure 6a). The N and P experiments were not significantly different, and therefore only the general relation is shown. The relation between dry-mass partitioning to the roots and leaf reduced-N concentration was much stronger (r^2=0.71, Figure 6b). In this case the slope of the relation was significantly different

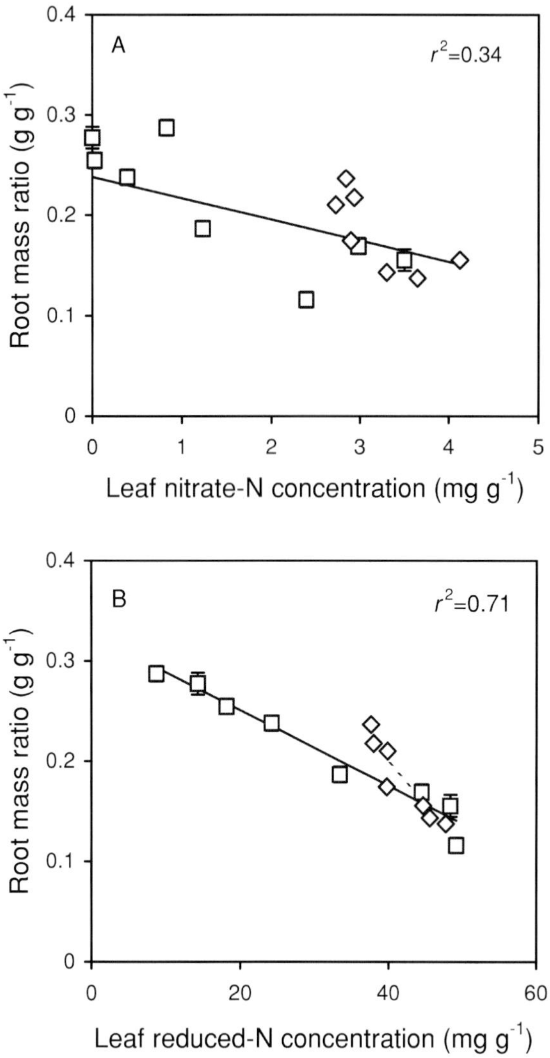

Figure 6. Dry-mass partitioning to the roots (root mass ratio) plotted against leaf nitrate-N concentration (A), and leaf reduced-N concentration (B). Squares=N experiment; diamonds=P experiment. Shown numbers for nitrate concentration are the average of the nitrate concentration at the end of the light and at the end of the dark period. Vertical bars indicate standard errors of means; only shown when larger than symbol (n = 6). Lines are fitted linear regressions, for all lines: $P<0.001$.

for the N and P experiment ($P<0.001$) due to the three most limiting P supply treatments.

Discussion

RGR response to phosphorus and nitrogen limitation

The differences in response of growth to N and P limitation (Figure 2; Burns et al., 1997; De Groot et al., 2001a) were not due to differences in inorganic phosphate and nitrate accumulation. P_i accumulated relatively more than nitrate in well-nourished plants (Figure 1a, b). When the increase in plant P concentration without an increase in RGR is due to the accumulation of P_i, RGR will show the same relation with N and P when only organic N and P are considered. However, correcting the leaf N and P concentration for accumulated nitrate and P_i did not change the shape of the curves (Figure 2a, b). This suggests that the depletion of accumulated nitrate or P_i cannot explain the differences in response of RGR to N or P limitation. Therefore the explanation for the differences in response is more likely due to the different roles of N and P in the machinery of the plant's energy metabolism (e.g., photosynthesis, respiration). In well nourished plants 94.5% of total N was reduced N (Figure 1b). A large portion of reduced N in a plant is associated with the machinery of the plant's energy metabolism, whereas a smaller portion is incorporated in structural cell components (e.g., structural proteins and nucleic acids; Mengel and Kirkby, 1987). In contrast, P remains in its oxidised form and a relatively large part is incorporated in structural cell components (e.g., phospholipids, nucleic acids). A smaller proportion of P plays a role in the machinery of the plant's energy metabolism (e.g., phosphorylated intermediates of glycolysis and the Calvin cycle; Mengel and Kirkby, 1987; Marschner, 1995). At N limitation a faster growth rate can only be achieved by increasing the plant's N concentration (more components required for instance for the photosynthetic machinery), whereas the phosphorylated intermediates can turn over more rapidly, with no need for a major increase in their pool size. Thus, an increase in N availability at low N will be used for building extra machinery. Consequently, the N concentration will increase, and with that extra machinery the plant can achieve a higher RGR. However, at low P, an increase in P availability will be used to make structural compounds (e.g., nucleic acids, phospholipids). Therefore, RGR increases because more structural material is produced, whilst the machinery is largely already there and does not have to be built first as was the case with N.

Leaf cytokinin concentration

The decrease in leaf cytokinin concentration (Table 2) was more distinct for N than for P supply as reported before (Horgan and Wareing, 1980; Kuiper et al.,

1988; Thorsteinsson and Eliasson, 1990). This was due to the weaker response of IPAR equivalents to P limitation than to N limitation, whereas the response of ZR type cytokinins tended to be similar for both experiments (Table 2). Though not significant, the decreased cytokinin concentration in the high N and P supply treatments compared with the free access treatments, without an effect on RGR (Figure 2b) might indicate an early influence of nutrient availability on cytokinin levels which in turn might affect growth (Thorsteinsson and Eliasson, 1990). The difference in cytokinin activity in the leaves grown at high N and P (320 mg g^{-1} day^{-1}) and at free access, might be caused by the differences in nitrate and phosphate concentration in the rooting medium. The roots of the plants in the free access treatment experienced a constant nitrate concentration of 12 m*M* and a constant phosphate concentration of 1 m*M*, while the high N and P plants were supplied daily with just enough nitrate or phosphate to grow at maximum rate, resulting in a low to zero nitrate or phosphate concentration in the rooting medium at the end of the day. This low concentration in the rooting medium of the high N and P supply treatments might have led to a reduction in cytokinin production in roots (Kuiper et al., 1988; Salama and Wareing, 1979; Thornsteinsson and Eliasson, 1990). Another possibility may be that, next to the cytokinin signal, other 'signals' may be needed to achieve a growth response. Our results suggest that the decrease in N concentration with P limitation may be mediated by cytokinins. However, further experiments are necessary to confirm this hypothesis.

Interaction between N and P uptake

The decrease in total plant N concentration with P limitation was due to a decrease in the N concentration of all plant organs and was thus not caused by a change in dry-mass partitioning (Figure 3a). Despite the limited role of leaf nitrate in reducing leaf N concentration at low P (Figure 4a), leaf nitrate concentration at the end of the light and at the end of the dark period showed a typical pattern (Figure 4b). The increase in leaf nitrate concentration over night at high P supply treatments may be due to a lack of nitrate reduction in the leaves at night, in combination with a continued nitrate uptake during the night as shown before for tomato plants (Cárdenas-Navarro et al., 1998; Le Bot and Kirkby, 1992). At low P supply rates (220 mg g^{-1} day^{-1} and lower) the leaf nitrate concentration did not increase overnight (Figure 4b). As mentioned above, in the dark the rate of nitrate reduction is very slow (Cárdenas-Navarro et al., 1998; Le Bot and Kirkby, 1992); therefore, we may assume that the lack of increase in leaf nitrate concentration over night reflects decreased uptake or translocation of absorbed nitrate from the root to the shoot.

It has been argued that effects of P supply on growth are mediated by N (Kuiper et al., 1988). However, we have shown that only 10 out of the 74% decrease in RGR with P limitation could be explained by a decrease in leaf N concentration, which suggests that effects of P supply on growth are not mediated by N. We suggest that the decrease in leaf N concentration with increasing P limitation (Figure 3a) is mediated by the decrease in cytokinin concentration upon P limitation (Table 2). This decrease in cytokinin concentration may have caused a decrease in nitrate reductase activity (Bueno et al., 1994; Gaudinová, 1990; Lu et al., 1992), a decrease in net protein synthesis (Klämbt, 1977), and hence a decrease in N incorporation in the shoot (Simpson et al., 1982). This is consistent with our data, which show that the decrease in leaf N concentration with increasing P limitation was mainly due to a decrease in leaf reduced-N concentration (Figure 4a), which is a measure for leaf protein content (Mengel and Kirkby, 1987). Furthermore, it has been shown that nitrate reductase activity and net protein synthesis may decrease with increasing P limitation (Andrews et al., 1999; Gniazdowska and Rychter, 2000; Jacob and Lawlor, 1991; Rufty et al., 1990, 1993). Low levels of cytokinins also inhibit cell expansion in leaves, due to their effects on cell wall extensibility (Rayle et al., 1982) thereby decreasing shoot growth. Decreased incorporation of N into proteins in the leaves as a result of low cytokinin production may explain increased N translocation (as aminoacids) from the shoot to the roots (Simpson et al., 1982) and decreased N transport from roots to shoot (Gniazdowska and Rychter, 2000; Jeschke et al., 1997; Rufty et al., 1990). This may result in N accumulation in the roots, which may cause feedback inhibition of N uptake, and thus a decreased plant N concentration (Gniazdowska and Rychter, 2000; Jeschke et al., 1997; Rufty et al., 1990).

An alternative explanation for the decreased N concentration with increasing P limitation may be decreased energy availability (ATP; Rufty et al., 1990). For both the uptake and the reduction of nitrate carbohydrates are needed. Starch concentration, measured at the end of the light and at the end of the dark period, decreased when P was severely limiting plant growth, which suggests that photosynthesis was restricted (De Groot et al., 2001a). Furthermore, net assimilation rate (NAR), which is largely the balance of photosynthesis and respiration, became more important in explaining effects of P limitation on RGR with increasing severity of P limitation (De Groot et al., 2001a). Also the amount of soluble sugars in the leaves measured at the end of the light period decreased with increasing P limitation, as did the difference in the concentration of sugars at the end of the day and night (Figure 5); this difference may largely be accounted for by export of sugars out of the leaves at night, and to a lesser extent due to respiration (Bouma et al., 1995; Ho and Thornley, 1978). These observed changes in carbon metabolism may all lead to a decreased availability of metabolic energy, and thus to a decreased N uptake and/or assimilation under P-limited conditions. However, the fraction of dry mass partitioned to the roots increased with increasing P limitation (Figure 6; De Groot et al., 2001a). Therefore it is less likely that the low nightly decrease of leaf sugar concentration at low P was due to decreased sugar translocation to the roots. Hence, decreased energy availability is a less likely cause of the reduced N uptake with P limitation.

In contrast to the clear decrease in plant N concentration under P-limiting conditions, there was not such a clear effect on plant P concentration in response to N limitation (Figure 3b). A low supply of N or P may decrease cytokinin levels in the shoots (Horgan and Wareing 1980; Kuiper et al., 1988; Wagner and Beck, 1993) and thus decrease protein synthesis (Klämbt, 1977), which may explain the decrease in plant N concentration under P-limiting conditions, as argued above. However, only a very small part of total P is incorporated into proteins as against 80–85% of total N (Mengel and Kirkby, 1987). Thus a decrease in protein synthesis with N limitation is not expected to decrease P concentration. This may explain the insensitivity of P concentration to N limitation.

Dry-mass partitioning to the roots

Dry-mass partitioning has been reported to correlate strongly with the nitrate level in the leaves (Scheible et al., 1997). We also found partitioning to be correlated with leaf nitrate concentration (Figure 6a); however, this relation was less strong than the correlation with reduced N (Figure 6b). Leaf reduced-N concentration is a measure for leaf protein content, because by far the largest fraction of leaf reduced-N is incorporated in proteins, only a small part being incorporated in nucleic acids or present as soluble amino N as already mentioned above (Mengel and Kirkby, 1987). Dry-mass partitioning correlates well with leaf soluble protein concentration (Andrews et al., 1999). These authors, therefore, concluded that effects of limitation of macronutrients on dry-mass partitioning are primarily mediated through decreases of protein synthesis. As argued previously, a decrease in protein synthesis may result from a decreased cytokinin level as a result of N or P limitation. Furthermore cytokinins have been involved in the regulation of dry-mass partitioning (Kuiper et al., 1988; Van der Werf and Nagel, 1996). However, the deviation of the points at severe P limitation, from the linear relation between reduced N concentration and dry-mass partitioning to the roots, shows that changes in cytokinin concentrations, and thus protein synthesis and reduced N concentration, cannot be the sole explanation for changes in dry-mass partitioning at severe P supply.

Conclusions

The different response of RGR to N compared to P limitation is most likely due to the different roles of N and P in the machinery used for energy metabolism (photosynthesis, respiration) of the plant.

A shift in dry-mass partitioning cannot explain the observed decrease in plant N concentration under P limiting conditions. We have shown that cytokinin concentrations (mainly ZR) tended to decrease upon N as well as upon P limitation. Our results suggest that the decrease in N concentration with P limitation may be mediated by cytokinins. However, further experiments are necessary to confirm this hypothesis. The relative increase in root growth as reflected in the increase in root mass ratio with increasing P limitation suggests that reduced energy availability might not play a role of importance in explaining the effects of P limitation on N uptake.

Acknowledgements

We are grateful to Eva Cabanes for her help with the experiments, to Eva Wirth for the anion analysis, to Marcel Visser for his help with the soluble sugars analysis, to Evert Davelaar for skilfully determining the cytokinins and to Anja Dieleman for reading the manuscript critically.

References

Andrews M, Sprent J I, Raven J A and Eady P E 1999 Relationships between shoot to root ratio, growth and leaf soluble protein concentration of *Pisum sativum*, *Phaseolus vulgaris* and *Triticum aestivum* under different nutrient deficiencies. Plant Cell Environ. 22, 949–958.

Bouma T J, De Visser R, Van Leeuwen P H, De Kock M J and Lambers H 1995 The respiratory energy requirements involved in nocturnal carbohydrate export from starch-storing mature source leaves and their contribution to leaf dark respiration. J. Exp. Bot. 46, 1185–1194.

Bueno M S, Alonso A and Villalobos N 1994 Nitrate reduction in cotyledons of *Cicer arietinum* L.: regulatory role of cytokinins. Plant Sci. 95, 117–124.

Burns I G, Walker R L and Moorby J 1997 How do nutrients drive growth? *In* Plant Nutrition – For Sustainable Food Production and Environment. Eds. T Ando et al. pp. 891–895. Kluwer Academic Publishers, Dordrecht.

Buwalda F and Warmenhoven M 1999 Growth-limiting phosphate nutrition suppresses nitrate accumulation in greenhouse lettuce. J. Exp. Bot. 50, 813–821.

Cárdenas-Navarro R, Adamowicz S and Robin P 1998 Diurnal nitrate uptake in young tomato (*Lycopersicon esculentum* Mill.) plants: test of a feedback-based model. J. Exp. Bot. 49, 721–730.

De Groot C C, Marcelis L F M, Van den Boogaard R and Lambers H 2001a Growth and dry-mass partitioning in tomato as affected by phosphorus nutrition and light. Plant Cell Environ. 24, 1309–1317.

De Groot C C, Marcelis L F M, Van den Boogaard R and Lambers H 2001b Regulation of growth by phosphorus supply in whole tomato plants. *In* Plant Nutrition – Food Security and Sustainability of Agro-ecosystems. Proceedings of XIV International Plant Nutrition Colloquium, Hannover, Germany. Eds. W J Horst et al. pp. 114–115. Kluwer Academic Publishers, Dordrecht.

De Pinheiro Henriques A R and Marcelis L F M 2000 Regulation of growth at steady-state nitrogen nutrition in lettuce (*Lactuca sativa* L.): interactive effects of nitrogen and irradiance. Ann. Bot. 86, 1073–1080.

Dhillon S S 1978 Influence of varied phosphorus supply on growth and xylem sap cytokinin level of sycamore (*Platanus occidentalis* L.) seedlings. Plant Phys. 61, 521–524.

Gaudinová A 1990 The effect of cytokinins on nitrate reductase activity. Biol. Plant. 32, 89–96.

Gniazdowska A and Rychter A M 1999 Low phosphate nutrition alters bean plants' ability to assimilate and translocate nitrate. J. Plant Nutr. 22, 551–563.

Gniazdowska A and Rychter A M 2000 Nitrate uptake by bean (*Phaseolus vulgaris* L.) roots under phosphate deficiency. Plant Soil 226, 79–85.

Ho L C and Thornley J H M 1978 Energy requirements for assimilate translocation from mature tomato leaves. Ann. Bot. 42, 481–483.

Horgan J M and Wareing P F 1980 Cytokinins and the growth response of seedlings of *Betula pendula* Roth. and *Acer pseudoplatanus* L. to nitrogen and phosphorus deficiency. J. Exp. Bot. 31, 525–532.

Humphries E C 1956 Mineral components and ash analysis. *In* Modern Methods of Plant Analysis, Volume 1. Eds. K Paech and M V Tracey. pp. 468–502. Springer, Berlin.

Jacob J and Lawlor D W 1991 Dependence of photosynthesis of sunflower and maize leaves on phosphate supply, ribulose-1,5-biphosphate carboxylase/oxygenase activity, and ribulose-1,5-bisphosphate pool size. Plant Physiol. 98, 801–807.

Jeschke W D, Peuke A D, Kirkby E A, Pate J S and Hartung W 1996 Effects of P deficiency on the uptake, flows and utilization of C, N and H_2O within intact plants of *Ricinus communis* L. J. Exp. Bot. 47, 1737–1754.

Jeschke W D, Kirkby E A, Peuke A D, Pate J S and Hartung W 1997 Effects of P deficiency on assimilation and transport of nitrate and phosphate in intact plants of castor bean (*Ricinus communis* L.). J. Exp. Bot. 48, 75–91.

Klämbt D 1977 Cytokinin and cell metabolism. *In* Plant Growth Regulation, proceedings of the 9th International Conference on Plant Growth Substances, Lausanne. Ed. P E Pilet. pp. 154–160. Springer, Berlin.

Kuiper D, Schuit J and Kuiper P J C 1988 Effects of internal and external cytokinin concentrations on root growth and shoot to root ratio of *Plantago major* ssp *pleiosperma* at different nutrient conditions. Plant Soil 111, 231–236.

Le Bot J and Kirkby E A 1992 Diurnal uptake of nitrate and potassium during the vegetative growth of tomato plants. J. Plant Nutr. 15, 247–264.

Lu J L, Ertl J R and Chen C 1992 Transcriptional regulation of nitrate reductase mRNA levels by cytokinin-abscisic acid interaction in etiolated barley leaves. Plant Physiol. 98, 1255–1260.

Marschner H 1995 Mineral Nutrition of Higher Plants. Academic Press, London.

Martin A C, del Pozo J C, Iglesias J, Rubio V, Solano R, de la Peña A, Leyva A and Paz-Ares J 2000 Influence of cytokinins on the expression of phosphate starvation responsive genes in *Arabidopsis*. Plant J. 24, 559–567.

Menary R C and Van Staden J 1976 Effect of phosphorus nutrition and cytokinins on flowering in the tomato, *Lycopersicon esculentum* Mill. Aust. J. Plant Physiol. 3, 201–205.

Mengel K and Kirkby E A 1987 Principles of plant nutrition. International Potash Institute Worblaufen, Bern.

Rayle L, Ross C W and Robinson N 1982 Estimation of osmotic parameters accompanying zeatin-induced growth of detached cucumber cotyledons. Plant Physiol. 70, 1634–1636.

Rufty T W Jr, MacKown C T and Israel D W 1990 Phosphorus stress effects on assimilation of nitrate. Plant Physiol. 94, 328–333.

Rufty T W Jr, Israel D W, Volk R J, Qiu J and Sa T 1993 Phosphate regulation of nitrate assimilation in soybean. J. Exp. Bot. 44, 879–891.

Salama A M S El-D A and Wareing P F 1979 Effects of mineral nutrition on endogenous cytokinins in plants of sunflower (*Helianthus annuus* L.). J. Exp. Bot. 30, 971–981.

Scheible W-R, Lauerer M, Schulze E-D, Caboche M and Stitt M 1997 Accumulation of nitrate in the shoot acts as a signal to regulate shoot-root allocation in tobacco. Plant J. 11, 671–691.

Schjørring J K 1986 Nitrate and ammonium absorption by plants growing at sufficient or insufficient level of phosphorus in nutrient solutions. Plant Soil 91, 313–318.

Simpson R J, Lambers H and Dalling M J 1982 Kinetin application to roots and its effect on uptake, translocation and distribution of nitrogen in wheat (*Triticum aestivum*) grown with a split root system. Physiol. Plant. 56, 430–435.

Steiner A A 1984 The universal nutrient solution. *In* Proceedings of the Sixth International Congress on Soilless Culture. pp. 633–649. International Society for Soilless Culture, Wageningen.

Thorsteinsson B and Eliasson L 1990 Growth retardation induced by nutritional deficiency or abscisic acid in *Lemna gibba*: The relationship between growth rate and endogenous cytokinin content. Plant Growth Regul. 9, 171–181.

Van der Werf A and Nagel O W 1996 Carbon allocation to shoots and roots in relation to nitrogen supply is mediated by cytokinins and sucrose: opinion. Plant Soil 185, 21–32.

Van Rhijn J A, Heskamp H H, Davelaar E, Jordi W, Leloux M S and Brinkman U A T 2001 Quantitative determination of glycosylated and aglycon isoprenoid cytokinins at sub-picomolar levels by microcolumn liquid chromatography combined with electrospray tandem mass spectrometry. J. Chromatogr. 929, 31–42.

Vonk C R, Davelaar E and Ribôt S A 1986 The role of cytokinins in relation to flower-bud blasting in Iris cv. Ideal: Cytokinin determination by an improved enzyme-linked immunosorbent assay. Plant Growth Regul. 4, 65–74.

Wagner B M and Beck E 1993 cytokinins in the perennial herb *Urtica dioica* L. as influenced by its nitrogen status. Planta 190, 511–518.

Implications of root architecture, root-soil interactions and mycorrhiza on plant P nutrition

Plant and Soil **248:** 271–283, 2003.
© 2003 Kluwer Academic Publishers. Printed in the Netherlands.

Phosphorus acquisition from soil by white lupin (*Lupinus albus* L.) and soybean (*Glycine max* L.), species with contrasting root development

Michelle Watt[1] & John R. Evans
Environmental Biology Group, Research School of Biological Sciences, Australian National University, GPO Box 475, Canberra, ACT 2601, Australia. [1]*Corresponding author**

Received 23 November 2001. Accepted in revised form 24 May 2002.

Key words: cluster roots, proteoid roots, root architecture, root development, soil phosphorus, soybean, white lupin

Abstract

White lupin and soybean have contrasting root morphologies: white lupin develops proteoid or cluster roots, roots with discreet clusters of short, determinate branch roots (rootlets) while soybean develops a more fibrous root system with evenly distributed, longer branch roots. Growth and P acquisition by white lupin and soybean were compared in a soil high in bound, total P, with or without additional inorganic P applied in solution. Additional P increased biomass by 25% and doubled total P in soybean. In contrast, white lupin did not respond to additional P in biomass or total P. However added P decreased cluster development on proteoid roots indicating that white lupin sensed the added P. The reduction in cluster weight per plant was exactly countered by an increase in dry weight of other roots. Soybean root development responded to P application, proliferating branch roots with active meristems in the upper portion of the soil profile where P was applied, and reducing root weight to plant weight by 13%. White lupin did not proliferate roots in response to P application. When P was not added to soil, soybean and lupin acquired similar P per unit root dry weight. However, white lupin accumulated 4.8 times more P per unit root length, suggesting that P acquisition in these plants involved other mechanisms such as the exudation of P solubilizing compounds. Soybean accessed P by developing more root length thus colonising more soil volume than white lupin and, therefore, was better able to take advantage of the added P. Pericycle and root tip meristem activities were critical to the differences in root development between white lupin and soybean, and therefore their responses to plant and soil P.

Introduction

Phosphorus (P) is partitioned between sorbed and organic forms of soil P (hereafter called bound P), and P in the soil solution. Most P in the bound fraction is non-labile and is tightly bound to mineral or organic compounds, rendering it unavailable to many plants. Studies in soils labelled with ^{32}P, which distinguish between different forms of soil P accessed by the plant, have shown that white lupin can access up to six times more P from this bound fraction compared to a range of crop plants including soybean (Braum and Helmke, 1995; Hocking et al., 1998). Superior acquisition of bound P by white lupin is attributed to its ability to develop proteoid roots, roots with clusters of closely spaced, determinate branch roots (rootlets) that exude P-solubilising compounds. These compounds include citrate that release P from P sorbed to mineral compounds (Gardner et al., 1983a) and acid phosphatase that release P from organic compounds (Gilbert et al., 1999). Soybean in contrast does not produce proteoid roots, but develops a more fibrous root system and exudes only trace levels of organic anions in response to P limitation (Ohwaki and Hirata, 1992). Interestingly, however, Braum and Helmke (1995) noted in ^{32}P studies that although white lupin was better at acquiring bound P, it was less effective than soybean at acquiring more readily available P. These studies suggest that lupin may be limited in its ability to access soluble P

* Present address: CSIRO Plant Industry, GPO 1600, Canberra, ACT 2601, Australia.
E-mail: michelle.watt@csiro.au

compared to soybean. Root development and P uptake per unit root were not quantified. In our study reported here white lupin and soybean were grown in a soil high in total P with and without additional inorganic P, applied in solution. The growth and P acquisition per unit root weight and length, by each species was compared in the two treatments. The ability to access P was related to root morphology and development.

Materials and methods

Soil

A medium grey clay soil, classified as Typic Pellusterts, was obtained from the top 20 cm of a plot at the Australian Cotton Research Institute, CSIRO Plant Industry, Narrabri, New South Wales. This plot had not received P fertiliser for at least 20 years. The air-dried soil had a pH ($CaCl_2$) of 7.0, 76 mg kg^{-1} extractable P, as measured by the Bray I technique, and 516 mg kg^{-1} total P. Approximately 80% of the total P was bound to mineral elements (calcium 50%, aluminium 20% and iron 10%), and 20% was bound to organic surfaces (C. Dorahy, pers. comm.). Three days prior to planting, the soil was mixed in a cement mixer 50:50 with river sand, and nutrient solution (described below) with or without P was added to give a final water content of 20% (v/v).

Plant growth

White lupin, *Lupinus albus* L. cv. Kiev mutant, and soybean, *Glycine max* L. cv. Lincoln, were grown in the soil/sand mix, with or without supplemented P in nutrient solution. Plants were grown in glasshouses matched for light and temperature, in Canberra, Australia, in December/January of 1998/1999. Day length was approximately 14 h, and average noon irradiance during that period was approximately 1300 μmol photons m^{-2} s^{-1}. Mean day and night temperatures in the glasshouses were 26 and 20 °C, respectively. Seeds were planted directly into 50 cm height$\times$9 cm diameter pots (five seeds per pot) filled with prepared soil, and shoots emerged 4 days after planting. Plants were thinned to one per pot 6 days after emergence, removing the two largest and smallest plants. At this stage, the soil surface was covered with 1-cm thick felt pads to minimise evaporative water loss from the pots. Pots were watered with nutrient solution two to three times per week to maintain 20% (v/v) soil water content. This was done by weighing a sample number of three pots per treatment and using mean weight loss. The nutrient solution was with or without 1.3 m*M* NaH_2PO_4, plus 4 m*M* KNO_3, 4 m*M* $Ca(NO_3)_2$, 1.5 m*M* $MgSO_4$, 0.05 m*M* EDTA FeNa, 0.001 m*M* $MnSO_4$, 0.001 m*M* $ZnSO_4$, 0.001 m*M* $CuSO_4$, 0.05 m*M* H_3BO_3, 0.0005 m*M* Na_2MoO_4, 0.1 m*M* NaCl, 0.0002 m*M* $Co(NO_3)_2$.

Harvest measurements

Whole plant

At 24 days after planting, a lupin plant from each treatment was harvested and examined for the presence of proteoid roots. All had developed proteoid roots, and the remaining plants were harvested at 25 or 37 days after planting (seven plants per treatment). The soybean plants were harvested 28 or 35 days after planting (four plants per treatment). Shoots were excised at soil level, separated from the leaves, and the leaf area determined using a leaf area meter (Li-Cor 3050A/3000A, Lincoln, NE, USA). The stems and leaves were oven dried at 70 °C for 3 days. Soil was gently washed from the roots. For lupin, for both harvests, roots were maintained on ice for further measurements and then dried at 70 °C for 3 days. For soybean at harvest 1, roots were immediately oven dried. At harvest 2, a subsample of first-order branch roots (approximately 50 mg dry weight) with any higher order branch roots were excised from the basal part of the root system of each plant (top 4 cm of the soil profile) and immediately fixed in 1.5% glutaraldehyde in 70% ethanol at 4 °C and maintained at 4 °C for further measurements (described below). Root pieces from individual plants were pooled per treatment.

Root measurements

For each lupin root system, the clusters (rootlets and axis) were cut from the proteoid roots, and counted (Figure 1A). In the second harvest, the tap root was divided into three equal lengths and the clusters developed on first-order branch roots of these parts were counted, to quantify clusters within the root system and estimate where the clusters predominated in the soil profile. After cutting out clusters at both harvests, a qualitative judgement of the nine to 12 largest clusters was made, and these were measured for rootlet length with a ruler. The occurrence of branch roots within the cluster that had extended twice the length of the rootlets was also scored (Figure 1A). These

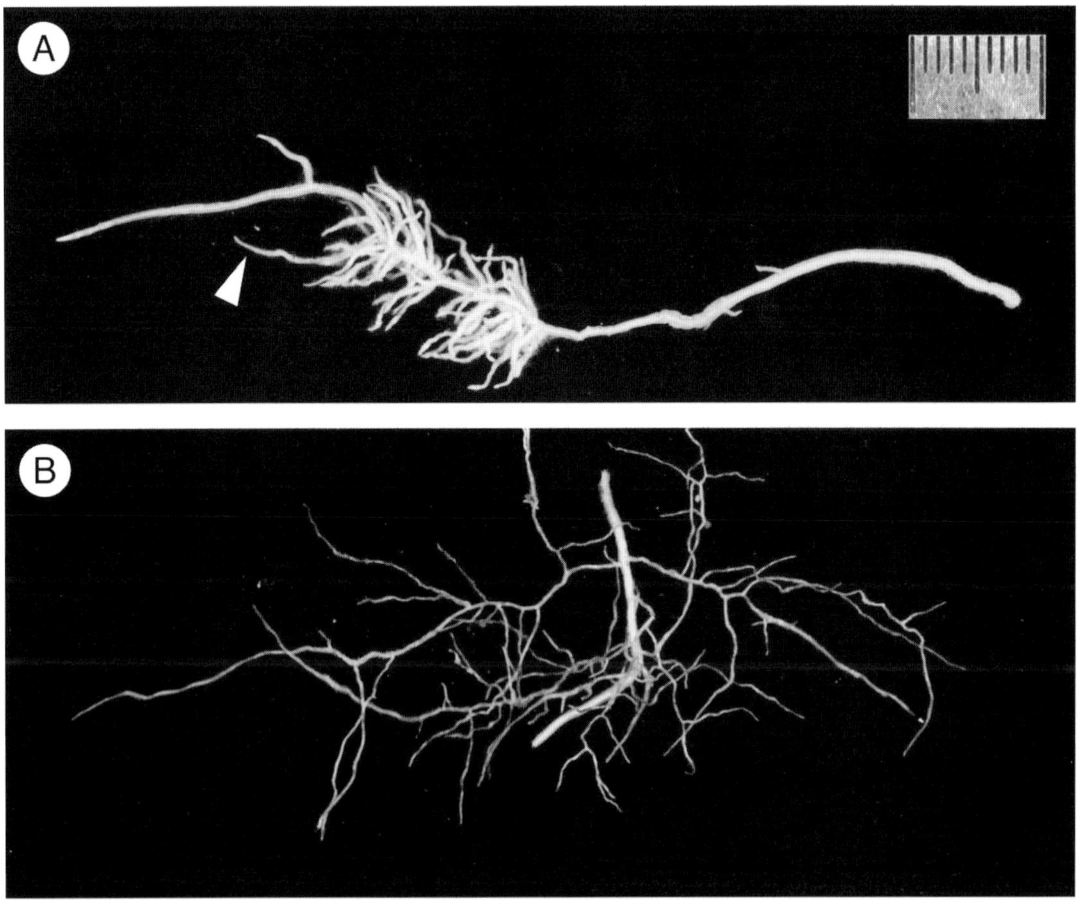

Figure 1. First-order branch roots of white lupin (oriented horizontally) and soybean (oriented vertically) grown in soil. (A) White lupin proteoid root with cluster of determinate branch roots (rootlets) and few other branch roots. Arrow indicates a second-order branch root within the cluster that has extended longer than the rootlets. (B) Soybean first-order branch root with numerous higher-order branch roots. Photos are at the same scale, smallest division=1 mm.

longer branch roots were considered to have a delayed onset of meristem determinacy. Following the measurements, the clusters and the rest of the root system were oven dried separately and weighed.

To measure specific root length (SRL) of lupin roots, proteoid roots were spread out in a shallow dish of water sitting over a ruler, and lengths of root axes and clusters measured under a dissecting microscope. Tissues were then blotted dry and fresh weight measured. A subset of root systems was used to estimate fresh weight to dry weight ratios. Clusters and axes from one root system from each treatment were separated, blotted dry, weighed for fresh weight, and then oven dried and weighed for dry weight. To estimate the SRL of the entire root system, root length of white lupin was divided into 70% axes and 30% clusters based on the dry weights of clusters and axes determined above.

The fixed soybean roots were observed under a dissecting microscope in a shallow tray of water and first photographed for subsequent diameter measurements. Then higher order branch roots were counted per length of first order branch (Figure 1B). These higher-order branch roots were then classed into three categories based on the appearance of their apical meristems and the proximity of root hair differentiation to the root tip. Roots in the first category had tips with large meristems and root hairs starting back from the tip, indicative of rapid growth; those of the second had reduced meristems with a small amount of swelling behind the tip, suggesting reduced division rates and elongation, and those of the third did not have a meristem, had swollen tips and root hairs to the tip, suggesting differentiation of meristem cells, determinacy and cessation of growth.

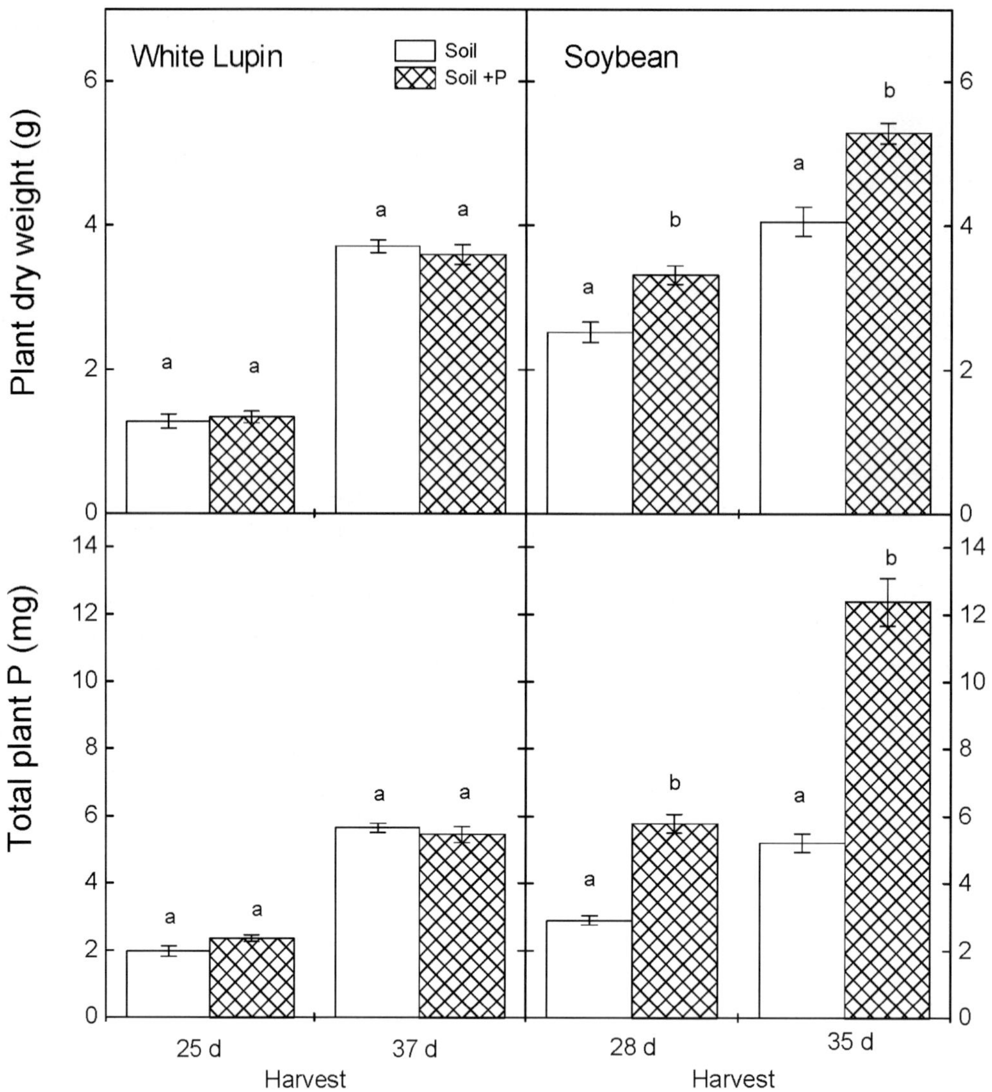

Figure 2. Total plant dry weight and total plant P (calculated as tissue dry weight×tissue P concentration) of white lupin and soybean at harvests 1 and 2. Plants were grown from seed in soil with or without 1.3 m*M* P (+P) added to soil in solution. Each value is the mean and S.E. of seven plants per treatment for white lupin and four plants per treatment for soybean. At each harvest, treatments with different letters are significantly different at $P = 0.005$.

To measure SRL of soybean, first order branch roots and their subtending branch roots were measured under a dissecting microscope, as described for white lupin. Tissues were then blotted to remove excess water and the fresh weight obtained. To estimate fresh weight to dry weight ratio of the soybean roots, and therefore estimate the SRL of the entire root system, root systems were blotted dry, weighed fresh and then oven dried for 48 h, and re-weighed.

Tissue P

After drying and weighing, the leaves, stems and roots (clusters plus rest of root system) were ground to a fine powder, and analyzed individually for total P. Approximately 0.03 g of powder was weighed into tubes, and 5 mL of digest mix (300 g K_2SO_4 in 1 L of 98% H_2SO_4) and 1 mL of hydrogen peroxide were added at room temperature. The tubes were then heated to 325 °C for 1.5 h, at which point 0.5 mL of hydrogen peroxide was added. The solutions were left for 1 h more, and then removed from the heat. When cool, they were diluted to 16 mL with distilled water. The

Table 1. Characteristics of clusters of white lupin

Treatment	Number of clusters[3] (plant^{-1})		Rootlet length (mm)		Number of longer roots[6] (cluster^{-1})		Number of longer[6] roots per mean cluster weight (mg^{-1})	
	h1[1]	h2[2]	h1[4]	h2[5]	h1[4]	h2[5]	h1	h2
Soil	33±3	115±3	9±0.6	10±0.3	6±1	13±1.7	3.2	4.0
Soil+P	25±4	92±4	10±0.4	11±0.4	10±0.5	18±2.8	4.1	4.8

[1]Harvest 1, at 25 days after planting.
[2]Harvest 2, at 37 days after planting.
[3]Mean and S.E., n=7.
[4]Mean and S.E., n=48 clusters from seven plants.
[5] Mean and S.E., n=70 from seven plants.
[6]Second-order branch roots within a cluster that extended at least twice the length of the rootlets (see Figure 1).

leaf samples were clear while the root samples had a small amount of white precipitate, and were spun for 5 min at 13 000 rpm prior to analysis. P in the solutions was detected using malachite green as a colour indicator (modified from Irving and McLaughlin, 1990). Aliquots (5, 10 and 20 μL) of digest solution were dispensed into a microtitre plate, diluted to 100 μL with 0.75 M H_2SO_4, and then 100 μL each of dH_2O and Malachite green indicator added to each well. The orange-coloured product was left to develop at room temperature for 1 h and the absorbances were determined at 660 nm. The P in the samples was determined from standard curves prepared in each microtitre plate.

Calculations

DW is dry weight; FW is fresh weight; h is harvest; RL is root length.

Total plant P

$$=(\text{shoot DW} \times \text{shoot [P]}) + (\text{root DW} \times \text{root [P]})\ (\text{mg})$$

P taken up as a percentage of P applied to soil

$$= \text{total plant P}/(\Sigma\text{P applied}) \times 100$$

This assumed total plant P was derived from total P applied in solution during plant growth.

P uptake by roots

(i) Per dry weight of total root system (mg g^{-1} day^{-1})

=daily change in total plant P per unit root dry weight, between harvest 1 and 2.

$$=(\text{plant P}_{h2} - \text{plant P}_{h1})/\exp[(\ln \text{root DW}_{h2} + \ln \text{root DW}_{h1})/2]/\text{days between harvest}$$

It was assumed that different parts of the root system contribute to P uptake at different times, and therefore the geometric mean between root dry weight at each harvest was used.

(ii) Per root length (mg m^{-1} day^{-1})

$$=(\text{P uptake root DW}^{-1}\ \text{day}^{-1}\ (\text{from (i)})) \times (\text{root DW FW}^{-1}) \times (\text{root FW RL}^{-1})$$

Results

Growth and P uptake in white lupin and soybean

Applying additional P did not alter the growth of lupin (Figure 2). There was also no effect of added P on total plant P or P concentrations in roots and shoots (Figures 2 and 3). In contrast to lupin, P addition increased soybean total plant dry weight by approximately 25% and additional P increased P concentration in both shoot and root tissue (Figures 2 and 3). This, combined with the stimulation in growth by the addition of P, doubled the P content per plant compared to the unamended soil treatment (Figure 2). The effects of P application on biomass were reflected in calculations of the proportion of P applied to soil that was taken up by plants. Assuming plants used all the more readily available applied P and did not take up the soil P, soybean used 2.5 times as much as lupin at each harvest, and by harvest 2 captured up to 30% of the applied P, while lupin captured 12%.

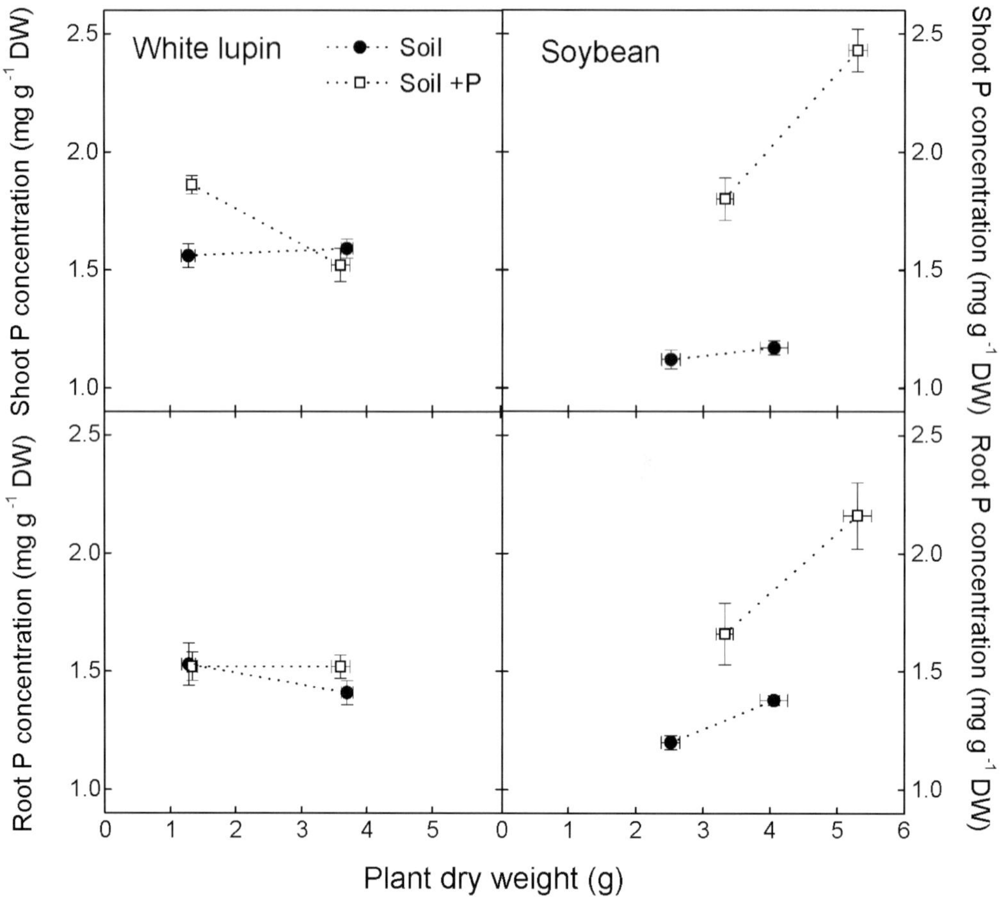

Figure 3. Tissue P concentrations of white lupin and soybean plotted against plant dry weight at harvests 1 and 2 (25 and 37 days for lupin; 28 and 35 days for soybean). Plants were grown in soil with or without 1.3 m*M* P (+P) added to soil in solution at watering. Each value is the mean and S.E. of seven plants per treatment for white lupin and four plants per treatment for soybean.

Effects of additional P on root development

White lupin

White lupin had a taproot system (Figure 4A), with first-order branch roots emerging from a dipolar stele (Figure 5A–C). Clusters of rootlets emerged from the first-order branch roots, with the very occasional emergence of a cluster from a second-order branch root. Across treatments, between harvests 1 and 2, the proportion of clusters relative to total root weight increased from 16 to 30% (Figure 6B), and 5.3 to 11.4% of plant dry weight. Allocation to roots increased with plant weight due to partitioning to clusters, which began after a lag (Figure 6A,B).

P supply affected cluster production independently of plant size by harvest 2 at 37 days (Figure 6B,C). Addition of P suppressed the percentage of root dry weight that was clusters by 5–6%, and the percentage of plant dry weight that was clusters by 2.2%. Cluster number per plant (Table 1) was affected by P supply earlier than cluster weight per plant (Figure 6). By harvest 1 at 25 days, plants supplied with P had 25% fewer clusters per plant but a similar total cluster weight per plant. By harvest 2, both cluster number and weight per plant were 20% lower in plants supplied with additional P.

The dry weight per cluster doubled between harvests 1 and 2 in all treatments, increasing from 2.1 to 4.2 mg. Mean cluster dry weight was greater with additional P at harvest 1, but was not affected by P at harvest 2. Rootlets were determinate with root hairs differentiated close to the tip (Figure 7A), except for rootlets of clusters recently emerged. Young, recently emerged clusters (with meristems on the rootlets) were observed only towards the tip of the proteoid root axis. The mean rootlet length at both harvests was 10 mm (Table 1). By harvest 2, mean rootlet length

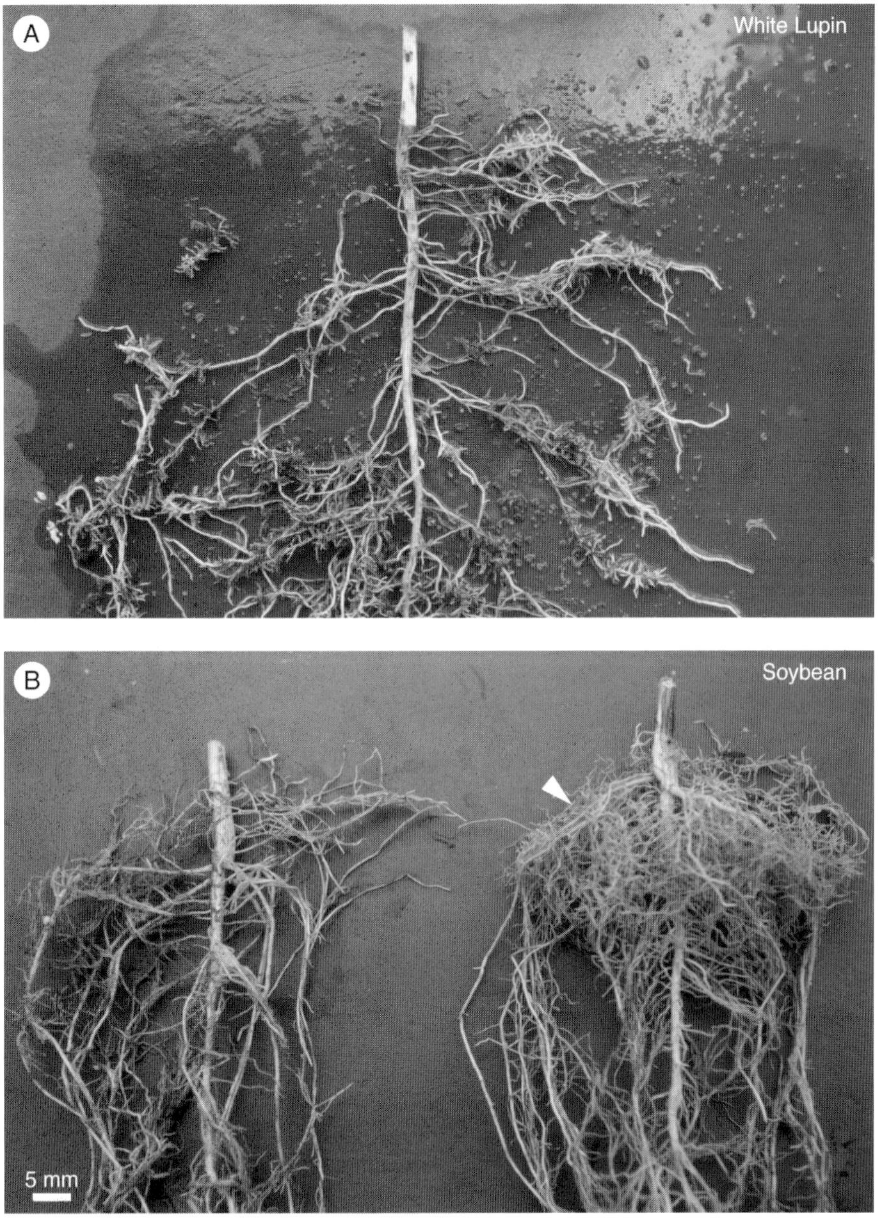

Figure 4. Basal portions of white lupin (A) and soybean (B) root systems harvested from soil at 35 and 37 days after planting, respectively. (A) Proteoid roots with clusters extend from the tap root. Root system from soil without applied P. (B) Soybean root systems. Right: 1.3 m*M* P applied in solution to the top of the pot (arrowhead indicates proliferation of roots); Left: no P applied.

was slightly longer with additional P, and the number of longer roots within a cluster was also greater with additional P, both per unit cluster and per cluster dry weight (Table 1). At harvest 2, clusters were distributed throughout the root system with the top third having 30% of the total number, the middle third 47%, and the bottom 22%. This distribution of clusters was not affected by application of P.

The SRL of the entire root system of white lupin was approximately half that of soybean and this was a reflection of thicker first-order and higher-order branch roots (Table 2). White lupin developed 3.4 times less root length per unit plant weight than soybean.

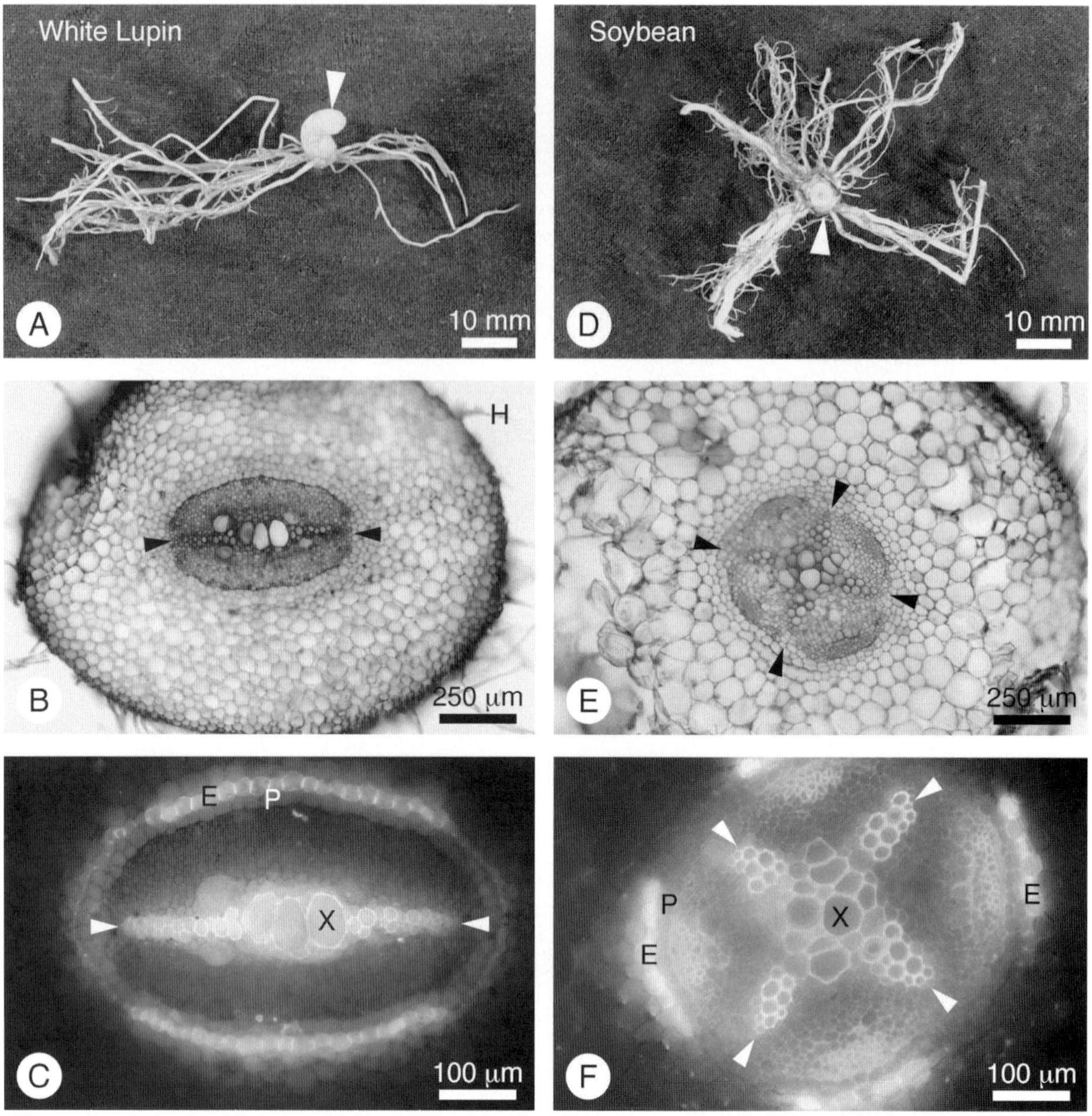

Figure 5. Roots of white lupin (A–C) and soybean (D–F) grown in soil and harvested 35 and 37 days after planting, respectively. (A,D) Tap roots (arrowheads) and extending branch roots viewed from above. Lupin branch roots extend in two directions (A), while soybean branch roots extend in four (D). (B,C,E,F) Hand-cut cross-sections through the tap roots, stained with toluidine blue, pH 4.4, and viewed with bright field optics (B,E) or UV-induced fluorescence (C,F). In lupin (B,C) the stele has two xylem poles (arrowheads), and in soybean (E,F), four (arrowheads). In white lupin (C) or soybean (F), branch roots would emerge from the pericycle (P) overlying the xylem poles, where the endodermis (E) has gaps in the fluorescing Casparian band formation and suberisation. H – root hair, X – xylem vessel.

Soybean

Similar to lupin, soybean had a taproot system (Figure 4B), however, first-order branch roots emerged from a tetrapolar stele (Figure 5D–F). The SRL was greater in soybean than white lupin (Table 2). Branching in soybean went up to the fourth-order compared to the third-order in lupin (Figure 1). The higher-order branch roots of soybean were 60% thinner than the cluster rootlets of lupin enabling soybean to explore a greater volume of soil per plant and per root dry weight (Table 2).

Soybean plants supplied with additional P developed a concentration of branch roots in the soil within the top of the profile where the nutrient solution was applied. This was evident upon observation of the whole root system at both harvests 1 and 2

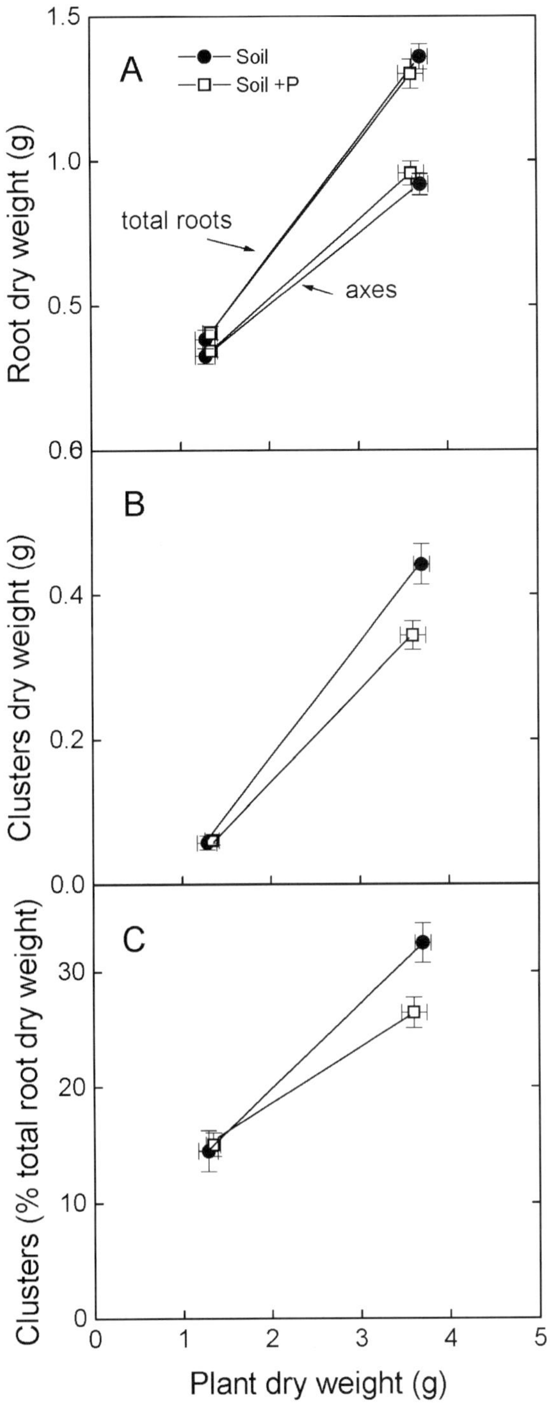

Figure 6. Root dry weights of white lupin plotted against plant dry weight. Plants were grown in soil with or without 1.3 m*M* P (+P) added to soil in solution. Each point is the mean and S.E. of roots of seven plants per treatment, at harvest 1, 25 days, and harvest 2, 37 days. A. Total roots: all roots of root system; Axes: all roots excluding clusters. (B,C) Clusters (rootlets plus axis) excised from proteoid roots.

(e.g., Figure 4B). The mean number of branch root tips growing from the first-order branch roots harvested from the top layer of the soil at harvest 2 was 13±3 when no P was added, and 44±28 when P was added to the soil. The addition of P to the soil also altered the meristems of these branch roots at the surface of the soil. Meristems were 5 to 7 times more likely to be large, indicative of rapid cell division, in the branch roots from plants grown with additional P (Figures 7 and 8). These plants had fewer tips with reduced or no meristems, compared to plants without additional P (Figure 8). The mean diameter of the surface branch roots was also altered by P application. P supply increased the mean branch diameter by 17% from 228±2 to 267±7 μm (n=25 to 31 roots from four plants per treatment). Application of additional P reduced biomass allocation towards soybean roots by 13% (mean±S.E. at harvests 1 and 2 of root weight/plant weight was 0.42±0.02 and 0.37±0.01 for the minus and plus P treatments, respectively).

P uptake per unit root in white lupin and soybean

In soil unamended with P, lupin and soybean obtained similar amounts of total P (Figure 2) and white lupin had 1.8 times higher uptake rate per unit root dry weight per day than soybean (Table 3). However, on a root length basis, white lupin obtained 5 times more P per day, compared to soybean (Table 3). When P was added to the soil, soybean obtained 1.7 times more P per unit dry weight and only 1.7 times less per unit length than lupin. The production of a more fibrous root system by soybean compared to white lupin, which had relatively few branch roots outside of the clusters (Figure 1) enabled soybean roots to colonise 3.3 fold more soil than those of white lupin, and thus capture the applied P.

Discussion

Soil P acquisition by white lupin

White lupin obtained as much P as soybean with one third of the root length (Figure 2, Table 2), at 5 times the rate per unit root length (Table 3). This indicates that white lupin acquired 66% of its P through mechanisms other than root extension. It is valid to assume that exudates from clusters contributed a large portion of the P acquired by white lupin, based on a number of direct measurements in the literature. Citrate and acid phosphatases (Adams and Pate,

Table 2. Root lengths for white lupin and soybean[1]

Species	Specific root length[2]					Root length per plant dry weight	Estimated soil volume colonised per plant	Estimated soil volume per root dry weight
	Entire root system	First order branch roots		≥second order branch roots				
	(m g^{-1} DW)[3]	(m g^{-1} DW)	(m g^{-1} FW)	(m g^{-1} DW)	(m g^{-1} FW)	(m g^{-1})[4]	(% soil volume)[5]	(cm^3 g^{-1})[6]
Lupin	79±12	43±9	2.3±0.5	169±9	6.5±0.3	22	8	62
Soybean	168±16	19±6	2.3±0.8	235±2	15.7±0.1	75	26	132

[1]Treatments with and without P combined.
[2]Means and S.E.; $n = 4$ root pieces pooled from different plants.
[3](Fresh weight/dry weight roots)×(length/fresh weight).
[4](Root weight ratio)×(fresh weight/dry weight roots)×(length/fresh weight).
[5]Both white lupin and soybean assigned a radius of 0.5 mm (root plus rhizosphere). Volume of root system=πr^2 (root length) and expressed as a percentage of soil volume per pot.
[6]Both white lupin and soybean assigned a radius of 0.5 mm (root plus rhizosphere). Volume of root system per root dry weight=πr^2 (specific root length of entire root system).

1992; Dinkelaker et al., 1989; Gerke et al., 1994; Li et al., 1997) and protons (Dinkelaker et al., 1989) have all been reported around white lupin clusters in soil. Labelling soil with ^{32}P enabled Braum and Helke (1995) to show that white lupin accessed up to 5 times more non-labile P from soil than soybean. Exudate compounds can release bound, non-labile P (Bar-Yosef, 1991; Hayes et al., 2000). The root measurements on both species can be combined with the Braum and Helmke data to show that 66% of white lupin's P was from the non-labile fraction, acquired by cluster development and exudation of P solubilising compounds. The cluster morphology likely enhances acquisition of non-labile P. Modelling of root P acquisition (Gardner et al., 1983b; Kirk et al., 1999) suggests that the closely spaced rows of rootlets create overlapping rhizospheres within a cluster, increasing the concentrations of exudates around a single rootlet. P mobilisation by organic anions is correlated positively with concentration of the anions (e.g., Fox et al., 1990); however, Dinkelaker et al. (1989) calculate that P acquisition by clusters is more efficient than can be explained by stochiometry of the soil chemistry.

Other mechanisms favouring acquisition of P by white lupin include root-hair development and activity and quantity of P transporters. Longer root hairs would extend the volume of soil exploited by a given length of root, and more dense hairs and transporters would enable more efficient extraction of P in solution. We did not measure hairs or transporter activity; however, longer, denser hairs would not necessarily enhance non-labile P uptake, which may not go into solution by diffusion alone.

Table 3. P-uptake rates of white lupin and soybean

Species	Treatment	P uptake rates	
		Per root dry weight[1] (mg P g^{-1} day^{-1})	Per root length[2] (μg P m^{-1} day^{-1})
White lupin	Soil	0.42	6.3
	Soil+P	0.35	5.3
Soybean	Soil	0.24	1.3
	Soil+P	0.6	3.1

[1]P-uptake per geometric mean dry weight of the entire root system between harvests 1 and 2, per day.
[2](P-uptake per dry weight per day, presented in adjacent column)×(dry weight roots/fresh weight roots)×(fresh weight roots/length roots).

Soil P acquisition by soybean

Soybean, conversely, acquired as much P per plant as white lupin by developing thinner roots that colonised 3 times the soil volume compared to white lupin roots (Figure 2, Tables 2 and 3). Continuous growth of roots into soil presumably enabled soybean to take up labile P, readily available in soil solution. Again the ^{32}P dilution studies of Braum and Helke (1995) support this finding since they showed that not only did white lupin access up to 5 times more non-labile P, but soybean used much more labile P than lupin. Soybean may have accessed a small portion of non-labile soil P via exudates including protons; however, this is not supported by the findings of Ohwaki and Hirata (1992) showing almost no organic anions in exudates of soybean subject to P starvation for 3 days in nutrient culture.

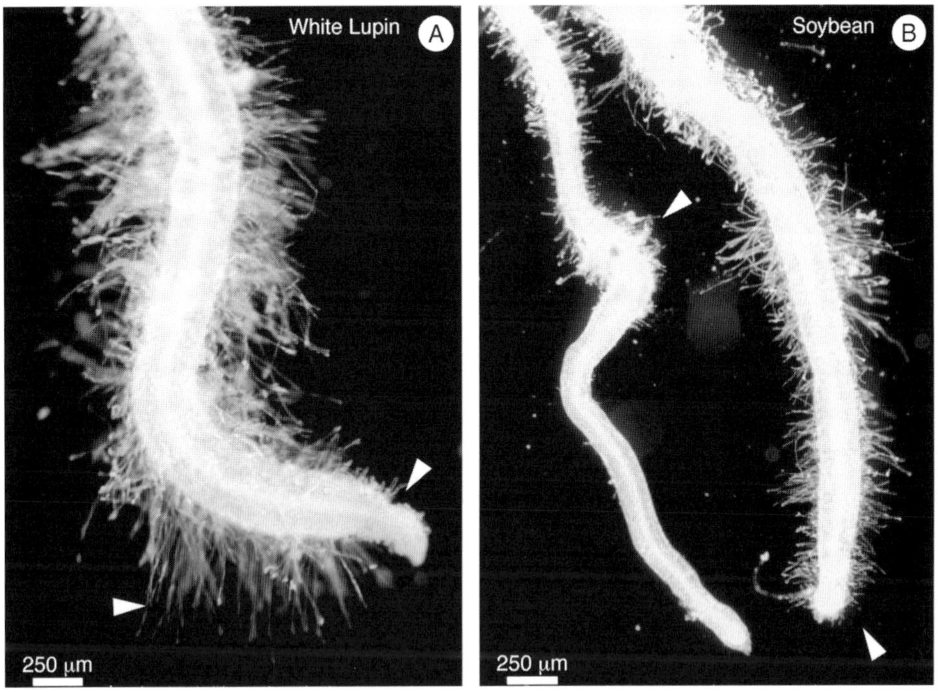

Figure 7. Root tips of white lupin rootlet (A) and soybean secondary branch roots (B). (A) Root hairs have developed to the tip, and there is no elongation zone, an indication that the root has stopped advancing. (B) Root on the left has a meristem and an elongation zone ahead of differentiating root hairs. The root on the right has differentiated root hairs around the tip, indicative of no active meristem or elongation. Arrowheads – root hairs.

Use of applied P by white lupin and soybean

P was applied to the soil at the top of the pot, in solution, and likely became bound to soil with time and movement through the soil profile. Roots need to intercept soil P as the roots grow through the soil to capture the solution P before it is soil-bound, or have the ability to unlock the bound fraction. The species examined here responded very differently to the applied P. Soybean readily captured and used the additional P to gain 25% more biomass, and double its total P (Figure 2). White lupin, however, had similar biomass and total P regardless of addition of P to soil. Cluster production may also have benefited the plant by mobilisation of micronutrients such as Fe and Mn (Braum and Helmke, 1995; Dinkelaker et al., 1989).

The reduction in cluster development by P addition independent of plant size shows that lupin sensed the extra P, and after a lag (perhaps due to seed P; approximately 1.2 mg per plant), responded by shifting allocation of root tissues from clusters to axes (Table 1, Figure 6). It could be argued that white lupin had access to the additional P but did not take it up, maintaining a constant homeostasis and P concentration in the tissues (Figure 3). However, studies in hydroponics (Keerthisinghe et al., 1998) show that P concentrations and biomass can increase in white lupin provided that the P is readily available in solution and is continually circulated over the roots. By increasing P in solution from 1 to 5 μM, plant dry weight increased approximately 4-fold, and P concentration in the tissues increased 40% to 2.1 mg g^{-1}. When 100 μM P was supplied, concentrations reached 8 mg g^{-1}.

The lack of response of white lupin to the amended P can be explained in terms of the inherent genetic differences in the development and thus the root architecture of the two species, and the responsiveness of soybean development to the applied P. Soybean colonised a much larger volume of soil per unit biomass, positioning its roots through the profile to capture applied P. Only one-third of the white lupin root system consisted of clusters of rootlets, and these had 58% less length per unit fresh weight compared to the thinner soybean fine branch roots (Figure 1, Table 2). The tetrarch organisation of the stele of soybean's tap root enabled soybean to position its basal primary branch roots more evenly within the soil compared to white lupin (Figure 5). Also, soybean developed branch roots to the fourth-order, was not constrained to the 'clustering' of branch roots as occurs in white

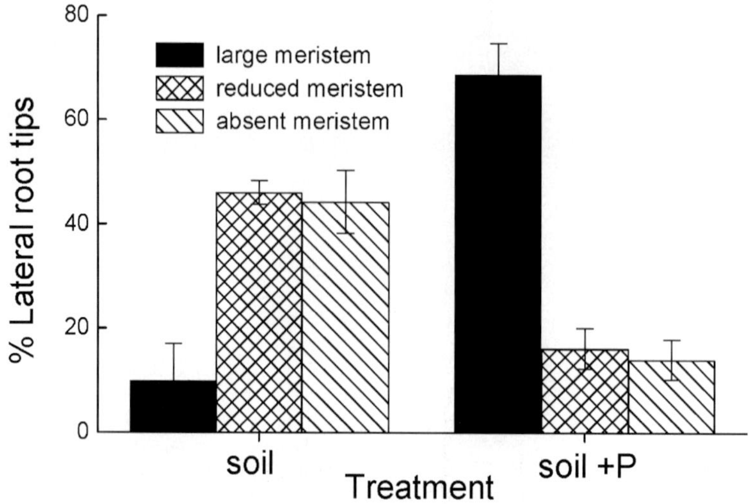

Figure 8. Soybean root tips. Frequency (percent of total tips scored per treatment) of soybean branch root tips harvested from the basal 10 cm of the taproot, with a large meristem, a reduced meristem, or no meristem. Plants were grown in soil with or without 1.3 m*M* P (+P) added to soil in solution at the tops of the pots. *n*=130–170.

lupin, and the branch roots were longer because their meristems were generally less determinate. Meristem determinacy in rootlets means that root tips cannot elongate and colonise new soil upon sensing of external P.

Soybean responded dramatically to local application of inorganic P (Figure 4). There was more root tissue with active meristems in the top of the soil profile in the plus P treatments (Figure 8). Del Castillo et al. (1989) also report active-looking tips around the basal portions of soybean root systems, corresponding to where complete nutrient fertilizer was applied. White lupin, in contrast, did not generate new roots in response to applied P. Lupin cluster distribution was maintained throughout the root system, increasing with plant size and decreasing with P application suggesting that cluster initiation is under a systemic signal within the plant (Watt and Evans, 1999), or that cluster initiation can be induced externally by non-labile P (throughout the soil), rather than labile P. New clusters in white lupin did not arise closer to the tip than the youngest existing cluster in response to inorganic P applied in solution, and new proteoid roots were not stimulated to develop in the basal part of the root where P was applied throughout the experiment. Clearly soybean and white lupin differ in the sensing and responsiveness to plant and soil P signals at the pericycle and the branch root meristems. It is not possible to tell from the data here what induced the formation of a greater number of clusters in white lupin with plant size (Figure 6C).

Speculation suggests that white lupin was also hampered in its response to the applied P because clusters have a short metabolic life span. Enzyme activities, citrate efflux, protein content, and mRNA levels decline sharply in white lupin clusters in hydroponics within 7 days (Neumann et al., 2000; Watt and Evans, 1999). P transporters at the plasma membrane of rootlets may also decline in activity, and the rootlets may have a limited time when they are connected to the transpiration stream if they senesce quickly. Older clusters at the base of the root system where P was continually applied throughout this experiment may not therefore have been able to capture P. Data of Keerthisinghe et al. (1998), also in hydroponics, would suggest that P uptake does not vary along a proteoid root with clusters up to 5 weeks old. ^{32}P was applied to segments of proteoid root and no difference in uptake was found with age of rootlets. The relative contribution, however, of axis and rootlets to uptake may change along the axis with age, masking any changes in rootlet activity alone. If P uptake by clusters declined with age, this may not appear in a hydroponics system where strong uptake and a diffusion gradient are not needed for contact with P.

Limitations and advantages of strategies for acquiring P

Data presented here highlight two strategies for P acquisition from soil: that of extensive root length and uptake of P in solution, employed by soybean, and that of cluster formation and bound P extraction, used by white lupin. While white lupin was very efficient at P acquisition from soil per unit root length, it was unable to capture the applied, inorganic P apparently because it did not develop roots to access the P. White lupin with proteoid roots appears adapted to soils with bound P, and soybean adapted to soils with labile P.

Studies of P acquisition strategies need to be placed in the context of the soils in which they will be used. Such studies can be extended to soils with different P availabilities and pH levels to test the abilities of different root strategies for acquiring P. Studies in soil that combine measurements of bound and labile P in the bulk soil and the rhizosphere, and P in the plant, with knowledge of root development and physiology, will be critical to understanding the key mechanisms determining species abilities to acquire P.

Acknowledgements

We are grateful to Ian Rochester for providing the soil and Chris Doherty for supplying details on the soil chemical composition, and to the Plant Culture Staff at the Research School of Biological Sciences for mixing the soil and sand, and maintaining the glasshouse facilities. Thanks also to Margaret McCully and Megan Ryan for helpful comments on the manuscript.

References

Adams M A and Pate J S (1992) Availability of organic and inorganic forms of phosphorus to lupins (*Lupinus* spp.). Plant Soil 145, 107–113.

Bar-Yosef B (1991) Root excretions and their environmental effects: influence on availability of phosphorus. *In* Eds. Waisel Y, Eshel A and Kafkafi U. Plant Roots: The Hidden Half (pp. 581–905). Marcel Dekker, New York.

Braum S M and Helmke P A (1995) White lupin utilizes soil phosphorus that is unavailable to soybean. Plant Soil 176, 95–100.

Del Castillo D, Acock B, Reddy V R and Acock M C (1989) Elongation and branching of roots on soybean plants in a carbon dioxide-enriched aerial environment. Agron. J. 81, 692–695.

Dinkelaker B, Romheld V and Marschner H (1989) Citric acid excretion and precipitation of calcium citrate in the rhizosphere of white lupin (*Lupinus albus* L.). Plant Cell Environ. 12, 285–292.

Fox T R, Comerford N B and McFee W W (1990) Phosphorus and aluminium release from a spodic horizon mediated by organic acids. Soil Sci. Soc. Am. J. 54, 1763–1767.

Gardner W K, Barber D A and Parbery D G (1983a) The acquisition of phosphorus by *Lupinus albus* L. III. The probable mechanism by which phosphorus movement in the soil/root interface is enhanced. Plant Soil 70, 107–124.

Gardner W K, Barber D A and Parbery D G (1983b) The acquisition of phosphorus by *Lupinus albus* L. V. The diffusion of exudates away from roots, a computer simulation. Plant Soil 72, 13–29.

Gerke J, Römer W and Jungk A (1994) The excretion of citric and malic acid by proteoid roots of *Lupinus albus* L.; effects on soil solution concentrations of phosphate, iron, and aluminum in the proteoid rhizosphere in samples of an oxisol and luvisol. Z. Pflanzenernähr. Bodenk. 157, 289–294.

Gilbert G A, Knight J D, Vance C P and Allan D L (1999) Acid phosphatase activity in phosphorus-deficient white lupin roots. Plant Cell Environ. 22, 801–810.

Hayes J E, Richardson A E and Simpson R J (2000) Components of organic phosphorus in soil extracts that are hydrolysed by phytase and acid phosphatase. Biol. Fertil. Soils 32, 279–286.

Hocking P J, Keerthisinghe G, Smith F W and Randall P J (1998) A comparison of the ability of different crop species to access poorly-available soil phosphorus. *In* Eds. Ando P, Fujita K, Mae T, Matsumoto H, Mori S and Sekiya J. Plant Nutrition for Sustainable Food Production and Environment (pp. 305–308). Kluwer, Tokyo.

Irving G C J and McLaughlin M J (1990) A rapid and simple field test for phosphorus in Olsen and Bray No. 1 extracts. Commun. Soil Sci. Plant Anal. 21, 2245–2255.

Keerthisinghe G, Hocking P J, Ryan P R and Delhaize E (1998) Effect of phosphorus supply on the formation and function of proteoid roots of white lupin (*Lupinus albus* L.). Plant Cell Environ. 21, 467–478.

Kirk G J D, Santos E E and Santos M B (1999) Phosphate solubilization by organic anion excretion from rice growing in aerobic soil: rates of excretion and decomposition, effects on rhizosphere pH and effects on phosphate solubility and uptake. New Phytol. 142, 185–200.

Li M, Shinano T and Tadona T (1997) Distribution of exudates of lupin roots in the rhizosphere under phosphorus deficient conditions. Soil Sci. Plant Nutr. 43, 237–245.

Neumann G, Massonneau A, Langlade N, Dinkelaker B, Hengeler, Römheld V and Martinoia E (2000) Physiological aspects of cluster root function and development in phosphorus-deficient white lupin (*Lupinus albus* L.). Ann. Bot. 85, 909–919.

Ohwaki Y and Hirata H (1992) Differences in carboxylic acid exudation among P-starved leguminous crops in relation to carboxylic acid contents in plant tissues and phospholipid level in roots. Soil Sci. Plant Nutr. 38, 235–243.

Watt M and Evans J R (1999) Linking development and determinacy with organic acid efflux from proteoid roots of *Lupinus albus* L. grown with low phosphorus and ambient or elevated atmospheric CO_2 concentration. Plant Physiol. 120, 705–716.

Plant and Soil **248:** 285–295, 2003.
© 2003 *Kluwer Academic Publishers. Printed in the Netherlands.*

Plantago lanceolata L. and *Rumex acetosella* L. differ in their utilisation of soil phosphorus fractions

Ann-Mari Fransson[1,3], Ingrid M. van Aarle[2], Pål Axel Olsson[2] & Germund Tyler[1]
[1]*Department of Ecology, Soil-Plant Research, Ecology Building, Lund University, SE-223 62 Lund Sweden.*
[2]*Department of Ecology, Microbial Ecology, Ecology Building, Lund University, SE-223 62 Lund Sweden.*
[3]*Corresponding author**

Received 4 December 2001. Accepted in revised form 21 March 2002

Key words: arbuscular mycorrhiza, phosphatase, phosphorus extraction, plant nutrition, unfertilised soil

Abstract

To establish relationships between soil phosphorus (P) fractions and leaf P, a mycorrhizal species (*Plantago lanceolata* L.) was compared with a typically non-mycorrhizal species (*Rumex acetosella* L.) in a glasshouse experiment. The plants were grown in 40 soils from non-fertilised, abandoned pastures or abandoned arable fields and leaf P concentration were found to be related to various soil P fractions after six weeks of growth. The differences in the P fractions in soil can account for a large share of the variation in leaf P concentration in both species, but the two species differed in their utilisation of P fractions. Leaf P concentration of *R. acetosella* was more related to extractable soil P than that of *P. lanceolata*. *Rumex acetosella* showed a higher maximum P concentration. The P fractions accounting for the largest share of the variation in leaf P concentration was the Bray 1 extractable and the weak oxalate (1 m*M*) extractable P, and for *P. lanceolata* also the Na_2SO_4+NaF extractable P fraction. P extracted with these methods accounted for up to 80% of the variation in P concentration in leaves of *R. acetosella* and 65% of the variation in leaves of *P. lanceolata*. More P extractable with weak oxalate, Na_2SO_4+NaF and strong oxalate (50 m*M*) was released from the soil than was taken up by the plants during the experimental period. The Bray 1 extractable P fraction, however, decreased in both unplanted and planted soils. Phosphatase release was not induced in any of the plants during the experimental period, indicating that they were not mobilising soil organic P. However, some of the methods extracted a large share of the organic P and still explained much of the variation in leaf P concentration. Mycorrhizal colonisation of *P. lanceolata* was inversely related to the extractable soil P. The consistently fast P uptake of *R. acetosella* indicates that this species have a high demand for P. The differences in P utilisation between *R. acetosella* and *P. lanceolata* could be caused by their different mycorrhizal status.

Abbreviations: ACP – acid phosphatase; ALP – alkaline phosphatase; AM – arbuscular mycorrhiza

Introduction

More than 170 phosphorus (P) containing compounds have been identified in soils (Holford, 1997). The fraction dominating a certain soil is determined by such factors as primary soil properties, soil acidity and rate of organic matter decomposition. Several of these compounds are in chemical equilibrium, replenishing those P pools most available to plants. Utilisation of different forms of soil P by plants is, therefore, governed primarily by soil factors influencing P solubility, such as pH, soil moisture, and texture, as well as the specific and differing capacities of the plants to render P fractions available for uptake.

Many studies have demonstrated fair to good relationships between plant P concentration and level of extractable soil P (e.g. Ziadi et al., 2001). Most, however, were conducted in agricultural systems and were designed to investigate responses of crop plants to fertilisation (Bray and Kurtz, 1945; Olsen et al.,

* FAX No: +46-46-222-44-23
E-mail: Ann-Mari.Fransson@planteco.lu.se

1954). As expected a large proportion of the P taken up in these cases originated from the easily solubilised P added as fertiliser. Few studies have considered relationships between plant P and soil P in undisturbed soils in order to determine which soil P fractions are available to plants.

A number of strategies used by plants for more efficient utilisation of soil P have been proposed, including morphological changes of the roots. These include formation of root hairs (Gahoonia et al., 1997), extensive lateral root formation, and proliferation of roots in response to nutrient patches (Hutchings and De Kroon, 1994). Physiological changes may also take place in the roots in response to low availability of P. Both experimental work and model simulations suggest that changes in uptake kinetics may play an important role in P uptake from fertile microsites (Caldwell et al., 1992). Plants may also alter the equilibrium between the different P compounds in the rhizosphere soil by exuding P-solubilising compounds (Marschner, 1995). Exudation of protons results in a decrease in rhizosphere pH, thereby changing the solubility of numerous P compounds. Many plants growing in acid soils exude oxalic acid and this may be a mechanism for solubilising P not only in calcareous soil (Ström, 1997; Tyler and Ström, 1995). Acid phosphatases exuded by the plant in the rhizosphere may hydrolyse soil organic P (Tarafdar and Claasen, 1988).

The organic P fraction represents a major pool of P in many soils. For example, up to 57% of the total P in acid grassland soils in Sweden was found to be organic (Tyler and Olsson, 1993). Associations between plants and microorganisms may influence P transformations in the rhizosphere and thereby alter the availability of P to plants. Arbuscular mycorrhiza (AM), present in most herbaceous plants, can increase the influx of P to roots (Fitter, 1995; Jakobsen, 1986; Sanders and Tinker, 1971; Smith and Read, 1997). A few studies have indicated that the AM fungi may have different access to pools of organic P, so that the availability of P could be higher to mycorrhizal than to non-mycorrhizal plants (Jayachandran et al., 1992; Tarafdar and Marschner, 1994). However, AM fungi are usually considered to take up P from similar fractions as plants (Bolan, 1991; Sanders and Tinker, 1971).

The aims of this study are to determine which soil P fractions, as determined by different extractants, are available to field-collected plants growing either in undisturbed unfertilised soil or in soil not recently fertilised and with P fractions in equilibrium. Species studied were *Plantago lanceolata* L., which typically forms arbuscular mycorrhizal association, and *Rumex acetosella* L., which typically is not colonised by mycorrhizal fungi. *Rumex acetosella* has previously been shown to be highly dependent on a high P status of soils (Tyler, 1996) and could be considered to be sensitive to variations in P availability. Both species occur frequently in the habitats where the soils were collected. Several methods for extraction of P are used to measure the different P fractions available to plants in the soil. By determining the relationship between leaf P and the concentration of differently extractable P fractions of soil, those fractions contributing to P uptake by plants are estimated.

Materials and methods

Soils

Forty different topsoils (0–10 cm) were collected from little used or abandoned pastures and abandoned arable fields in a non-calcareous area of southern Sweden (ca 56° 05′ N, 13° 50′ E). Soils were selected due to vegetation characteristics to be of low to moderate P status. They were assumed to have P fractions in equilibrium since none had been fertilised in the last 5 years and many of the soils had probably never been fertilised. Soils were sieved (6 mm) in the field and water holding capacity determined after water saturation from below. Loss on ignition (600 °C) and water content (105 °C) were determined gravimetrically. Acidity (0.1 *M* $BaCl_2$ extractable) was determined after titration with NaOH, and 0.1 *M* $BaCl_2$ extractable cations were measured by ICP-ES in all soils prior to experimental set-up. Soil pH, using the same extractant as for acidity and cat ions (0.1 *M* $BaCl_2$), was determined both initially and after the experimental period. Soils were kept at 4 °C and all extractions were performed within one week after collection.

P extraction methods

Prior to commencing the glasshouse studies, each of the soils (at field moisture) were extracted by the following methods:

(a) *Weak oxalate extractable P*: An aliquot of soil was extracted with a solution containing 0.8 m*M* Na-oxalate and 0.2 m*M* oxalic acid at pH 4.4, soil/solution ratio 1:20 w/v with a 30 min extraction time (Fransson, 2001). This extraction was used to simulate realistic concentrations of P exuded close to the root surface in the soil (Jones, 1998). Both total P (measured by ICP-ES) and molybdate reactive P was determined. Organic weak oxalate extractable P was calculated as the difference between molybdate reactive and total P in the extract (Tyler and Olsson, 1993).

(b) *Na_2SO_4+NaF extractable P*: An aliquot of soil was extracted with a solution containing 0.05 *M* Na_2SO_4 and 0.02 *M* NaF, soil/solution ratio 1:10 w/v for 20 min (Tyler and Olsson, 1993). This method is assumed to extract phosphate exchangeable with SO_4^{2-} and some surface Al-phosphates solubilised by formation of Al–F-complexes at ambient soil pH. The extract was analysed spectrophotometrically using the molybdate blue method of Murphy and Riley (1962), thereby excluding organic P.

(c) *Bray 1 extractable P*: An aliquot of soil was extracted with a mixture of 0.03 *M* NH_4F and 0.025 *M* HCl, soil/solution ratio 1:10 w/v, 1 min extraction time. This extraction is widely used since it is known to correlate well to P uptake in agricultural systems (Bray and Kurtz, 1945). The Bray 1 extraction releases Al-phosphates by the complex formation between Al and F-ions as in extractant (a) in combination with P solubilisation as a result of a decrease in pH. It should be noted that the P extracted may be underestimated as some P is incorporated into organic matter aggregating as a result of the pH decrease. Only molybdate reactive P was determined.

(d) *H_2SO_4 extractable P*: An aliquot of soil was extracted with 0.5 *M* H_2SO_4, soil/solution ratio 1:25 w/v, with a 16 h extraction time. This was used to compare the P released by strong oxalate and the Bray 1 extraction with a strong mineral acid extraction H_2SO_4 extraction is considered to extract all Ca-phosphates and as it is a part of the total organic P determination (according to Kuo, 1996). Only molybdate reactive P was determined.

(e) *Strong oxalate extractable P*: An aliquot of soil was extracted with a solution containing 40 m*M* Na-oxalate and 10 m*M* oxalic acid at pH 4.8, soil/solution ratio 1:20 w/v with a 30 min extraction time. This extraction was used to determine the potentially oxalate soluble P in the soil (Fransson, 2001). Only total P was determined. It was not possible to analyse molybdate reactive P because of the high background colour of dissolved organic compounds.

(f) *Total organic P* was determined after ignition of the soil and extraction of both ignited and non-ignited soil with H_2SO_4. The difference in molybdate reactive P was considered as total organic P (Tyler and Olsson, 1993).

(g) *Total P* was analysed spectrophotometrically according to Murphy and Riley (1962) after hot digestion of soil with concentrated HNO_3 and subsequent dilution.

All extracts were filtered through Whatman OOK filters before analysis.

Experimental design

Plants of *Plantago lanceolata* L. and *Rumex acetosella* L. were collected (21 April 1998) from a slightly acid sandy unfertilised soil. Plants about the same size were selected, washed carefully, and roots and shoots were cut, leaving a similar amount of biomass. For each of the 40 soils, three *P. lanceolata* individuals were planted a pot (1.5 L). Each of 40 similar pots was planted with one *R. acetosella* individual. Another 40 pots were left unplanted. All pots were placed in a completely randomised design in the glasshouse. The mean glasshouse temperature was 20.5 °C with a maximum temperature of 29.1 °C and a minimum of 14.3 °C recorded. High-pressure sodium lamps provided additional light (150 μmol m^{-2} s^{-1} PAR) for 12 h d^{-1} (06.00–18.00 h). Soil moisture was maintained at 60–65% of the water holding capacity by watering the pots with distilled water every day. After 6 weeks of growth plants were harvested and the soil collected. Roots had grown to such an extent that they were present throughout the whole soil volume. Plant tissue was dried to constant weight at 70 °C and the weight recorded. Total P concentration in leaves was determined according to Murphy and Riley (1962) after total digestion of the entire leaf fraction with concentrated HNO_3.

After the completion of the experiment, soil P was extracted with: (a) Na_2SO_4+NaF, (b) weak oxalate, (c) strong oxalate, or (d) the Bray 1 extractant. Total P, total organic P and H_2SO_4 extractable P were not measured after the experiment since these were either closely correlated to one of the chosen extractants (Fransson, 2001) or were not expected to be changed as a result of the plant growth. The pH of the extracts was determined. Concentrations of NO_3^- and NH_4^+ in $BaCl_2$ extract of the soils were determined using flow injection analysis according to Ruzika and Hansen (1981).

Phosphatase activity assay

Alkaline phosphatase (ALP) and acid phosphatase (ACP) activity in the soils were determined spectrophotometrically using *p*-nitrophenyl phosphate (*p*NPP, Sigma Chemical Co.) as substrate (Tabatabai and Bremner, 1969). Two g of soil was added to 40 mL of distilled water and shaken at 500 rpm for 15 min and aliquots of the suspension were used to determine ALP (pH 10.3) and ACP (pH 4.8) activity (Van Aarle et al., 2001). Absorbance in three blanks of heated extracts from each of the three treatments was determined and a mean value was calculated. Absorbance was measured at 420 nm after filtration through a 0.5 μm hydrophilic PTFE filter (Millipore). Enzyme activity expressed as Units g^{-1} soil (1 unit of phosphatase activity being the amount of enzyme activity, which would liberate 1 nmol of *p*-nitrophenol s^{-1}) was calculated after correction for soil blanks.

AM fungal root colonisation

Roots of *P. lanceolata* and *R. acetosella* were stained according to a modification of the method described by Phillips and Hayman (1970). A subsample of fine roots grown during the experimental period was washed free of soil, cut into pieces and heated at 80 °C in 10% KOH until sufficiently cleared. After rinsing in water, roots were stained at 80 °C in 0.1% Trypan blue in lactic acid, glycerol and water (1:1:1, v/v/v), rinsed in water and destained in 50% glycerol. Percentage of root length colonised was determined by the grid-line intersect method under a dissection microscope using a Petri dish with lines (Giovanetti and Mosse, 1980).

Calculations and statistical methods

Soil P and leaf P concentrations were calculated as μmol g^{-1} dry mass. The variation in leaf P concentration was hypothesised to be a function of the extractable P concentration in the soil. Since the distribution of data was somewhat skewed, the data were $\log_{10}$ transformed in order to obtain a more even distribution of the data over the measured range. In this way the contribution of soils with low soil P concentration was better demonstrated. Further, a reciprocal transformation was made in order to increase the influence of soils with low soil P concentration. The amount of soil P and above ground biomass P per pot was also calculated. Differences between extractable P before and after the experiment were determined in all soils using paired students' *t*-test (where each soil represented a block) and Bonferroni correction for multiple testes (Sokal and Rohlf, 1995). Each pot was considered to be independent of the others.

Table 1. Median, maximum and minimum P concentrations (μmol g^{-1} soil) extracted from the soils with the different methods at the commencement of the experiment

Extraction method[a]	Initial soil P concentrations		
	Median	Min	Max
Weak Ox, mo	0.03	0.002	0.77
Weak Ox, org	0.07	0.02	0.18
Weak Ox, total	0.12	0.03	0.91
Na_2SO_4+NaF	0.30	0.001	2.16
Bray 1	1.18	0.14	13.70
H_2SO_4	5.67	1.14	14.12
Strong Ox	7.61	2.42	26.71
Total organic P	9.30	2.30	24.00
Total P	30.28	14.46	54.26

[a] Abbreviations: Weak Ox=weak oxalate; mo=molybdate reactive; org=organic; Strong Ox=strong oxalate.

Table 2. Average chemical properties of all soils at the commencement of the experiment. Mean, maximum and minimum values for $BaCl_2$ extractable pH, acidity, base saturation, cation concentrations, and organic matter content

Soil variables	Mean	Min	Max
Initial pH	4.09	3.54	4.63
Acidity (μmol g^{-1}dry mass)	11.63	0.35	33.3
Base saturation (%)	69.76	23.2	99.7
Organic matter (% dry mass)	8.06	2.35	14.2
Cations (μmol g^{-1} dry mass)			
Al	3.18	0.06	8.89
Ca	15.73	2.02	49.5
Fe	0.01	<0.01	0.04
K	1.65	0.30	5.94
Mg	2.60	0.44	9.58
Mn	0.09	<0.01	0.54
Na	0.35	0.03	1.40

Results

General characteristics of the soils

The efficiency of the different methods to extract P varied considerably between the 40 soils. Median extractable P ranged from 0.03 μmol g^{-1} soil of molybdate-reactive weak oxalate-extractable P to 7.6 μmol g^{-1} soil of strong oxalate-extractable P (Table

Table 3. Mean, minimum and maximum leaf P concentration (μmol g^{-1} dry mass), total plant biomass (g dry mass pot^{-1}), pH ($BaCl_2$) and $BaCl_2$ extractable N (NH_4^++NO_3^-) concentration in the 40 different soils where either *Plantago lanceolata* or *Rumex acetosella* had been grown for 6 weeks

	P. lanceolata			*R. acetosella*		
	Mean	Min	Max	Mean	Min	Max
Leaf P (μmol g^{-1} dry mass)	40.6	15.5	97.3	88.9	30.2	194.5
Biomass (g dry mass pot^{-1})	3.19	2.03	5.83	4.70	2.04	8.56
pH ($BaCl_2$)[a]	4.05	3.51	4.46	4.08	3.49	4.72
N (μmol g^{-1} dry mass)[b]	0.14	<0.04	0.25	0.06	<0.04	0.13

[a] Mean pH in the unplanted soil was 4.09 after the experiment.
[b] Mean $BaCl_2$ extractable N mineralised in the unplanted soil during the experiment was 4.60 μmol g^{-1} dry mass.

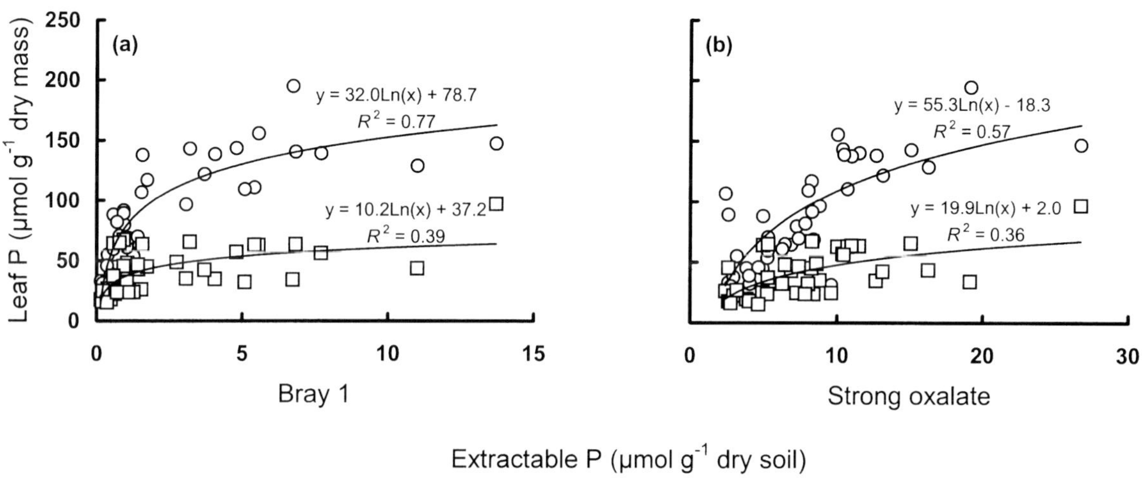

Figure 1. The relationship between concentration of P in leaves of *Rumex acetosella* (circles) and *Plantago lanceolata* (squares) (μmol g^{-1} dry mass) and (a) Bray 1 and (b) strong oxalate extractable soil P concentration (μmol g^{-1} dry soil).

1). Soil organic matter content ranged from 2 to 14%, and mean soil pH ($BaCl_2$) was 4.1, with a range from 3.5 to 4.6 (Table 2). Soil in which either of the plant species had been grown had slightly lower soil pH values than unplanted soil (paired *t*-test, *P. lanceolata*, *P*=0.024 and *R. acetosella* *P*=0.012, Table 3). Base saturation ($BaCl_2$) and pH were not correlated within the pH range of the soils. The plants used >95% of the N mineralised during the experiment. At the end of the experiment the mean $BaCl_2$ extractable N concentration was less than 0.25 μmol g^{-1} in the soil after *P. lanceolata* growth and less than 0.13 μmol g^{-1} in the soil for *R. acetosella* (Table 3).

Relation between soil P and leaf P

For both species, all the transformations tested improved regression values when used as a description of the relationship between leaf P and the various soil P fractions (Table 4). This shows that there was a stronger relationship for soils with low P than for soils with high P. There was, however, a clear difference between the two species in that the reciprocal transformation of both variables improved regression most beneficially for *P. lanceolata*, while logarithmic transformation of the independent variable improved regression most for *R. acetosella* (Table 4). The relations between soil P and leaf P were similar regardless of whether calculations were made on initial, final or unplanted final soil variables. Leaf P concentration in leaves were well correlated to the P fractions extracted by a range of methods. Some relationship patterns could be recognised. For example, leaf P in *R. acetosella* was generally better accounted for by P extracted by weak methods. Leaf P of *P. lanceolata* generally correlated better to soil P than leaf P of *R. acetosella* in the soils with low P (Figure 1). There was no obvious soil P fraction that marked a border for the P availability, but the organic P frac-

Table 4. Regression coefficients (R^2) for the relation between concentrations of P in 40 different field-collected soils, measured using different methods and concentrations of P in leaves of *Plantago lanceolata* and *Rumex acetosella* cultivated in these soils. Linear regression was applied to both non-transformed data (lin–lin) as well as after three types of data transformations (log10–lin, log10–log10, lin^{-1}–lin^{-1}). All relationships are positive. Levels of significance: ^{2} = 0.25 corresponds to P=0.001 (other levels of significance is not considered; R^2 > 0.25 shown in bold). For abbreviations of extraction methods see Table 1

Extraction method	*P. lanceolata* lin–lin	log10–lin	log10–log10	lin^{-1}–lin^{-1}	*R. acetosella* lin–lin	log10–lin	log10–log10	lin^{-1}–lin^{-1}
Calculations on initial soil concentrations								
Weak Ox, mo	0.21	**0.33**	**0.34**	0.04	**0.49**	**0.75**	**0.64**	**0.43**
Weak Ox, org	**0.30**	**0.33**	**0.37**	**0.41**	**0.51**	**0.54**	**0.40**	**0.61**
Weak Ox, total	**0.31**	**0.43**	**0.46**	**0.57**	**0.54**	**0.73**	**0.60**	**0.74**
Na_2SO_4+NaF	0.17	**0.34**	**0.44**	**0.59**	**0.55**	**0.70**	**0.64**	**0.54**
Bray 1	**0.34**	**0.39**	**0.45**	**0.47**	**0.55**	**0.77**	**0.70**	**0.72**
Strong Ox	**0.38**	**0.36**	**0.38**	**0.38**	**0.58**	**0.57**	**0.48**	**0.49**
H_2SO_4	0.23	**0.27**	**0.35**	**0.46**	**0.68**	**0.58**	**0.60**	**0.52**
Total organic P	0.21	0.17	0.14	0.21	0.02	0.07	0.07	0.06
Total P	**0.40**	**0.40**	**0.47**	**0.55**	**0.58**	**0.57**	**0.51**	**0.69**
Calculations on final soil concentrations								
Weak Ox, mo	**0.26**	**0.44**	**0.46**	**0.39**	0.22	**0.44**	**0.33**	**0.26**
Weak Ox, org	0.07	**0.32**	0.07	0.04	**0.51**	**0.54**	**0.30**	**0.30**
Weak Ox, total	**0.27**	**0.29**	**0.30**	0.22	**0.39**	**0.50**	**0.39**	**0.58**
Na_2SO_4+NaF	0.17	**0.36**	**0.41**	**0.65**	**0.49**	**0.72**	**0.62**	**0.53**
Bray 1	**0.34**	**0.41**	**0.47**	**0.56**	**0.49**	**0.80**	**0.66**	**0.73**
Strong Ox	**0.29**	**0.29**	**0.32**	**0.34**	**0.53**	**0.50**	**0.41**	**0.41**
Calculations on unplanted final soil concentrations								
Weak Ox, mo	**0.27**	**0.43**	**0.47**	**0.37**	**0.47**	**0.61**	**0.57**	**0.36**
Weak Ox, org	**0.31**	**0.35**	**0.36**	**0.38**	**0.51**	**0.58**	**0.52**	**0.44**
Weak Ox, total	**0.29**	**0.40**	**0.41**	**0.48**	**0.50**	**0.63**	**0.57**	**0.49**
Na_2SO_4+NaF	0.17	**0.40**	**0.46**	0.18	**0.50**	**0.63**	**0.62**	0.19
Bray 1	**0.43**	**0.47**	**0.52**	**0.61**	**0.57**	**0.80**	**0.79**	**0.77**
Strong Ox	**0.42**	**0.45**	**0.48**	**0.50**	**0.61**	**0.62**	**0.60**	**0.49**

tion gave the weakest relation for both plant species. Leaf P concentration also correlated well with total soil P, however, differences between the species were detected. The soil P fraction extracted with the Bray 1 method accounted for most of the variation in leaf P concentration of *R. Acetosella* and up to 80% of the leaf P variation could be accounted for by this extractant after logarithmic transformation. The variation in leaf P of *P. lanceolata* was in some cases best accounted for by the Na_2SO_4+NaF extractable P.

The extent to which leaf P was related to soil P differed between the species. Leaf P concentration in *R. acetosella* was more closely correlated to soil P concentration than it was for *P. lanceolata*, for all different soil P fractions examined in this study (paired *t*-test, P<0.001). The relationship was best described by a hyperbolic curve indicating a saturation of leaf P (Figure 1). Leaf P concentration in *R. acetosella* increased more than in *P. lanceolata* as the extractable soil concentration increased. The regression for leaf P concentrations in *R. acetosella* was better accounted for by soil P concentration, while leaf P amounts per pot were better accounted for by soil P amounts per pot (paired *t*-test P<0.001). There were no differences between regression coefficients based on soil P concentrations of soil or amounts per pot for *P. lanceolata*.

Changes in soil P during the experiment

Phosphorus extracted using weak oxalate (paired *t*-test, P<0.001), Na_2SO_4+NaF (paired *t*-test, P<0.001),

Table 5. The change in soil phosphorus concentration (μmol g^{-1} soil dry mass) during the experiment estimated by different extraction methods and calculated as the mean difference±95% confidence interval between initial soil and either unplanted soil in which *P. lanceolata* was grown or soil in which *R. acetosella* was grown (*n*=40). Negative values denote a decrease in P concentration during the experiment. *P* values were calculated from paired *t*-tests. For abbreviations of extraction methods see Table 1

	Extraction method				
	Na_2SO_4+NaF	Bray 1	Weak Ox mo	Weak Ox total	Strong Ox
Unplanted soils					
mean	0.12±0.05	−0.12±0.23	0.03±0.01	0.13±0.03	1.20±0.47
% change[a]	35.1	−4.5	27.1	67.5	14.7
	P<0.001	n.s.	*P*<0.001	*P*<0.001	*P*<0.001
P. lanceolata soils					
mean	0.03±0.04	−0.41±0.30	−0.005±0.007	0.03±0.02	1.32±0.38
% change	9.0	−16.8	−4.8	13.4	16.5
	n.s.	*P*=0.016	n.s.	*P*=0.01	*P*<0.001
R. acetosella soils					
mean	−0.005±0.04	−0.33±0.34	−0.02±0.04	0.01±0.03	0.78±0.60
% change	−1.5	−12.5	−16.3	6.2	9.5
	n.s.	n.s.	n.s.	n.s.	*P*<0.001

[a] Mean% concentration changes.

and strong oxalate ($P<0.001$) increased in the unplanted pots during the experiment (Table 5). The Bray 1 extractable P fraction did not differ between unplanted and initial soil (P=0.32). Growth of *P. lanceolata*, however, decreased the Bray 1 extractable P concentration in the soil by 0.41 μmol g^{-1} compared to the initial soil (paired *t*-test, P=0.016), and by 0.34 (μmol g^{-1}) compared to the unplanted pots (paired *t*-test, P=0.004). The strong oxalate extractable P concentration, however, increased compared to the initial soil (Table 5).

AM fungal root colonisation

In *P. lanceolata* the fine roots that developed during the experimental period were colonised by AM fungi to some extent in all soils. The mean level of colonisation was 34% but the amount of colonisation differed widely according to the soil type. In some soils up to 80% of *P. lanceolata* root length was colonised, while in others less than 5%. Low colonisation was notably in those soils with the highest P concentration (Table 6). Colonisation showed a slight correlation to initial P concentration of the soil. Strong oxalate extractable P ($R^2 = 0.26$) and total P ($R^2 = 0.32$) were the fractions accounting for most variation. The relationship between colonisation and initial strong oxalate extractable P was stronger after removing one outlier soil from the calculation, when 47% of the variation in root colonisation could be accounted for (Figure 2). The soil excluded was the one with the highest P concentration, as measured by all methods. No colonisation was observed in 80% of the soils with *R. acetosella*. The occasional colonisation was not related to low soil P concentration but seemed random and could not

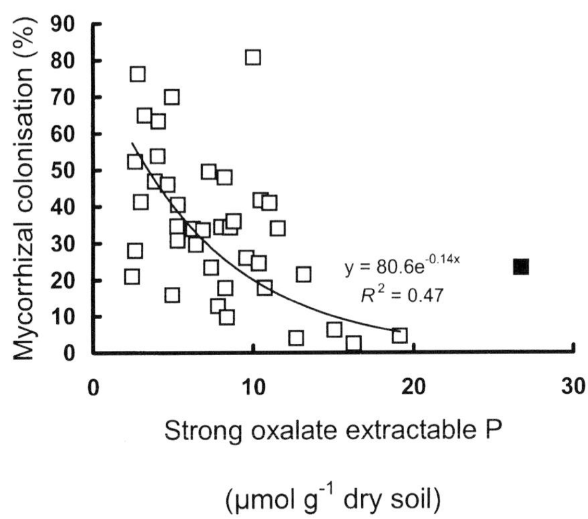

Figure 2. The relationship between strong oxalate extractable P concentration (μmol g^{-1} dry soil) and arbuscular mycorrhizal colonisation in *Plantago lanceolata* (%; open squares). One outlying soil (closed square) was excluded from the calculation.

Table 6. Acid and alkaline phosphatase (ACP and ALP) activities and arbuscular mycorrhizal (AM) fungal root colonisation presented as means for all soils at the end of the experiment (n=40, mean±SD). ACP activity was calculated as units both per g dry soil and per g organic matter content (OM). The values for ACP activity in treatments with plants were compared with the unplanted soil using paired t-test (**denotes P<0.01 and ***denotes P<0.001)

Treatment	ACP (units g^{-1} soil)	ACP (units g^{-1} OM)	ALP (units g^{-1} soil)	AM colonisation (%)
Unplanted pots	14.4±3.1	172±35	<1	–
P. lanceolata	10.7±2.4**	118±19***	3.2±0.7	34±6
R. acetosella	10.1±2.6***	114±21***	<1	<2

explain the variation in leaf P concentration (R^2 < 0.14).

Soil phosphatase activity

Differences in phosphatase activity did not account for the variability in leaf P concentration. The highest ACP activity was found in the unplanted soil; soil in which either *P. lanceolata* or *R. acetosella* had been grown had lower ACP activity (Table 6) whether calculated per g organic matter or on a per dry soil basis. Detectable ALP activity was found only in the soil where *P. lanceolata* had grown (Table 6).

Discussion

We have described the relationships between leaf P and soil P for two plant species over a range of acid non-cultivated soils. We have shown that the two species differ in this relationship and that P fractions in soil are able to account for a large share of this variation for both species as well as for part of the variation of AM colonisation in *P. lanceolata*. The tested transformations of the data showed that the regression was similar for most of the extraction methods, while there was a difference between the species in which transformation improved regression most. The higher correlation between leaf P and soil P using reciprocally transformed data than non-transformed data shows that the plant P uptake is more affected by the availability of P on soils low in P than on those with higher P. This applied especially for *P. lanceolata*. On soils richer in P, the soil appeared to fulfil plant needs for P more readily and factors other than P availability modified P uptake and concentration in the plant. In contrast, for *R. acetosella*, P uptake was still well related to soil P at high soil P levels (Figure 1). This result provides experimental support for the observations from field studies that *R. acetosella* depends on high soil P (Tyler, 1996). Leaf P concentration in *P. lanceolata* was to a larger extent determined by other factors. Different demand for P by the two species could possibly explain this. Further, differences in P uptake dynamics and kinetics might also explain the differences between the two plant species. Phosphorus uptake in spring wheat may for example differ between life stages (Schilling et al., 1998). Although adult individuals were studied in both species, *P. lanceolata* could have reached a state where P needed for new growth was re-translocated within the plant rather than taken up from the soil. Any such re-translocation would result in leaf P being less well correlated to soil P (Brundrett and Kendrick, 1990). Studies showing differences in utilisation of added P have been reported within and between other species. For example different strains of *Trifolium repens* and *Lotus pedunculatus* may use different soil P fractions (Trolove et al., 1996). Furthermore, uptake of P in corn seemed better correlated to Bray 1 extractable P than in soybean, even though the variation among soils was even greater (Borges and Mallarino, 1998). The high demand for P by *R. acetosella* is, however, difficult to explain since both P concentration and biomass produced were higher in *R. acetosella* than in *P. lanceolata*. In any case higher P depletion of the soil P fractions determined was not found in the *R. acetosella* pots.

Both species decreased or maintained the N concentration of most of the soils to levels that might have been growth limiting. However, the difference between the species could not be accounted for by differences in N limitation, as the N limitation to growth appeared more or less the same for both species (Table 3). It is, however, possible that the two species have a different tolerance of N deficiency.

The less close relationship between soil P and shoot P of *P. lanceolata* may be due to the higher AM colonisation at low soil P concentrations. A high AM colonisation at low soil P probably also means that there is more external AM mycelium (Olsson et al., 1997) that can take up P. AM colonisation might also be a reason why *P. lanceolata* used partly other P fractions than *R. acetosella*. However, data so far indicate that AM fungi are not important in the turnover of organic P in soil (Joner et al., 2000) and experiments with labelled P in agricultural soil indicate that the AM fungal mycelium utilises similar P fractions to plants (Sanders and Tinker, 1971). Though, studies using ^{32}P may be difficult to interpret since added ^{32}P may exchange with less available P fractions given sufficient time (Bolan, 1991). A negative influence of increased soil P on AM colonisation has been reported a number of times in fertiliser experiments (e.g. Bååth & Spokes, 1989; Jasper et al., 1979; Olsson et al., 1997). It is apparently not the soil P concentration itself that inhibits AM colonisation, but a high P concentration in the roots (Menge et al., 1978; Sanders, 1975). This suggests that the negative influence of high soil P concentration on colonisation of *P. lanceolata* may be due to higher uptake by the roots. In contrast to *P. lanceolata*, *Rumex* species are typically not colonised by mycorrhiza. We also found that *R. acetosella* was rarely colonised by AM fungi, as reported by others (Harley and Harley, 1987), and the occasional colonisation showed no relation to soil P.

Release of acid phosphatases to the soil may be a plant response to low P conditions (Helal and Dressler, 1989; Marschner, 1995). However, for both plant species, the soil had decreased acid phosphatase activity compared to unplanted soil and the remaining activity showed no correlation with any extractable P fraction. This indicates that regardless of any phosphatase exudation by the plants, a net immobilisation, or decomposition by the enzyme took place in the soil when roots were present. The pools of soil enzyme could also have been inactivated by some factor induced by the presence of the plants. It can be assumed that the soil acid phosphatase activity largely originated from microorganisms and from the relatively stable pool of acid phosphatases known to be present in soils (Helal and Dressler, 1989; Tyler, 1974). Alkaline phosphatase activity was only detected in soils where *P. lanceolata* had grown, probably originating either from the AM fungi or from saprophytic microorganisms in the rhizosphere surrounding *P. lanceolata* (see Joner et al., 2000).

When correlating extractable P with P uptake in *Lolium rigidum* grown in 34 soils collected in Denmark, the more easily soluble soil P fractions gave higher correlation coefficients than the less soluble fractions (Sibbesen, 1978). This could be interpreted as *L. rigidum* having a poor ability to utilise less soluble soil P fractions. However, in natural soils less soluble P fractions are important for plant uptake and this is supported by the results of our study. Many of the methods tested extracted P fractions that seem available to the plants. P extracted by different methods is very well correlated in these soils and partly 'overlapping' as determined on the data material used in this study (Fransson, 2001). This could explain why even the strong oxalate extractable P fraction was fairly well correlated with plant P concentration. The soluble P fractions in soils are continuously, though sometimes slowly replenished. The soils used in our study showed high P replenishment ability (Table 5). The only fraction actually decreasing due to plant growth was the Bray 1 extractable P fraction. This could indicate that even on rather acid soils, Ca phosphates play an important role in P availability, even though their concentrations are relatively low in such soils. Several of the growth conditions may have influenced P availability and plant P uptake. Soil temperature in the pots was not much different from mean air temperature in the greenhouse, but probably higher than under conditions prevailing in the field. Increasing the temperature may increase the Olsen-P extracted (Saleque et al., 1996), such that more P was mobilised than utilised by the plants.

In conclusion, *R. acetosella* and *P. lanceolata* differ in their utilisation of soil P. This difference is most probably due to differences in utilisation of P fractions in the soil and could be caused by different mycorrhizal status. Another factor may be due to the consistently higher demand for P and therefore higher uptake by *R. acetosella* (Tyler, 1996) than for *P. lanceolata*. The large P fraction extractable only with strong oxalate/oxalic acid (50 m*M*) explains a high proportion of the variation in leaf P concentration. Less soluble P fractions in the soil may, therefore, be relevant for plant uptake in natural soils.

Acknowledgement

We wish to thank the CF Lundström Foundation and the Swedish Council for Forestry and Agricultural Research for the financial support, Mrs Anita Balogh for her excellent technical assistance and Anne Ashford for critical comments on the manuscript.

References

Bååth E and Spokes J 1989 The effect of added nitrogen and phosphorus on mycorrhizal growth response and infection in *Allium schoenoprasum*. Can. J. Bot. 67, 3227–3232.

Bolan N S 1991 A critical review on the role of mycorrhizal fungi in the uptake of phosphorus by plants. Plant Soil 134, 189–207.

Borges R and Mallarino A P 1998 Variation of early growth and nutrient content of no-till corn and soybean in relation to soil phosphorus and potassium supplies. Commun. Soil Sci. Plant Anal. 29, 2589–2605.

Bray R H and Kurtz L T 1945 Determination of total, organic, inorganic and available form of phosphorus in soil. Soil Sci. 59, 39–45.

Brundrett M and Kendrick B 1990 The roots and mycorrhizas of herbaceous woodland plants. I. Quantitative aspects of morphology. New Phytol. 114, 457–468.

Caldwell M M, Dudley L M and Lilieholm B 1992 Soil solution phosphate, root uptake kinetics and nutrient acquisition: implications for a patchy soil environment. Oecologia 89, 305–309.

Fitter A H 1995 Interpreting quantitative and qualitative characteristics in comparative analyses. J. Ecol. 83, 730.

Fransson A-M 2001 Evaluation of oxalate/oxalic acid for extracting plant available phosphorus in unfertilised acid soils. Commun. Soil Sci. Plant Anal. 32, 2469–2484.

Gahoonia T S, Care D and Nielsen N E 1997 Root hair and phosphorus acquisition of wheat and barley cultivars. Plant Soil 191, 181–188.

Giovanetti M and Mosse B 1980 An evaluation of techniques for measuring vesicular-arbuscular mycorrhizal infection in roots. New Phytol. 84, 489–500.

Harley J L and Harley E L 1987 A check-list of mycorrhiza in the British flora. New Phytol. (suppl.) 105, 1–102.

Helal H M and Dressler A 1989 Mobilization and turnover of soil phosphorus in the rhizosphere. Z. Pflanzenernähr. Bodenk. 152, 175–180.

Holford I C R 1997 Soil phosphorus: its measurement, and its uptake by plants. Aust. J. Soil Res. 35, 227–239.

Hutchings M J and De Kroon H 1994 Foraging in plants: the role of morphological plasticity in resource acquisition. Adv. Ecol. Res. 25, 159–238.

Jakobsen I 1986 Vesicular arbuscular mycorrhiza in field-grown crops. 3. Mycorrhizal infection and rates of phosphorus inflow in pea-plants. New Phytol. 104, 573–581.

Jasper D A, Robson A D and Abbott L K 1979 Phosphorus and formation of vesicular-arbuscular mycorrhizas. Soil Biol. Biochem. 11, 501–505.

Jayachandran K, Schwab A P and Hetrick B A D 1992 Mineralisation of organic phosphorus by VAM fungi. Soil Biol. Biochem. 24, 897–903.

Joner E J, Van Aarle I M and Vosatka M 2000 Phosphatase activity of extra-radical arbuscular mycorrhizal hyphae: A review. Plant Soil 226, 199–210.

Jones D L 1998 Organic acids in the rhizosphere – a critical review. Plant Soil 205, 25–44.

Kuo S 1996 Phosphorus. *In* Methods of Soil Analysis. Part 3: Chemical Methods. Ed. D L Sparks. pp. 875–876. Soil Science Society of America Inc., Madison, Wisconsin.

Marschner H 1995 Mineral Nutrition of Higher Plants. 2nd ed. Academic Press, London.

Menge J A, Steirle D, Bagyaraj D J, Johnsson E L V and Leonard R T 1978 Phosphorus concentrations in plants responsible for inhibition of mycorrhizal infection. New Phytol. 80, 575–578.

Murphy J and Riley J P 1962 A modified single solution method for the determination of phosphate in natural waters. Anal. Chim. Acta 27, 31–36.

Olsen S R, Cole C V, Watanabe F S and Dean L A 1954 Estimation of available phosphorus in soils by extraction with sodium bicarbonate. USDA Circular no. 939. 1–19.

Olsson P A, Bååth E and Jakobsen I 1997 Phosphorus effects on mycelium and storage structures of an arbuscular mycorrhizal fungus as studied in the soil and roots by fatty acid signatures. Appl. Environ. Microbiol. 63, 3531–3538.

Phillips J M and Hayman D S 1970 Improved procedures for clearing roots and staining parasitic and vesicular-arbuscular mycorrhizal fungi for rapid assessment of infection. Trans. Brit. Mycol. Soc. 55, 158–162.

Ruzika J and Hansen E H 1981 Flow injection analysis. Chemical Analysis. John Wiley & Sons, New York.

Saleque M A, Abedin M J and Bhuiyan N I 1996 Effects of moisture and temperature regimes on available phosphorus in wetland rice soils. Commun. Soil Sci. Plant Anal. 27, 2017–2023.

Sanders F E 1975 The effect of foliar-applied phosphate on the mycorrhizal infections of onion roots. *In* Endomycorrhizas. Eds. F E Sanders, B Mosse and P B Tinker. pp. 261–276. Academic Press, London.

Sanders F E and Tinker P B 1971 Mechanisms of absorption of phosphate from soil by *Endogone* mycorrhizas. Nature 233, 278–279.

Schilling G, Gransee A, Deubel A, Lezovic G and Ruppel S 1998 Phosphorus availability, root exudates, and microbial activity in the rhizosphere. Z. Pflanzenernähr. Bodenk. 161, 465–478.

Sibbesen E 1978 An investigation of the anion-exchange resin method for soil phosphate extraction. Plant Soil 50, 305–321.

Smith S E and Read D J 1997 Mycorrhizal Symbiosis, 2nd ed. Academic Press, San Diego. pp. 605.

Sokal R R and Rohlf F J 1995 Biometry. 3rd ed. W.H Freeman and Company, New York. pp. 356, 451–539.

Ström L 1997 Root exudation of organic acids: importance to nutrient availability and the calcifuge and calcicole behaviour of plants. Oikos 80, 459–466.

Tabatabai M A and Bremner J M 1969 Use of *p*-Nitrophenyl phosphate for assay of soil phosphatase activity. Soil. Biol. Biochem. 26, 301–307.

Tarafdar J C and Claasen N 1988 Organic phosphorus compounds as a phosphorus source for higher plants through the activity of phosphatases produced by plant roots and microorganisms. Biol. Fertil. Soils 5, 308–312.

Tarafdar J C and Marschner H 1994 Dual inoculation with *Aspergillus fumigatus* and *Glomus mosseae* enhances biomass production and nutrient uptake in wheat (*Triticum aestivum* L.) supplied with organic phosphorus as Na-phytate. Plant Soil 173, 97–102.

Trolove S N, Hedley M J, Caradus J R and Mackay A D 1996 Uptake of phosphorus from different sources by *Lotus pedunculatus* and three genotypes of *Trifolium repens*. 2. Forms of phosphate utilised and acidification of the rhizosphere. Aust. J. Soil Res. 34, 1027–1040.

Tyler G 1974 Heavy metal pollution and soil enzymatic activity. Plant Soil 41, 303–311.

Tyler G 1996 Soil chemistry and plant distribution in rock habitats of southern Sweden. Nordic J. Bot. 16, 609–635.

Tyler G and Olsson P A 1993 The calcifuge behaviour of *Viscaria vulgaris*. J. Veg. Sci. 4, 29–36.

Tyler G and Ström L 1995 Differing organic acid exudation pattern explains calcifuge behaviour of plants. Ann. Bot. 75, 75–78.

Van Aarle I M, Olsson P A and Söderström B 2001 Microscopic detection of phosphatase activity of saprophytic and arbuscular mycorrhizal fungi using a fluorogenic substrate. Mycologia 93, 17–24.

Ziadi N, Simard R R, Tran T S and Allard G 2001 Soil-available phosphorus as evaluated by desorption techniques and chemical extractions. Can. J. Soil Sci. 81, 167–174.

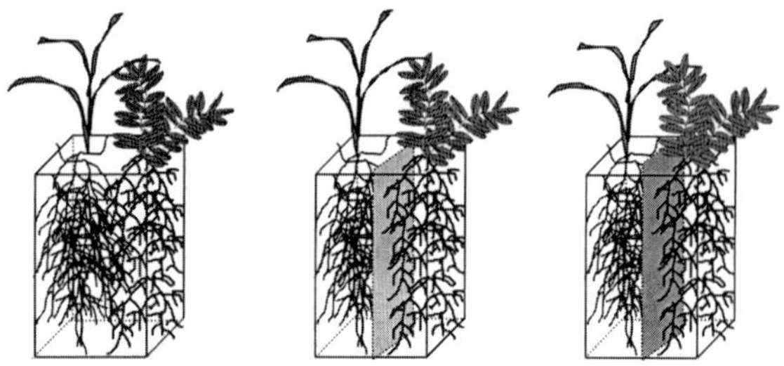

Figure 1. Schematic diagram of separating the root systems of wheat and chickpea by the solid root barrier, nylon mesh and no root barrier.

Interspecific facilitation of *P* uptake by intercropped species in agro-ecosystems has been reported for inorganic *P* sources in a number of studies. For example, white lupin (*Lupinus albus* L.) increased *P* uptake of intercropped wheat via chelation of Ca^{2+} by citrate exuded from roots and subsequent release of *P* from Ca–*P* complexes (Gardner and Boundy, 1983; Horst and Waschkies, 1987). Pigeonpea facilitated *P* uptake by sorghum (Ae et al., 1990), and faba bean facilitated *P* uptake by maize in field conditions (Li et al., 2002).

Organic forms of *P* can account for 50–80% of total soil *P* (Dalal, 1977). Some plant species release large amounts of acid phosphatase and phytase from roots under *P* stress (Li et al., 1997), allowing mobilization and utilization of organic *P* in soil (Grierson and Adams, 2000). In intercropped agro–ecosystems at least two species grow together sharing the same resources. If some species can mobilize organic *P* in soil and leave more inorganic *P* available to the other species, interspecific competition for *P* resources will be reduced (or interspecific facilitation will be enhanced). The objective of the present study was to test whether such interspecific interaction occurred between intercropped wheat and chickpea when supplied with organic (phytate) and sparingly soluble inorganic *P* ($FePO_4$).

Materials and methods

Experiment 1

A factorial experiment, with 4 replicates, was conducted in a glasshouse. The treatments consisted of 2 *P* sources and 3 types of interspecific root contacts between wheat and chickpea. The two *P* sources, phytate ($C_6H_6O_{24}P_6Na_{12}$, SIGMA P-3168) and $FePO_4$(Merck, 1.03923.0500), were added at a rate of 60 mg *P* kg^{-1} soil. Pots were made up of two 2-L juice cartons (25 cm (height) × 10 cm × 10 cm) combined side by side to create two compartments. Wheat plants were grown in one compartment and chickpea in the other. There were three interspecific root contacts between the compartments: (i) a solid root barrier between wheat and chickpea to prevent root interactions, (ii) a 30-μM nylon mesh between wheat and chickpea to prevent direct root contact but permit interactions through mass flow and diffusion, and (iii) no root barrier to allow complete interaction between the root systems of species (Figure 1).

An undisturbed sandy soil (top 0.20-m layer) of virgin brown sand (Uc4.22, Northcote 1971) was collected from a bushland site 15 km south-east of Lancelin, Western Australia (31.56 S, 115.20 E). It contained 2 mg NO_3^-–N, 1 mg NH_4^+–N, 8.4 g organic C, 5 mg $NaHCO_3$–extractable *P* and 21 mg K per kg of soil, and had pH ($CaCl_2$) of 5.70. The soil was air-dried and sieved through a 2-m*M* sieve. Each compartment in the pot was filled with 3.5 kg soil. Basal nutrients in solution were added to soil at the following rates (mg kg^{-1} soil): urea-N (30), K_2SO_4 (133), $CaCl_2{\cdot}2H_2O$ (167), $MgSO_4{\cdot}7H_2O$ (43), $MnSO_4{\cdot}H_2O$ (7), $ZnSO_4{\cdot}7H_2O$ (10), $CuSO_4{\cdot}5H_2O$ (2), H_3BO_3 (0.67) and $Na_2MoO_4{\cdot}2H_2O$ (0.17). After the soil was re-dried, *P* was added in a solid form, and the soil was thoroughly mixed by shaking. Twenty germinated seeds of wheat (*Triticum aestivum* L. cv. Westonia) and nine of chickpea (*Cicer arietinum* L. cv. Sona) were sown in the pot (one species in each compartment). Plants were thinned to 16 per compartment for wheat and to 6 for chickpea 10 days after sowing. Soil was watered daily to field capacity (11%, w/w).

Plants were harvested 42 days after sowing. The whole roots of wheat and chickpea were separated from soil. The root length was measured by a Comair® Root Scanner (Commonwealth Aircraft Corporation Ltd., Melbourne). The dry weight of shoots and roots were measured after plants were dried at 60 °C for 72 h. Phosphorus concentrations in plant tissues were determined using the vanadomolybdate method after plant materials were digested in concentrated nitric acid. Soil pH was measured after soil extraction in 0.01 M $CaCl_2$ by shaking for 17 h at the soil:extractant ratio of 1:5.

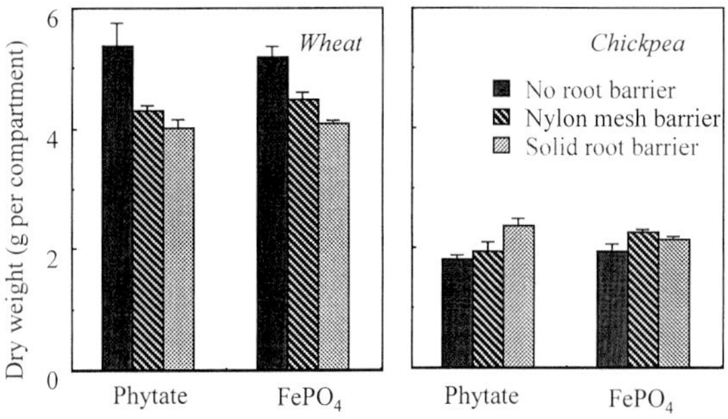

Figure 2. Dry weight of shoots of wheat and chickpea supplied with either phytate or $FePO_4$ and grown in various treatments of root contact between the two species for 42 d. Bars are the standard error of means of four replicates (Experiment 1).

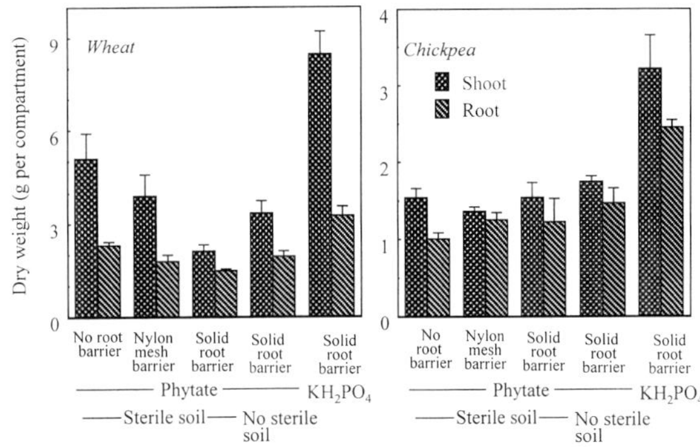

Figure 3. Dry weights of root and shoots of wheat and chickpea grown in a sterilized soil or a non-sterile soil and supplied with either phytate or KH_2PO_4 for 28 d. Bars are the standard error of means of four replicates (Experiment 2).

Experiment 2

The soil was the same as Experiment 1, but was steam-sterilized at 90 °C for 60 min twice in 2 days. There were five treatments: (1) No root barrier with phytate, (2) Nylon mesh separation with phytate, (3) Solid root barrier with phytate, (4) Solid root barrier with KH_2PO_4 as a control, and (5) Solid root barrier with phytate but using non-sterile soil. Thirty mg *P* kg^{-1} soil was applied as phytate or KH_2PO_4. Other procedures were the same as Experiment 1.

Plants were grown for 4 weeks. Root and shoot weights, root length, soil pH and *P* concentration in shoots were measured with the same methods as in Experiment 1.

Statistical analysis

The experiments were set up in the completely randomized design. Statistical significance of differences between treatments was analyzed by analysis of variance (ANOVA) and LSD (least significant deviation) multiple comparison (SAS Institute, 1985).

Results and discussion

Biomass

In Experiment 1, the biomass of wheat was significantly increased by the root intermingling of intercropped wheat and chickpea compared with the treatments with either the root barrier or the nylon mesh, regardless of *P* supply (phytate or $FePO_4$) (Figure 2). However, the biomass of chickpea was decreased by interspecific root interactions (Figure 2). The combined biomass of wheat and chickpea was significantly greater when intermingling of roots was allowed than in the treatment with the nylon mesh or the solid root barrier when *P* was supplied as phytate–*P*. The combined biomass of plants grown with complete (no root barrier) and partial root contacts (nylon mesh) was greater than that with the solid root barrier when *P* was supplied as $FePO_4$–*P*. Intercropping of wheat and chickpea thus exhibited a positive interspecific interaction in growth.

In Experiment 2, the biomass of wheat and chickpea was significantly greater with KH_2PO_4 than with phytate, indicating that *P* supply from phytate was inadequate for optimal plant growth. While soil sterilization reduced the biomass of wheat, it did not reduce biomass of chickpea compared to non-sterilization when roots of the two crops were separated by the solid root barrier. These findings suggested that wheat was more dependent on microorganisms than chickpea for mobilization and utilization of the organic *P* in soil. When chickpea was supplied with phytate, there was no significant difference in shoot biomass of chickpea between root partition treatments or between sterilization and non-sterilization. Shoot biomass of wheat was significantly greater when its roots were intermingled with chickpea roots compared to them being separated from chickpea roots in the sterilized soil. This study demonstrated that the intermingling of chickpea and wheat roots did improve biomass of wheat, and did not reduce biomass of chickpea under sterile condition (Figure 3). Root biomass (Figure 3) and length (data not shown) of wheat and chickpea showed trends similar to those for the shoot biomass.

Phosphorus

In Experiment 1 where phytate–*P* was applied, the root intermingling of wheat and chickpea increased the *P* concentration in wheat, but decreased the *P* concentration in chickpea compared with either the root separation by nylon mesh or the solid root barrier (Table 1). Where inorganic *P* ($FePO_4$) was applied, root separation treatments did not affect *P* concentration in either species (Table 1).

The intermingling of root systems of the two species significantly increased *P* uptake by wheat from both organic and inorganic *P* sources compared to the treatment with the solid root barrier. In addition, *P* uptake by chickpea in the treatment with the solid root barrier and organic *P* source was greater than that in the treatment without a solid root barrier, suggesting that organic *P* mobilized by chickpea probably contributed to *P* uptake by wheat. There was no difference in total *P* uptake by both species between inorganic and organic *P* sources. When phytate–*P* was supplied, however, the intermingling of wheat and chickpea roots significantly increased *P* uptake by the two species (Table 1). Phosphorus concentration and total *P* uptake in wheat plants grown with the inorganic *P* source, in general, were greater than in plants grown with the organic *P* source. However, such a difference was not observed in chickpea (Table 1). The results indicate that wheat is less able to use organic *P* than inorganic *P*, whereas chickpea is able to use both *P* sources to a similar extent.

In Experiment 2, in the treatments with the sterilized soil, *P* concentration in shoots of wheat grown with phytate was greater when roots of the two species intermingled, or were separated by the nylon mesh as compared with the roots of two species being separated by the solid root barrier. In contrast, *P* concentration in chickpea was not affected by root separation. Phosphorus concentration (and uptake) of both species was much higher when supplied with KH_2PO_4 than phytate (Table 2). There was a good relationship between biomass production and *P* concentration in shoot of wheat, irrespective of treatments (Figure 4), confirming that the difference in growth at various treatments was related to *P* nutrition. Phosphorus uptake of wheat was 2.5 and 1.9 times greater with the root intermingling of two species and the root separation by nylon mesh, respectively, than the root separation by the solid root barrier (Table 2). The effect of root separation on wheat *P* uptake was much greater than that in Experiment 1 using a non-sterile soil. Although root separation did not significantly affect *P* uptake by chickpea, total *P* uptake by the two species was increased by 70% with the intermingling and by 39% with nylon-mesh separation between roots of the two species, compared to the solid root barrier (Table 2), implying that an increase in *P* uptake by intercropped wheat was a consequence of chickpea effects. Furthermore, total *P* uptake by wheat and chickpea was less in sterilized than in non-sterilized soil.

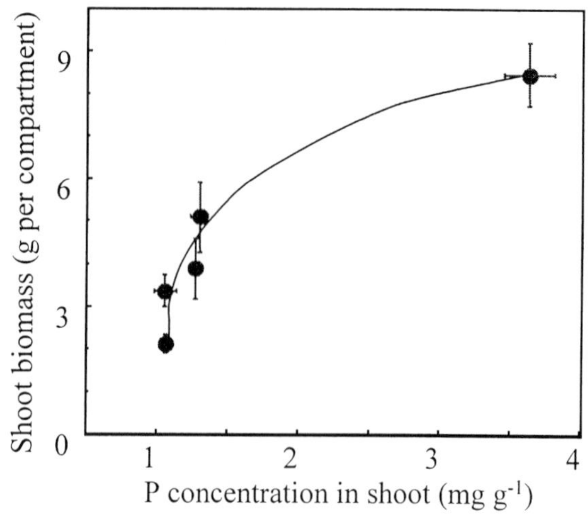

Figure 4. Relationship between wheat shoot biomass production and shoot *P* concentration in various treatments after 28 d. Vertical and horizontal bars are the standard errors of means (n=4) for shoot biomass and *P* concentration, respectively (Experiment 2).

In another study, Rao et al. (1999) found that acquisition of *P* by the legume, *Arachis pintoi* CIAT 17434, was markedly greater than that of the grass, *Brachiaria dictyoneura* CIAT 6133, regardless of the *P* supply in inorganic or organic form. Richardson et al. (2000) found that a low intrinsic level of phytase activity in wheat roots was a critical factor limiting the ability of wheat to obtain *P* from phytate when grown in agar under non-limiting conditions. Similar interspecific differences in *P* source utilization were observed between sorghum and pigeonpea intercropping, where pigeonpea was more efficient in utilizing iron-bound *P* in an Alfisols compared with sorghum, soybeans, *Panicum americanum* and maize (Ae et al., 1990).

Organically-bound *P* has to be hydrolyzed before plant roots can absorb *P*. Both microbes and plant roots possess phosphatases that could be of importance, especially in soils with low concentrations of inorganic *P* in the soil solution. In the present study,

Table 1. Phosphorus concentration in shoot (g P kg^{-1} d.wt.) and uptake by the whole plant (mg P per compartment) of wheat and chickpea grown with two P sources and different intercropping treatments (Experiment 1)

Treatment		Wheat		Chickpea		Total P uptake by two species (mg pot^{-1})
		Concentration	Uptake	Concentration	Uptake	
Phytate	Solid root barrier	1.88b[1]	9.6b[1]	1.96a	6.6a	16.2b
	Nylon mesh	1.92b	9.7b	1.80ab	5.7ab	15.4b
	No root barrier	2.88a	14.6a	1.51b	5.2b	19.8a
	Average[2]	*2.22*	*11.3*	*1.76*	*5.8*	*17.9*
$FePO_4$	Solid root barrier	2.51a	12.6ab	1.54a	5.2a	17.8a
	Nylon mesh	2.26a	11.5b	1.67a	5.8a	17.3a
	No root barrier	2.51a	13.2a	1.70a	5.4a	18.5a
	Average	*2.42*	*12.4*	*1.63*	*5.4*	*17.1*

[1]Mean values in each column of P source followed by the same letter are not significantly different at the 0.05 level.
[2]Mean values for three intercropping treatments with the same P source.

Table 2. Phosphorus concentration in shoot (g P kg^{-1} d.wt.) and uptake by the whole plant (mg P per compartment) of wheat and chickpea grown with phytate or KH_2PO_4 in sterilized and non-sterile soils (Experiment 2)

Treatment		Sterilised	P concentration		P uptake		Total P uptake by two species (mg pot^{-1})
			Wheat	Chickpea	Wheat	Chickpea	
Phytate	No root barrier	Yes	1.30	1.05	8.23	2.61	10.84
Phytate	Nylon mesh	Yes	1.27	1.08	6.18	2.67	8.84
Phytate	Solid root barrier	Yes	1.07	1.01	3.23	3.21	6.44
KH_2PO_4	Solid root barrier	Yes	3.63	2.14	34.84	11.96	46.81
Phytate	Solid root barrier	No	1.06	1.00	4.63	3.26	7.89
LSD (P=0.05)[1]							
P treatment			*0.22*		*1.49*		*2.26*
Species × *P treatment*			*0.31*		*2.11*		

[1]Effects of P treatment and species, and the interaction are significant at P<0.001.

P nutrition of wheat receiving organic P was significantly improved by the intermingling of root systems with the chickpea root system. One possible mechanism was that chickpea could hydrolyze organic phosphorus by phosphatase released from roots (Gupta, 1998). As a result, chickpea used more organic P in soil and left more inorganic P to wheat, or the hydrolyzed P might be absorbed directly by wheat. This might explain why root intermingling enhanced P uptake by the chickpea and wheat association when supplied with organic P, especially in the sterilized soil. It is concluded that intercropping species inefficient in utilizing organic P with species efficient in utilizing this P source could improve P nutrition of inefficient plant species. This has an important implication for advantages of intercropping in agroecosystems as well as co-existence of species in plant communities.

Specific absorption rate of phosphorus

The specific absorption rate of phosphorus (SARP) by wheat in the treatment without the solid root barrier was significantly increased compared to that with the nylon mesh or the solid root barrier when plants were supplied with organic P. In contrast, there was no difference in SARP by wheat regardless of the intercropping treatment when wheat and chickpea received inorganic P (Table 3).

There was no significant difference in SARP by chickpea receiving organic P regardless of the root separation treatment. Root intermingling of wheat and chickpea increased SARP by chickpea receiving inorganic P. In addition, there were no significant differences in SARP for wheat and chickpea between organic and inorganic P supplies (Table 3).

Plant and Soil **248:** 305–312, 2003.
© 2003 *Kluwer Academic Publishers. Printed in the Netherlands.*

Using competitive and facilitative interactions in intercropping systems enhances crop productivity and nutrient-use efficiency

Fusuo Zhang[1] & Long Li
Department of Plant Nutrition, China Agricultural University, Key Laboratory of Plant-Soil Interactions, Ministry of Education, Key Laboratory of Plant Nutrition, Ministry of Agriculture, Beijing 100094, P. R. China. [1]*Corresponding author*

Received 18 December 2001. Accepted in revised form 12 March 2002

Key words: intercropping, interspecific competition, interspecific facilitation, recovery, yield advantage

Abstract

This paper reviews recent research on the processes involved in the yield advantage in wheat (*Triticum aestivum* L.)/maize (*Zea mays* L.), wheat/soybean [*Glycine max* (L.) Merr.], faba bean (*Vicia faba* L.)/maize, peanut (*Arachis hypogaea* L.)/maize and water convolvulus (*Ipomoea aquatica* Forsk.)/maize intercropping. In wheat/maize and wheat/soybean intercropping systems, a significant yield increase of intercropped wheat over sole wheat was observed, which resulted from positive effects of the border row and inner rows of intercropped wheat. The border row effect was due to interspecific competition for nutrients as wheat had a higher competitive ability than either maize or soybean had. There was also compensatory growth, or a recovery process, of subordinate species such as maize and soybean, offsetting the impairment of early growth of the subordinate species. Finally, both dominant and subordinate species in intercropping obtain higher yields than that in corresponding sole wheat, maize or soybean. We summarized these processes as the 'competition-recovery production principle'. We observed interspecific facilitation, where maize improves iron nutrition in intercropped peanut, faba bean enhances nitrogen and phosphorus uptake by intercropped maize, and chickpea facilitates P uptake by associated wheat from phytate-P. Furthermore, intercropping reduced the nitrate content in the soil profile as intercropping uses soil nutrients more efficiently than sole cropping.

Introduction

In China, one-third of all the cultivated land area is used for multiple cropping and half of the total grain yield is produced with multiple cropping (Tong, 1993). Intercropping, as one of the multiple cropping systems, has been practiced by farmers for many years in various ways and most areas, and has played a very important role in China. Most studies on intercropping have focused on the legume-cereal intercropping, a productive and sustainable system, its resource utilization (water, light, nutrients), and its effect on N input from symbiotic nitrogen fixation into the cropping system and reduction of negative impacts on the environment (Jensen, 1996; Willey, 1979). However, a cereal–cereal association, e.g., wheat (*Triticum aestivum* L.)/maize (*Zea mays* L.) intercropping where two crops grow together for about 70 to 80 days, with a yield of more than 12,000 kg ha^{-1}, has become increasingly popular in the irrigated area of the Hexi Corridor in the Gansu province, irrigated areas along the Huanghe river in Ningxia, and in Inner Mongolia Autonomous regions, China. The area under wheat/maize intercropping was 75,100 ha in Ningxia in 1995, producing 43% of the total grain yield for the area. In Gansu the area is 200,000 ha annually. Wheat/soybean [*Glycine max* (L.) Merr.] and faba bean (*Vicia faba* L.)/maize intercropping systems are popular in irrigated areas of northwest China, and peanut (*Arachis hypogaea* L.)/maize intercropping in the northern part of China. There

* FAX No: +86-10-6289-1016
E-mail: zhangfs@mail.cau.edu.cn

have been few published research publication on interspecific interactions, especially interspecific below-ground interactions between intercropped species in wheat/maize, wheat/soybean, peanut/maize and faba bean/maize intercropping systems. In the past 10 years, many studies, including field trials in Gansu and Henan provinces and greenhouse experiments, were conducted to determine interspecific below-ground interactions between intercropped species. This paper reviews the main results and progress of our research works along with closely related work done elsewhere.

A significant contribution of below-ground interactions to border-row yields and nutrient uptake by intercropped wheat

In intercropping systems, two or more crops grow simultaneously on the same field such that the period of overlap is long enough to include the vegetative stage (Gomes and Gomez, 1983). In maize/soybean strip intercropping, West and Griffith (1992) observed a 26% increase in maize yield and a 27% reduction in soybean yield of border rows located at the outside of 8-row alternating strips in Indiana, USA. Ghaffarzadeh et al. (1994) found that strip intercropping led to 20–24% greater maize yields and 10–15% lower soybean yields in adjacent border rows in a maize/soybean intercropping system in Iowa, USA. In our study, yield advantage of intercropped wheat resulted from both the border-row and inner-row effect in wheat/maize and wheat/soybean intercropping systems (Li et al., 2001a). The yields in border rows of intercropped wheat were significantly higher than those in inner rows, both in wheat/maize and wheat/soybean intercropping systems (Figure 1). The 74% increase of grain yield in wheat intercropped with maize, 47 and 27% was attributed to above-ground and below-ground interaction, respectively. The 53% increase of grain yield in wheat intercropped with soybean, was due to a 30 and 23% above- and below-ground interaction between the two species, respectively. The contributions of above-ground and below-ground interactions to the increase of N uptake by wheat were 50 and 59%, respectively, for wheat/maize intercropping, and 23 and 19%, respectively, for wheat/soybean intercropping. Similarly, the contribution of above-ground and below-ground interactions to the increase of P uptake were 56 and 42%, respectively, for wheat/maize intercropping, and 26 and 28%, respectively, for wheat/soybean intercropping (Zhang et al., 2001). Interspecific below-ground interactions and rhizosphere effects between intercropped species played an important role in yield advantage of intercropping.

Competitive – recovery production principle

Yield advantage of dominant species from interspecific competition

Interspecific competition may occur when two crops are grown together (Van der Meer, 1989). Such competition usually decreases survival, growth or reproduction of at least one species (Crawley, 1997). In the wheat/maize or wheat/soybean intercropping systems in north China, there is a 70 to 80-day overlapping growth period that causes intense interspecific interactions between the intercropped species. The interactions frequently occurred at the interface between two crop species where they were nearest in distance, and resulting in an increase or decrease of growth, development and even yields. The yield advantage in the border row of intercropped wheat probably derived from the differences in interspecific competitiveness, and wheat proved to be more competitive than maize and soybean during overlapping growth periods in the wheat/soybean and wheat/maize intercropping systems. Aggressivity measures the interspecies competition in intercropping by relating the yield changes of the two component crops (Willey and Rao, 1980).

$$A_{ab} = Y_{ia}/(Y_{sa} \times F_a) - Y_{ib}/(Y_{sb} \times F_b), \qquad (1)$$

where Y_{ia} and Y_{ib} are yields of crops a and b in intercropping, Y_{sa} and Y_{sb} are yields of crops a and b in sole cropping. F_a and F_b are the proportion of the area occupied by crop a and crop b in the intercropping. When A_{ab} is greater than 0, the competitive ability of crop a exceeds that of crop b in intercropping.

In a study by Li et al. (2001a), aggressivities of wheat relative to either maize (0.26–1.63 of A_{wm}) or soybean (0.35–0.95 of A_{ws}) revealed the greater competitive ability of wheat compared with either maize or soybean. The nutrient competitive ratio, 1.09–7.54 for wheat relative to maize, and 1.2–8.3 for wheat relative to soybean, showed that wheat had a greater capability of acquiring nutrients compared with soybean and maize. However, the mechanism of the difference in competitive abilities between intercropped species is rarely analyzed. The determination of the mechanism will be helpful in controlling the interspecific

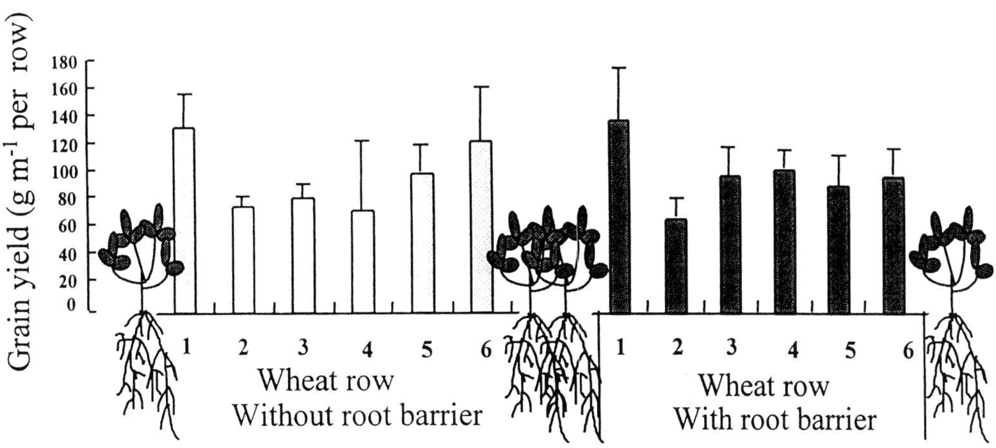

Figure 1. Effect of interspecific root interaction on border row yield of intercropped wheat in a wheat/soybean intercropping system, six rows of wheat and two rows of soybean. Row spaces for wheat and soybean were 0.12 and 0.20 m, respectively. Bars are the standard error of means. The experimental site (37°05′N, 104°40′E) is located in Jingyuan county, Gansu Province, at 1645 m above sea level. Annual mean temperature is 6.6 °C. Annual precipitation is 259 mm, and potential evaporation is 2369 mm. This site is on the edge of a loess plateau. The soils at the site are classified as Aridisol.

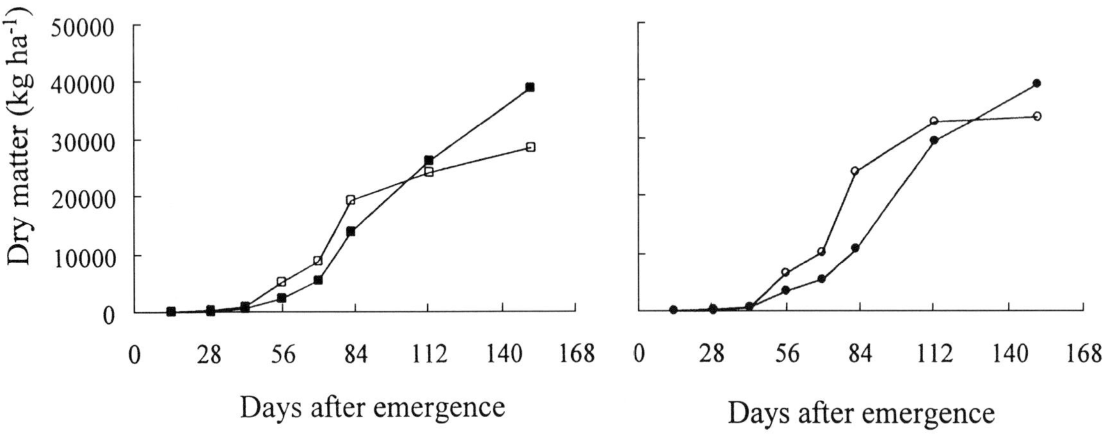

Figure 2. The difference in dry matter accumulation between intercropped maize (IM) and sole maize. (■, intercropped maize without P fertilizer, □, sole maize without P fertilizer, ●, intercropped maize receiving P fertilizer at 53 kg P ha^{-1}, and ○, sole maize receiving P fertilizer at 53 kg P ha^{-1}). The experimental site (38°37′N, 102°40′E) is located 15 km north of Wuwei City, Gansu Province at 1504 m above sea level. Annual mean temperature is 7.7 °C. Annual precipitation is 150 mm, and potential evaporation is 2021 mm. The soils at the site are classified as Aridisol.

competition and in gaining higher yield advantage of intercropping.

Recovery or compensation of subordinate species at a late growth stage

In the wheat/maize or wheat/soybean intercropping, at wheat harvesting, the growth and nutrient uptake by the subordinate species maize or soybean are reduced due to interspecific competition. There is recovery of nutrient uptake and growth after harvest of the earlier-maturing species, which made the later-maturing species compensate for impaired early growth once the early-maturing species was harvested. In a field trial conducted at the Baiyun site in 1997, the rates of dry matter accumulation in the intercropped maize (10–20 g m^{-2} day^{-1}) were significantly lower than those in the sole maize (17–35 g m^{-2} day^{-1}) during the early stage from 7 May to 3 August, while mostly intercropped with wheat. After 3 August, however, the rates of dry matter accumulation for intercropped maize, increasing from 59 to 70 g m^{-2} day^{-1}, were significantly greater than in sole maize (23–52 g m^{-2} day^{-1}) (Figure 2); nutrient acquisition showed the same trend as growth. At the Jingtan site in 1998, the disadvantage of the border row of intercropped maize resulting from interspecific competition diminished after wheat harvest, and had disappeared at maize maturity (Li et al., 2001b). However, the recovery was limited to the treatment without N and P fertilizer application (Li et al., 2001b). The recovery needs further study in understanding the mechanism involved, which will contribute to understanding yield advantage of intercropping where short-season associated with long-season species.

Interspecific below-ground facilitation – nutrients mobilized by efficient species contribute to inefficient species

Maize improved iron nutrition of peanut

Peanut (*Arachis hypogaea* L.), a major oilseed crop in China, accounts for 30% of the cropped area and 30% of the total oilseed production in the country. However, iron deficiency chlorosis frequently occurs on calcareous soils, especially in north China. Interestingly, iron deficiency chlorosis of the crop is more severe in sole cropping systems than in the intercropping system of peanut/maize on these soils. There is, therefore, considerable interest in devising practical approaches for the correction or prevention of Fe deficiency in crops in Chinese agriculture. The results demonstrated that the improvement in the Fe nutrition of peanut intercropped with maize was mainly caused by rhizosphere interactions between peanut and maize (Zuo et al., 2000).

The severity of iron deficiency chlorosis in young peanut leaves in the intercropping systems was closely related to the distance of the peanut plants from the maize roots when treatments were assessed during the peanut flowering period. In the unrestricted intercropping treatment, where neighboring roots of peanut and maize intermingled fully, the young leaves of peanut plants in rows 1–3 from the maize grew without visible symptoms of iron deficiency, while those in rows 5–10 showed variable degrees of chlorosis. In the treatment where the roots were physically separated between peanut and maize, the peanut plants in all rows were chlorotic, albeit that most of the peanut plants in the row closest to maize showed only slight signs of chlorosis (80% of young leaves were green). In contrast, in the treatment where a 400 mesh nylon mesh separation (nominal aperture 37 μm) was inserted into the ground between the two crop species to prevent direct root contact but allow interactions through mass flow and diffusion, the plants in rows 1 and 2 remained completely green, while those in rows 3–10 were chlorotic to varying extents. In the monocropping treatment, about 90% of the peanut plants showed severe Fe deficiency chlorosis in the young leaves.

Peanut is a species with strategy I for iron uptake and frequently showed Fe deficiency chlorosis on calcareous soil, whereas maize with strategy II is an Fe efficient species (Marschner, 1995). When peanut and maize grow together, phytosiderphore released from maize roots may mobilize Fe(III) and benefit the iron nutrition of peanut plant. However, there is no direct evidence to support this contention.

Maize benefits from associated faba bean

Interspecific root interactions between intercropped faba bean and maize played an important role in the yield advantage and nitrogen and phosphorus acquisition by the intercropping system (Li et al., 1999, 2002a). When the roots of two species intermingled, the land equivalent ratio (Wiley, 1979), N uptake and P uptake were 1.21, 38.4 g N m^{-2} and 4.2 g P m^{-2}, respectively. However, when the roots of the two species were separated completely, they were reduced to 1.06, 31.8 g N m^{-2} and 3.3 mg P m^{-2}, respectively (Li et al., 2002a). This study, however, does not provide information on the mechanism of increased N and P uptake. Possible mechanisms for interspecific facilitation are: (1) the phosphorus that is mobilized by faba bean may have been made available for maize; (2) nitrogen that was fixed by faba beans may have been transferred to maize. More research, such as on the difference in root distribution, in the ability to mobilize sparingly soil P between faba bean and maize, needs to be done.

Chickpea-facilitated P nutrition in associated wheat

In a pot experiment with no root barrier, a nylon mesh root barrier, and a solid root barrier, biomass of wheat was significantly increased by the root intermingling of intercropped wheat and chickpea (*Cicer arietinum* L.) compared with the treatments with either the solid root barrier or the nylon mesh, regardless of whether P was supplied as phytate or $FePO_4$.

Where phytate was applied, the root intermingling of wheat and chickpea increased the P concentration in wheat and P uptake by wheat, compared with the treatment in which the roots were separated by nylon mesh or a solid barrier. In addition, P uptake by chickpea in the treatment with the root barrier and an organic P source was greater than that in the treatment without root barrier, suggesting that organic P mobilized by chickpea probably contributed to P uptake by wheat. In another study, Rao et al. (1999) found that acquisition of P by the legume was markedly greater than that by the grass, regardless of the P form being inorganic or organic. The results indicate that wheat is less able to use organic P than inorganic P, whereas chickpea is able to use both P sources equally effectively. When wheat and chickpea were grown together, the following may have happened. Firstly, chickpea can mobilize and absorb some organic P by releasing phosphatase into soil, and also leave some inorganic P for wheat. Secondly, wheat with a greater competitive ability acquires more P from the root zone of both wheat and chickpea, resulting in P depletion in the chickpea rhizosphere (Li et al., 2002b). The P depletion in the chickpea rhizosphere then induces the release of substance that mobilizes organic P. Therefore, furthermore study should focus on the difference in phytase and phosphatase between chickpea and wheat.

Intercropping reduced nitrate accumulation in soil profile

Wheat/maize and faba bean/maize intercropping

There has been a rapid increase of fertilizer application in recent years to achieve high yields, especially in the wheat/maize intercropping systems. The average rate of N application in China is 180 kg N ha^{-1} $year^{-1}$. In the irrigated area of Northwest China, application rates of N fertilizers average 450 kg N $ha^{-1}year^{-1}$ for wheat/maize intercropping, which increases the risk of ground-water nitrate pollution. Applying N up to the economic optimum rate (estimated from yield and N rate data from individual trials) was associated with small increases in soil nitrate-N after harvest (the mean increase was 4 kg N ha^{-1} $year^{-1}$). Where the optimum N rate was exceeded, soil nitrate-N levels increased to a greater extent (Richards et al., 1996). Intercropping faba bean with oats or spring wheat could reduce the nitrate accumulation in soil profiles (Stuelpnagel, 1993). Intercropping maize with ryegrass was also an effective way for increasing N uptake under conditions of high N application (Zhou et. al., 2000). Nitrogen uptake by intercropping of wheat and maize was greater than that by corresponding sole cropping under same N supply (Li et al., 2001a). Therefore, we propose that wheat and maize intercropping could reduce the nitrate accumulation in soil compared with sole maize and wheat. In an irrigated area of Hexi corridor, northwest China, the amounts of NO_3^- in the soil profile after wheat harvest was greater for sole wheat than for wheat intercropped with maize, and was greater for sole faba bean than for faba bean intercropped with maize supplied with 200 and 400 kg N ha^{-1} of application. The decrease was about 0–41% for wheat and 0–31% for faba bean. Amounts of NO_3^- in soil after maize harvest was in the order of sole wheat and faba bean, intercropping wheat and faba bean, maize intercropped with faba bean and wheat. The results indicate that intercropping decreases the accumulation of nitrate in soil profiles.

Water convolvulus/maize intercropping

A field study was conducted in Beijing, China to test the hypothesis that intercropping could reduce nitrate in the soil profile. There were three treatments: (1) sole maize with 0.40 m row space, (2) sole water convolvulus (*Ipomoea aquatica* Forsk.) with 0.20 m row space, and (3) water convolvulus/maize intercropping with 2 rows of vegetable with 0.20 m row space and 2 rows of maize with 0.40 m row space. The results demonstrated that nitrate in the soil profile (0–1 m) where grew both vegetable and maize was reduced by intercropping (Figure 3).

Interspecific competition and facilitation are two aspects of the same interaction

Geno and Geno (2001) concluded that interspecific competition and facilitation occurs at the same time. Van der Meer (1989) noted that both competition and facilitation take place in many intercropping systems, and that it is possible to obtain the net result of land equivalent ratio (LER), an indicator of intercropping advantage, >1 where the complementary facilitation is contributing more to the interaction than the competitive interference. Thus, an LER>1 could result from low interspecific competition or strong facilitation. A pot experiment was conducted to investigate the interspecific competition and facilitation in a wheat/soybean association. The pots were separated into two compartments either by a plastic sheet to eliminate root contact and any solute movement, or by a nylon mesh (30 μm) to prevent root contact but permitting solute exchange between the compartments. In addition, there was a control treatment without any separation. Phosphorus was added at 0, 150, and 300 μg P g^{-1} soil as KH_2PO_4. Wheat

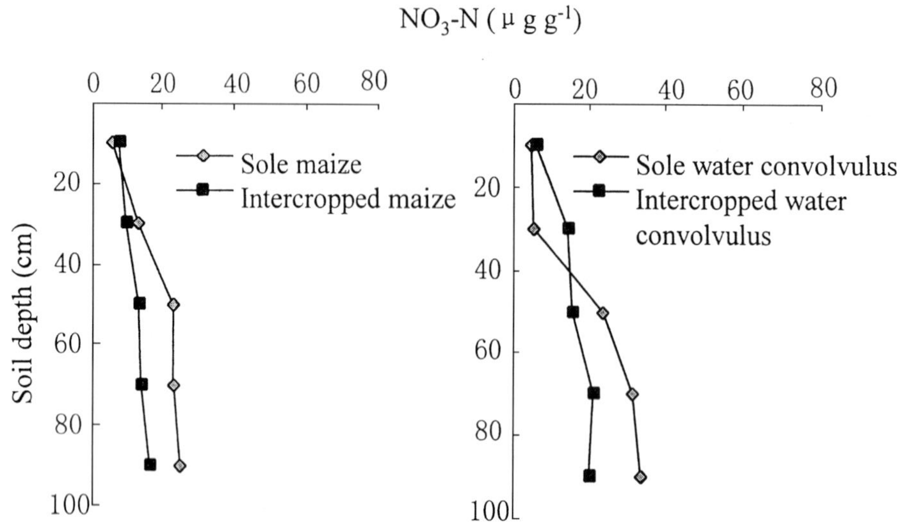

Figure 3. Nitrate concentrations in the soil profile after water convolvulus and maize harvesting (October 10, 2001) in an intercropping system and a corresponding sole cropping system. Experimental site (39°56′N, 116°17′E), located in the campus of China Agricultural University, Beijing at 54 m above sea level. Annual mean temperature is 11.8 °C. Annual precipitation is 577 mm.

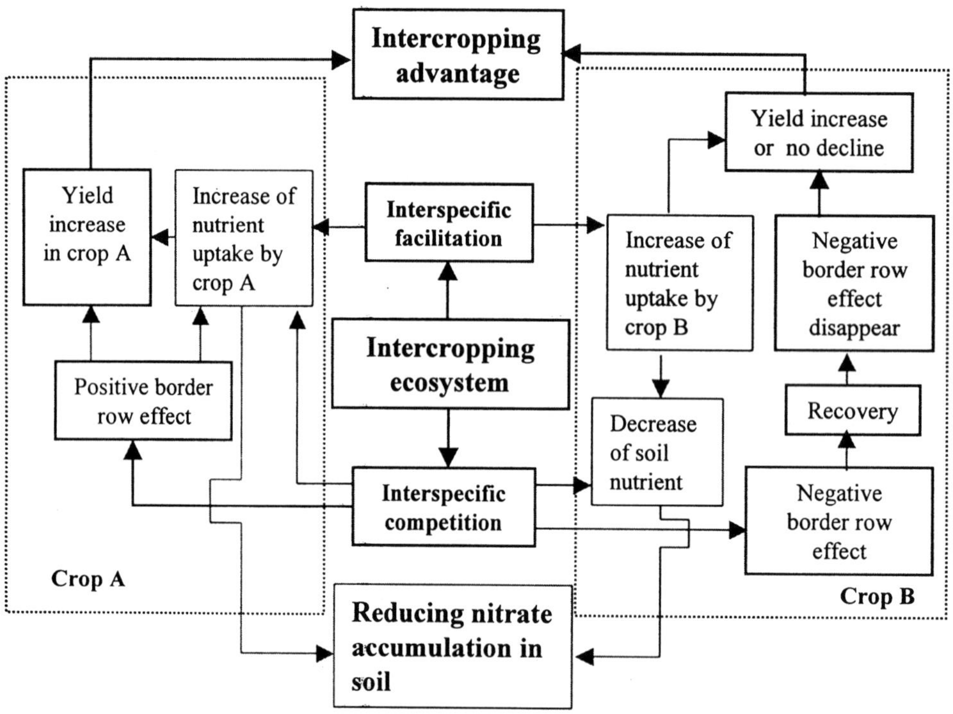

Figure 4. The relationship between yield advantage of intercropping and interspecific competition, facilitation, recovery and soil nitrate content (modified from Li et al., 2001b).

plants were grown in one compartment and soybean in the other. The obvious facilitation of wheat to phosphorus uptake by soybean was observed in the pot with the mesh separation, where competition of wheat for phosphorus was eliminated. The biomass and P uptake of wheat in the absence of a separation were increased by 25 and 7% compared with those with the plastic sheet partition. The biomass and phosphorus uptake of soybean at the nylon mesh separation treatment were 150 and 134% higher, respectively, than those with the plastic sheet separation. In the pots with the nylon net separation, more roots of soybean were distributed near the nylon mesh on which wheat formed a dense root mat at the other side. In contrast, soybean roots were distributed more or less evenly in the whole compartment in the treatment with the plastic sheet. It is suggested that the beneficial effect of soybean resulted from a rhizosphere effect. At the same time, biomass and P uptake of soybean were significantly reduced by the intermingling of roots of the two species, indicating that there was an interspecific competition between wheat and soybean P. This study also proved that interspecific competition and facilitation existed together in the wheat/soybean association.

Conclusion

A 'competition-recovery production principle' is proposed based on several years of studies on intercropping. It means interspecific interaction increases growth, nutrient uptake and yield of dominant species, but decreases growth and nutrient uptake of the subordinate species during the co-existence stage of two crop species. After the dominant species is harvested, the subordinate species has a recovery or complementary process so that the final yields remain unchanged or even increase compared with corresponding sole species. The principle is suitable to intercropping of short-season/long-season species. Interspecific competition and facilitation exist simultaneously. Interspecific facilitation has been observed in some intercropping systems. These include maize improving iron nutrition in peanut, faba bean enhancing nitrogen and phosphorus uptake by maize, and chickpea-facilitated P uptake by associated wheat. Yield advantage of intercropping and its relationship to interspecific interactions are summarized in Figure 4.

Acknowledgements

We thank Prof. H. Lambers for his support for F. S. Zhang to attend the Workshop on Structure and Functioning of Cluster (proteoid) Roots and Plant Responses to Phosphate Deficiency. The authors are also grateful to Dr. Tang for his comments on a draft of this paper. The Major State Basic Research Development Programme (Project No. G1999011707) and The Major Research programme of Education of China (Project No. 0112) are acknowledged for fiancial support.

References

Crawley M J 1997 Plant Ecology. Blackwell Science, Cambridge.

Geno L and Geno B 2001 Polyculture Production – Principles, Benefits and Risks of Multiple Cropping Land Management Systems for Australia. A report for the rural industries research and development corporation. CIRDC Publication No 01/34.

Ghaffarzadeh M, Prechac F G and Cruse R M 1994 Grain yield response of corn, soybean, and oat grown in a strip intercropping system. Am. J. Altern. Agric. 9, 171–177.

Gomez A A and Gomez K A 1983 Multiple cropping in the Humid Tropics of Asia, IDRC 176E, IDRC, Ottawa.

Jensen E S 1996 Grain yield, symbiotic N_2 fixation and interspecific competition for inorganic N in pea-barley intercrops. Plant Soil 182, 25–38.

Li L, Sun J H, Zhang F S, Li X L, Yang S C and Rengel Z 2001a Wheat/maize or wheat/soybean strip intercropping. I. Yield advantage and interspecific interactions on nutrients. Field Crops Res. 71, 123–137.

Li L, Sun J H, Zhang F S, Li X L, Rengel Z and Yang S C 2001b Wheat/maize or soybean strip intercropping. II. Recovery or compensation of maize and soybean after wheat harvesting. Field Crops Res. 71, 173–181.

Li L, Yang S C, Li X L and Zhang F S and Christie P 1999 Interspecific complementary and competitive interaction between intercropped maize and faba bean. Plant Soil 212, 105–114.

Li L, Zhang F S, Li X L, Christie P, Sun J H, Yang S C and Tang C 2002a Interspecific facilitation of nutrient uptakes by intercropped maize and faba bean. Nutr. Cycl. Agroecosyst. (in press).

Li L, Tang C, Rengel Z and Zhang F S 2002b Chickpea facilitates phosphorous uptake by intercropped wheat from an organic phosphorus source. Plant Soil 248, 297–303.

Marschner H 1995 Mineral Nutrition of Higher Plants. Academic Press, London.

Rao I M, Borrero V, Ricaurte J and Garcia R 1999 Adaptive attributes of tropical forage species to acid soils. V. Differences in phosphorus acquisition from less available in organic and organic sources of phosphate Plant Nutri. 22: 1175–1196.

Richards I R, Wallace P A and Paulson G A 1996 Effects of applied nitrogen on soil nitrate-nitrogen content after harvest of winter barley. Fert. Res. 45, 61–67.

Stuelpnagel R 1993 Intercropping of faba beans (*Vicia faba L.*) with oats or spring wheat. *In* Proceedings of the International Crop Science Congress, 14–44, July, 1992. Iowa State University, Ames, Iowa.

Tong P Y 1994 Achievements and perspectives of tillage and cropping systems in China (in Chinese), Cropping System and Cultivation Technology (Genzuo Yu Zaipei). 77, 1–5.

Van der Meer J 1989 The Ecology of Intercropping. Cambridge University Press, New York.

West T D and Griffith D R 1992 Effect of strip-intercropping corn and soybean on yield and profit. J. Prod. Agric. 5, 107–110.

Willey R W 1979 Intercropping – its importance and research needs. Part 1. Competition and yield advantages. Field Crop Abstr. 32, 1–10.

Willey R W and Rao M R 1980 A competitive ratio for quantifying competition between intercrops. Exp. Agric. 16, 117–125.

Zhang F S, Li L and Sun J H 2001 Contribution of above- and below-ground interactions to intercropping. *In* Plant Nutrition – Food Security and Sustainability of Agro-ecosystems. Ed. Horst et al. pp. 979–980. Kluwer Academic Publishers, Doirdrecht.

Zhou X M, Chandra A M, Angus F M, Kaluli J W and Smith D L 2000 Corn yield and fertilizer N recovery in water-table-controlled corn-rye-grass. Eur. J. Agron. 12, 83–92

Zuo Y M, Zhang F S, Li X L and Cao Y P 2000 Studies on the improvement in iron nutrition of peanut by intercropping with maize on a calcareous soil. Plant Soil 220, 13–25.

Plant and Soil **248:** 313–320, 2003.
© 2003 *Kluwer Academic Publishers. Printed in the Netherlands.*

Phosphorus uptake by a community of arbuscular mycorrhizal fungi in jarrah forest

M. Zakaria Solaiman[1] & Lynette K. Abbott[2,3]
[1]*Soil Science and Plant Nutrition, School of Earth and Geographical Sciences, The University of Western Australia, Crawley, 6009, Australia.* [2]*Soil Science and Plant Nutrition, School of Earth and Geographical Sciences, The University of Western Australia, Crawley, 6009, Australia.* [3]*Corresponding author**

Received 18 January 2002. Accepted in revised form 17 May 2002

Key words: arbuscular mycorrhizas, community, forest, morphotypes, phosphate, soil

Abstract

Communities of indigenous arbusuclar mycorrhizal (AM) fungi are expected to alter phosphorus uptake and biomass productivity of plants according to characteristics of the life cycles of the fungi present and the way they interact with each other inside roots and with host plants. Differences in the relative abundance of AM fungi inside roots could influence P uptake if the fungi present differ in effectiveness at accessing P and transferring it to the plant. However, it is difficult to assess the contribution of AM fungi under field conditions. We investigated P uptake, from point sources of P placed 2, 4 and 6 cm from roots, by plants colonised by a community of AM fungi in jarrah forest soil. Roots were retained within a mesh bag to prevent them from growing towards the point source of P. The relative abundance of morphotypes of fungi inside roots and the P status of plants were assessed after 12 and 16 weeks. First, a bioassay was carried out in undisturbed forest soil cores using two host plants, a forest understorey plant *Phyllanthus calycinus* Labill and the annual pasture species subterranean clover (*Trifolium subterraneaum* L.), to assess the infectivity of the indigenous community of AM fungi. Roots of both bioassay host plants were colonised in similar proportions by morphotypes of AM fungi resembling *Glomus*, *Acaulospora*, *Scutellospora* and fine endophytes. In this bioassay, there were positive correlations between the proportion of root length colonised and plant biomass and P uptake for *P. calycinus*, but not for subterranean clover. In the experiment assessing the capacity of *P. calycinus* to access P placed at increasing distances from the root, shoot P content and concentration in *P. calycinus* were greater when P was placed 2 cm compared with 4 and 6 cm from roots. The length of hyphae in the vicinity of the point source of P decreased with increasing distance from the plant. The extent to which the individual AM fungi were involved in P uptake is not known. The *Glomus* morphotype was dominant at both times of sampling.

Introduction

Jarrah (*Eucalyptus marginata* Donn ex Smith) forest is unique to the south-western corner of Western Australia. The ecosystem has a diverse flora that is strongly influenced by topography, available moisture, soil acidity, and the limited availability of nutrients such as P and N (Bell and Heddle, 1989). The soil typically has a high phosphate-fixing capacity and the infectivity of AM fungi in this soil can influence mycorrhiza response depending on the amount of P added to the soil and the time the soil is collected from this area (Bolan and Abbott, 1983; Bolan et al., 1983).

Arbuscular mycorrhizal (AM) associations are assumed to be important in soil with low chemical fertility, such as that in the jarrah forest. Research carried out on the abundance and diversity of AM fungi in this natural ecosystem near Jarrahdale (Brundrett and Abbott, 1994, 1995) demonstrated seasonal variation in inoculum levels and considerable heterogeneity in distribution of AM fungi in soil. Several techniques

* FAX No. +61-8-9380-1050.
E-mail: labbott@cyllene.uwa.edu.au

have been used to isolate the AM fungi that are dominant within roots of plants grown in this region (based on methods reported by Brundrett et al., 1999). Representatives of AM fungi established in pot culture included a *Glomus* sp. for which no spores have been found (F. Jayasundara, pers. comm.). Three dominant morphotypes of AM fungi found in roots at this site were from the genera *Glomus*, *Acaulospora* and *Scutellospora* (Brundrett and Abbott, 1994). Additional morphotypes resembling another *Glomus* sp., another *Acaulospora* sp. and fine endophytes have also been observed (F. Jayasundara, pers. comm., based on descriptions of Abbott (1982)). Fine endophyte colonisation may have been formed by more than one species, but this could not be confirmed (Thippayarugs et al., 1999). This represents a low diversity of species of AM fungi (Bever et al., 2001; Cuenca. et al., 1998; Stutz and Morton, 1996).

The diversity of AM fungi is relevant because high species diversity may help stabilise plant community structure and ecosystem processes (Klironomos et al., 2000; Tilman, 1996). However, high diversity can also lead to competitive exclusion and cause a reduction in the number of coexisting species (Huston, 1994). For AM fungi, this may only reduce their abundance below levels of detection (Bever et al., 2001). The competitive ability of species of AM fungi in roots is probably an important factor in determining the dominance of AM fungi inside roots as well as in soil. This is compounded by seasonal changes in infectivity of AM fungi (Merryweather and Fitter, 1998a) which is influenced by fungal life cycles (Abbott and Gazey, 1994) such as sporulation and associated changes in infectivity of the hyphae (Pearson and Schweiger, 1993). The benefit to the host plant through enhanced P uptake is expected to be markedly altered according to the effectiveness of dominant AM fungi in roots. Species diversity of AM fungi has been examined from the perspective of presence or absence of fungi in soil, not in roots (Bever et al., 2001; Van der Heijden et al., 1998a). The effectiveness of fungi in taking up P in these studies was not considered, but it is not easy to determine, particularly as effectiveness can change for the same fungus depending on the P status of the soil and plant (Thomson et al., 1986). There is little evidence of a simple relationship between the relative abundance of morphotypes of AM fungi inside roots and spores in soil (Merryweather and Fitter, 1998a; Scheltema et al., 1987). Therefore, diversity of AM fungi needs to be investigated further, in relation to both the relative abundance of fungi in the roots (Abbott and Gazey, 1994) and the particular species of AM fungi present (Wardle, 1999).

As AM fungi occur as communities in soil and in roots, they are likely to collectively contribute to P uptake (Jakobsen et al., 2001) depending on the mycorrhizal dependency of the host plant (Van der Heijden et al., 1998b). Although there have been intensive studies on single-AM fungus function, there have been few investigations of the contributions of communities of AM fungi. Reconstructed communities of AM fungi in soil can promote plant growth (Daft, 1983; Daft and Hogart, 1983), or have no effect on plant growth (Sylvia et al., 1993). These studies did not quantify the relative abundance of different fungi within the roots. However, Merryweather and Fitter (1998b) demonstrated a correlation between the occurrence of AM fungal morphotypes and seasonal P uptake in a natural ecosystem. AM fungi can differ in their capacity to take up P which may be associated with hyphal characteristics (Smith et al., 2000). Therefore, information about the relative abundance and activity of fungi inside roots could be more useful for explaining variation in the contributions that communities of AM fungi make to plant growth than a measure of fungal diversity based on spores in soil.

The accessibility of a localised source of P to hyphae of a community of AM fungi in a pasture soil was related to hyphal length density in the vicinity of the P source (Jakobsen et al., 2001). We investigated this further in a highly P-deficient forest soil. First, we assessed the infectivity of the AM fungi present in jarrah forest soil using two host plants. We then determined the extent to which plants colonised by this indigenous community of AM fungi could access P from a point source placed at increasing distances (2, 4 and 6 cm) from the roots.

Materials and methods

Experiment 1

The infectivity of AM fungi present in jarrah forest soil was assessed in mycorrhizal bioassays conducted in undisturbed soil cores (Scheltema et al., 1987). *Phyllanthus calycinus* Labill, a native forest understorey plant, and subterranean clover, *Trifolium subterraneum* L., were grown in undisturbed soil cores under glasshouse conditions. We used these two plant species to trap the AM fungi; clover because it is routinely used as a mycorrhizal host, and *P. calycinus*

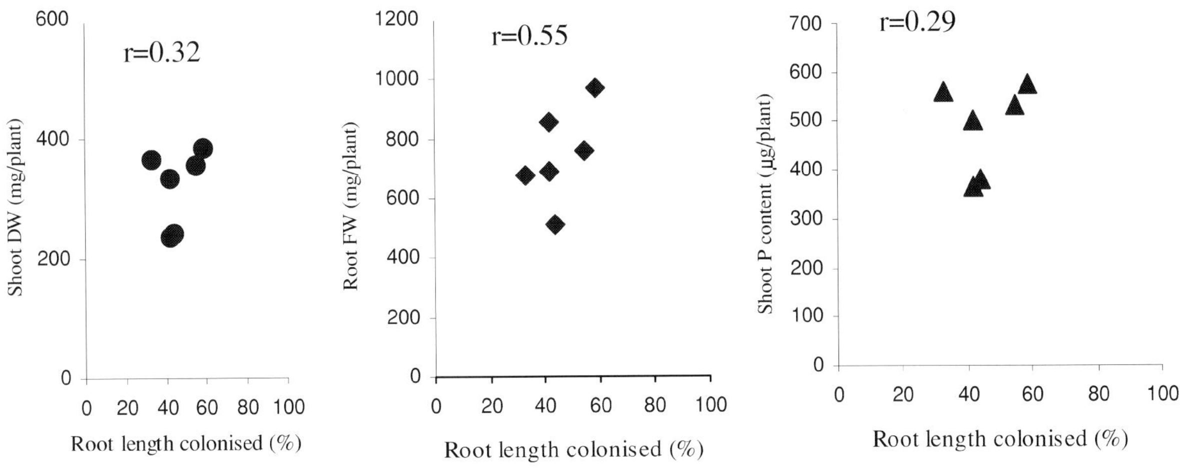

Figure 1. Correlation between % root length colonised, biomass and shoot P content in subterranean clover (Expt. 1).

because it was highly mycorrhizal when inoculated with forest soil (Brundrett and Abbott, 1991). Soil cores were collected near Jarrahdale, Western Australia. The intact cores were collected using a soil corer 12 cm in diameter and 12 cm high. They were transferred to 1.5 kg capacity plastic-lined pots, then stored in a cold room for about 2 weeks. Seeds of *P. calycinus* were immersed in Seed Starter (smoked water routinely used for germinating seeds of Australian native plants), left overnight, rinsed and aerated in distilled water for 1 day before sowing them in seedling trays. The seedlings were grown in the tray filled with steamed Lancelin sand for one month. Three seedlings were transplanted into each soil core. Subterranean clover seeds were immersed in aerated water overnight; then six seeds were planted in different soil cores and thinned to three after emergence. Six replicate pots were prepared for each plant species. Pots were placed in a temperature-controlled water tank at 20 °C in a glasshouse. The plants were watered daily to maintain soil moisture at field capacity.

Plants were harvested 12 weeks after transplanting and dry weight (DW) of shoots, fresh weight (FW) of roots, and root length colonised by different morphotypes of AM fungi were recorded. The roots were cut into 1 cm lengths then about 0.5 g was subsampled at random. Sub-samples of roots were cleared and stained with Trypan Blue (Abbott and Robson, 1981). Root length was estimated using a line intersect method (Newman, 1966), and the percentage of root length colonised by AM fungi was estimated simultaneously. P concentrations of shoots were determined using the molybdovanado-phosphate method (Boltz and Lueck, 1958) after acid digestion (Johnson and Ulrich, 1959).

Experiment 2

Soil cores were collected from the same area of the jarrah forest as in the bioassay described above. Point sources of P were placed at increasing distance from plants. KH_2PO_4 (1.14 mg P) was added to 100 g soil in root–free mesh bags (37 μm) which were inserted 2, 4 and 6 cm away from plants constrained in additional mesh bags. Additional P and basal nutrients were added as a solution of $KH_2PO_4$to the whole of each core at the rate of 11.4 μg P g^{-1} soil to support measurable plant growth within the time of the experiment (Jasper et al., 1989). Basal nutrients were added in solution to the whole core at the following rates ($\mu g\ g^{-1}$ soil): 50 K_2SO_4. 5 H_2O, 0.15 $CuSO_4$. $5H_2O$, 0.075 $ZnSO_4.7H_2O$, and 0.01 $NaMoO_4$. $2H_2O$. Basal nutrients were added in solution to the surface of the soil in each pot prior to watering. All treatments were replicated three times.

The native understorey plant *P. calycinus* was used in this experiment. Seeds were germinated and plants grown as in the mycorrhizal bioassay. One plant was transplanted per soil core, which was watered to field capacity and kept within a temperature-controlled water tank at 20 °C in a glasshouse.

Plants were harvested 12 and 16 weeks after transplanting as in the mycorrhizal bioassay. The parameters recorded were dry weight (DW) of shoots, fresh weight (FW) of roots, root length colonised and % root length colonised by morphotypes of AM fungi, length of external hyphae, and P content of shoots. The

Table 1. Plant biomass, mycorrhizal colonisation, and shoot P status 12 weeks after transplanting (Expt. 1)

Host plant	Shoot DW (mg/plant)	Root FW (mg/plant)	Root length colonised (%)	Shoot P conc. (mg/g)	Shoot P content (μg/plant)
Trifolium subterraneum	319	744	46	1.5	486
Phyllanthus calycunus	174	250	55	2.0	340
LSD ($P < 0.05$)	36	54	NS	0.5	62

Table 2. Percent root length colonised in two host plants by AM fungal community 12 weeks after transplanting (Expt. 1)

Host plant	Percentage of root length colonised by morphotypes of AM fungi				
	Glomus sp.	*Acaulospora*[1]	*Scutellospora*[1]	Fine endophyte	Total
Trifolium subterraneum	30	4	7	5	46
Pyllanthus calycinu	31	8	13	3	55
LSD ($P < 0.05$)	NS	3	4	NS	NS

[1]These morphotypes were probably those of *A. laevis* and *S. aurogloba* (F. Jayasundara and C. Walker, pers. comm. based on species in soil).

length of hyphae in soil was measured by an aqueous extraction and membrane filtration technique (Jakobsen et al., 1992). The Newman formula for calculating root length (Newman, 1966) was used to calculate the hyphal length on each filter.

Treatment effects were detected by separate one-way analysis of variance at each harvest (GENSTAT 5). Means were compared by Least Significance Difference (LSD) test at $P < 0.05$. Paired t-test was done to compare the differences in colonisation between two sampling periods.

Results

Mycorrhizal bioassay (Experiment 1)

In the mycorrhizal bioassay conducted using two host plants (*P. calycinus* and subterranean clover) in forest soil cores, both species had a similar proportion of root length colonised by indigenous AM fungi (Table 1). The mycorrhizas were formed by a community dominated by three morphotypes. There were no correlations between root length colonised and plant biomass and shoot P content in subterranean clover (Figure 1). In contrast, for *P. calycinus,* root length colonisation by the community of AM fungi was correlated with both plant biomass and shoot P content (Figure 2). Subterranean clover produced more biomass than did *P. calycinus*, but *P. calycinus* had a higher P concentration in shoots. Both host species had a similar proportion of each morphotype of AM fungi (Table 2). *Glomus* sp. was the dominant morphotype. Morphotypes resembling *Acaulospora*, *Scutellospora* and fine endophytes were also present.

Effect of an AM fungal community on plant growth and P uptake (Experiment 2)

Root colonisation by distinct morphotypes of AM fungi was similar to that in the mycorrhizal bioassay. *Glomus* sp. was the dominant AM fungus in roots after both 12 and 16 weeks (Table 3). The percentage root length colonised by the *Glomus* morphotype increased with time ($P < 0.05$) and that of the *Scutellospora* morphotype decreased ($P < 0.01$). Colonisation by the *Acaulospora* morphotype and fine endophytes remained at a relatively low level from 12 to 16 weeks (Table 3). The hyphal length in the mesh bag de-

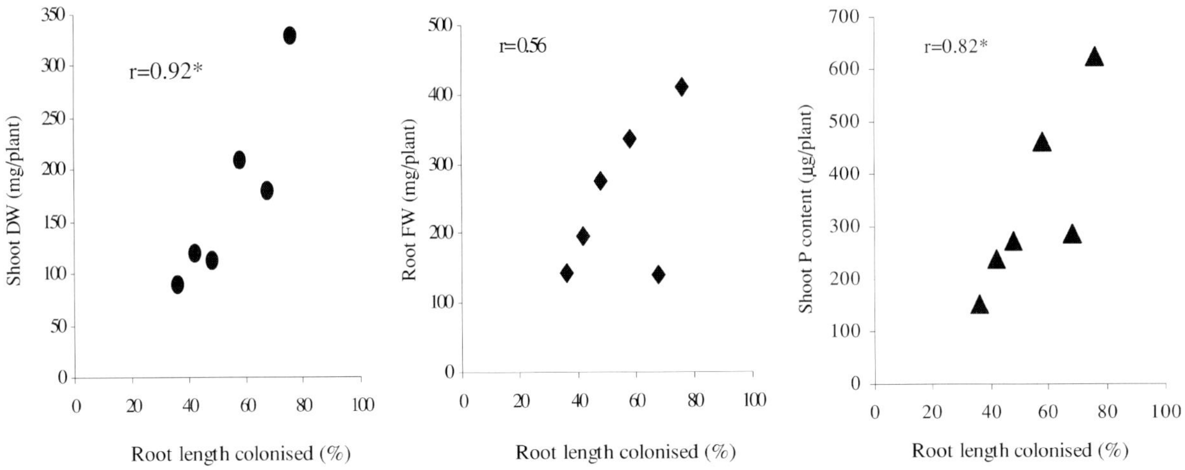

Figure 2. Correlation between % root length colonised, plant biomass and shoot P content in *Phyllanthus calycinus* (Expt. 1).

Table 3. Percent root length colonised in *P. calycinus* by AM fungal community 12 and 16 weeks after transplanting (Expt. 2). Data in parentheses are total root length colonised (cm/plant)

Distance from plant	Percentage of root length colonised by morphotypes of AM fungi				
	Glomus sp.	*Acaulospora*[1]	*Scutellospora*[1]	Fine endophyte	Total
12 weeks					
2 cm	29 (127)	8 (35)	16 (70)	3 (13)	56 (246)
4 cm	27 (113)	8 (34)	15 (63)	2 (8)	52 (218)
6 cm	30 (121)	9 (36)	17 (68)	4 (16)	60 (241)
LSD ($P < 0.05$)	NS	NS	NS	NS	5 (21)
16 weeks					
2 cm	55 (257)	8 (37)	3 (14)	4 (19)	70 (327)
4 cm	52 (246)	9 (43)	4 (19)	3 (14)	68 (322)
6 cm	57 (258)	7 (32)	2 (9)	3 (14)	69 (312)
LSD ($P < 0.05$)	NS	NS	NS	NS	NS

[1]These morphotypes were probably those of *A. laevis* and *S. aurogloba* (F. Jayasundara and C.Walker, pers. comm., based on species in soil).

creased with distance from the plant for samples taken after both 12 and 16 weeks (Table 4).

Shoot P uptake was greatest when the point source of P was placed 2 cm away from the root, but there was no corresponding difference in plant growth (Table 4).

Discussion

It has commonly been assumed that species of AM fungi have a similar function because of their common occurrence and the absence of clearly defined host specificity (see Dodd et al., 2000). Changes in plant growth induced by AM fungi vary for different fungal species, but both growth promotion and reduction depend upon particular plant-fungal combinations (Johnson et al., 1997).

We used two bioassay host plants to determine the extent of root length colonisation by morphotypes of AM fungi in a forest soil in relation to plant growth. The relative proportion of root colonised by the morphotypes was similar in both plants. There was a positive correlation between the proportion of root length colonised, plant biomass and shoot P content for *P. calycinus*, but not for subterranean clover. The different response of the two hosts was unrelated to differences in the dominant morphotype of AM fungus inside the roots. A difference in the rate of mycorrhiza formation prior to the first sampling time may be responsible, but this was not assessed.

Table 4. Plant biomass, hyphal length of all AM fungi and shoot P uptake in *P. calycinus* by AM fungal community 12 and 16 weeks after transplanting (Expt. 2)

Distance from plant	Shoot DW (mg/plant)	Root FW (mg/plant)	Hyphal length (m/g dry soil)	Shoot P Conc. (mg/g)	Shoot P content (μg/plant)
12 weeks					
2 cm	558	932	21	2.0	1095
4 cm	441	890	17	1.7	747
6 cm	353	854	15	1.5	514
LSD ($P < 0.05$)	NS	NS	1	0.3	347
16 weeks					
2 cm	794	992	25	1.4	1112
4 cm	670	1005	20	1.2	775
6 cm	646	959	16	0.9	587
LSD ($P < 0.05$)	NS	NS	3	0.2	267

The high variability in colonisation of bioassay plants grown in individual soil cores demonstrated spatial variability in infectivity of AM fungi at the forest site, as has been shown previously (Brundrett and Abbott, 1995). In our study, a species of *Glomus* was the most infective fungus for both species, with no evidence of preferential colonisation as may sometimes occur (Bever et al., 2001).

The naturally occurring AM fungal community at this site in the jarrah forest was able to access P when the P source was placed 2 and 4 cm from the root, but not when it was placed 6 cm away from the roots. This observation corresponded with changes in the length of hyphae in the soil in the vicinity of the P source.

In previous work, it was reported that seasonal fluctuation in mycorrhiza formation by propagules of AM fungi in a similar part of the jarrah forest was not substantial (Brundrett and Abbott, 1994). Thus, hyphal networks and other mycorrhizal fungus propagules in the jarrah forest soil probably maintained a similar capacity to colonise roots throughout the year, even during extended periods of summer drought. In our study over a 4-week period, colonisation by *Glomus* from soil cores increased with time, while that of *Scutellospora* decreased. Changes in relative abundance of the fungi inside roots could be due to competitive interactions between the fungi (Juniper et al., 1998) or to a decline in infectivity of *Scutellospora* (presumably *S. aurogloba* based on spore identification) if it has a similar life cycle to that of *S. calospora* (Pearson and Schweiger, 1993). A comprehensive study of seasonal dynamics of a community of AM fungal morphotypes in roots of three grassland species has been published recently (Smilauer, 2001), and variation in the relative abundance of morphotypes of AM fungi were observed. Earlier work also reported seasonal variation of infectivity of morphotypes of AM fungi in annual pastures (e.g., Scheltema et al., 1987).

Jakobsen et al. (2001) showed that ^{32}P uptake by subterranean clover planted in undisturbed cores of soil from pastures in Western Australia was correlated with the quantity of hyphae formed in soil in the vicinity of the P source. The mycorrhizas in the different soils were dominated by different morphotypes of AM fungi. Differences in function of AM fungi studied separately have often been demonstrated and may be related to the growth and activity of their hyphae (Jakobsen et al., 1992), but not necessarily the distance that the hyphae grow away from roots (Smith et al., 2000). However, when the same AM fungi are present in the field, these differences in function may be of less importance because their relative abundance is controlled by environmental factors and interactions between the fungi (Juniper et al., 1998). Therefore, P-related contributions of AM fungi under field conditions depend both on their relative abundance and on their effectiveness in transferring P to the plant.

AM fungi in this forest soil have the potential to contribute to P uptake of an understorey plant from point sources of P placed within 4 cm of the root surface. There was little possibility that P diffusion occurred in this soil due to its high phosphate adsorption characteristics (Bolan and Abbott, 1983). P is extremely limiting in this environment, and AM fungi have the capacity to access isolated sources of P that might not otherwise be available to the roots.

Acknowledgements

This research was supported by the Australian Research Council (ARC). We thank Dr I. Jakobsen and Professor H. Lambers for discussion in the developmental stages of this research.

References

Abbott L K 1982 Comparative anatomy of vesicular-arbuscular mycorrhizas formed on subterranean cover. Aust. J. Bot. 30, 485–490.

Abbott L K and Gazey C 1994 An ecological view of the formation of VA mycorrhizas. Plant Soil 159, 69–78.

Abbott L K and Robson A D 1981 Infectivity and effectiveness of five endomycorrhizal fungi: competition with indigenous fungi in field soils. Aust. J. Agric. Res. 32, 621–630.

Bell D T and Heddle E M 1989 Floristic, morphologic and vegetational diversity. *In* The Jarrah Forest. Eds. B Dell, J J Havel and N Malajczuk. pp. 53–66. Kluwer Academic Publishers, Dordrecht.

Bever J D, Schultz P A, Pringle A and Morton J B 2001 Arbuscular mycorrhizal fungi: more diverse than meets the eye, and the ecological tale of why. BioSci. 51, 923–931.

Bolan N S and Abbott L K 1983 Seasonal variation in infectivity of vesicular-arbuscular mycorrhizal fungi in relation to plant response to applied phosphorus. Aust. J. Soil. Res. 21, 195–198.

Bolan N S, Robson A D and Barrow N J 1983 Plant and soil factors including mycorrhizal infection causing sigmoidal response of plants to applied phosphorus. Plant Soil 73, 187–203.

Boltz D F and Lueck C H 1958 Phosphorus. *In* Colorimetric Determination of Non-metals. Ed. D F Boltz. pp. 29–46. Interscience Publishers, New York.

Brundrett M C and Abbott L K 1991 Roots of jarrah forest plants. I. Mycorrhizal associations of shrubs and herbaceous plants. Aust. J. Bot. 39, 445–457.

Brundrett M C and Abbott L K 1994 Mycorrhizal fungus propagules in the jarrah forest. I. Seasonal study of inoculum levels. New Phytol. 127, 539–546.

Brundrett M C and Abbott L K 1995 Mycorrhizal fungus propagules in the Jarrah forest. II Spatial variability in inoculum levels. New Phytol. 131, 461–469.

Brundrett M C, Abbott L K and Jasper D A 1999 Glomalean mycorrhizal fungi from tropical Australia. I. Comparison of the effectiveness and specificity of different isolation procedures. Mycorrhiza 8, 305–314.

Cuenca G, De Andrale Z and Escalante G 1998 Diversity of Glomalean spores from natural, disturbed and revegetated communities growing on nutrient-poor tropical soils. Soil Biol. Biochem. 30, 711–719.

Daft M J 1983 The influence of mixed inocula on endomycorrhizal development. Plant Soil 71, 331–337.

Daft M J and Hogart B G 1983 Competitive interactions amongst four species of *Glomus* on maize and onion. Trans. Br. Mycol. Soc. 80, 339–345.

Dodd J C, Boddington CL, Rodriguez A, Gonzalez-Chavez C and Mansur I 2000 Mycelium of arbuscular mycorrhizal fungi (AMF) from different genera: form, function and detection. Plant Soil 226: 131–151.

Huston M A 1994 Biological diversity: The Coexistence of Species on Changing Landscapes. Cambridge University Press, Cambridge.

Jakobsen I, Abbott L K and A D Robson 1992 External hyphae of vesicular-arbuscular mycorrhizal fungi associated with *Trifolium subterraneum* L. 1. Spread of hyphae and phosphorus inflow into roots. New Phytol. 120, 371–380.

Jakobsen I, Gazey C and Abbott L K 2001 Phosphate transport by communities of arbuscular mycorrhizal fungi in intact soil cores. New Phytol. 149, 95–103.

Jasper D A, Abbott L K and Robson A D 1989 Hyphae of a VA mycorrhizal fungus maintains infectivity in dry soil, except when the soil is disturbed. New Phytol. 112, 101–107.

Johnson C M and Ulrich A 1959 Analytical methods for use in plant analysis. Calif. Agric. Exp. Sta. Bull. 767, 25–78.

Johnson N C, Graham J H and Smith F A 1997 Functioning of mycorrhizal associations along the mutualism-parasitism continuum. New Phytol. 135, 575–585.

Juniper S, Abbott L K and Jayasundara F 1998 Approaches to the study of interactions between arbuscular mycorrhizal fungi. *In* Mycorrhiza Manual. Ed. Varma A., pp. 25–39. Springer-Verlag, Heidelberg.

Klironomos J N, McCune J, Hart M, Neville J 2000 The influence of arbuscular mycorrhizae on the relationship between plant diversity and productivity. Ecol. Lett. 3, 137–141.

Merryweather J W and Fitter A H 1998a The arbuscular mycorrhizal fungi of *Hyacinthoides non-scripta* – I. Diversity of fungal taxa. New Phytol. 138, 117–129.

Merryweather J W and Fitter A H 1998b The arbuscular mycorrhizal fungi of *Hyacinthoides non-scripta* – II. Seasonal and spatial patterns of fungal populations. New Phytol. 138, 131–142.

Newman E I 1966 A method of estimating the total length of root in a sample. J. Appl. Ecol. 3, 139–145.

Pearson J N and Schweiger P 1993 *Scutellospora calospora* (Nicol. And Gerd.) Walker and Sanders associated with subterranean clover: dynamics of colonisation, sporulation and soluble carbohydrates. New Phytol. 124, 215–219.

Scheltema M A, Abbott L K and Robson A D 1987 Seasonal variation in infectivity of VA mycorrhizal fungi in annual pastures in a Mediterranean environment. Aust. J. Agric. Res. 38, 707–715.

Smilauer P 2001 Communities of arbuscular mycorrhizal fungi in grassland: seasonal variability and effects of environment and host plants. Folia Geobotanica 36, 243–263.

Smith F A, Jakobsen I and Smith S E 2000 Spatial differences in acquisition of soil phosphate between two arbuscular mycorrhizal fungi in symbiosis with *Medicago truncatula*. New Phytol. 147, 357–366.

Stutz J C and Morton J B 1996 Successive pot cultures reveal high species richness of arbuscular endomycorrhizal fungi in arid ecosystems. Can. J Bot. 74, 1883–1889.

Sylvia D M, Wilson D O, Graham J H, Maddox J J, Millner P, Morton J B, Skipper H D, Wright S F and Jarstfer A G 1993 Evaluation of vesicular arbuscular mycorrhizal fungi in diverse plants and soils. Soil Biol. Biochem. 25, 705–713.

Thippayarugs S, Bansal M and Abbott L K 1999 Morphology and infectivity of fine endophyte in a Mediterranean environment. Mycol. Res. 103, 1369–1379.

Thompson B D, Robson A D and Abbott L K 1986 Effects of phosphorus on the formation of mycorrhizas by *Gigaspora margarita* and *Glomus fasciculatum* in relation to root carbohydrates. New Phytol. 103, 751–765.

Tilman D 1996 Biodiversity: population versus ecosystem stability. Ecology 77, 350–363.

Van der Heijden M G A, Klironomos J N, Ursic M, Moutoglis P, Streitwolf-Engel, Boller T, Wiemken A and Sanders I R 1998a Mycorrhizal fungal diversity determines plant biodiversity, ecosystem variability and productivity. Nature 396, 69–72.

Van der Heijden M G A, Boller T, Wiemken A and Sanders I R 1998b Different arbuscular mycorrhizal fungal species are potential determinants of plant community structure. Ecology 79, 2082–2091.

Wardle D A 1999 Is 'sampling effect' a problem for experiments investigating biodiversity-ecosystem function relationships? Oikos 87, 403–407.

Plant and Soil **248:** 321–333, 2003.
© 2003 *Kluwer Academic Publishers. Printed in the Netherlands.*

Relationships between cluster root-bearing taxa and laterite across landscapes in southwest Western Australia: an approach using airborne radiometric and digital elevation models

W. H. Verboom[1] & J. S. Pate[2]

[1]*Department of Agriculture, 10 Doney Street, Narrogin, Western Australia 6312. (E-mail at: wverboom@agric.wa.gov.au for reprints and correspondence on soil and geomorphological issues).* [2]*Faculty of Plant Sciences (Botany) and Co-operative Research Centre for Plant-based Management of Dryland Salinity, University of Western Australia, Western Australia 6009 (Contact address for correspondence on botanical issues).*

Received 19 April 2002. Accepted in revised form 17 July 2002

Key words: carboxylate, cluster roots, gamma-ray, phosphate, spiral waves, thorium

Abstract

As part of a general soil and regolith mapping exercise across cleared and remnant bush land, radiometric data for distribution of potassium (K), uranium (U) and thorium (Th) were examined alongside relief models and floristic and soil surveys in test catchments at Elashgin and Toolibin in the Western Australian wheat-belt. The Elashgin survey showed that highly weathered low K soils co-concentrated U and Th and were vegetated mainly by cluster root-bearing Proteaceae and Casuarinaceae. In granitic soils ratios of U to Th were higher and cluster root bearing taxa much less prominent, except where ferricrete gravels were concentrated. Draping of radiometric imagery over a digital elevation model showed spiral waveforms of high and low U and Th signal which were largely independent of topography but demarcated different oligotrophic communities. General observations and a detailed 900-m transect along an aeolian sand plume at Toolibin showed very high U and Th in ferricrete gravels where Proteaceae were dominant, but failed to separate proteaceous versus myrtaceous shrublands on deep sands due to truncation of signal. Augering along the transect and examination of floristic, soil and signal composition of 32 sites in the Lake Toolibin catchment confirmed presence and degree of development of ferricretes in the rhizosphere of Proteaceae-dominated communities and showed high reactivity scores for Al in 'Bs' horizons in profiles carrying such vegetation. Highly specific associations between Proteaceae and very high U and Th were generally evident on exposed ferricrete gravels. The data are discussed in relation to the effects of root exudates on ferricrete formation and destruction and how the broader spatial pattern of such processes might relate to competition for soil phosphorous.

Introduction

The landscapes of the Western Australian wheatbelt are characterised by widespread occurrences of variously developed intact and denuded laterites vegetated by flora displaying extraordinary endemism, species diversity and specificity for soil type. The intact laterites have much in common with Western Australian podzols. Both tend to be vegetated by flora rich in Proteaceae and other taxa known to bear carboxylate-secreting cluster roots (Pate et al., 2001; Verboom and Galloway, 2000) and both are leached and highly differentiated and show zonal accumulations of iron and aluminium sesquioxides in their respective subsoil (Bs) layers. However laterites distinguish themselves by having ferricretes embracing structures, ranging from discrete pisoliths to continuous reticulate or vermiform skeletons (see McFarlane, 1983 for a fuller description and illustrations of the relevant structures). In a recent publication, Pate et al. (2001) examined (a) the possible origins of laterites (b) the extent of the coincidences worldwide in past and present times between Proteaceae and oligotrophic soil types of lateritic character including podzols and (c) evidence of functioning of cluster (proteoid) roots of Proteaceae in formation of ferricrete layers due to their copious exudation of low molecular weight carboxylates. They speculated generally that phosphate acquisition strategies of plants in oligotrophic settings might apply as much to conifers as well as to angiosperms and thus be deep seated in the genetic make up of land plants.

Within ferric-rich layers of these laterites one encounters concentrations of chelatable metallic elements including thorium (Th) and uranium (U). Thorium, occurring only in valence state (Th^{4+}), is highly insoluble in most surface soil environments and available data for natural waters suggests that its mobility is greatly influenced by formation of complexes with organic ligands (Langmuir and Herman, 1980). The bulk of dissolved organic matter in natural waters consists of fulvic-type acids which have long been implicated in the dissolution of Fe and Al and other chelatable metals in soil environments, particularly podzols (see Farmer and Lumsden, 2001). However, it is now increasingly evident that readily metabolised low molecular weight carboxylates may play a major role in the dissolution of Fe, Al and other cations from primary minerals. The resulting complexes may be microbially decomposed to yield ferrihydrite and imogolite type materials in Bs horizons of podzols (Farmer and Lumsden, 2001; Lundstrom et al., 2000) and laterites (Pate et al., 2001). The main chemical difference between laterites and podzols, is that in podzols one usually encounters a Bh overlying the Bs horizons. However, the Bs horizon is a dominant feature of both and fulvic type acids may simply recycle metals such as Fe and Al between the O and Bh horizons (Farmer and Lumsden, 2001).

Scott and Dickson's (1991) observation that Th and to a lesser extent the more mobile U attach readily to iron oxides and possibly co-precipitate with Fe as cutans on lateritic pisoliths, taken together with the susceptibility of ferric iron to chelation, suggest that these two elements might be mobilised and precipitated together with Fe in such situations.

Over the last decade, airborne geophysical techniques and digital elevation models (DEM's) have been applied in soil and regolith mapping in agricultural regions in south west Australia (Cook et al., 1996; George et al., 2002; Pracilio et al., 1998). In the process, high resolution maps of natural distribution of U, Th and potassium (K) and associated magnetic and relief structures have very recently become available for parts of the Western Australian wheatbelt. Where remnant vegetation occurs in the areas surveyed, opportunities accordingly arise for examining predictions regarding our earlier hypotheses relating to biogenic origins of laterite (see Pate et al., 2001). In particular, if the mode of laterite formation is as we have suggested, one would expect to find evidence of U and Th concentrating alongside Fe in oligotrophic settings, and for such features to specifically distinguish soil types carrying vegetation dominated by cluster root-bearing taxa.

This paper tests the above predictions in two landscape settings, and assembles detailed geophysical and floristic evidence further supporting the concept of a strong influence of biological processes in soil formation and distribution in landscapes of the region.

Materials and methods

Description of study areas

Major bioclimatic and laterite zones for the two study areas are indicated in the location map shown in Figure 1. As shown, both study areas lie within the Extra-dry mediterranean zone (see Beard, 1984). The bedrock at Elashgin and Toolibin is dominated by granite and gneiss variously laced by parallel swarms of dolerite, diorite and quartz-feldspar dykes. Both areas have been extensively cleared for agriculture leaving confined areas of remnant vegetation in near pristine condition (see Figure 1).

Study Area 1, at Elashgin, embraces residual Yellow Sandplain (see Pate et al., 2001) consisting of ferricrete gravels and yellow sands interspersed with granite outcrops and their associated soils, and flat valleys of red clay loams. Judging from evidence from blocks of remnant vegetation and road-side reserves, gravels and yellow sand were probably vegetated originally by kwongan communities of casuarinas, acacias and Proteaceae. Granitic soils originally contained *Acacia acuminata*, *Eucalyptus loxophleba* and *Allocasuarina* spp. while the valley bottoms principally comprised woodlands of *E. salmonophloia* and *E. salubris* (see vegetation maps and descriptions of Beard, 1980, 1990).

Study Area 2, at Toolibin, straddles the Younger Laterite Zone (Mulcahy, 1960, 1967) and the Grey Sandplain Zones (see Pate et al., 2001). High level 'older' laterites of the former zone, are confined to the peripheries of isolated mesas vegetated by eucalypts with an understorey of mostly proteaceous shrubs and small trees. Soil landscapes of the Grey Sandplain Zone are similar to those of the Yellow Sandplain,

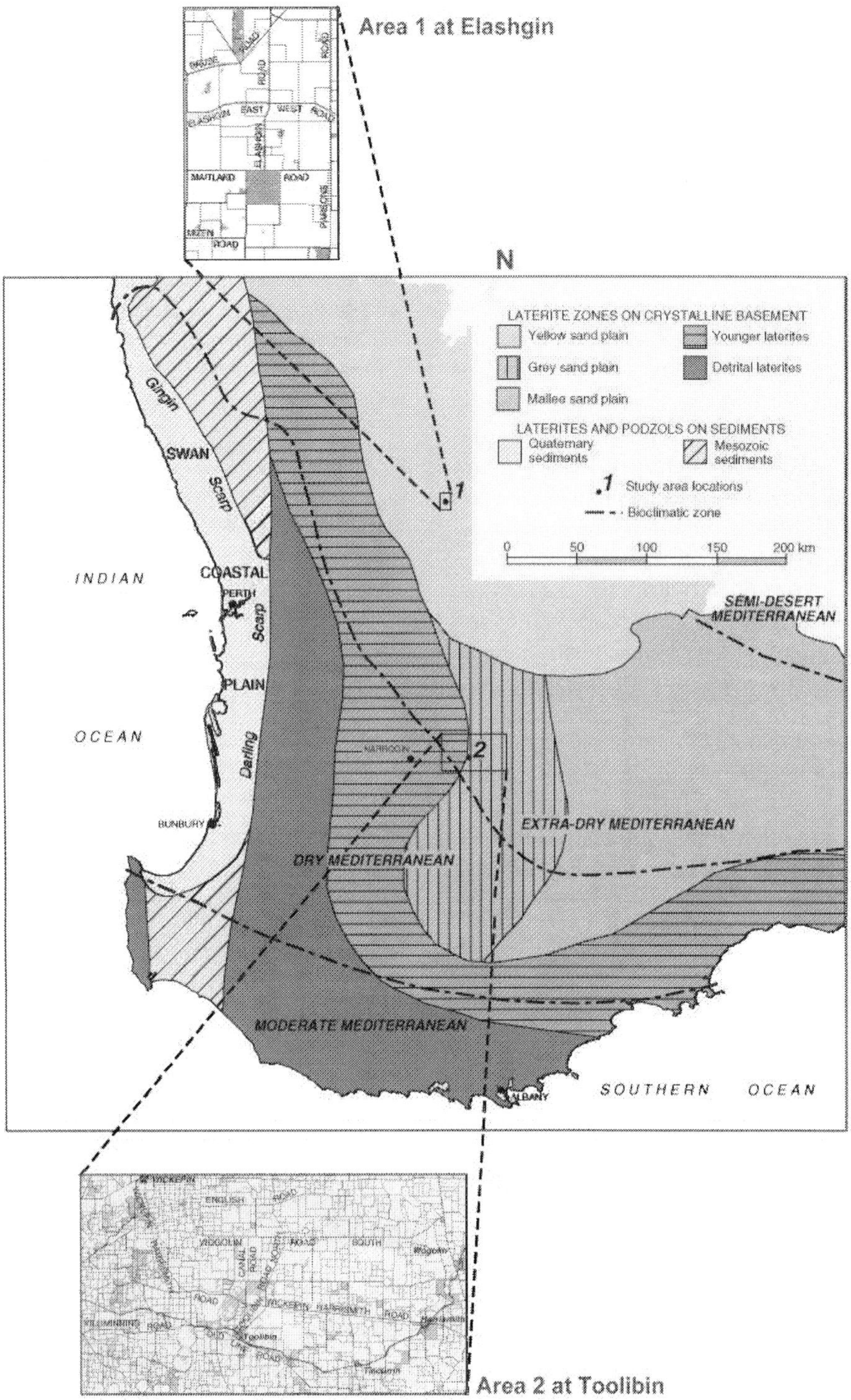

Figure 1. Major bioclimatic and laterite zones of south-west Western Australia (modified from Pate et al., 2001). Study areas showing location of remnant vegetation (shaded grey) are shown as insets for Elashgin and Toolibin.

but its soils are typically better segregated with respect to iron. Highly indurated laterites in the area are dominated by species of *Dryandra*, *Banksia* and *Allocasuarina* and their flanking 'deposits' often contain pedogenetic ferricrete set in a grey sandy matrix.

Digital elevation and radiometric data

Relatively high-resolution digital elevation models (DEMs) from the LandMonitor project, http://www.landmonitor.wa.gov.au, are available for both areas. Elevations are generally accurate to 1–2 m sampled on a 10 m easting/northing grid.

Study Area 1 has been surveyed by high-resolution airborne radiometrics over an area of 8000 ha at Elashgin (Figure 1). These data, acquired during the mid-summer dry season during the year 2000, provide estimates of concentrations of K, Th and U in the upper 30 cm of soil surfaces, with minimal attenuation from vegetation cover and soil moisture. Values for K concentration derive directly from gamma-rays released during the decay of ^{40}K while concentrations of Th and U are inferred from the characteristic gamma-ray emissions of their respective daughter elements ^{208}Tl and ^{214}Bi. The air survey was conducted by UTS Geophysics (2002) using their latest acquisition system at a flying height of 20 m and a line spacing of 25 m. Radiometric data were collected with 32 l of thallium-doped sodium iodide crystals, using half second count times and processed using the enhanced Maximum Noise Fraction methodology described by Dickson and Taylor (1998).

Lower resolution radiometric data for Study Area 2 cover the 46 700 ha area of the Lake Toolibin Catchment (Figure 1). The survey was flown in 1997 by the World Geoscience Corporation Limited at a nominal terrain clearance of 60 m and a flight line spacing of 150 m. Radiometric data were collected with 33.5 l of thallium doped sodium iodide crystals, using 1 s count times. These data were processed using Spectra+Plus (Pracilio et al., 1998).

Soil and vegetation studies

Data on soil properties in the Elashgin catchment are available for 132 auger sites (Taylor et al., 2002). Surveys of the sparsely distributed remnant vegetation were conducted during the course of this study.

Detailed information on remnant vegetation at a number of monitoring sites is available in the Lake Toolibin catchment (see Brown & Root Asia Pacific Pty Ltd, 2002). A transect of vegetation was selected for detailed investigation during the present study to examine transitions in proteaceous and other vegetation in relation to changing profiles of topography, underlying soil type and radiometric signals. Soil pits dug at intervals along the transect provided information on soil morphology and distribution of cluster roots where present. The profile in each pit was subjected to spot testing for reactive aluminium (Fieldes and Perrott, 1966; Isbell, 1992) to distinguish regions of imogolite type enrichment, betraying the presence of podzolic 'Bs' horizons. The authors also examined the floristic composition of 32 upland monitoring sites in the Lake Toolibin catchment and revisited these to test the strength of the above mentioned relationships.

Data analysis

The comprehensive geophysical and soil data for Elashgin permitted investigations of interrelationships of native vegetation, habitat soil, landscape relief and radiometric patterns. Firstly relationships were considered between ground observation and airborne sensor measurements in an attempt to separate the components of sensor data into recognisable geomorphic features, choosing K signals and large scale topographic patterns to partition the landscape. Using these broadly interpreted geomorphic units, relationships between the natural clustering of soil–vegetation and the sensor data for K, U and Th were then defined. Secondly, the observed patterns of soil and vegetation delineated by the distributed radiometric data were evaluated in a topographical context using data from high resolution DEM.

Radiometric data for the Toolibin catchment were analysed in topographical context and in relation to vegetation, to occurrences of deep sand and ferricrete and to soil profile development. The data were also examined in respect of broad taxonomic classes of vegetation and soil at the 32 monitoring sites. Reactive aluminium tests were scored according to the five reactivity classes (0=non-reactive up to 5=very strongly reactive) as listed by Isbell (1992). Monitoring site scores were bulked into low (0–2) and high (3–5) classes to effect meaningful χ^2 testing.

Results

Elashgin results

Figure 2 shows the radiometrics-derived K map of

Area 1 draped over a digital elevation model. Immediately apparent is a range of single and repeating structures conforming to a range of K patterns. Pattern A, in this figure, coincides with flat-planated valley floors variously containing clays, loams and shallow sandy duplexes, all dominated by kaolinite in the clay fraction and vegetated by woodlands of salmon gum (*Eucalyptus. salmonophloia*) and gimlet (*E. salubris*). The uniform (low variance) K signal is interpreted to represent reworking and homogenising effect of valley waters on potassic feldspars. Potassium in clay fractions across the area is relatively low (Taylor et al., 2002). The uplands divide naturally into two contrasting regions namely, Pattern B with relatively stable low K landforms where remnant vegetation is principally Proteaceae, Casuarinaceae and *Acacia* spp., and Pattern C where it is interpreted that eroding granitic uplands show K feldspars coalescing into channels and terminating as fan deposits. Low woodlands of *Acacia acuminata* and *Allocasuarina huegeliana* in C regions occur around granitic outcrops with patches of *Allocasuarina campestris* in places.

Plots of aerial sensor measurements for U and Th are illustrated in Figure 3 for the ground observation sites on upland areas B (closed symbols) and C (open symbols) indicated in Figure 2. Soil types encountered at each site are designated by symbols and where remnant vegetation is present communities are listed in the caption to the Figure.

When comparing the point studies with the aerial values, caution needs to be exercised in two respects. Firstly each airborne measurement at Area 1 was averaged over a swath of ground of around 60 m width and 90 m long which of course is much larger than the area which is examined by a single auger boring. Secondly, correlation between signals for Th and U are exaggerated by the methodology used to remove noise from the raw spectral data. Even so, it is still apparent that a linear relationship is more strongly evident between Th and U for the oligotrophic uplands (B) than for corresponding granitic areas (C), where ratios of Th to U are noticeably lower and more variable.

The relief ternary radiometric image of the surveyed portion of the Elashgin catchment, shown in Figure 4, merges the maps of potassium, uranium and thorium into a single element ratio map and drapes this over a digital elevation model. Brightest colours are used for highest levels of Th (green), U (blue) and K (red) grading to black where lowest levels of all three are recorded and to white where all elements are at high concentration. It shows that Th and U occur at high concentration in rock and 'fresh' granite soil (R and G, white to violet), and in the clay size fraction of soil (yellow pink and white) where K is also high. Highest levels of U and Th (brightest blue/green) denote exposed ferricretes (L) where potassium is low.

Figure 4 shows surprisingly complex patterns for signalling of Th and U on areas B in Figure 2. Here, mosaics of exposed ferricretes (L, greeny-blue) and yellow sands (Y, brown) conform to spiral and wave-like configurations swirling up, down and across slopes, largely independent of topography. Elsewhere, rocky outcrops (R, pinkish to white), flow into feldspar-rich sandy channels (C, red) which in turn terminate in clayey fan deposits (F, orange). In the flat-floored valleys, faint vein like patterns mark shallow channels of non-granitic surface sand (red coloring Figure 4) against a background of loams and clays (pink yellow and white areas) rich in all three elements. For further information on the mineralogical interpretation of airborne radiometrics see Dickson and Scott (1997).

Toolibin results

Lake Toolibin is a seasonally wet playa on the south-west margin of the Australian arid zone. This region is sensitive to climatic oscillations (Mabbutt 1977) and the lake itself has continuously experienced changes in hydrologic, sediment and salinity regimes. Figure 5A views the lakes catchment from the south-west by means of a ternary radiometric image draped over a DEM. Interfluve crests to the right (east) of the line A–B are of similar elevation to those in the western area but the main trunk valleys are longer and down-stream gradients are less. These gradients influence the distribution of landforms and soils. The irregularly undulating granitic hills to the left of the line A–B are interpreted as showing active erosion and geological effects on interfluve topography. The smoothly undulating terrain to the right of the line A–B is taken to indicate deeply-weathered regolith. The soil materials are generally residual or colluvial and because of weathering, U and Th concentrations are distorted relative to those expected of the underlying granites and gneiss. As at Elashgin (cf. Figures 3 and 4) chelatable metals such as U (blue) and Th (green) concentrate to considerable degree in the ferricretes and emissions are strong where these outcrop in the landscape.

Paleo-lunettes to the south east of Lake Toolibin are thought to have formed in response to the wet season winds from the north-west during the Pleistocene

(Bowler, 1976). These lunettes, easily visible on the inset relief radiometric model shown in Figure 5B, are disrupted by uplands rising from the valley floors 1.5 km southeast of the lake. The 900 m transect selected for detailed study (location as shown in Figure 5B) traversed the sand plume blown from the lake onto the uplands.

The study transect at Toolibin encompassed a range of vegetation types and soil profiles and Figure 6 designates these along side intensities in airborne U and Th signals. Numerical scores for reactive aluminium are annotated for zones of the profiles of pits dug at stations along the transect. Vegetation cover and composition in respect of Proteaceae, soil profile characteristics and presence and abundance of cluster roots are also given for each sampling station.

Stations 1–3 lay in Banksia woodlands carrying a predominantly myrtaceous understorey. Stations 1 and 2 in *B. prionotes* woodland were on bleached white over yellow sand with discontinuous soft nodulated 'ferricrete' above placic (thin ironpan) layers overlying pockets of similar nodules and ortstein (pockets of pale brown soil weakly cemented by imogolite and organic aluminium type material). *B. attenuata* dominated at Station 3 where the profile showed bleached sand over a thick 'ferricrete' layer. The latter consisted of small irregular semi-indurated nodules. Cluster roots on both species of Banksia were prolific in A horizons and at lower frequency where roots intermixed with the 'ferricrete' gravels in the Bs layers. Extension of the zone of cluster root development into sand layers below the A horizon was particularly marked in *B. attenuata*. All three Stations scored strong (4) for reactive aluminium in their respective 'Bs' horizons contrasting with zero scores in superficial horizons. The above features collectively indicate a podzolic character.

Station 4, located in mixed heath with 15% cover of Proteaceae, showed a thin layer of relatively well developed ferricrete below bleached sand, again with evidence of cluster roots of Proteaceae in the A horizon and at depth in the ferricretes. The reactive Al score was 5 for the 'Bs' horizon, 0 for overlying parts of the profile.

Station 5 contrasted with the above in lacking Proteaceae and ferricrete and carrying vegetation consisting almost entirely of Myrtaceae. The soil profile comprised deep sand over mottled clay. Scores for reactive Al were non-reactive (0) to weak (2) throughout the profile.

Moving further up slope through Stations 6–8, ferricrete increased progressively in layer thickness and nodule size, concomitant with increasing proportional representation of Proteaceae and associated cluster roots and increasing shallowness of the overlying sand. Scores for reactive Al in all three profiles were strong (4) to very strong (5) for 'Bs' horizons but in these cases induration would disqualify classification as a podzol. As in Stations 1–3, reactivity scores in the overlying horizons were close to zero.

Station 9 was located in the region of the transect richest in Proteaceae in terms of both ground cover and taxonomic diversity. The profile showed a very shallow layer of bleached sand over strongly indurated laterite and cluster roots were prominent throughout the sand and wherever roots penetrated the ferricrete. Situations such as this, grading to fully exposed ferricrete, consistently comprised prime habitat for communities dominated by Proteaceae. However, this Station appeared to differ from all others in showing somewhat elevated K and containing coarse angular quartz grains, features together suggesting influence from underlying geology. Magnetic data (unpublished) indicated that a thin dolerite dyke coincides

Figure 2. Combined potassium-DEM image of Area 1 with coarse scale K/topographic patterns delineated. Vertical exaggeration is 15×. Variation in potassium concentration is represented by a spectrum with high potassium bright red and very low potassium dark blue. (A) Low variance K signal from flat valley floors; (B) relatively stable low K landforms; and (C) 'high' K granitic uplands and associated fan deposits.

Figure 3. Uranium and thorium concentrations at ground observation sites on Elashgin catchment's oligotrophic upland areas (B; closed symbols) and granitic upland areas (C; open symbols) indicated in Figure 2. The number in the plot designates remnant vegetation where present. Principal Kwongan communities on oligotrophic soils are: (1) *Allocasuarina campestris*, *Allocasuarina acutivalvis* and various short shrub Proteaceae; (2) tall shrub Banksia, *Grevillea* and *Acacia* spp. Communities on granitic soils are: (3) *E. loxophleba*, *Acacia acuminata* and *Allocasuarina hugleana*; (4) *Allocasuarina campesteris*, *Acacia acuminata* and *Melaleuca uncinata*; and (5) *Allocasuarina campestris*, *Isopogon divergens* and *Acacia lasiocaylyx*.

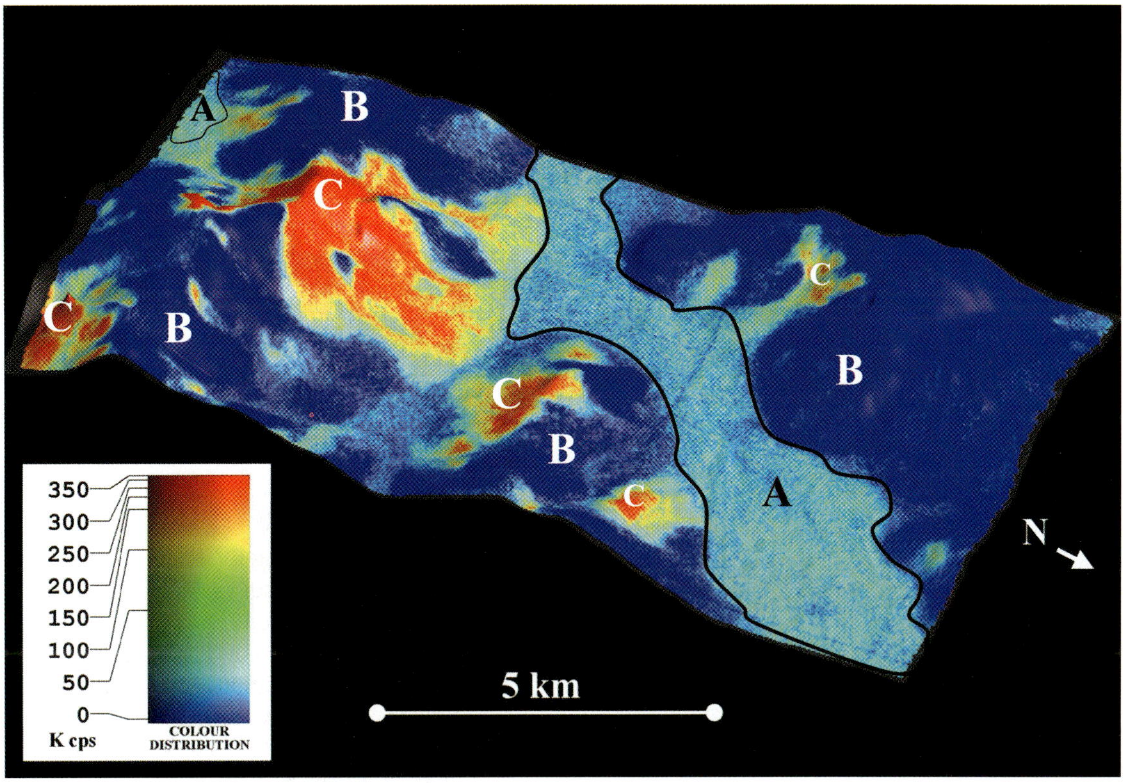

Figure 2.

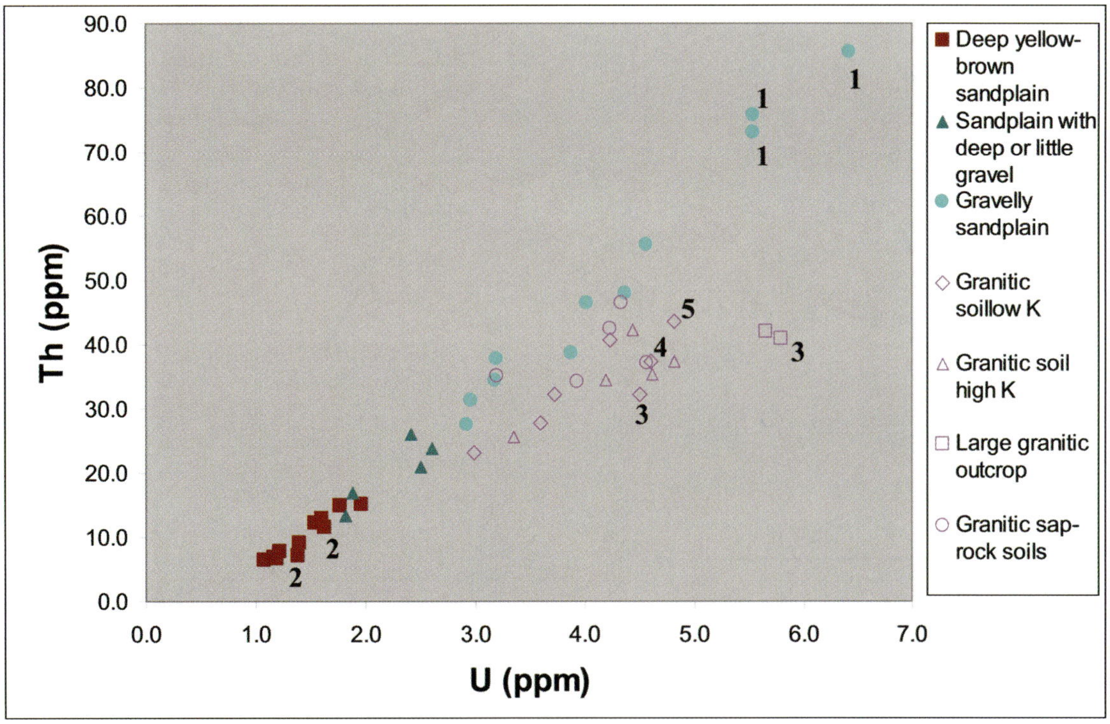

Figure 3.

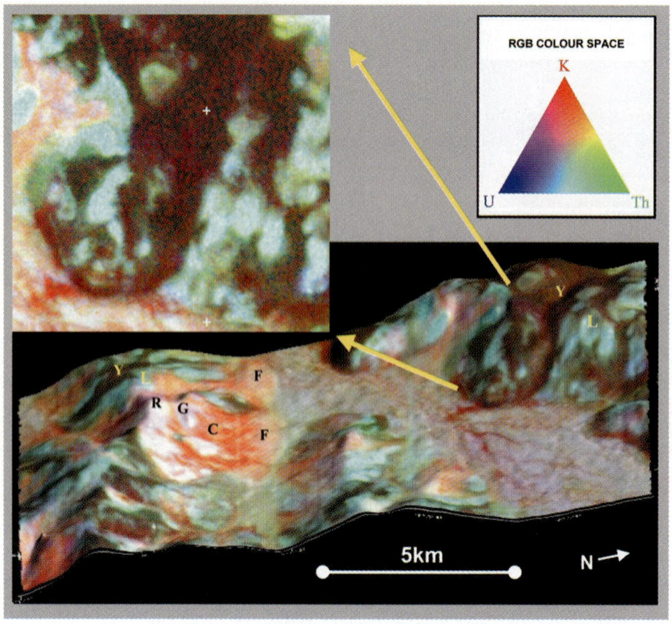

Figure 4.

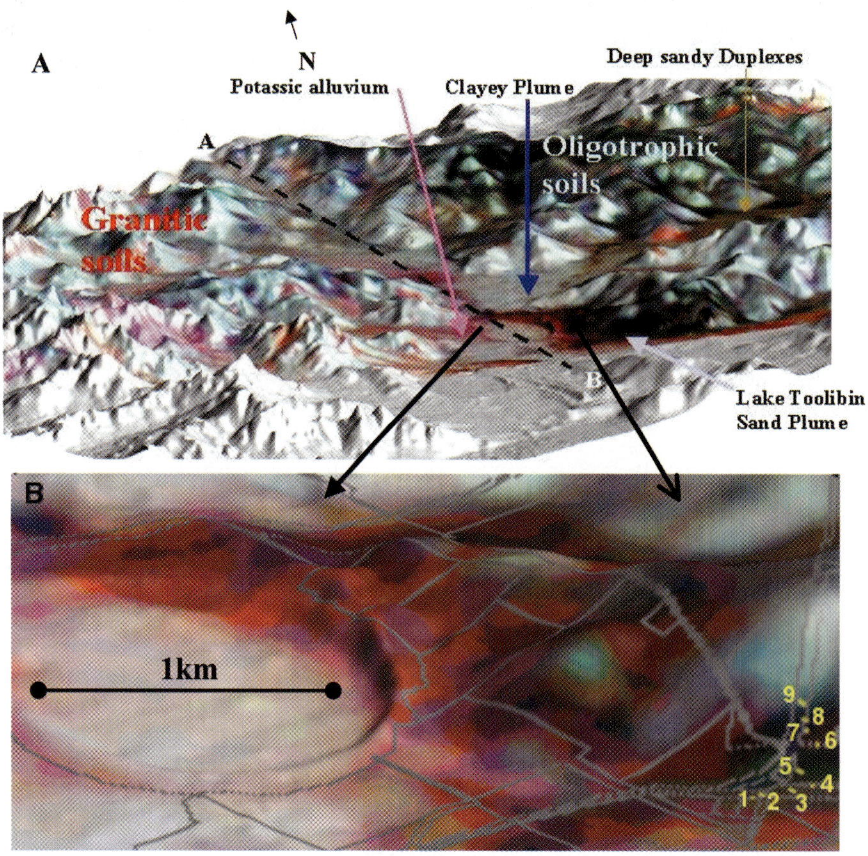

Figure 5.

←

Figure 4. DEM aligned ternary radiometric image of the Elashgin catchment (×15 vertical exaggeration) with inset expanded view of spiral wave structures and ternary colour scale. (R) Granite rock outcrops; (G) granitic soils; (C) granitic sandy channels; (F) clayey fan deposits; (L) exposed lateritic ferricretes; (Y) yellow sands. Note that the colour coding of the site symbols, in the plot, Figure 3, is selected to correspond as closely as possible to the ternary colour gradings at those sites.

←

Figure 5. (A) DEM aligned ternary radiometric image of a portion the Lake Toolibin catchment viewed from the south-west with inset expanded view (B) of the Lake Toolibin sand plume showing location of roads (grey) and sampling stations (yellow dots annotated numerically) along transect selected for detailed study. Vertical exaggeration is ×15. Respective WGS 84 coordinates for Stations 1 and 9 are S32.94286 E117.62124 and S32.93960 E117.62783.

Table 1. Frequency of soil type under proteaceous and non-proteaceous communities

Vegetation community	Non-lateritic soil	Lateritic soil
Proteaceous	0	16
Non-proteaceous	14	2

** χ^2 probability <0.01.

with the ferricrete close to the surface and may be supplying Fe. This would partly account for the degree of development of the ferricrete layer. Reactivity scores for Al were essentially as in Stations 6–8.

On the basis of field visits and matching floral descriptions for monitor sites given by Brown and Roots (2002), the robustness of the association between Proteaceae and lateritic soil, and lateritic soils and radiometric signal and the reactive Al test is shown in Tables 1 and 2. The relationship between lateritic profiles and proteaceous communities, lateritic soils and low K signal and lateritic soils and reactive Al were all highly significant. The significant relationship between soil and Th is mainly attributable to low Th signalling from duplex soils.

Discussion

The data presented in this paper for a series of sites on the oligotrophic and granitic uplands at Elashgin reveal reproducible relationships between aerial gamma-ray measurements of U and Th and the already well-documented associations between soil types and vegetation classes. Uranium and Th tend to remain closely associated in many soil profiles including granitic areas (Dickson and Scott, 1997). So it is reasonable to suppose that granitic areas ranging from bare rock to soil end-members should collectively exhibit patterns of weathering in which small increases in the ratio of Th toU would indicate differential loss of U. This is consistent with the general ordering of U and Th signals in Figure 3 and conforms generally to the expected order of element leaching noted by other authors (Bierwirth et al., 1996; Martz and de Jong, 1990). In contrast ferricretes commonly exhibit concentration of Th relative to U as evidenced here for both Elashgin and Toolibin and reported elsewhere for other areas of Australia (see Dickson and Scott, 1997). This implies that chemical transport has been involved in formation of Th-rich laterites and not a residual concentration of resistates as suggested by some models of laterite formation. However, one must bear in mind that the apparent propensity at Elashgin for Th to concentrate in ferricretes to several-fold greater extents than U, might be explained in terms of higher solubility of U, signal truncation effects by sand layers or simply to artefacts of signal processing procedures. Furthermore, Pate et al. (2001) have implied that Fe and hence U and Th might be cycled by plant-based chelating agents within 40–400 thousand years. On this time scale differential movement between U and its gamma-emitting daughter (^{214}Bi) might corrupt current estimates of U since the time taken for concentrations of U to equilibrate with those of its daughters takes more than 1.5 million years versus a much shorter time frame for equilibration of Th with ^{208}Tl. However, expected disequilibrium in the U decay series is a little more complex since ^{230}Th in the U decay series will preferentially move with the ^{232}Th. ^{230}Th has a half-life of 80 000 years, so Th-rich ferricretes would be expected to show quite a large excess gamma radiation relative to initial U until decaying down to equilibrium with ^{234}U/^{238}U at around 300 000 years. A measure of Th230/U234 disequilibrium in ferricretes with Th/U>5 might accordingly comprise a means of dating the ferricretes (B.L. Dickson, pers. comm.).

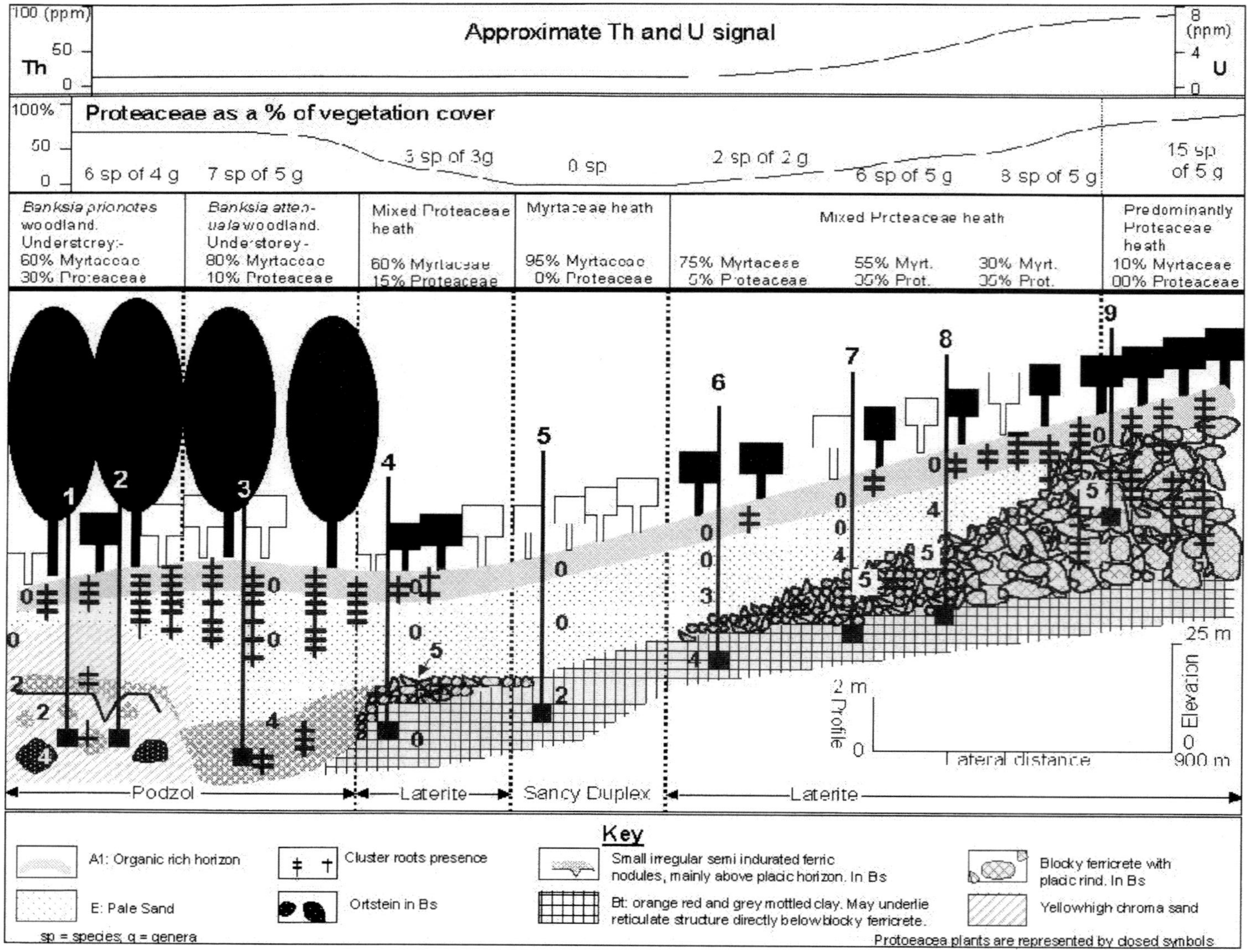

Figure 6. Schematic cross-section of vegetation types soil profiles and intensities in U and Th signal at Toolibin along the study transect located in Figure 4B with Proteaceae represented by closed symbols. Scores for reactive aluminium, for zones of the profile along the transect, refer to the following classes: (0) non-reactive; (1) very weak; (2) weak; (3) moderate; (4) strong; (5) very strong.

Table 2. Frequency of radiometric and Al reactivity classes for lateritic and non-lateritic soils

Soil type	K signal**		Th signal*		Al reactivity in B horizon**	
	Low	High	Low	High	Low	High
Lateritic	15	2	8	9	1	17
Non-Lateritic	5	9	12	2	13	1

*χ^2 probability <0.05; **χ^2 probability <0.01.

Observations here and in earlier work (Pate et al., 2001) show that ferricretes of various types and associated cluster root-bearing taxa are generally encountered in low K settings. They also occur in soil pockets within granitic profiles devoid of a pallid clay zone and where K rich-feldspars would remain relatively intact but where more-weatherable plagioclase and mafic minerals would disappear. The latter situation clearly diverges from standard models of laterite formation embodying humid tropical deep weathering, loss of K and complementary and synchronous development of laterite, mottled and pallid zones. However it is fully consistent with our earlier-presented biological model of laterite formation (see Pate et al., 2001) in which leaching and loss of fertility is not necessarily related to depletion of K and presence of pallid zones. In such cases, we would contend that the main driver for colonisation by oligotrophic species is competition for P. Benefit would then accrue to cluster root-forming species capable of accessing intractable forms of the element and immobilising P resources within ferricrete layers (see discussion in Pate et al., 2001).

The general comments of Beard (1980) on the vegetation of the Elashgin area are that the mixed Kwongan communities on ferricrete gravel do include a number of members of the Proteaceae but are typically dominated by *Allocasuarina*. Thickets dominated by species of *Acacias*, *Banksia* and *Grevillea* occur on associated yellow sands, while low woodlands comprising *Acacia acuminata* and *Allocasuarina huegliana* surround granitic outcrops, with patches of *Allocasuarina campestris* in places. It is interesting to note that certain species of *Allocasuarina* and *Acacia* form cluster roots akin to those in Proteaceae (see review of Pate and Watt, 2002). Refering to the spatial arrangement of soil and plant associations in the Elashgin and adjacent areas Beard (1990) writes – "the soils are not usually arranged in catena. Sand and laterite may occur on low ground, and red soil on high ground. Soil and vegetation types merge into one an other", and "while therefore the aerial photography shows that there is a broad division of the landscape into sandplains and valleys and salt lake systems, each is a mosaic of vegetation too intricate to be mapped at 1:250 000 even where it still exists".

Observations consistent with the above statements are evident in the earlier mentioned spiral waveforms of radiometric signals depicted in Figure 4. Spiralling configurations of a similar nature, supposedly rotating over time around meandering or stationary cores, are generally viewed as emerging from the self-organisation of dynamic, coherent systems. These are evident for example at very large scale in hurricanes and galaxies, at intermediate scale in ecological interactions and at small scale in chemical reactions. Many plausible mathematical models of ecological interactions, including observations on cultured microorganisms, are supportive of such dynamics but as Rohani et al. (1997) have noted, many ecologists believe that such symmetries would be broken in real-world situations. In our data for Elashgin, the radiometrically defined waveforms are profuse and insensitive to topography, while also conforming closely to distribution of different plant communities. An alternative explanation is that the spiral waves denote areas where weathering-resistant granitic core materials intrude into more easily weathered gneiss. Weathering profiles with upper surfaces of sand and laterite would then develop over the areas concerned. Furthermore, gneisses can show complex folding and differential weathering and erosion which would then account for the observed spiralling of radiometric signals. Were this the case, plant communities would be expected to follow closely the surface geochemistry (B.L. Dickson, pers. comm.). This suggestion can not be discounted on the basis of present evidence from Elashgin.

Additional evidence suggesting an underlying biological dynamic comes from the spatial relationships of the two major oligotrophic soils and their respective vegetation communities to the granitic areas at Elashgin. Drawings by Bettenay and Hingston (1964) of lateritised landscapes in the nearby Merredin area, suggested that yellow lateritic sandy earths, usually topped by paler sands, are typically contained by exposed ferricretes. Such sandy earths are said to be largely colluvial in origin, and indeed they often yield to paler lateritic sands on lower slopes. However, we find this view partly inconsistent with our radiometric images from Elashgin in which yellow sands (brown colour code, Figure 4) often sweep over crests while still bordered by ferricretes (blue-green code) and distinctly separated from granitic areas (pink and red).

Ground observations at Elashgin suggest that vegetation dominated by *Allocasuarina campestris* and underlying ferricretes, originate within and spread out from the granitic areas. Eventually, following colonisation by competing oligotrophic communities, mature ferricretes would break down to yellow sands. Hence their containment within mature ferricretes. Such a scenario is partly consistent with views of Mulcahy (1960) and Prescott (1931), namely, that the sands in question are formed by the breakdown of ferricrete materials and are largely residual. Such *in situ* development of yellow sand, in places of acid reaction, would entertain the idea of different groupings of cluster root bearing taxa having subtly different mechanisms for P uptake depending on soil pH the types of organic anions secreted and whether or not protons accompany such secretion (see Lambers et al., 1998; Roelofs et al., 2001).

As well as providing radiometric data fully consistent with conclusions drawn from Elashgin, the Toolibin catchment provided special opportunities for comparing soil profiles, vegetation and radiometric signals within a Quaternary sand plume. Here, the associations between Proteaceous vegetation, podzols and laterites depicted in Figure 6 are highly specific and provide particularly vivid examples of targeting of cluster root distributions in A horizons and podzolic and lateritic Bs horizons parallel to those already observed elsewhere (Pate et al., 2001). In marked contrast, within the transect and elsewhere in the catchment, Proteaceae are replaced by non-cluster root-bearing taxa where sand overlies a clayey Bt in duplex soils largely free of ferricrete gravel. The clayey (Bt) horizons in non-proteaceous areas score low in the reactive aluminium test, whereas similarly located 'Bs' horizons of adjacent Proteaceae-rich areas carrying podzols and laterites score high in the same test. The later is to be expected from conclusions drawn by Farmer and Lumsden (2001) and ourselves (Pate et al., 2001) regarding complexation of Al with low molecular weight organic anions and subsequent precipitation of the Al following metabolic utilisation of the anion. In our hypothesis this would refer specifically to organic anion secretion by P-absorbing cluster roots.

Because of attenuation of signalling in areas of deep sand, radiometrics failed to clearly distinguish contrasting soil types and associated proteaceous and non-proteaceous communities across the Lake Toolibin catchment. However, the prominent relationships between regions where Proteaceae dominate, the occurrence of near surface gravel and intense Th and U signals are already well borne out by the vegetation surveys of Beard (loc cit) and the radiometric surveys of Cook et al. (1996) and Pracilio et al. (1998). Counterparts of this were also encountered in uncleared areas at Elashgin and the same correlation would supposedly apply to similar profiles alluded to in our earlier studies (Pate et al., loc cit).

Conclusion

This paper combines radiometric, elevation, soil and botanical data to reveal patterns of association consistent with our earlier hypothesis. When drawn together with other field evidence, we are led to believe that laterite production and reworking has a strong biogenic component. As the study progressed it became apparent that airborne radiometrics would provide a particularly useful tool for delineating where cluster root-bearing taxa such as Proteaceae currently exist or were likely to have existed prior to land clearing. Radiometrics also provide subtleties within their imagery likely to be relevant to understanding of pedogenic processes, particularly those mediated biologically along the lines suggested here and in our earlier publications (Pate et al., 2001; Pate and Watt, 2002; Verboom and Galloway, 2000). Despite limitations due to attenuation of signals, the technique highlights faithful association between high U and Th and low K and cluster root bearing taxa at least where lateritic profiles are intact. However the full picture cannot be obtained unless soil profiles are also examined, whereupon the presence of equally strong associations between cluster root bearing taxa and oligotrophic laterites and podzols are consistently revealed. To this extent radiometric surveys should be considered complementary to rather than substituting directly for ground based surveys.

Acknowledgements

We thank Miranda Taylor for some of the data collection and analysis, Gregor Broekmann and Phil Goulding for preparing some of the figures, Paul Galloway for comments on the manuscript and Shaun Molloy for some of the plant community identification at Elashgin. Contributions to the manuscript from Bruce Dickson and two anonymous referees were of great value. We are also much obliged to Rob Fitzpatrick for his suggestion that we include testing for reactive Aluminium at Toolibin.

References

Cook S E, Corner R J, Groves P R and Grealish G J 1996 Use of airborne gamma-radiometric data for soil mapping. Aust. J. Soil Res. 34, 183–194.

Beard J S 1980 The Vegetation of the Kellerberrin Area, Western Australia. Explanatory memoir, 1:250 000 map series. Vegmap Publications, Perth, Australia. 43 pp.

Beard J S 1984 Biogeography of the Kwongan. *In* Kwongan, plant life of the sandplain. Eds. J S Pate and J S Beard. pp. 1–26. University of Western Australia Press, Perth.

Beard J S 1990 Plant life of Western Australia. Kangaroo Press, Kenthurst, NSW, Australia. 319 p.

Bettenay E and Hingston F J 1964 Development and distribution of soil in the Merredin area, Western Australia. Aust. J. Soil Res. 2, 173–186.

Bierwirth P N, Gessler, P and McKane D 1996 Empirical investigations of airborne gamma-ray images as an indicator of soil properties. *In* Proceedings of the 8th Australasian remote sensing conference. pp. 81–89. Wagga Wagga, Australia.

Bowler J M 1976 Aridity in Australia: Age origins and expression in Aeolian landforms and sediments. Earth Sci. Rev. 12, 279–310.

Brown & Roots Services Asia Pacific Pty. Ltd. 2002 Flora and vegetation survey, Tooliban catchment. Perth, Australia. 46 pp.

Dickson B and Taylor G 1998 Noise reduction on aerial gamma-ray surveys. Exp. Geophys. 23, 324–329.

Dickson B L and Scott K M 1997 Interpretation of aerial gamma-ray surveys – adding the geochemical factors. J. Aust. Geol. Geophys. 17, 187–200.

Farmer V C and Lumsden D G 2001 Interactions of fulvic acid with aluminium and a proto-imogolite sol: the contribution of E-horizon eluates to podzolization. Eur. J. Soil Sci. 52, 177–188.

Fieldes M and Perrott K W 1966 Rapid field and laboratory test for allophane. NZ J. Sci. 9, 623–629.

George R J, Verboom W H and Green A 2002 Airborne geophysics meets dryland salinity: Great expectations 'v' realistic applications. *In* Proceedings of the International Groundwater Conference: Balancing the ground water budget. Darwin, Australia.

Hunt P A, Mitchell P B and Paton T R 1977 'Laterite profiles' and 'lateritic ferricretes' on the Hawkesbury Sandstone, Australia. Geoderma 19, 105–121.

Langmuir D and Herman J S 1980 The mobility of thorium in natural waters at low temperatures. Geochemica et Cosmochemica Acta 44, 1753–1766.

Lambers H, Chapin F S III and Pons T L 1998 Plant Physiological Ecology. Springer-Verlag, New York.

Isbell R F 1996 The Australian Soil Classification. CSIRO Publishing, pp. 143, Collingwood, Australia.

Lundstrom U S, van Breemen N, Bain D C, van Hees P A W, Giesler R, Gustafsson J P, Ilvesniemi H, Karltun E, Melkerud P A, Olsson M, Riise G, Wahlberg O, Bergelin A, Bishop K, Finlay R, Jongmans A G, Magnusson T, Mannerkoski H, Nordgren A, Nyberg L, Starr M and Tau Strand L 2000 Advances in understanding the podzolization process resulting from a multidisciplinary study of three coniferous forest soils in the Nordic Countries. Geoderma 94, 335–353.

Martz L W and de Jong E 1990 Natural radionuclides in soil of a small agricultural basin in the Canadian praries and their association with topography soil properties and erosion. Catena 17, 85–96.

Mabbutt J A 1977 Desert Landforms. ANU press, Canberra, Australia. 340 pp.

McFarlane M J 1983 Laterites. *In* Chemical sediments and geomorphology: precipitates and residua in the near-surface environment. Eds. A S Goudie and K Pye. pp. 7–58. Academic Press, London.

Mulcahy M J 1960 Laterites and lateritic soils in south-western Aust. J. Soil Sci. 11, 206–225.

Mulcahy M J 1967 Landscapes, Laterites and Soils in Southwestern Australia. *In* Landform Studies from Australia & New Guinea. Eds. J N Jennings J A and Mabbutt. pp. 211–230. Australian National University Press, Canberra.

Pate J S and Watt M 2002 Roots of *Banksia* spp. (Proteaceae) with special reference to functioning of their specialized root clusters. pp. 989–1006. *In* Plant Roots: The Hidden Half (Third edition). Eds. Y Waisel, A Eshel and U Kafkafi U. Marcel Dekker, New York, USA.

Pate J S, Verboom W H and P D Galloway 2001 Co-occurrence of Proteaceae, laterite and related oligotrophic soils: Coincidental associations or causative inter-relationships? Aust. J. Bot. 49, 529–560.

Prescott J A 1931 The soils of Australia in relation to vegetation and climate. Bulletin No 52, Council for Scientific and Industrial Research, Australia.

Pracilio G, Street G J, Nallan Chakravatula P, Angeloni J R, Sattel D, Owers M and Lane R 1998 Airborne geophysical surveys to assist planning for salinity control, Report No 3. Lake Toolibin saltmap survey: National Airborne Geophysics Project Report for the National Dryland Salinity Program. World Geoscience Corporation. Perth, Australia.

Roelofs R, Rengel Z, Cawthray G R, Dixon K W and Lambers H 2001 Exudation of carboxylates in Australian Proteaceae: Chemical composition. Plant Cell Envir. 24, 891–903.

Rohani P, Lewis T J, Grunbaum D and Ruxton G D 1997 Spatial self-organized in ecology: pretty patterns or robust reality? Trends Ecol. Evol. 12, 70–74.

Scott K M and Dickson B L 1991 Study of pisolith-bearing samples from the Cobar Region, Western NSW. CSIRO Aust. Div. Exploration Geoscience. Restricted report 207R.

Taylor J T, Smettem K, Pracilio G and Verboom W H 2002 Investigation of the relationships between soil properties and high-resolution radiometrics. Exp. Geophys. 32 (in press).

UTS Geophysics 2002 Logistics Report: Detailed airborne magnetic, radiometric and digital elevation survey for the Elashgin catchment.

Verboom W H and Galloway P D 2000 Hypothetical effects of rhizosphere associates of Proteaceae and their lateritic products on landscape evolution: Explanatory descriptions from south-western Australia. *In* Proceedings of the Australian Society of Soil Science, Soils 2000 Conference. Eds. C Tang and D R Williamson. pp. 24–35. University of Western Australia, Perth, Western Australia.